T0207489

Lecture Notes in Bioinformatics 13278

Subseries of Lecture Notes in Computer Science

More information about this subseries at https://link.springer.com/bookseries/5381

Itsik Pe'er (Ed.)

Research in Computational Molecular Biology

26th Annual International Conference, RECOMB 2022
San Diego, CA, USA, May 22–25, 2022
Proceedings

 Springer

Editor
Itsik Pe'er 🆔
Columbia University
New York, NY, USA

ISSN 0302-9743   ISSN 1611-3349 (electronic)
Lecture Notes in Bioinformatics
ISBN 978-3-031-04748-0   ISBN 978-3-031-04749-7 (eBook)
https://doi.org/10.1007/978-3-031-04749-7

LNCS Sublibrary: SL8 – Bioinformatics

This Springer imprint is published by the registered company Springer Nature Switzerland AG
The registered company address is: Gewerbestrasse 11, 6330 Cham, Switzerland

Preface

This volume contains 17 extended abstracts and 23 short abstracts representing a total of 40 proceedings papers presented at the 26th International Conference on Research in Computational Molecular Biology (RECOMB 2022), hosted by the University of California, San Diego. The conference took place at La Jolla, California, USA, during May 22–25, 2022. These 40 contributions were selected by a rigorous peer-review process from 188 submissions to the conference. Each of these 188 submissions received reviews from at least three members of the Program Committee (PC) members or their designated sub-reviewers. Following an initial process of independent reviewing, all submissions were opened for discussion by their reviewers and the conference program chair through the EasyChair Conference Management System. Final decisions were made based on reviewer assessments with some adjustment to ensure the technical diversity of the conference program.

RECOMB 2022 allowed authors an option to publish their full extended papers in the conference proceedings or to provide short abstracts for the proceedings and pursue alternative arrangements for publishing the full paper. In addition, the authors of a select set of accepted papers were invited to submit revised manuscripts for consideration for publication in the partner journals, Cell Systems and Genome Research. All papers that appear as extended abstracts in the proceedings were invited for submission to the RECOMB 2022 special issue of the Journal of Computational Biology.

RECOMB 2022 also featured highlight talks of computational biology papers that were published in journals during the previous 18 months. Of the 38 submissions to the highlights track, ten were selected for oral presentation at RECOMB.

In addition to presentations of these contributed papers, RECOMB 2022 featured seven invited keynote talks given by leading scientists:

- Bing Ren (University of California, San Diego), "Single-cell analysis of epigenome in health and disease".
- Howard Chang (Stanford University), "Personal regulome navigation".
- Lenore Cowen (Tufts University), "Pathways for Learning from Structure and Organization of Protein Interaction Networks".
- John Chodera (Sloan Kettering Institute), "The COVID Moonshot: Open science discovery of a novel oral SARS-CoV-2 antiviral".
- John Marioni (European Bioinformatics Institute), "Computational challenges in single-cell genomics".
- Wenyi Wang (MD Anderson Cancer Center), "Deciphering cancer cell evolution and ecology".
- Regina Barzilai (Massachusetts Institute of Technology), "Infusing Biology into Molecular Models for Drug Discovery".

RECOMB also featured a special invited workshop on genomic privacy, organized by Gamze Gürsoy (Columbia University).

In addition, five topical RECOMB satellite meetings took place in parallel directly preceding the main RECOMB meeting:

- The 19th RECOMB Satellite Conference on Comparative Genomics (RECOMB-CG) co-chaired by Siavash Mirarab (University of California, San Diego), Lingling Jin (University of Saskatchewan), and Dannie Durand (Carnegie Mellon University).
- The 6th RECOMB Satellite Conference on Bioinformatics Education (RECOMB-BE) chaired by Niema Moshiri (University of California, San Diego).
- The 10th RECOMB Satellite on Computational Methods in Genetics (RECOMB-Genetics) co-chaired by Anna-Sapfo Malaspinas (Swiss Institute of Bioinformatics), Sriram Sankararaman (University of California, Los Angeles), and Gillian Belbin (Mount Sinai Institute for Genomics Health).
- The 12th RECOMB Satellite Workshop on Massively Parallel Sequencing (RECOMB-Seq) co-chaired by Can Alkan (Bilkent University), Benjamin Langmead (Johns Hopkins University), Paul Medvedev (Pennsylvania State University), and Stefano Tonzani (Cell Press)
- The RECOMB Satellite Workshop on Computational Cancer Biology (RECOMB-CCB) co-chaired by Hannah Carter (University of California, San Diego) and Simone Zaccaria (University College London).

Two additional RECOMB satellite meetings were held in honor of members of our community: Mike Waterman's 80th birthday was celebrated in a satellite meeting co-chaired by Remo Rohs and Fengzhu Sun (University of Southern California). Benny Chor, who our community lost this year, was commemorated in a satellite meeting co-chaired by Sagi Snir (University of Haifa) and Zohar Yakhini (Israel Institute of Technology). We thank them for organizing these great companion meetings and they and their Program Committees for their hard work in making them possible.

The organization of this conference was the work of many colleagues contributing their time, effort, and expertise. I am especially grateful to the local organizing committee, particularly Conference Chair Vineet Bafna (University of California, San Diego) and co-organizers Vikas Bansal (University of California, San Diego), Jocelyn Bernardo (University of California, San Diego), Melissa Gymrek (University of California, San Diego), Siavash Mirarab (University of California, San Diego), Glenn Tesler (University of California, San Diego), and Kaiyuan Zhu (University of California, San Diego). I am grateful to the many others who volunteered their time and work, including those whose names were not yet known to us at the time of this writing. I also want to thank the Poster Chair, Yaron Orenstein (Ben-Gurion University, Israel), Keynotes Chair, Ewa Szczurek (University of Warsaw), Satellites Chair, Sriram Sankararaman (University of California, Los Angeles), and Highlights Chair, Sushmita Roy (University of Wisconsin, Madison) for their efforts in ensuring a high-quality technical program. I am further grateful to all of those PC members and sub-reviewers who took time out of their busy schedules to review and discuss submissions on a very tight schedule. I also thank the authors of the proceedings papers, the highlights, and the posters for contributing their work to the meeting and for their attendance at the conference.

Final thanks go to all our conference sponsors for their support, who at press time for this volume included Akamai Technologies, Illumina, the University of California, San Diego, and the Department of Computer Science and Engineering (University of California, San Diego), and especially to the sponsors of our student travel awards, the National Science Foundation (NSF) and the International Society for Computational Biology (ISCB).

Beyond the formal details about the conference, and thanking all the direct contributors, RECOMB 2022 represents a pivotal moment for our community, coming together in person for the first time in three years. RECOMB 2020 was canceled in physical form with only a few weeks' notice - the 2020 template I was using for this preface document had no idea a pandemic was coming, squeezing the conference into a Zoom window. We have since all learned the extent to which in-person meetings provide deeper engagement, greater value, and a more immersive sense of community. One of the highlighted talks in RECOMB 2022 even made this point quantitatively. I would thus like to express how honored I am to be ushering back participation in RECOMB in person.

May 2022   Itsik Pe'er

To Benny, to the RECOMB Community in Memory of Benny

This past year, 2021, sadly marked the passing away of Prof. Benny Chor, a scientist, a pillar in the bioinformatics community and in RECOMB, a colleague and a friend. Benny made significant contributions in cryptography, in computational biology, and to the teaching of computer science at all levels.

Benny introduced all his students and many of his colleagues to seeing science in the context of culture, of friendship and human interactions, as a lifestyle more than as a vocation. For Benny, a student or a colleague was first a friend, a person with a full life, maybe with a family, maybe with habits and ideas to explore and to learn from. He was proud, for example, of a PhD student from the Technion who took his complexity course and upon graduation opened a falafel shop. Benny was an avid traveler and sailor, taking his students on challenging adventures. He was an evangelist of high standards and zero compromise morality in science. An outstanding characteristic of Benny was his modesty. Few people know that he won the ACM award for his PhD, due to its seminal founding contributions in cryptography. Only some colleagues in the RECOMB community know of his theoretical computer science career, and his remarkable achievements. To the privileged people who worked with him he conveyed a culture of rigorous science in a pleasant atmosphere. Science with a smile.

In 1985 Benny and colleagues established the concept of verifiable secret sharing and developed related methods and definitions that are still used in the field. In 1998 he started his work on private information retrieval (PIR), a fundamental topic of high importance in cryptography. His contribution to cryptography continued in further developing techniques and theories related to PIR and other privacy primitives. Among Benny's most prominent contributions to computational biology is the introduction of highly rigorous approaches from computer science and math to the study of evolution, specifically to phylogenetics. This includes his dynamic programming algorithm for quartet-based phylogenetics, the introduction of analytical, symbolic algebra-based approaches to maximum likelihood phylogenetics, and his proof of the NP-hardness of maximum likelihood phylogenetics. In recent years Benny also worked on analyzing gene expression and genomics data, including studying techniques and complexity results related to order-preserving submatrices, as well as developing methods for analyzing HiC data and producing biologically interesting results. Benny greatly contributed to computer science education. Benny once likened the use of Python in teaching CS to the use of Toyota in driving lessons. One doesn't learn how to drive a Toyota – one learns the art and skills of driving.

We are grateful to have had the privilege to work with Prof. Benny Chor and to have learned so much from him. We are grateful for having spent many fun science hours with Benny on trails, in pubs and cafes, and on sailboats. He will always be remembered for his great science and personality. He will always be remembered with a smile.

Sagi Snir
Zohar Yakhini

Organization

General Chair

Vineet Bafna   University of California, San Diego, USA

Program Chair

Itsik Pe'er   Columbia University, USA

Steering Committee

Vineet Bafna	University of California, San Diego, USA
Bonnie Berger (Chair)	Massachusetts Institute of Technology, USA
Eleazar Eskin	University of California, Los Angeles, USA
Teresa Przytycka	National Institutes of Health, USA
Cenk Sahinalp	National Institutes of Health, USA
Roded Sharan	Tel Aviv University, Israel
Martin Vingron	Max Planck Institute for Molecular Genetics, Germany

Program Committee

Derek Aguiar	University of Connecticut, USA
Tatsuya Akutsu	Kyoto University, Japan
Can Alkan	Bilkent University, Turkey
Srinivas Aluru	Georgia Institute of Technology, USA
Mukul S. Bansal	University of Connecticut, USA
Niko Beerenwinkel	ETH Zurich, Switzerland
Bonnie Berger	Massachusetts Institute of Technology, USA
Sebastian Böcker	Friedrich Schiller University Jena, Germany
Christina Boucher	University of Florida, USA
Dongbo Bu	Chinese Academy of Sciences, China
Sha Cao	Indiana University School of Medicine, USA
Cedric Chauve	Simon Fraser University, Canada
Brian Chen	Lehigh University, USA
Rayan Chikhi	CNRS, France
Lenore Cowen	Tufts University, USA
Barbara Di Camillo	University of Padua, Italy

Anthony Gitter	University of Wisconsin-Madison, USA
Mohammed El-Kebir	University of Illinois at Urbana-Champaign, USA
Nadia El-Mabrouk	University of Montreal, Canada
Travis Gagie	Diego Portales University, Chile
Faraz Hach	University of British Columbia and Vancouver Prostate Centre, Canada
Xin Gao	King Abdullah University of Science and Technology, Saudi Arabia
Iman Hajirasouliha	Cornell University, USA
Bjarni Halldorsson	deCODE genetics and Reykjavik University, Iceland
Fereydoun Hormozdiari	University of Washington, USA
Tao Jiang	University of California, Riverside, USA
Daisuke Kihara	Purdue University, USA
Carl Kingsford	Carnegie Mellon University, USA
Gunnar W. Klau	Heinrich Heine University Düsseldorf, Germany
Tal Korem	Columbia University, USA
Mehmet Koyuturk	Case Western Reserve University, USA
Gregory Kucherov	CNRS and LIGM, France
Jens Lagergren	KTH Royal Institute of Technology, Sweden
Benjamin Langmead	Johns Hopkins University, USA
Jingyi Jessica Li	University of California, Los Angeles, USA
Wei Vivian Li	Rutgers University, USA
Stefano Lonardi	University of California, Riverside, USA
Jian Ma	Carnegie Mellon University, USA
Jianzhu Ma	Purdue University, USA
Wenxiu Ma	University of California, Riverside, USA
Veli Mäkinen	University of Helsinki, Finland
Salem Malikic	National Institutes of Health, USA
Florian Markowetz	University of Cambridge, UK
Bernard Moret	Ecole Polytechnique Fédérale de Lausanne, Switzerland
Onur Mutlu	ETH Zurich, Switzerland
Luay Nakhleh	Rice University, USA
William Stafford Noble	University of Washington, USA
Ibrahim Numanagic	University of Victoria, Canada
Layla Oesper	Carleton College, USA
Yaron Orenstein	Ben-Gurion University, Israel
Laxmi Parida	IBM, USA
Cinzia Pizzi	University of Padua, Italy
Teresa Przytycka	National Institutes of Health, USA
Ben Raphael	Princeton University, USA

Knut Reinert	FU Berlin, Germany
Sushmita Roy	University of Wisconsin – Madison, USA
Cenk Sahinalp	National Cancer Institute, USA
Sriram Sankararaman	University of California, Los Angeles, USA
Michael Schatz	Cold Spring Harbor Laboratory, USA
Alexander Schoenhuth	Bielefeld University, Germany
Russell Schwartz	Carnegie Mellon University, USA
Mingfu Shao	Carnegie Mellon University, USA
Roded Sharan	Tel Aviv University, Israel
Ritambhara Singh	Brown University, USA
Sagi Snir	University of Haifa, Israel
Jens Stoye	Bielefeld University, Germany
Fengzhu Sun	University of Southern California, USA
Wing-Kin Sung	National University of Singapore, Singapore
Krister Swenson	CNRS and Université de Montpellier, France
Haixu Tang	Indiana University Bloomington, USA
Glenn Tesler	University of California, San Diego, USA
Tamir Tuller	Tel Aviv University, Israel
Fabio Vandin	University of Padua, Italy
Martin Vingron	Max Planck Institute for Molecular Genetics, Germany
Jerome Waldispuhl	McGill University, Canada
Sheng Wang	University of Washington, USA
Wei Wang	University of California, Los Angeles, USA
Tandy Warnow	University of Illinois at Urbana-Champaign, USA
Sebastian Will	Ecole Polytechnique, France
Jinbo Xu	Toyota Technological Institute at Chicago, USA
Min Xu	Carnegie Mellon University, USA
Yuedong Yang	Sun Yat-sen University, China
Yun William Yu	University of Toronto, Canada
Jianyang Zeng	Tsinghua University, China
Chi Zhang	Indiana University School of Medicine, USA
Louxin Zhang	National University of Singapore, Singapore
Xiuwei Zhang	Georgia Institute of Technology, USA
Jie Zheng	ShanghaiTech University, China

Additional Reviewers

Gad Abraham	Matthew Aldridge
Alyssa Adams	Mohammed Alquraishi
Ben Adcock	Bayarbaatar Amgalan
Jeffrey Adrion	Ulzee An
Jarno Alanko	Juan Andrade

Sandro Andreotti
Maria Anisimova
Yoann Anselmetti
Hrayer Aprahamian
Jesus Eduardo Rojo Arias
Peter Arndt
Jasmijn Baaijens
Xin Bai
Peter Bailey
Dipankar Ranjan Baisya
Alexander Baker
Daniel Baker
Giacomo Baruzzo
C. J. Battey
Brittany Baur
Fritz Bayer
Michael Beckstette
Giora Beller
Chandrima Bhattacharya
Jeremy Bigness
Wout Bittremieux
Luc Blassel
Alexander Bockmayr
Richard Border
Nico Borgsmüller
Vincent Brault
Maria Brbic
Laurent Brehelin
Matthew Brendel
Sharon Browning
Diyue Bu
Volker Busskamp
Patrick Cahan
Stefan Canzar
Marco Cappellato
Marcello Carazzo
Bastien Cazaux
Giulia Cesaro
Mark Chaisson
Sakshar Chakravarty
Addie Chambers
Wennan Chang
Samrat Chatterjee
Ke Chen
Ken Chen

Mandi Chen
Runsheng Chen
Xing Chen
Liang Cheng
Benjamin Chidester
Leonid Chindelevitch
Julien Chiquet
Philippe Chlenski
Kil To Chong
Julie Chow
Simone Ciccolella
Phillip Clarke
Alejandro Cohen
Matteo Comin
Helen Cook
Raghuram Dandinasivara
Anjali Das
Bu Dechao
James Degnan
Mattéo Delabre
Pinar Demetci
Emilie Devijver
Meleshko Dmitrii
Norbert Dojer
Monica Dragan
Ruben Drews
Yuxuan Du
Timothy Durham
J. N. Eberhardt
Ghazal Ebrahimi
Gudmundur Einarsson
Baris Ekim
Michael Epstein
Michael Ford
Anat Fuchs
Philippe Gambette
Shan Gao
Lana Garmire
Mathieu Gascon
Alexander Gawronski
Paul Geeleher
Sam Gelman
Vanesa Getseva
Ali Ghaffaari
Maryam Ghareghani

Parham Ghasemloo Gheidari
Mazyar Ghezelji
Manoj Gopalkrishnan
Olivier Gossner
Ilan Gronau
Barak Gross
Mario Guarracino
Alessandro Guazzo
Pawel Górecki
Georg Hahn
Spencer Halberg
Gisli Halldorsson
Michiaki Hamada
Renmin Han
Wenkai Han
Marteinn Hardarson
Ananth Hari
Shabbeer Hassan
Verena Heinrich
Hadas Hezroni
Minh Hoang
Ermin Hodzic
Christian Hoener zu Siederdissen
Steve Hoffmann
Jan Hoinka
Jill Hollenbach
Yuhui Hong
Farhad Hormozdiari
Borislav Hristov
Jian Hu
Jiawei Huang
Xingfan Huang
Michael Hudgens
Pelin Icer
Anastasia Ignatieva
Elham Jafari
Mansoureh Jalilkhany
Joseph Janizek
Aleksander Jankowski
Joulien Jarroux
Cangzhi Jia
Xiaofang Jiang
Suraj Joshi
Ben Kaufman
Yasin Kaymaz

David Kelley
Raiyan Khan
Parsoa Khorsand
Kengo Kinoshita
Florian Klimm
Maren Knop
Liang Kong
Alexey Kozlov
Sandeep Krishna
Semih Kurt
Chee Keong Kwoh
Lukas Käll
Da-Inn Lee
Heewook Lee
Peter Lee
Seokho Lee
Jure Leskovec
Haipeng Li
Xiang Li
Qi Liao
Xingyu Liao
Ran Libeskind-Hadas
Antoine Limasset
Jie Liu
Kaiyuan Liu
Yining Liu
Yuelin Liu
Yu-Chen Lo
Enrico Longato
Xiaoyu Lu
Yang Lu
Shishi Luo
Xiang Ge Luo
Xiao Luo
Yunan Luo
Cong Ma
Feiyang Ma
Jianzhu Ma
Chris Magnano
Lauren Mak
Bastien Mallein
Ion Mandoiu
Camille Marchet
Parmita Mehta
Cassidy Mentus

David Merrell
Ivana Mikocziova
Ryan Mills
Hossein Moeinzadeh
Mohammadhossein Moeinzadeh
Mohammadreza Mohaghegh Neyshabouri
Amirsadra Mohseni
Matthieu Muffato
Ghulam Murtaza
Veli Mäkinen
Sheida Nabavi
Ozkan Nalbantoglu
Rami Nasser
Wilfred Ndifon
Tin Nguyen
Pavlos Nikolopoulos
Macha Nikolski
Mor Nitzan
Eirini Ntoutsi
Ziad Obermeyer
Baraa Orabi
Soumitra Pal
Xiaoyong Pan
Enrico Paolini
Luca Parmigiani
Ali Pazokitoroudi
Leonardo Pellegrina
Matteo Pellegrini
David Pellow
Sitara Persad
Johan Persson
Nico Pfeifer
Marina Pinskaya
Daniel Pirak
Petr Popov
Gunnar Pálsson
Yutong Qiu
Yoav Ram
Vincent Ranwez
Nadav Rappoprt
Farid Rashidi Mehrabadi
Anat Reiner-Benaim
Vladimir Reinharz
Stefano Rensi
Mina Rho

Julie Rodor
Caroline Ross
Oscar Rueda
Jean-Francois Rupprecht
Timothy Sackton
Armita Safa
Negar Safinianaini
Yana Safonova
Tiziana Sanavia
Rebecca Santorella
Diego Santoro
Roman Sarrazin-Gendron
Itay Sason
Gryte Satas
Johannes Schlüter
Sebastian Schmeier
Jacob Schreiber
Tizian Schulz
Celine Scornavacca
Bengt Sennblad
Ekaterina Seregina
Saleh Sereshki
Bar Shalem
Tal Shay
Ron Sheinin
Hong-Bin Shen
Yihang Shen
Noam Shental
Qian Shi
Shinsuke Shigeto
Ilan Shomorony
Tomer Sidi
Brynja Sigurpalsdottir
Nasa Sinnott-Armstrong
Pavel Skums
Haris Smajlovic
Ruslan Soldatov
Reza Soltani
Yan Song
Yun Song
Vsevolod Sourkov
Alexandros Stamatakis
Leen Stougie
Alexander Strzalkowski
Pascal Sturmfels

Shiwei Sun
Yuqi Tan
Jing Tang
Tianqi Tang
Eric Tannier
Zahin Tasfia
Erica Tavazzi
Fabian Theis
Marketa Tomkova
Cole Trapnell
Isotta Trescato
Xinming Tu
Laura Tung
Eline van Mantgem
Jan H. van Schuppen
Thomas Varley
Isana Veksler
M. P. Verzi
Davide Verzotto
Riccardo Vicedomini
Gian Marco Visani
Pieter-Jan Volders
Dong Wang
Jihua Wang
Liangjiang Wang
Tianyang Wang
Weixu Wang
Xuran Wang
Ying Wang
Leah Weber
April Wei
Ethan Weinberger
Mathias Weller
Daniel Westreich
J. White Bear
Roland Wittler
Jochen Wolf
Aiping Wu
Chih Hao Wu
Lani Wu
Yi-Chieh Wu
Yufeng Wu
Chencheng Xu

Hanwen Xu
Jie Xu
Junyan Xu
Xiaopeng Xu
Muyu Yang
Yang Yang
Moran Yassour
Esti Yeger-Lotem
Melih Yilmaz
Keren Yizhak
Kaan Yorgancioglu
Serhan Yılmaz
Tristan Zaborniak
Xiaofei Zang
Ron Zeira
Chao Zeng
Xiangrui Zeng
Qing Zhan
Bin Zhang
Chuanyi Zhang
Louxin Zhang
Michael Zhang
Qimin Zhang
Ran Zhang
Ruochi Zhang
Sai Zhang
Shilu Zhang
Wen Zhang
Xiang Zhang
Xining Zhang
Xiuwei Zhang
Yang Zhang
Zidong Zhang
Zijun Zhang
Ziqi Zhang
Yi Zhao
Fan Zheng
Hongyu Zheng
Qinghui Zhou
Tianming Zhou
Kaiyuan Zhu
Zifan Zhu

Contents

Short Papers

Extended Abstracts

Unsupervised Integration of Single-Cell Multi-omics Datasets with Disproportionate Cell-Type Representation

Pınar Demetçi[1,2], Rebecca Santorella[3], Björn Sandstede[3], and Ritambhara Singh[1,2(✉)]

[1] Center for Computational Molecular Biology, Brown University, Providence, RI 02912, USA
{pinar_demetci,ritambhara}@brown.edu
[2] Department of Computer Science, Brown University, Providence, RI 02912, USA
[3] Division of Applied Mathematics, Brown University, Providence, RI 02912, USA
{rebecca_santorella,bjorn_sandstede}@brown.edu

Abstract. Integrated analysis of multi-omics data allows the study of how different molecular views in the genome interact to regulate cellular processes; however, with a few exceptions, applying multiple sequencing assays on the same single cell is not possible. While recent unsupervised algorithms align single-cell multi-omic datasets, these methods have been primarily benchmarked on co-assay experiments rather than the more common single-cell experiments taken from separately sampled cell populations. Therefore, most existing methods perform subpar alignments on such datasets. Here, we improve our previous work Single Cell alignment using Optimal Transport (SCOT) by using unbalanced optimal transport to handle disproportionate cell-type representation and differing sample sizes across single-cell measurements. We show that our proposed method, SCOTv2, consistently yields quality alignments on five real-world single-cell datasets with varying cell-type proportions and is computationally tractable. Additionally, we extend SCOTv2 to integrate multiple ($M \geq 2$) single-cell measurements and present a self-tuning heuristic process to select hyperparameters in the absence of any orthogonal correspondence information.

Available at: http://rsinghlab.github.io/SCOT.

Keywords: Single-cell sequencing · Multi-omics · Data integration · Unsupervised learning · Optimal transport · Unbalanced alignment

1 Introduction

The ability to measure multiple aspects of the single-cell offers the opportunity to gain critical biological insights about cell development and diseases. However, many existing single-cell sequencing technologies cannot be simultaneously

© The Author(s), under exclusive license to Springer Nature Switzerland AG 2022
I. Pe'er (Ed.): RECOMB 2022, LNBI 13278, pp. 3–19, 2022.
https://doi.org/10.1007/978-3-031-04749-7_1

applied to the same cell, resulting in multi-omics datasets sampled from distinct cell populations. While these measurements can be analyzed separately, integrating them prior to analysis can help explain how different molecular views interact and regulate cellular functions. Unfortunately, single-cell assays that measure different molecular aspects in separately sampled cell populations lack direct sample–sample and feature–feature correspondences across these measurements. This lack of correspondences makes it hard to use integration methods that require some shared information to perform single-cell alignment [4]. Therefore, *unsupervised* single-cell multi-omics data alignment methods are crucial for integrative single-cell data analysis.

Several unsupervised methods [4,10,12,15], including our previous work, SCOT [9], have shown state-of-the-art performance for integrating different single-cell measurement domains. Since these methods were mainly evaluated on real-world co-assay datasets (with 1–1 correspondence between cells across domains), our understanding of their performance on datasets obtained from experiments that are not co-assays is limited. Such experiments perform separate sampling to measure distinct genomic features, like gene expression and 3D chromatin conformation. Therefore, their datasets can consist of varying proportions of cell-types across different measurements, creating cell-type imbalance and lacking 1–1 cell correspondences. We hypothesize that alignment methods that perform well on co-assay datasets may not effectively handle the differences in cell-type proportions of the commonly available non-co-assay datasets. Indeed, a recent method, Pamona [5], extended our SCOT framework and used partial Gromov-Wasserstein (GW) optimal transport to allow for missing or underrepresented cell-types in one domain when performing alignment. It showed that current integration methods [4,9,12,15] tend to perform worse under such settings.

We present SCOTv2, a novel extension of SCOT that can effectively align both co-assay and non-co-assay datasets using a single framework. It uses *unbalanced* GW optimal transport to align datasets with disproportionate cell-types while only introducing one additional hyperparameter. This unbalanced framework relaxes the constraint that each point must be mapped with its original mass during the optimal transport. Specifically, an underrepresented cell-type in one domain can be transported with more mass to match the proportion of that cell-type in the other domain and vice-versa. The SCOTv2 framework is summarized in Fig. 1. We demonstrate that SCOTv2 aligns datasets with imbalance in cell-type representations better than state-of-the-art baselines and computationally scales as well as the fastest methods. Furthermore, we extend SCOTv2 to integrate single-cell datasets with more than two measurements, making it a multi-omics alignment tool. We perform alignments of five real-world single-cell datasets, with both simulated and natural cell-type imbalance as well as two and more than two domains ($M \geq 2$), demonstrating SCOTv2's applicability across a wide range of scenarios. Finally, similar to the previous version, we present a self-tuning heuristic process to select hyperparameters for SCOTv2 without any corresponding information like cell-type annotations or matching cells or features in truly unsupervised settings.

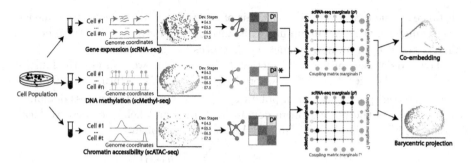

Fig. 1. Overview of SCOTv2 on scNMT-seq dataset [8], which contains unbalanced cell-type representation across three domains - RNA expression, chromatin accessibility, and DNA methylation. SCOTv2 selects an anchor domain (denoted with *) and aligns other measurements to it. First, it computes intra-domain distances matrices D^m for $m = 1, 2, 3$, which are used to solve for correspondence matrices between the anchor and other domains. The circle sizes in the matrices depict the magnitude of the correspondence probabilities or how much mass to transport. Unbalanced GW relaxes the mass conservation constraint, so the transport map does not need to move each point with its original mass. Finally, it either co-embeds the domains into a common space or uses barycentric projections to project them onto the anchor domain.

2   Method

Optimal transport finds the most cost-effective way to move data points from one domain to another. One can imagine it as the problem of moving a pile of sand to fill in a hole through the least amount of work. Our previous framework SCOT [9] uses Gromov-Wasserstein optimal transport, which preserves local geometry when moving data points from one domain to another. The output of SCOT is a matrix of probabilities that represent how likely it is that data points from one modality correspond to data points in the other.

Here, we reintroduce the SCOT formulation to integrate M domains (or single-cell measurements) $X^m = (x_1^m, x_2^m, \ldots x_{n_m}^m) \in \mathbb{R}^{d_m}$ for $m = 1, \ldots M$ with n_m data points (or cells) each. For each dataset, we define a marginal distribution p^m, which can be written as an empirical distribution over the data points:

$$p^m = \sum_{i=1}^{n_m} p_i^m \delta_{x_i}. \tag{1}$$

Here, δ_{x_i} is the Dirac measure. For SCOT, we choose these distributions to be uniform over the data.

Gromov-Wasserstein optimal transport performs the transport operation by comparing distances between samples rather than directly comparing the samples themselves [2]. Therefore, for each dataset, we compute the intra-domain distance matrix D^m. Next, we construct k-NN graphs based on correlations between data points and use Dijkstra's algorithm to compute the shortest path

distance on the graph between each pair of nodes. Finally, we connect all uncon-
nected nodes by the maximum finite distance in the graph and set D^m to be the
matrix resulting from normalizing the distances by this maximum.

For two datasets and a cost function $L : \mathbb{R} \times \mathbb{R} \to \mathbb{R}$, we compute the fourth-
order tensor $\mathbf{L} \in \mathbb{R}^{n_x \times n_x \times n_y \times n_y}$, where $\mathbf{L}_{ijkl} = L(D^1_{ik}, D^2_{jl})$. Intuitively, L quan-
tifies how transporting a pair of points x^1_i, x^1_k onto another pair across domains,
x^2_j, x^2_l, distorts the original intra-domain distances and helps to preserve local
geometry. Then, the discrete Gromov-Wasserstein problem is,

$$GW(p^1, p^2) = \min_{\Gamma \in \Pi(p^1, p^2)} \sum_{i,j,k,l} \mathbf{L}_{ijkl} \Gamma_{ij} \Gamma_{kl}, \tag{2}$$

where Γ is a coupling matrix from the set:

$$\Pi(p^1, p^2) = \{\Gamma \in \mathbb{R}^{n_1 \times n_2}_+ : \Gamma \mathbb{1}_{n_2} = p_1, \ \Gamma^T \mathbb{1}_{n_1} = p_2\}. \tag{3}$$

One of the advantages of using optimal transport is the probabilistic interpretation of the resulting coupling matrix Γ, where the entries of the normalized row
$\frac{1}{p_i}\Gamma_i$ are the probabilities that the fixed data point x_i corresponds to each y_j.
Each entry Γ_{ij} describes how much of the mass of x_i should be mapped to y_j.

To make this problem more computationally tractable, we solve the entrop-
ically regularized version:

$$GW_\epsilon(p^1, p^2) = \min_{\Gamma \in \Pi(p^1, p^2)} \langle \mathbf{L}(D^1, D^2) \otimes \Gamma, \Gamma \rangle - \epsilon H(\Gamma). \tag{4}$$

where $\epsilon > 0$ and $H(\Gamma)$ is the Shannon entropy defined as $H(\Gamma) = \sum_{i=1}^{n_x} \sum_{j=1}^{n_y} \Gamma_{ij} \log \Gamma_{ij}$. Larger values of ϵ make the problem more convex but
also lead to a denser coupling matrix, meaning there are more correspondences
between samples. In SCOT, we use the cost function $L = L_2$.

2.1   Unbalanced Optimal Transport of SCOTv2

Our proposed solution to align datasets with different numbers of samples or proportions of cell-types is to use unbalanced Gromov-Wasserstein optimal transport, which adds divergence terms to allow for mass variations in the marginals
[11,16]. We follow Séjourné et al. [16], and use the Kullback-Leibler divergence,

$$\mathrm{KL}(p||q) = \sum_x p(x) \log \left(\frac{p(x)}{q(x)} \right), \tag{5}$$

to measure the difference between the marginals of the coupling Γ and the input
marginals p^1 and p^2. Thus, we solve the unbalanced GW problem:

$$GW_{\epsilon,\rho}(p^1, p^2) = \min_{\Gamma \geq 0} \langle \mathbf{L}(D^1, D^2) \otimes \Gamma, \Gamma \rangle - \epsilon H(\Gamma) + \rho \mathrm{KL}(\Gamma \mathbb{1}_{n_2} || p^1) + \rho \mathrm{KL}(\Gamma^T \mathbb{1}_{n_1} || p^2), \tag{6}$$

where $\rho > 0$ is a hyperparameter that controls the marginal relaxation. When ρ
is large, the marginals of Γ should be close to p^1 and p^2, and when ρ is small,
the marginals of Γ may differ more, allowing each point to transport with more
or less mass than it originally had. We detail the optimization procedure for
unbalanced Gromov-Wasserstein optimal transport (UGWOT) in Algorithm 3.

2.2   Extending SCOTv2 for Multi-domain Alignment

We provide the details of SCOTv2 in Algorithm 1. To align more than two datasets ($M > 2$), we use one domain as an anchor to align the other domains. The anchor should be the domain with the clearest biological structures, for example, a dataset with the best-defined cell-type clusters. We propose selecting the anchor via the kNN graph computed. For every node x_i^m in the graph, we calculate the average of the k neighboring node values $\mathcal{N}_k(x_i^m)$. We measure the difference between this average and the true value of the node. This difference reflects how well the averaged neighborhood represents the given node. We then average these differences across the graph and select the domain with the lowest averaged difference as the anchor. Intuitively, we select the anchor whose kNN graph best reflects its dataset. Suppose X^1 is the anchor dataset. Then, for $m = 2, 3, \ldots, N$, we compute the coupling matrix Γ^m according to Eq. 4.

To have all of the datasets aligned in the same domain, we can either use barycentric projection to project each X^m for $m = 2, 3, \ldots, M$ onto X^1 or find a shared embedding space as described in Sect. 2.3. In the first iteration of SCOT, we used a barycentric projection to align and project one dataset onto the other. Due to the marginal relaxation, we now search for a non-negative $n_1 \times n_m$ dimensional matrix Γ instead of $\Gamma \in \Pi(p^1, p^m)$. Because of this change, the adjusted barycentric projection is:

$$x_i^m \mapsto \frac{\sum_{j=1}^{n_1} \Gamma_{ij}^m x_j^1}{\sum_{j=1}^{n_1} \Gamma_{ij}^m}. \tag{7}$$

2.3   Embedding with the Coupling Matrix

Other methods such as MMD-MA and UnionCom align datasets by embedding them into a common latent space of dimension $p \le \min_{m=1,\ldots,M} d_m$. Here d_m represents the original dimension size of measurement (or domain) m. Embedding the datasets in a new space often leads to a better alignment as it introduces the additional benefits of dimension reduction, allowing more meaningful structures in the datasets such as cell-types to be more prevalent. Due to these benefits, we also enable the embedding option through a modification of the t-SNE method proposed by UnionCom [4]. For each domain m, we compute P^m, an $n_m \times n_m$ cell-to-cell transition matrix; each entry $P_{j|i}^m$ is the conditional probability that a data point x_i^m would pick x_j^m as its neighbor when chosen according a Gaussian distribution centered at x_i^m. Similarly, for the lower-dimensional embeddings, we compute a cell-to-cell probability matrix $Q^{m'}$ through a Student-t distribution. The full descriptions of P^m and $Q^{m'}$ are given in Appendix.

Then, to jointly embed all domains through the anchor domain X^1, the optimization problem is:

$$\min_{X^{1'},\ldots,X^{M'}} \sum_{m=1}^{M} \mathrm{KL}(P^m \| Q^{m'}) + \beta \sum_{m=2}^{M} \|X^{1'} - X^{m'}(\Gamma^m)^T\|_F^2, \tag{8}$$

Algorithm 1: Pseudocode for SCOTv2 Algorithm

Input: Datasets $X^1, \ldots, X^M$, number of graph neighbors k, entropic regularization coefficient ϵ, mass relaxation coefficient ρ.

for $m = 1, \ldots, M$ **do**

    // Initialize marginal probabilities: $p^m \leftarrow$ Uniform(X^m);

    //Construct G^m, a $k-$NN graph based on pairwise correlations

    // Compute intra-domain distances D^m with Dijsktra's algorithm.

    $c^m = \frac{1}{n_m} \sum_{i=1}^{n_m} \frac{1}{k} \sum_{x_j^m \in \mathcal{N}_k(x_i^m)} \text{corr}(x_j^m, x_i^m)$ //"neighborbood corr.

end

// Select an anchor domain X^{m*}: $m^* = \arg\max_{m=1,\ldots M} c^m$

for $m = 1, \ldots, M \; (m \neq m^*)$ **do**

    $\Gamma^m \leftarrow UGWOT_{\epsilon,\rho}(p^m, p^{m*})$ // Compute pairwise couplings w/ X^{m*}:

    if *Barycentric projection* **then**

        $x_i^{m'} \leftarrow \frac{\sum_{j=1}^{n_1} \Gamma_{ij}^m x_j^{m*}}{\sum_{j=1}^{n_1} \Gamma_{ij}^m}$

    end

    else

        $X^{1'} \ldots X^{M'} \leftarrow$

        $\min_{X^{m'}, \ldots, X^{M'}} \sum_{m=1}^{M} \text{KL}(P^m || Q^{m'}) + \beta \sum_{m \neq m*} ||X^{m*'} - X^{m'}(\Gamma^m)^T||_F^2$

        // Find shared embedding

    end

end

Return: Aligned datasets, $X^{1'} \ldots X^{M'}$.

---

where $X^{m'}$ is the lower dimensional embedding of X^m, and Γ^m is the coupling matrix from solving Eq. 6 for $m = 2, \ldots, M$. These two terms seek to find an embedding that both preserves the local geometry in the original domain and aligns the domains according to the correspondence found by GW. The intuition behind the term $\text{KL}(P^m || Q^{m'})$ is very similar to that of GW; if two points have a high transition probability in the original space, then they should also have a high transition probability in the latent space. The term $||X^{1'} - X^{m'}(\Gamma^m)^T||_F^2$ measures how well aligned the new embeddings $X^{1'}$ and $X^{m'}$ are according to the prescribed coupling matrix Γ^m. Finally, $\beta > 0$ controls the trade-off between preserving the original geometry with the KL term and enforcing the alignment found with GW. We solve this optimization problem using gradient descent from UnionCom with a default latent space dimension size $p = 3$ [4].

2.4 Heuristic Process for Self-tuning Hyperparameters

SCOTv2 has three hyperparameters: (1) k for the number of neighbors to consider in nearest neighbor graphs, (2) the weight of the entropic regularization term, ϵ, and (3) the coefficient of the mass relaxation constraint, ρ. The barycentric projection of one domain onto another does not require any hyperparameters. However, jointly embedding the domains in a latent space requires selecting the dimension p.

Ideally, orthogonal correspondence information such as 1–1 correspondences and cell-type labels can guide hyperparameter tuning as validation. However, such information is hard to obtain in most cases. First, no validation data on cell-to-cell correspondences exists for non-co-assay datasets. Second, it is challenging to infer cell-types for certain sequencing domains such as 3D chromatin conformation. Lastly, the cell-type annotations may not always agree across single-cell domains.

We provide a heuristic to self-tune hyperparameters in the completely unsupervised setting. We first choose a k for the neighborhood graphs that yields a high average correlation value between the neighborhood predicted values and measured genomic values of the graph nodes. This step is the same as the one used to select the anchor domain for multi-omics alignment in Sect. 2.2. Next, we choose ϵ and ρ values that minimize the Gromov-Wasserstein distance between the aligned datasets. Algorithm 2 gives the details of this procedure.

Algorithm 2: Unsupervised hyperparameter search procedure

Input: Datasets $X^1, \ldots, X^M$.
for $m = 1, \ldots, M$ **do**
$\quad k^m = \underset{k \in \{10,20,\ldots,150\}}{\mathrm{argmax}} \ \frac{1}{n_m} \sum_{i=1}^{n_m} \frac{1}{k} \sum_{x_j^m \in \mathcal{N}_k(x_i^m)} \mathrm{corr}(x_j^m, x_i^m)$ // Find k_m's
$\quad$ // Use k^m to compute D^m
end
for $m = 2, \ldots, M$ **do**
$\quad \epsilon^m, \rho^m = \arg\min_{\epsilon, \rho} GW_{\epsilon, \rho}(\mathbb{1}_{n_1}, \mathbb{1}_{n_m})$ // Use GW distance to pick ρ, ϵ
end
Return: k^m, ϵ^m, ρ^m.

---

3   Experimental Setup

3.1   Datasets

We evaluate SCOTv2 on single-cell datasets with disproportionate cell-types using two schemes. (1) We subsample different cell-types in co-assay datasets to simulate cell-type representation disparities between sequencing modalities. (2) We select real-world separately sequenced single-cell multi-omics datasets, which lack 1–1 cell correspondences and have different cell-type proportions across modalities due to the sampling procedure. Additionally, we present results on the original co-assay datasets with 1–1 cell correspondence to demonstrate the flexibility of SCOTv2 across balanced and unbalanced single-cell datasets.

Co-assay Single-Cell Datasets with 1–1 Cell Correspondence. We use three co-assay datasets to validate our model, sequenced by SNARE-seq, scGEM, and scNMT technologies. SNARE-seq is a two-modality sequencing technology that simultaneously captures the chromatin accessibility and transcriptional profiles of cells [6]. This dataset contains a total of 1047 cells from four cell lines: BJ (human fibroblast cells), H1 (human embryonic cells), K562 (human erythroleukemia cells), and GM12878 (human lymphoblastoid cells) (Gene Expression Omnibus access code: GSE126074). We follow the same data preprocessing steps outlined by Chen *et al.* [6]. The scGEM technology is a three-modality sequencing technology that profiles the genetic sequence, gene expression, and DNA methylation states in the same cell [7]. The dataset we use is derived from human somatic cell samples undergoing conversion to induced pluripotent stem cells (Sequence Read Archive accession code SRP077853) [7]. We access the preprocessed data provided by Welch *et al.* [17], which only contains the gene expression and DNA methylation modalities.[1] The dataset sequenced by scNMT-seq method [3] contains three modalities of genomic data: gene expression, DNA methylation, and chromatin accessibility, from mouse gastrulation samples, going through the Carnegie stages of vertebrate development (Gene Expression Omnibus access code: GSE109262). We access the preprocessed data through GitHub[2]. While the SNARE-seq and scGEM datasets contain the same number of cells across measurements, scNMT-seq modalities contain different cell-type proportions after preprocessing due to varying noise levels in measurements (Table 1).

Single-Cell Datasets with Simulated Cell-Type Imbalance. To test alignment performance sensitivity to different levels and types of cell-type proportion disparities across modalities, we generate simulation datasets by subsampling SNARE-seq and scGEM co-sequencing datasets in two ways. (1) We remove a cell-type from one modality. (2) We reduce the proportion of a cell-type in one modality by subsampling it at 50% and another cell-type in the other modality by subsampling it at 75%. We simulate this setting to test how the alignment methods will behave when multiple cell-types have disproportionate representation at different levels (for example, half or quarter percentage of cell-types missing) across modalities. For these cases, we uniformly pick at random which cell-type to subsample or remove. Specifically, for scGEM in simulation case (1), we remove "d16T+" cells in the DNA methylation domain while retaining the original gene expression domain, and remove the "d24T+" cells in the gene expression domain while retaining the original DNA methylation domain. For the SNARE-seq dataset, we remove "GM" cells in the gene expression domain and "K562" in the chromatin accessibility domain. In simulation case (2), we subsample the "d8" cluster of the scGEM dataset at 75% in the gene expression modality and the "d16T+" cluster at 50% in the DNA methylation modality.

---

[1] Preprocessed data for the scGEM dataset accessed here: https://github.com/jw156605/MATCHER.

[2] Dimensionality reduced data, used by Pamona and us, here: https://github.com/caokai1073/Pamona/tree/master/scNMT. Preprocessing scripts for the raw data provided by the authors here: https://github.com/PMBio/scNMT-seq/.

Table 1. Number of cells in (and percentages of) each cell-type across different modalities in the scNMT-seq co-assayed dataset after quality control procedures and the non-coassay datasets.

	Modality #1 (Gene Expression)	Modality #2 (Chromatin Accessibility)	Modality #3 (DNA Methylation or 3D chrom. conform.)
scNMT dataset	(n = 579) **E4.5:** 76 (12.73%) **E5.5:** 104 (17.42%) **Day6.5:** 146 (24.46%) **E7.5:** 271 (45.39%)	(n = 647) **E4.5:** 63 (9.73%) **E5.5:** 89 (13.76%) **E6.5:** 220 (34.00%) **E7.5:** 175 (42.50%)	(n =725) **E4.5:** 65 (8.96%) **E5.5:** 91 (12.55%) **E6.5:** 278 (38.34 %) **E7.5:** 291 (40.14%)
sciOmics dataset	(n = 1,058) **Day0:** 489 (46.22%) **Day3:** 127 (12.00%) **Day7:** 78 (7.37%) **Day11:** 145 (13.71%) **NPC:** 219 (20.70%)	(n = 1,296) **Day0:** 164 (12.65%) **Day3:** 702 (54.17%) **Day7:** 77 (5.94%) **Day11:** 175 (13.50%) **NPC:** 178 (13.73%)	(n =2,154) **Day0:** 987 (45.82 %) **Day3:** 435 (20.19 %) **Day7:** 243 (11.28 %) **Day11:** 164 (7.61 %) **NPC:** 325 (15.09 %)
MEC dataset	(n = 26,273) **Basal:** 11,138 (42.39 %) **L-Sec (Prog):** 7,683 (29.24 %) **L-HR:** 3,439 (13.09 %) **L-Sec (Mat):** 2,869 (10.92 %) **L-Sec (Prolif):** 758 (2.89 %) **Stroma:** 386 (1.47 %)	(n = 21,262) **Basal:** 13,353 (62.80 %) **L-Sec (Prog):** 3,343 (15.72 %) **L-HR:** 2,624 (12.34 %) **L-Sec (Mat):** 1,165 (5.48 %) **L-Sec (Prolif):** 7 (0.033 %) **Stroma:** 770 (3.62 %)	N/A

For SNARE-seq, we subsample the "H1" cluster at 75% and the "K562" cluster at 50% in the gene expression and chromatin accessibility domains, respectively.

Single-Cell Datasets Without 1–1 Correspondences. We also align non-co-assay datasets, containing separately sequenced single-cell -omic measurements. Bonora *et al.* generated the first dataset we use, "sciOmics" [1]. This dataset consists of sciRNA-seq, sciATAC-seq, and sciHiC measurements, capturing gene expression, chromatin accessibility, and 3D chromosomal conformation profiles of mouse embryonic stem cells undergoing differentiation. The measurements were taken at five stages: days 0, 3, 7, 11, and as fully differentiated neural progenitor cells (NPCs). The second non-co-assay dataset, "MEC," contains gene expression and chromatin accessibility measurements taken using the 10X Chromium scRNA-seq and scATAC-seq technologies on mouse mammary epithelial cells (MEC). Since each modality consists of separately sampled cell populations, these contain disparate cell-type proportions across modalities (Table 1).

3.2 Evaluation Metrics and Baseline Methods

Although most of the datasets lack 1–1 cell correspondences, we can evaluate alignment using cell-type labels through label transfer accuracy (LTA) as in [4,5,9]. This metric assesses the clustering of cell-types after alignment by training a kNN classifier on a training set (50% of the aligned data) and then evaluates its predictive accuracy on a test dataset (the other 50% of the aligned data). Higher values correspond to better alignments, indicating that cells that belong to the same cell-type are aligned close together after integration. We benchmark our method against the current unsupervised single-cell multi-omic alignment methods, Pamona [5], UnionCom [4], MMD-MA [14], bindSC [10], Seuratv4 [15], and the previous version of SCOT, which performs alignment without the KL term [9]. Pamona [5], as previously discussed, uses partial Gromov-Wasserstein (GW) optimal transport to align single-cell datasets. UnionCom [4] performs unsupervised topological alignment through a two-step procedure that first finds a correspondence between the domains, considering both global and local geometries with a hyperparameter to control the trade-off between them, and then embeds them in a new shared space. MMD-MA [14] uses the maximum mean discrepancy (MMD) measure to align and embed two datasets in a new space. BindSC [10] requires the users to bring input datasets to the gene expression feature space by constructing a gene activity score matrix for the epigenomic domains, then finds a correspondence matrix between samples through bi-order canonical correspondence analysis (bi-CCA), and jointly embeds the domains into a new space. Finally, Seuratv4 [15] also requires gene activity score matrices for epigenomic domains and then identifies correspondence anchors via CCA. Based on these anchors, it imputes one genomic domain based from the other domain and co-embeds them into a shared space using UMAP.

Since bindSC and Seurat v4 require the creation of gene activity score matrices for epigenomic datasets, they might be more difficult to use with certain sequencing domains. For instance, gene activity scoring is challenging for 3D chromosomal conformation. Of all the selected baselines, only Pamona and UnionCom can align more than two domains, so we only use them as baselines for experiments with multiple domains ($M > 2$). For each benchmark, we define a hyperparameter grid of similar granularity and perform extensive tuning (see Appendix). We report the alignment results with the best performing hyperparameter combinations in Sect. 4.1.

4   Results

4.1   SCOTv2 Gives High-Quality Alignments Consistently Across All Single-Datasets

We first present the alignment results for real-world co-assay datasets with simulated cell-type imbalance. The results we present are obtained by the best performing hyperparameter combinations for all methods compared in this study. Figure 2(A) visualizes the barycentric projection alignments performed by SCOTv2 plotted as 2D PCA for SNARE-seq and scGEM datasets, respectively. We use barycentric projection for visualizations because we set this to be the default projection method of our method since it does not require additional hyperparameters. Here, we integrate datasets under three different settings described in the previous section: (1) Balanced datasets (or "full datasets" with no subsampling), (2) Missing cell-type in the epigenomic domains, and (3) Subsampled cells

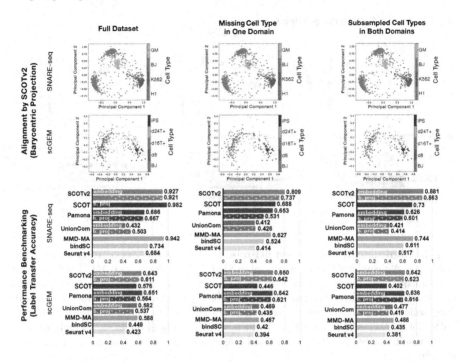

Fig. 2. Alignment results for simulations and balanced co-assay datasets.
A visualizes the barycentric projection alignment on SNARE-seq and scGEM for the
full co-assay datasets, simulations with a missing cell-type in the epigenomic domain,
and subsampled cell-types in both domains. **B** compares the alignment performance of
SCOTv2 to the benchmarks through LTA. For SCOTvs, Pamona, and UnionCom, we
report results on both embedding into a shared space (solid bars) and the barycentric
projection (dotted bars).

in both domains (one cell-type at 50% in the epigenomic domains and another cell-
type at 75% in the gene expression domains). We include alignment results on the
full datasets with 1–1 sample correspondences to ensure that SCOTv2 performs
well for balanced cases as well.

Qualitatively, we see that SCOTv2 preserves the cell-type annotations after
alignment for all three settings. In Fig. 2(B), we report the quantitative performance of SCOTv2 and all the other state-of-the-art baselines using the Label
Transfer Accuracy (LTA) scores. MMD-MA, UnionCom, Seurat, and bindSC fail
to reliably align datasets with disproportionate cell-type representation across
modalities. While Pamona tends to yield high-quality alignments for cases with
cell-type disproportion, it fails to perform well on the SNARE-seq balanced
dataset as well as its subsampling simulation.

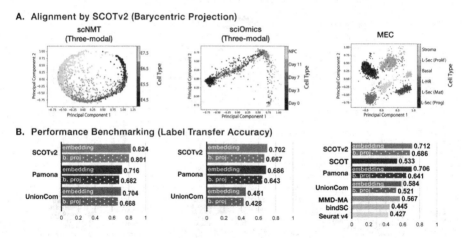

Fig. 3. **Alignment results for multi-modal** ($M > 2$) **and separately sequenced datasets. A** visualizes the alignment of scNMT-seq, sciOmics, and MEC. All datasets have unequal sample sizes and cell-type proportions across domains. **B** benchmarks alignment performance through LTA. As in Fig. 2, we report results both by embedding (solid bars) and barycentric projection (dotted bars) for the methods that allow for both. For scNMT-seq and sciOmics, which are three-modal datasets, we only demonstrate results for SCOTv2, Pamona, and UnionCom, which can handle more than two modalities.

Among all methods tested, SCOTv2 consistently gives more high-quality alignments across different scenarios of cell-type representation. It also demonstrates a ~22% average increase in LTA over the previous version of the algorithm (SCOT) when comparing the barycentric projection results and ~27% for the embedding results. UnionCom, Pamona, and SCOTv2 allow us to perform both barycentric projections and embed the single-cell domains in a lower-dimensional space. Overall, we observe that embedding yields higher LTA values than barycentric projection. Since the barycentric projection projects one domain onto another, the separation of the domain being projected onto (or anchor domain) limits the clustering separation after alignment. In contrast, the embedding utilizes t-SNE to enhance cell-type separation, allowing for better-separated clusters after alignment.

Next, we report the alignment performance of SCOTv2 on single-cell datasets with disparities in cell-type representation due to sampling during experiments. We include scNMT, a co-assay with varying levels of cells across domains due to quality control procedures, along with sciOmics and MEC for this experiment. Note that scNMT and sciOmics have three different modalities, and hence, we can only report the baselines for methods that can align datasets with $M > 2$. Figure 3(A) presents the qualitative alignment results for SCOTv2 with PCA. SCOTv2 performs well on all three datasets, including the ones with three modalities. The LTA scores in Fig. 3(B) demonstrate that SCOTv2 consistently yields the best alignments on the three real-world datasets. These results highlight its ability to reliably integrate separately sampled with disproportionate cell-type representation and multiple ($M > 2$) modalities simultaneously.

4.2 Hyperparameter Self-tuning Aligns Well Without Depending on Orthogonal Correspondence Information

The benchmarking results above present the alignment performance of each algorithm at its best hyperparameter setting; however, users may not have 1–1 correspondences to validate alignments, for the purpose of hyperparameter selection, in real-world applications. While users may have access to cell-type labels, inferring cell-types is highly difficult in specific modalities of single-cell sequencing, such as 3D chromatin conformation. Additionally, different sequencing modalities might disagree on cell-type clustering (as is often the case with scRNA-seq and scATAC-seq datasets). In these situations, users might not have sufficient validation data for tuning hyperparameters.

Table 2. Alignment performance benchmarking in the fully unsupervised setting. We run SCOTv2 and SCOT using their heuristics to approximately self-tune hyperparameters. We use default parameters for other methods due to a lack of similar procedures for unsupervised self-tuning.

	SNARE (full dataset)	SNARE (missing cell-type)	SNARE (subsam. dataset)	scGEM (full dataset)	scGEM (missing cell-type)	scGEM (subsam. dataset)	scNMT	sciOmics	MEC
SCOTv2	0.826	0.653	0.751	0.509	0.521	0.415	0.727	0.537	0.584
SCOT	0.852	0.572	0.588	0.423	0.323	0.314	N/A	N/A	0.466
Pamona	0.554	0.423	0.419	0.385	0.414	0.308	0.588	0.329	0.417
MMD-MA	0.523	0.407	0.431	0.360	0.296	0.287	N/A	N/A	0.233
UnionCom	0.411	0.406	0.422	0.332	0.315	0.276	0.474	0.306	0.349
bindSC	0.713	0.584	0.475	0.387	0.254	0.262	N/A	N/A	0.412
Seurat	0.428	0.517	0.503	0.408	0.377	0.329	N/A	N/A	0.387

We design a heuristic process (described in Sect. 2.4), as done previously for SCOT, that allows SCOTv2 to select hyperparameters in a completely unsupervised manner. Other alignment methods do not provide an unsupervised hyperparameter tuning procedure. Therefore, without validation data, a user would have to use the default parameters. In Table 2, we compare alignment performance for our heuristic against the default parameters of other methods. While our heuristic does not always yield the optimal hyperparameter combination, it does give more favorable results over the default settings of the other methods. Thus, we recommend using it in cases that lack orthogonal information for hyperparameter tuning.

4.3   SCOTv2 Scales Well with Increasing Number of Samples

We compare the runtime of SCOTv2 with the top performing methods: Pamona, MMD-MA, UnionCom, and the previous version of SCOT by subsampling various numbers of cells from the MEC dataset. MMD-MA, UnionCom, and SCOTv2 have GPU versions, while Pamona and SCOT only have CPU versions. We run MMD-MA and UnionCom on a single NVIDIA GTX 1080ti GPU with VRAM of 11 GB and Pamona and SCOT on Intel Xeon e5-2670 CPU with 16 GB memory. We also run SCOTv2 on the same CPU to give comparable results to Pamona's run-times. Figure 4 depicts that SCOT, MMD-MA, Pamona, and SCOTv2 show similar computational scaling.

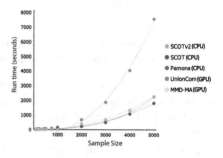

Fig. 4. Runtimes for SCOTv2, SCOT, Pamona, UnionCom, and MMD-MA as the number of samples increases.

5   Discussion

We present SCOTv2, an improved unsupervised alignment algorithm for multi-omics single-cell alignment. It extends the alignment capabilities of SCOT to datasets with cell-type representation disproportions across different sequencing measurements. It also performs alignment for single-cell datasets with more than two measurements ($M > 2$). Experiments on real-world subsampled co-assay datasets and separately sampled and sequenced single-cell datasets demonstrate that SCOTv2 reliably yields high-quality alignments for a wide range of cell-type disproportions without compromising its computational scalability. Furthermore, SCOTv2's flexible marginal constraints enable it to consistently give good alignments results for both balanced and unbalanced single-cell datasets. In addition to effectively handling cell-type imbalances and multi-omics alignment, SCOTv2 can self-tune its hyperparameters making it applicable in complete unsupervised settings. Therefore, SCOTv2 offers a convenient way to align multiple single-cell measurements without requiring any orthogonal correspondence information.

In this second iteration of SCOT, we have utilized the coupling matrix in a new way to find a latent embedding space. While this dimension reduction improves cell-type separation, using the coupling matrix directly may offer even more insights into interactions between the aligned domains. Future work will consider how to use the probabilities in the coupling matrix directly for downstream analysis like improved clustering and pseudo-time inference. Though SCOTv2 has runtimes that scale with other methods, it requires $O(n^2)$ memory storage for the distance matrices, which may be an issue for especially large datasets. One way to address this limitation would be to develop a procedure to align a representative subset of each domain that can be extended to the entire dataset. Therefore, we will explore this direction to further improve the scalability of SCOTv2.

Appendix

Embedding Method Details

The full details of t-SNE can be found in [13]. For each domain m, we compute P^m, an $n_m \times n_m$ cell-to-cell transition matrix; each entry $P^m_{j|i}$ is the conditional probability that a data point x^m_i would pick x^m_j as its neighbor when chosen according a Gaussian distribution centered at x^m_i:

$$P^m_{j|i} = \frac{\exp(-||x^m_i - x^m_j||^2/2\sigma^2_i)}{\sum_{k \neq i} \exp(-||x^m_i - x^m_k||^2/2\sigma^2_i)}. \qquad (9)$$

The bandwidth σ_i is chosen according to the density of the data points through a binary search for the value of σ_i that achieves the user-supplied perplexity value. P^m is computed by averaging $P^m_{i|j}$ and $P^m_{j|i}$ to give more weight to outlier points:

$$P^m_{ij} = \frac{P^m_{i|j} + P^m_{j|i}}{2n_m} \qquad (10)$$

Then, to jointly embed all domains through the anchor domain X^1, the optimization problem is:

$$\min_{X^{1'},\dots,X^{M'}} \sum_{m=1}^{M} KL(P^m || Q^{m'}) + \beta \sum_{m=2}^{M} ||X^{1'} - X^{m'}(\Gamma^m)^T||^2_F, \qquad (11)$$

where $X^{m'}$ is the lower dimensional embedding of X^m, P^m is defined as in Eq. 9, and Γ^m is the coupling matrix from solving Eq. 6 for $m = 1, 2, \dots, M, X^{m'}$. The probability matrix Q^m is computed through a Student-t distribution with one degree of freedom:

$$Q^{m'}_{ij} = \frac{(1 + ||x^{m'}_i - x^{m'}_j||)^{-1}}{\sum_{k \neq l} 1 + (||x^{m'}_k - x^{m'}_l||)^{-1}}. \qquad (12)$$

The intuition behind the cost $KL(P^m || Q^{m'})$ is very similar to that of GW; if two points have a high transition probability in the original space, then they should also have a high transition probability in the latent space.

Hyperparameter Tuning Procedure Details

For each alignment method, we define a grid of hyperparameters and choose the best performing combination for each experiment. If methods share similar hyperparameters in their formulation, we keep the range defined for these consistent across all algorithms. We refer to the publication and the code repository for each method to choose a hyperparameter ranges whenever possible.

For Pamona, we search the number of neighbors in the cell neighborhood graphs, $k \in \{20, 30, \ldots, 150\}$, the entropic regularization coefficient, $\epsilon \in \{5e{-}4, 3e{-}4, 1e{-}4, 7e{-}3, 5e{-}3, \ldots, 1e{-}2\}$, geometry preservation trade-off coefficient, $\lambda \in \{0.1, 0.5, 1, 5, 10\}$, and lastly, embedding dimensionality, $p \in \{3, 4, 5, 10, 30, 32\}$, the output dimension for embedding. For UnionCom, we search the trade-off parameter $\beta \in \{0.1, 1, 5, 10, 15, 20\}$, the regularization coefficient $\rho \in \{0, 0.1, 1, 5, 10, 15, 20\}$, the maximum neighborhood size permitted in the neighborhood graphs, $k_{max} \in \{40, 100, 150\}$, and embedding dimensionality $p \in \{3, 4, 5, 10, 30, 32\}$. For MMD-MA:, we tune the weights λ_1 and $\lambda_2 \in \{1e{-}2, 5e{-}3, 1e{-}3, 5e{-}4, \ldots, 1e{-}9\}$, and the embedding dimensionality, $p \in \{3, 4, 5, 10, 30, 32\}$. For bindSC, we choose the coefficient that assigns weight to the initial gene activity matrix $\alpha \in \{0, 0.1, 0.2, \ldots 0.9\}$, the coefficient that assigns weight factor to multi-objective function $\lambda \in \{0.1, 0.2, \ldots, 0.9\}$, and the number of canonical vectors for the embdedding space $K \in \{3, 4, 5, 10, 30, 32\}$. Lastly, for Seuratv4, we tune the number of neighbors to consider when finding anchors, $k \in \{5, 10, 15, 20\}$, co-embedding dimensionality, $p \in \{3, 4, 5, 10, 30, 32\}$ and the choice of the reference and anchor domains when finding anchors.

Algorithm 3: Pseudocode for Unbalanced GW Optimal Transport (UGWOT)

Input: Marginal probabilities p^1 and p^2, intra-domain distance matrices D^1 and D^2, relaxation coefficient ρ, regularization coefficient ϵ

Initialize the coupling matrix: $\Gamma = \pi = p^1 \otimes p^2$

while Γ *not converged* **do**

$\quad \Gamma_{(mass)} \leftarrow \sum_{i,j} \Gamma_{i,j} \quad \tilde{\epsilon} \leftarrow \Gamma_{(mass)}\epsilon, \quad \tilde{\rho} \leftarrow \Gamma_{(mass)}\rho$

$\quad$ // Compute cost C:

$\quad \Gamma^1 \leftarrow \Gamma \mathbb{1}_{n_2}, \; \Gamma^2 \leftarrow \Gamma^T \mathbb{1}_{n_1}$

$\quad A \leftarrow (D^1)^{\circ 2} \Gamma^1, \; B \leftarrow (D^2)^{\circ 2} \Gamma^2$

$\quad D \leftarrow D^1 \Gamma D^2$

$\quad E \leftarrow \epsilon \sum_{ij} \log\left(\frac{\Gamma_{i,j}}{p_i^1 p_j^2}\right) \Gamma_{i,j} + \rho\left(\sum_i \log\left(\frac{\Gamma_i^1}{p_i^1}\right) \Gamma_i^1 + \sum_j \log\left(\frac{\Gamma_j^2}{p_j^2}\right) \Gamma_j^2\right)$

$\quad C \leftarrow A + B - 2D + E$

$\quad$ **while** (u, v) *not converged* **do**

$\qquad u \leftarrow -\frac{\epsilon\tilde{\rho}}{\tilde{\epsilon}+\tilde{\rho}} \log\left[\sum_{i,j} \exp(v_j - C_{ij})/\tilde{\epsilon} + \log p^2\right]$

$\qquad v \leftarrow -\frac{\epsilon\tilde{\rho}}{\tilde{\epsilon}+\tilde{\rho}} \log\left[\sum_{i,j} \exp(u_i - C_{ij})/\tilde{\epsilon} + \log p^1\right]$

$\quad$ // Update: $\pi_{ij} \leftarrow \exp[u_i + v_j - C_{ij}] p_i^1 p_j^2$

$\quad$ // Rescale: $\pi \leftarrow \sqrt{\Gamma_{(mass)}/\pi_{(mass)}} \pi$ and set $\Gamma \leftarrow \pi$

Return: Γ

References

1. Bonora, G., et al.: Single-cell landscape of nuclear configuration and gene expression during stem cell differentiation and x inactivation. Genome Biol. **22**(1), 279 (2021). https://doi.org/10.1186/s13059-021-02432-w

2. Alvarez-Melis, D., Jaakkola, T.S.: Gromov-wasserstein alignment of word embedding spaces. arXiv preprint arXiv:1809.00013 (2018)
3. Argelaguet, R., Clark, S.J., Mohammed, H., Stapel, L.C., Krueger, C., Kapourani, C.A., et al.: Multi-omics profiling of mouse gastrulation at single-cell resolution. Nature **576**(7787), 487–491 (2019). https://doi.org/10.1038/s41586-019-1825-8
4. Cao, K., Bai, X., Hong, Y., Wan, L.: Unsupervised topological alignment for single-cell multi-omics integration. Bioinformatics **36**(Suppl._1), i48–i56 (2020)
5. Cao, K., Hong, Y., Wan, L.: Manifold alignment for heterogeneous single-cell multi-omics data integration using Pamona. Bioinformatics **38**(1), 211–219 (2021). https://doi.org/10.1093/bioinformatics/btab594
6. Chen, S., Lake, B.B., Zhang, K.: High-throughput sequencing of transcriptome and chromatin accessibility in the same cell. Nat. Biotechnol. **37**(12), 1452–1457 (2019)
7. Cheow, L.F., Courtois, E.T., Tan, Y., Viswanathan, R., Xing, Q., Tan, R.Z., et al.: Single-cell multimodal profiling reveals cellular epigenetic heterogeneity. Nat. Methods **13**(10), 833–836 (2016)
8. Clark, S.J., Argelaguet, R., Kapourani, C.A., Stubbs, T.M., Lee, H.J., et al.: scNMT-seq enables joint profiling of chromatin accessibility DNA methylation and transcription in single cells. Nat. Commun. **9**(1), 1–9 (2018)
9. Demetci, P., Santorella, R., Sandstede, B., Noble, W.S., Singh, R.: Gromov-wasserstein optimal transport to align single-cell multi-omics data. BioRxiv (2020)
10. Dou, J., Liang, S., Mohanty, V., Cheng, X., Kim, S., Choi, J., et al.: Unbiased integration of single cell multi-omics data. bioRxiv (2020). https://doi.org/10.1101/2020.12.11.422014. https://www.biorxiv.org/content/early/2020/12/11/2020.12.11.422014
11. Liero, M., Mielke, A., Savaré, G.: Optimal entropy-transport problems and a new hellinger-kantorovich distance between positive measures. Invent. Math. **211**(3), 969–1117 (2018)
12. Liu, J., Huang, Y., Singh, R., Vert, J.P., Noble, W.S.: Jointly embedding multiple single-cell omics measurements. BioRxiv, p. 644310 (2019)
13. Van der Maaten, L., Hinton, G.: Visualizing data using t-SNE. J. Mach. Learn. Res. **9**(11) (2008)
14. Singh, R., Demetci, P., Bonora, G., Ramani, V., Lee, C., Fang, H., et al.: Unsupervised manifold alignment for single-cell multi-omics data. In: Proceedings of the 11th ACM International Conference on Bioinformatics, Computational Biology and Health Informatics, pp. 1–10 (2020)
15. Stuart, T., Butler, A., Hoffman, P., Hafemeister, C., Papalexi, E., III, W.M.M., et al.: Comprehensive integration of single-cell data. Cell **77**(7), 1888–1902 (2019)
16. Séjourné, T., Vialard, F.X., Peyré, G.: The unbalanced gromov wasserstein distance: Conic formulation and relaxation. arXiv (2021)
17. Welch, J.D., Hartemink, A.J., Prins, J.F.: Matcher: manifold alignment reveals correspondence between single cell transcriptome and epigenome dynamics. Genome Biol. **18**(1), 138 (2017)

Semi-supervised Single-Cell Cross-modality Translation Using Polarbear

Ran Zhang[1], Laetitia Meng-Papaxanthos[2], Jean-Philippe Vert[3], and William Stafford Noble[1,4]($\boxtimes$)

[1] Department of Genome Sciences, University of Washington, Seattle, USA
[2] Google Research, Brain Team, Zurich, Switzerland
[3] Google Research, Brain Team, Paris, France
[4] Paul G. Allen School of Computer Science and Engineering, University of Washington, Seattle, USA
william-noble@uw.edu

Abstract. The emergence of single-cell co-assays enables us to learn to translate between single-cell modalities, potentially offering valuable insights from datasets where only one modality is available. However, the sparsity of single-cell measurements and the limited number of cells measured in typical co-assay datasets impedes the power of cross-modality translation. Here, we propose Polarbear, a semi-supervised translation framework to predict cross-modality profiles that is trained using a combination of co-assay data and traditional "single-assay" data. Polarbear uses single-assay and co-assay data to train an autoencoder for each modality and then uses just the co-assay data to train a translator between the embedded representations learned by the autoencoders. With this approach, Polarbear is able to translate between modalities with improved accuracy relative to state-of-the-art translation techniques. As an added benefit of the training procedure, we show that Polarbear also produces a matching of cells across modalities.

Keywords: Single cell · Cross-modality translation · Cross-modality alignment · Semi-supervised learning

1 Introduction

Single-cell measurements are immensely valuable for quantifying the variance of certain forms of molecular activity within the cell and for identifying and characterizing cell subpopulations within complex tissues. However, a weakness of most single-cell assays is that only a single form of activity can be measured for each cell.

Consequently, a variety of machine learning methods have been proposed to translate between different types of single-cell measurements [1–6]. Typically, co-assay data is used to train such models in a fully supervised fashion. However, in practice co-assay measurements are often more challenging to produce and lower

© The Author(s), under exclusive license to Springer Nature Switzerland AG 2022
I. Pe'er (Ed.): RECOMB 2022, LNBI 13278, pp. 20–35, 2022.
https://doi.org/10.1007/978-3-031-04749-7_2

throughput than standard single-cell measurements. Furthermore, co-assays have only been developed relatively recently and thus are less abundant than single-assay data. Although the recently released Cobolt [7] and MultiVI [6] methods incorporate single-assay data into model training, neither directly assessed whether adding single-assay data from unrelated public datasets improves cross-modality translation. We hypothesize that training a translation model using both labeled data (co-assay data) and unlabeled data (single-assay data) from unrelated studies will result in more accurate translation performance than training from co-assay data alone.

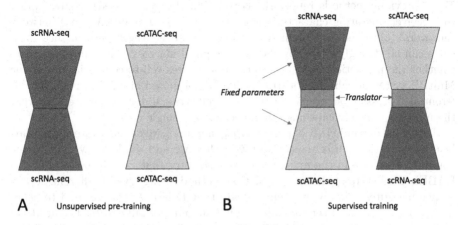

Fig. 1. Polarbear's semi-supervised framework. A In stage 1, Polarbear trains an autoencoder for each data modality, using both single-assay and co-assay data. **B** In stage 2, the encoder from one modality is stitched together with a decoder from the other modality (and vice versa), and the translation layers are trained in a supervised fashion using co-assay data.

Here we propose Polarbear, a semi-supervised approach that learns to translate between single-cell modalities by leveraging both single-assay and co-assay data. We focus on translation between scRNA-seq and scATAC-seq; i.e., given a scRNA-seq profile of a cell, Polarbear will produce as output the scATAC-seq profile of that same cell, and vice-versa. Polarbear is applicable to several types of co-assays that measure expression and chromatin accessibility within single cells, including CAR-seq [8], SNARE-seq [9], and Paired-seq [10]. Polarbear operates in two phases (Fig. 1). In the first phase, we train two deep variational autoencoder (VAE) neural networks that learn, in an unsupervised fashion, to reduce each given type of data to a latent representation (the encoder) and then expand that representation to recover the original data (the decoder). (V)AEs have already been successfully applied to scRNA-seq and scATAC-seq data, primarily for the purpose of de-noising [11–17]. Polarbear trains one VAE for each type of data, while taking into consideration sequencing depth and batch factors [15,16]. In phase two, we stitch together the encoder for one data type with the decoder of a second data type, interposing between them a single, fully connected "translator" layer. During this phase, the parameters of the encoder and decoder are frozen, and the translator parameters are trained in a supervised fashion using co-assay data. Repeating this procedure in reverse, Polarbear allows for bidirectional translation between scRNA-seq and scATAC-seq data. In principle, our method can also be applied to co-assays operating on other data modalities.

In order to evaluate the performance of Polarbear, we propose a set of evaluation metrics for single-cell translation tasks, with the aim of teasing out several important aspects of translation performance. A drawback of current methods lies in the choice of evaluation metrics used for cross-modality profile prediction. Many previous methods report the correlation (for scRNA-seq) or area under the receiver operating characteristic curve (AUROC, for scATAC-seq) between the overall observed and predicted profiles. However, these performance measures can be strongly driven by the average profiles across cells, failing to reflect whether the prediction method accurately captures cell-to-cell variation. While MultiVI systematically demonstrates that the proposed method can predict differential expression between cell clusters or cell types, it does not address whether the model accurately captures differences among single cells.

Using our expanded set of performance measures, we demonstrate that Polarbear's translation performance improves when we add single-assay data to the training procedure in the first phase. We also show that Polarbear outperforms BABEL [2], a state-of-the-art translation method, using several different performance measures. Finally, we demonstrate that Polarbear can be used to accurately match cells between modalities. Overall, our work illustrates the utility of exploiting single-assay data to aid in the prediction of cross-modality profiles.

1.1   Related Work

Several previous methods have been developed for cross-modality prediction (Table 1). TotalVI builds a VAE that takes as input the concatenation of gene and protein expression profiles from the CITE-seq co-assay. The autoencoder learns to impute missing protein expression profiles based on scRNA-seq profiles [1]. BABEL translates between single-cell modalities by joining two autoencoders, one from each data domain [2]. The scMM method uses a mixture-of-experts multimodal deep generative model to learn a joint embedding between modalities and predict missing modalities [3]. Seurat predicts the missing domain profile of a cell by identifying neighboring cells in co-assay data and then computing the average profile of those neighbor cells in the second modality [4]. Multigrate jointly embeds data from two or more modalities and uses the joint embedding to infer profiles in each domain [5]. MultiVI embeds scRNA-seq and scATAC-seq profiles into a shared space by joining two VAEs. The trained model is then able to take single-assay data as input and predict the missing modality [6].

Polarbear differs from previous translation models in several ways. First, Polarbear uses a stepwise optimization approach to first learn embeddings with both single and co-assay data and then translate between embeddings across

Table 1. Method comparison

	totalVI	BABEL	scMM	Seurat	Multigrate	MultiVI	Polarbear
scRNA → scATAC		✓	✓			✓	✓
scATAC → scRNA		✓	✓			✓	✓
Batch correction	✓		✓	✓	✓	✓	✓
Uses single-assay data						✓	✓
Neural network model	✓	✓	✓		✓	✓	✓
Uses a joint or shared embedding for prediction	✓	✓	✓		✓	✓	
Translates between embeddings							✓
Peak-wise evaluation						✓*	✓
Gene-wise evaluation						✓*	✓
Cluster matching evaluation			✓	✓	✓	✓	
Cell matching evaluation						✓	✓

*Evaluation is done across cell types per-peak or per-gene, rather than across individual cells.

modalities based on co-assays, whereas other methods optimize the entire model jointly with a weighted sum of losses for each task. The separate optimization steps make Polarbear less likely to be biased toward optimization of a specific task and requires less hyperparameter tuning. Second, previous models generate predictions based on a joint or shared embedding of both modalities. Polarbear does not require a shared or joint embedding; instead, it adds a translation layer between the embeddings across modalities based on co-assay data. Thus Polarbear is more flexible at leveraging single-assay data and incorporating pre-trained models from new data modalities.

More importantly, Polarbear is able to use single-assay data collected from public datasets to improve its translation performance. Although MultiVI can learn from single-assay data, the question of whether adding single-assay data from unrelated datasets improves translation performance has not been addressed.

In this study, we choose to compare our method with BABEL, for several reasons. First, BABEL directly addresses the task of translating between scRNA-seq and scATAC-seq, and it has been applied to the SNARE-seq co-assay data we use in this study, so it is most likely we can make a fair comparison with BABEL. We attempted to run multiVI, which was published on bioRxiv very recently, but it ran out of memory when trained on the SNARE-seq data. Second, because the focus and novelty of Polarbear is the semi-supervised framework that leverages single-assay data from unrelated studies, instead of comparing extensively with current methods that are not specifically designed for this task, we demonstrate the power of our semi-supervised framework leveraging single-assay data ("Polarbear") by comparing it with a Polarbear model that is only trained on co-assay data ("Polarbear co-assay"). We foresee that Polarbear's semi-supervised framework could be adapted to other existing architectures to boost their translation performance.

2 Methods

2.1 Polarbear Model

Polarbear's autoencoder model adopts ideas from scVI and peakVI [15,16], which take into account sequencing depth and batch factors. Specifically, Polarbear has the following architecture (Fig. 2). One of Polarbear's encoders takes gene expression raw counts as input and assumes they follow zero-inflated negative binomial (ZINB) distributions. The other encoder takes in binarized scATAC counts and assumes they follow Bernoulli distributions. To save memory for the scATAC parts of the model, we only allow for within-chromosome connections in the first two encoder layers and the last two decoder layers. We first specify the latent dimension of each autoencoder, and then we define the dimension of each hidden layer as half of the geometric mean between the input and latent dimensions. To correct for batch effects, we one-hot encode batch factors and concatenate them to the input and embedding layers. To correct for sequencing depth differences across cells, Polarbear includes a sequencing-depth factor as the sum of counts per cell, which is used to calculate the ZINB loss (for scRNA-seq) and binary cross entropy loss (for scATAC-seq), together with other distributional variables learned from the embedding layers in the corresponding modality. The loss of each VAE is the sum of the reconstruction loss and a weighted KL divergence loss.

After the autoencoders in both domains are optimized, we learn a single translation layer between the embedding layers of the scRNA-seq and scATAC-seq, supervised by co-assay data, to minimize the translation loss on each modality.

In both translation directions, since the distributional variables are independent of sequencing depth, the size-normalized expectations of the distributions (i.e. the "norm estimation" in Fig. 2) can be directly used for subsequent tasks such as differential expression analysis. Because the true scATAC-seq profiles used for evaluation are unnormalized binarized counts, we further generate an "unnormalized" prediction so that we can evaluate on the true profiles and make a fair comparison to methods that do not take into account sequencing depth. Since there is a clear correlation between the scATAC-seq and scRNA-seq depth factor when both modalities are observed, which may capture both technical (e.g. batch effect) and biological (e.g. cell cycle) effects, we predict the scATAC-seq depth factor in the translation task based on the known scRNA-seq profile. Specifically, we use ridge regression for this prediction task, with a penalty term determined by cross-validation within the training set. Finally, we multiply the learned sequencing depth with the normalized predictions to generate unnormalized predictions in the test set.

2.2 Hyperparameter Tuning

The Polarbear neural network architecture has two primary hyperparameters: the latent dimensions of the autoencoders, and the weight of the KL divergence term in each VAE. In this study, we use the validation set to choose hyperparameters, selecting the number of latent dimensions ($\{10, 25, 50\}$) and the KL divergence weight ($\{0, 0.5, 1, 2\}$). In the random test set setup, we randomly split the SNARE-seq dataset, assigning 60% of cells to the training set, 20% to the validation set, and 20% to the test set. In the unseen cell type scenario, we use the same validation and test set as BABEL, where the validation and test set are the largest two cell clusters based on the SNARE-seq scRNA-seq dataset. The rest of cells are used as the training set.

We downloaded BABEL's scripts and followed the instructions to generate predictions on SNARE-seq [2]. We verified that we are able to reproduce the performance reported in the paper. In all scenarios, we make sure that BABEL's train/validation/test splits are the same as Polarbear's. BABEL has proposed a set of default parameters (latent dimension: 16, weight factor: 1.13); however, for a fair comparison we tune the following 2D grid of hyperparameters: number of latent dimensions in $\{10, 16, 25, 50\}$ and weight factor to balance scATAC loss in $\{0.2, 1.33, 5\}$. We then select BABEL's best performing model based on each task's performance on the validation set.

2.3   Performance Measures

In designing performance measures for cross-modality translation, we tried to place ourselves in the shoes of a prospective end user of our predictive model. Imagine a scenario in which we are interested in leveraging an existing scRNA-seq dataset to predict chromatin accessibility in a particular biological system, applying our trained model to the scRNA-seq matrix to yield a predicted matrix of scATAC values. Given the predicted peak activations, we can imagine trying to solve two different problems.

In the first setting, we begin by identifying cell types using the original scRNA-seq data or identifying cell groups based on the experimental design (e.g. disease and control groups). We may then be interested in the pattern of predicted chromatin peak activations within each cell type or group. In this setting, a classifier-based measure such as AUPR or AUROC, computed separately for each peak, would accurately capture the per-peak predictive behavior across single cells and thus be indicative of per-peak predictive power across clusters or groups. Each of these two measures has advantages. The AUPR emphasizes enriching the top of the ranked list of predictions with positives. On the other hand, AUROC explicitly corrects for differences in "skew" (i.e., differences in the number of non-zero values) for each peak. To correct for the skew in AUPR measurement toward peaks with large positive proportions (PP defined as #(cells with peak expressed)/#cells), we calculate AUPRnorm = (AUPR-PP)/(1-PP), where 0 represents the behavior or a random predictor and 1 indicates perfect predictor. In this work, we use AUROC and AUPRnorm as performance measurements.

In the second setting, we can use the profile of predicted peak activations across each cell to match scRNA-seq profiles to corresponding scATAC-seq profiles. In this setting, we identify these matches based on Euclidean distance. We want to ensure that each predicted profile's nearest neighbor is the correct match; hence, we can use the fraction of samples closer than the true match (FOSCTTM) as a performance measure [18].

Based on these scenarios, we report here the average per-peak AUROC and AUPRnorm, as well as the FOSCTTM. Similarly, when predicting gene expression from scATAC-seq, we report the average per-gene Pearson correlation (on log-scaled expression) and the FOSCTTM. We do not foresee a scenario in which the overall "flattened" performance of the model, in which we treat all values in the matrix as a single list and compute a single score (Pearson correlation, MSE, AUPR, or AUROC), will be of primary interest to an end user.

2.4   Single-Cell Data Pre-processing

Table 2. Data sets

Data set	Cells	Assay	Platform
SNARE-seq	~10k	Co-assay	Illumina HiSeq 2500/4000
Li *et al.*	~800k	snATAC-seq	Illumina HiSeq 2500
Fang *et al.*	~55k	scATAC-seq	Illumina HiSeq 2500
Zeisel *et al.*	~160k	scRNA-seq	10× Genomics

For co-assay data, we use SNARE-seq data from mouse adult brains (~10k cells) [9]. We filter out peaks that occur in fewer than 5 cells or more than 10% of cells, and we filter out genes that are expressed in fewer than 200 cells or more than 2500 cells, as in the original SNARE-seq paper. To learn robust representations of each domain, we first train the autoencoders using the SNARE-seq data combined with publicly available scRNA-seq and scATAC-seq profiles from adult mouse brains [19–21]. We collected scRNA-seq profiles from ~160k cells and scATAC-seq profiles from ~855k cells, and we randomly downsampled the latter dataset to ~170k cells for use in training (Table 2).

For scRNA profiles, we use SNARE-seq genes as a reference, map genes in the other datasets to the gene symbols in the SNARE-seq data, and further filter out non-protein-coding genes based on Gencode annotations [22]. We remove sex chromosome genes for consistency across datasets. In this way, 17,271 genes are maintained for input to Polarbear. For the scATAC profile, we first lift all peaks to mm10 [23]. Because the ATAC-seq peak locations vary across datasets, we use the SNARE-seq peaks as a reference and map features from other datasets to SNARE-seq peaks if there is an overlap of 1 bp or more. Peaks from sex chromosomes are again filtered out. In the end, 220,526 peaks are input to the Polarbear model.

2.5   Cluster-Level Analysis

A common task in single-cell analysis is to cluster the cells according to the similarity of their scRNA-seq or scATAC-seq profiles; accordingly, a good translator should be able to produce predicted profiles that yield clusters similar to the clusters produced by the observed data.

An important task for our predicted profile is to be able to predict cell-type specific marker genes. To evaluate this, we use the clustering analysis performed in the original SNARE-seq study, which yielded an expert-curated list of marker genes for each cell cluster [9]. This analysis also identified genes and peaks that are significantly highly expressed in each cell type. To calculate whether a gene or a peak is specifically expressed in a specific cell type, we label cells in the corresponding cell type as positive and cells in other cell types as negative. We then calculate the AUROC of the predicted gene/peak-wise profile relative to these labels. A high AUROC score suggests the gene/peak is specifically expressed in the corresponding cell type.

To calculate differentially expressed genes for a cell type, we perform a one-sided Wilcoxon rank-sum test between the expression pattern in the corresponding cell type and that in unrelated cell types, and we control the FDR using the Benjamini-Hochberg procedure. To validate differential expression predictions, we label differentially expressed genes derived from the true profile (FDR $\leq$ 0.01) as positive and other genes as negative, and we calculate a precision-recall curve using the predicted differential expression p-value.

3   Results

3.1   Polarbear Accurately Translates Between Single-Cell Data Domains

We begin by testing Polarbear's ability to translate between scRNA and scATAC profiles in a SNARE-seq adult mouse brain co-assay dataset. To learn robust representations of each domain, we train the autoencoders with large-scale, publicly available scRNA and scATAC single-assay profiles, also derived from adult mouse brains (see Sect. 2.4). We then train Polarbear's translator layer in a supervised fashion using a training set of 80% of the cells from the SNARE-seq dataset, evaluating translation performance on the test set comprised of the remaining 20%.

To evaluate the performance of our model, we measure how well the predicted scRNA-seq profiles allow us to recapitulate gene expression differences across cells. To do this, we calculate the gene-wise correlation between the predicted profile and the true normalized profile [24]. In this analysis, Polarbear outperforms BABEL, yielding an improved correlation for 1067 out of 1205 genes

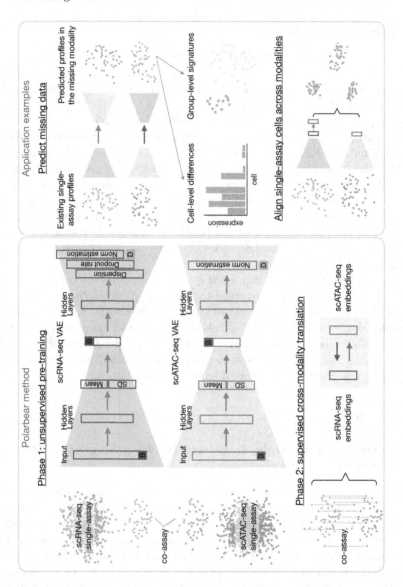

Fig. 2. Polarbear's semi-supervised framework and applications. Left: Polarbear's semi-supervised framework. In Phase 1, Polarbear trains a variational autoencoder for each data modality, using both single-assay and co-assay data. In Phase 2, the encoder from one modality is stitched together with a decoder from the other modality (and vice versa), and the translation layers are trained in a supervised fashion using co-assay data. Specifically, the VAEs take into account sequencing depth (D) and batch effects (B). The scRNA-seq VAE assumes that counts are drawn from a zero-inflated negative binomial distribution, and the scATAC-seq VAE assumes a Bernoulli distribution. Right: Sampled applications of Polarbear. Polarbear can predict missing domain profiles based on the known domain, capturing individual cell-level differences and group-level signatures in the missing domain. Furthermore, Polarbear can match single cell profiles across modalities.

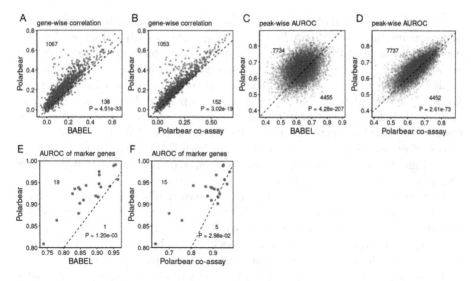

Fig. 3. Cross-modality prediction on the random test set. A,B: Gene-wise correlation between the true and predicted profile, comparing Polarbear with BABEL (A) or with Polarbear co-assay (B), only showing genes that are differentially expressed across cell types. BABEL performance is reported based on the best performing model in each task after a hyperparameter grid search. Each dot is a gene, and numbers indicate the number of dots above and below the diagonal line. P-values are calculated by one-sided Wilcoxon rank-sum test. "Polarbear co-assay" only uses co-assay data to train the model. **C,D**: Peak-wise AUROC, comparing Polarbear with BABEL (C) or with Polarbear co-assay (D). Each dot represents a peak, only peaks differentially expressed across cell types are shown. **E,F**: Gene-wise AUROC. For each marker gene, we calculate the AUROC of its prediction to be higher in cells in corresponding cell type compared to unrelated ones. Polarbear with BABEL (E) or with Polarbear co-assay (F).

(Wilcoxon rank-sum test p-value 4.51×10^{-33}, Fig. 3A, Table 3). We also observe that Polarbear strongly outperforms a Polarbear variant ("Polarbear co-assay") that is trained only with co-assay data (Fig. 3B). For translation in the opposite direction (scRNA-seq → scATAC-seq), we calculate the peak-wise area under the receiver operating characteristic curve (AUROC) of the predicted profile relative to the observed, binarized scATAC-seq profile. Polarbear outperforms both competing methods in this scenario (Fig. 3C–D, Fig. 6A–B, Table 3).

Polearbear will be particularly useful if its predictions can be used to derive biological insights, such as predicting genes specific to each cell ltype. Accordingly, we use a predefined sets of cell-type signature genes that are annotated in the SNARE-seq study [9], as a gold standard, and ask whether Polarbear's predictions allow us to rediscover these gene to the corresponding cell types. To do so, we calculate the AUROC for a gene's predicted expression in a one-vs-all fashion for one cell type versus all others. Polarbear is able to predict the gold standard cell type markers correctly with a median AUROC of 0.936. This performance is significantly better than both BABEL (median AUROC = 0.869; Fig. 3E) and the co-assay variant of Polarbear (median AUROC = 0.910; Fig. 3F).

Polarbear also predicts cell-type specific genes that are not captured based on the true scRNA-seq profiles. Here we focus on the microglia cell type, which consists of only 91 cells in the SNARE-seq dataset. Based on scATAC-seq profiles in the test set, we predict microglia specific genes by calculating AUROC for each gene's predicted expression in microglia cells against all other cells. Based on Polarbear's prediction, the *Sall1* gene is specifically expressed in microglia (AUROC = 0.851), but this gene is not highly expressed in microglia based on the observed scRNA-seq profiles (AUROC = 0.498). Interestingly, *Sall1* has been found previously to be a microglia signature gene, and it encodes the transcription factor, Sall1, that maintains microglia identity [25].

Table 3. Translation performance represented by median and interquartile range (IQR).

Test set	Evaluation metric	BABEL	Polarbear co-assay	Polarbear
		Median (IQR)	Median (IQR)	Median (IQR)
Random 20%	Gene-wise correlation	0.131 (0.141)	0.146 (0.147)	0.195 (0.185)
	Peak-wise AUROC	0.636 (0.0872)	0.647 (0.0839)	0.663 (0.0845)
	Peak-wise AUPRnorm	0.0136 (0.0153)	0.0156 (0.0167)	0.0190 (0.0199)
Unseen cell type	Gene-wise correlation	0.0701 (0.0667)	0.0561 (0.0985)	0.122 (0.0999)
	Peak-wise AUROC	0.560 (0.0871)	0.564 (0.0784)	0.592 (0.0876)
	Peak-wise AUPRnorm	0.00795 (0.0128)	0.00896 (0.0133)	0.0132 (0.0183)

3.2   Polarbear Generalizes to New Cell Types

Because the amount of available co-assay data is limited relative to single-assay data, a common challenge for models such as Polarbear is to translate between modalities in cell types for which no training data is available. The authors of the BABEL model simulated this scenario by creating a train/test split in which an entire SNARE-seq cell cluster is held out for testing. Accordingly, we also investigate this setting, using BABEL's train/test split.

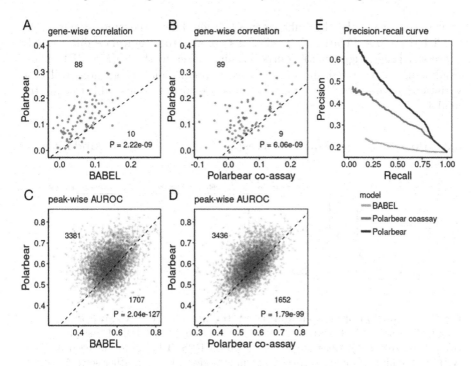

Fig. 4. Cross-modality prediction on an unseen cell type. A,B: Gene-wise correlation between the true and predicted profile, comparing Polarbear with BABEL (A) or with Polarbear co-assay (B). Each dot is a gene, and numbers indicate the number of dots above and below the diagonal line. P-values are calculated by one-sided Wilcoxon rank-sum test. "Polarbear co-assay" only uses co-assay data to train the model. **C,D**: Peak-wise AUROC, comparing Polarbear with BABEL (C) or with Polarbear co-assay (D). **E**: Precision-recall curve on prioritizing the true set of differentially expressed genes based on differential expression pattern on the predicted profiles.

First, we investigate whether Polarbear predictions can capture variations within this unseen population. Because the scRNA-seq and scATAC-seq profiles within a cell type are expected to be relatively homogeneous, successfully translating across modalities in this scenario requires the model to capture differences between individual cells, not just differences across cell types. We observe that Polarbear consistently outperforms BABEL and Polarbear co-assay in translation in both directions, suggesting that Polarbear predictions are able to recapitulate meaningful variations across cells within a cell type (Fig. 4A–D, Fig. 6C–D, Table 3).

Next, we test Polarbear's ability to predict gene expression signatures of the unseen cell population, based only on the scATAC-seq profiles. For this step, we derive the gene signatures for the unseen cell type by identifying genes with significantly higher expression in the unseen cell type compared to cells in the SNARE-seq training set (see Sect. 2.5) We then calculate a precision-recall curve relative to these cell labels, ranking genes by their differential expression from the predicted profiles. Polarbear's predictions are able to recapitulate differentially expressed genes, significantly outperforming other methods (Fig. 4E). These results suggest that Polarbear can correctly predict intra- and inter- cell type variations, even for cell types for which no co-assay data is available to train the model.

3.3   Polarbear Can Match Corresponding Cells Across Modalities

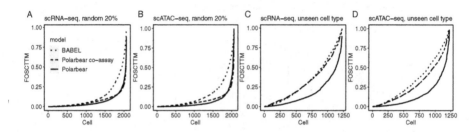

Fig. 5. Evaluation of cross-modality matching. The FOSCTTM score for Polarbear (solid line), Polarbear trained only with co-assay data (dashed line), as well as BABEL (dotted line). Cells are sorted based on FOSCTTM score for each method. **A,B**: Matching performance on the random 20% test set, using either scRNA-seq (A) or scATAC-seq (B) as queries. **C,D**: Matching performance on the unseen cell type, using either scRNA-seq (C) or scATAC-seq (D) as queries.

Polarbear can also be used to match corresponding cells from different modalities. Given unpaired single-assay profiles in each modality, we can use Polarbear to match those cells between modalities, supervised by the co-assay data. To simulate this setting, we project the scRNA-seq and scATAC-seq profiles in the held out test set to the bottleneck layer of the Polarbear model, and we match cells from different modalities in a greedy fashion based on Euclidean distance in the latent space. To assess the matching performance, we calculate for each cell the fraction of samples closer than the true match (FOSCTTM) [18]. Polarbear achieves a lower FOSCTTM score than the competing methods in matching cells in the random test set (Fig. 5A–B) as well as matching cells within the unseen cell type (Fig. 5C–D), suggesting that adding single-assay data improves cross-modality matching.

4   Discussion

We propose Polarbear, a semi-supervised framework that leverages both co-assay and publicly available single-assay data to translate between scRNA-seq and scATAC-seq profiles. We demonstrate that Polarbear improves upon methods that only train on co-assay data. Polarbear predictions are able to capture cell-type and individual cell-level differences, and can predict missing domain knowledge for cell types without any co-assay data available. Polarbear can be used to generate biological hypotheses in the missing domain, such as inferring differentially expressed genes/peaks between cell types or experimental groups. We expect Polarbear to be used to facilitate biological discoveries on uncharacterized domains at the single-cell level, such as identifying individual cell or subclone specific regulatory elements based on scRNA-seq profiles in tumor samples.

Currently, Polarbear predictions do not improve scATAC-seq predictions as much as scRNA-seq predictions. Possible reasons for this difference are that scATAC-seq profiles are sparse and noisy, and scATAC-seq potentially contains more information than scRNA-seq because a single gene can be regulated by multiple scATAC-seq peaks. We foresee that models taking into account prior knowledge (e.g. DNA-sequence features or regulatory region annotations) may further improve scATAC-seq predictions.

Polarbear can also match single cells across modalities with improved accuracy. We envision that our semi-supervised matching framework could be adapted for aligning the large compendium of publicly available single-assay profiles, so that we can generate new hypothesis (e.g. gene-peak relationships and cell clustering based on joint features) based on the predicted paired scRNA-seq and scATAC-seq profiles.

Thanks to the flexible training framework, the current Polarbear model could, in the future, be combined with pre-trained models from other data domains by learning the translation layer based on a limited number of co-assay data, and thus be generalized to translate between multi-modalities.

Polarbear code and data used in this study can be found on https://github.com/Noble-Lab/Polarbear.

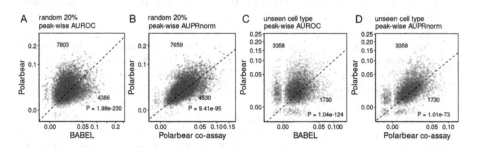

Fig. 6. **Evaluation of predicted scATAC-seq profile by peak-wise AUPRnorm. A,B**: Peak-wise AUPRnorm on the random test set, comparing Polarbear with BABEL (A) or with Polarbear co-assay (B). Each dot represents a peak, only peaks differentially expressed across cell types are shown. **C,D**: Gene-wise AUPRnorm on an unseen cell type, comparing Polarbear with BABEL (C) or with Polarbear co-assay (D). Only peaks highly expressed in the unseen cell type compared to other cell types are shown.

Acknowledgements. We would like to thank Noble Lab members, especially Yang Lu, Gang Li, Ayse Dincer, Anupama Jha and Dejun Lin, for valuable discussions. This work was funded in part by National Institutes of Health award UM1 HG011531.

References

1. Gayoso, A., Steier, Z., Lopez, R., Regier, J., Nazor, K.L., Streets, A., Yosef, N.: Joint probabilistic modeling of single-cell multi-omic data with totalVI. Nat. Methods **18**(3), 272–282 (2021)
2. Wu, K.E., Yost, K.E., Chang, H.Y., Zou, J.: Babel enables cross-modality translation between multiomic profiles at single-cell resolution. Proc. Natl. Acad. Sci. **118**(15) (2021)
3. Minoura, K., Abe, K., Nam, H., Nishikawa, H., Shimamura, T.: A mixture-of-experts deep generative model for integrated analysis of single-cell multiomics data. Cell Rep. Methods **1**, 100071 (2021)
4. Hao, Y., Hao, S., Andersen-Nissen, E., Mauck III, W.M., Zheng, S., Butler, A., et al.: Integrated analysis of multimodal single-cell data. Cell **184**, 573–3587 (2021)
5. Lotfollahi, M., Litinetskaya, A., Theis, F.: Multigrate: single-cell multi-omic data integration (2022).
6. Ashuach, T., Gabitto, M.I., Jordan, M.I., Yosef, N.: Multivi: deep generative model for the integration of multi-modal data. bioRxiv (2021)
7. Gong, B., Zhou, Y., Purdom, E.: Cobolt: joint analysis of multimodal single-cell sequencing data. bioRxiv (2021)
8. Cao, J., Cusanovich, D.A., Ramani, V., Aghamirzaie, D., Pliner, H.A., et al.: Joint profiling of chromatin accessibility and gene expression in thousands of single cells. Science **361**(6409), 1380–1385 (2018)
9. Chen, S., Lake, B.B., Zhang, K.: High-throughput sequencing of the transcriptome and chromatin accessibility in the same cell. Nat. Biotechnol. **37**(12), 1452–1457 (2019)
10. Zhu, C., Yu, M., Huang, H., Juric, I., Abnousi, A., Hu, R., et al.: An ultra high-throughput method for single-cell joint analysis of open chromatin and transcriptome. Nat. Struct. Mol. Biol. **26**, 1063–1070 (2019)
11. Talwar, D., Mongia, A., Sengupta, D., Majumdar, A.: AutoImpute: Autoencoder based imputation of single-cell RNA-seq data. Sci. Rep. **8**, 16329 (2018)
12. Trong, T.N., Mehtonen, J., González, G., Kramer, R., Hautamäki, V., Heinäniemii, M.: Semisupervised generative autoencoder for single-cell data. J. Comput. Biol. **27**(8), 1190–1203 (2020)
13. Eraslan, G., Simon, L.M., Mircea, M., Mueller, N.S., Theiss, F.J.: Single-cell RNA-seq denoising using a deep count autoencoder. Nat. Commun. **10**, 390 (2019)
14. Wang, D., Gu, J.: VASC: dimension reduction and visualization of single-cell RNA-seq data by deep variational autoencoder. Genom. Proteom. Bioinform. **16**(5), 320–331 (2018)
15. Lopez, R., Regier, J., Cole, M.B., Jordan, M.I., Yosef, N.: Deep generative modeling for single-cell transcriptomics. Nat. Methods **15**(12), 1053–1058 (2018)
16. Ashuach, T., Reidenbach, D.A., Gayoso, A., Yosef, N.: PeakVI: a deep generative model for single cell chromatin accessibility analysis. bioRxiv (2021)
17. Xiong, L., Xu, K., Tian, K., Shao, Y., Tang, L., Gao, G., et al.: SCALE method for single-cell ATAC-seq analysis via latent feature extraction. Nat. Commun. **10**(1), 1–10 (2019)

18. Liu, J., Huang, Y., Singh, R., Vert, J.-P., Noble, W.S.: Jointly embedding multiple single-cell omics measurements. In: Huber, K.T., Gusfield, D. (eds.) 19th International Workshop on Algorithms in Bioinformatics (WABI 2019), vol. 143 of Leibniz International Proceedings in Informatics (LIPIcs), pp. 10:1–10:13, Dagstuhl, Germany. Schloss Dagstuhl-Leibniz-Zentrum fuer Informatik (2019). PMC8496402

19. Zeisel, A., Hochgerner, H., Lönnerbergg, P., Johnsson, A., Memic, F., Van Der Zwan, J., et al.: Molecular architecture of the mouse nervous system. Cell **174**(4), 999–1014 (2018)

20. Fang, R., Preissl, S., Li, Y., Hou, X., Lucero, J., Wang, X., et al.: Comprehensive analysis of single cell ATAC-seq data with SnapATAC. Nat. Commun. **12**(1), 1–15 (2021)

21. Li, Y.E., Preissl, S., Hou, X., Zhang, Z., Zhang, K., Fang, R., et al.: An atlas of gene regulatory elements in adult mouse cerebrum. Nature. **598**(7879), 129–136 (2021)

22. Harrow, J., Denoeud, F., Frankish, A., Reymond, A., Chen, C.K., Chrast, J., et al.: GENCODE: producing a reference annotation for ENCODE. Genome Biol. **7**(Suppl 1), S4 (2006)

23. Hinrichs, A.S., Karolchik, D., Baertsch, R., Barber, G.P., Bejerano, G., Clawson, H., et al.: The UCSC genome browser database: update 2006. Nucleic Acids Res. **34**(suppl 1), D590–D598 (2006)

24. Lun, A.T.L., Bach, K., Marioni, J.C.: Pooling across cells to normalize single-cell RNA sequencing data with many zero counts. Genome biol. **17**(1), 75 (2016)

25. Buttgereit, A., Lelios, I., Yu, X., Vrohlings, M., Krakoski, N.R., Gautier, E.L., et al.: Sall1 is a transcriptional regulator defining microglia identity and function. Nat. Immunol. **17**(12), 1397–1406 (2016)

Transcription Factor-Centric Approach to Identify Non-recurring Putative Regulatory Drivers in Cancer

Jingkang Zhao[1,2], Vincentius Martin[1,3], and Raluca Gordân[1,3,4,5](✉)

[1] Center for Genomic and Computational Biology, Duke University, Durham,
NC 27708, USA
[2] Program in Computational Biology and Bioinformatics, Duke University, Durham,
NC 27708, USA
[3] Department of Computer Science, Duke University, Durham, NC 27708, USA
[4] Department of Biostatistics and Bioinformatics, Duke University, Durham,
NC 27708, USA
[5] Department of Molecular Genetics and Microbiology, Duke University, Durham,
NC 27708, USA
raluca.gordan@duke.edu

Abstract. Recent efforts to sequence the genomes of thousands of matched normal-tumor samples have led to the identification of millions of somatic mutations, the majority of which are non-coding. Most of these mutations are believed to be passengers, but a small number of non-coding mutations could contribute to tumor initiation or progression, e.g. by leading to dysregulation of gene expression. Efforts to identify putative regulatory drivers rely primarily on information about the recurrence of mutations across tumor samples. However, in regulatory regions of the genome, individual mutations are rarely seen in more than one donor. Instead of using recurrence information, here we present a method to prioritize putative regulatory driver mutations based on the magnitude of their effects on transcription factor-DNA binding. For each gene, we integrate the effects of mutations across all its regulatory regions, and we ask whether these effects are larger than expected by chance, given the mutation spectra observed in regulatory DNA in the cohort of interest. We applied our approach to analyze mutations in a liver cancer data set with ample somatic mutation and gene expression data available. By combining the effects of mutations across all regulatory regions of each gene, we identified dozens of genes whose regulation in tumor cells is likely to be significantly perturbed by non-coding mutations. Overall, our results show that focusing on the functional effects of non-coding mutations, rather than their recurrence, has the potential to prioritize putative regulatory drivers and the genes they dysregulate in tumor cells.

Keywords: Non-coding mutations · Regulatory drivers ·
Transcription factors · DNA-binding specificity · Enhancers and
promoters · Combining p-values · Liptak's method

J. Zhao and V. Martin—The authors contributed equally to this work.

© The Author(s), under exclusive license to Springer Nature Switzerland AG 2022
I. Pe'er (Ed.): RECOMB 2022, LNBI 13278, pp. 36–51, 2022.
https://doi.org/10.1007/978-3-031-04749-7_3

1 Introduction

Studies of somatic mutations in cancer genomes are generally focused on mutations that alter the amino acid sequences of protein-coding genes. However, whole-genome sequencing of human tumors has revealed that the vast majority of somatic mutations in cancer are non-coding [1], suggesting that they could play a role in cancer initiation and development. Tumorigenesis is thought to be due to the accumulation of multiple drivers that confer a growth advantage to the tumor cells; some of these driver mutations may be non-coding [2]. But only a small proportion of the mutations present in cancer cells are drivers, so it is important to accurately identify them and distinguish them from the much larger number of passenger mutations [3].

Given that driver mutations are expected to be under positive selection, their identification is generally based on patterns of recurrence among tumor samples. Several recent studies have attempted to discover non-coding driver mutations in regulatory DNA sites using recurrence information (e.g. [4–7]). Such studies usually involve the identification of genomic regions with high mutational frequency (i.e. hotspots) by comparing the mutation rate within a DNA window to a background distribution. However, it is generally challenging to precisely estimate the background mutation rate in small genomic regions, given the heterogeneity across different patients and across the genome [8]. A recent meta-analysis of methods for predicting regulatory driver mutations reported that hotspot-based methods can generate large sets of candidate drivers, many of which are false positives [9]. To narrow down the list of candidates, one can also incorporate information on the functional impacts of putative non-coding driver mutations, in particular their effect on transcription factor (TF) binding. One of the most widely used approaches to prioritize mutations in regulatory regions involves the identification of TF binding sites created or disrupted by the mutations, which can be predicted using position weight matrices and motif prediction algorithms [10–12]. However, these methods are limited by the high false positive and false negative rates of binding site prediction algorithms.

In addition, Rheinbay et al. [9] have recently reported that non-coding regulatory driver mutations are much less frequent than protein-coding drivers, with the only notable exception being driver mutations in the *TERT* gene promoter [13,14]. Moreover, some non-coding drivers identified in previous studies were found to be the result of poorly-understood localized hypermutation processes such as mutations originating from differential DNA damage [15] or differential DNA repair [16,17]. On the other hand, recent studies of cancer drivers have shown that mutations do *not* have to be highly recurrent in order to be true drivers [18]; in fact, even mutations that occur in individual tumor samples can drive tumorigenesis.

Here, we describe a new method for analyzing non-coding cancer mutations in regulatory genomic regions (i.e. promoters and enhancers) with the goal of prioritizing mutations based on their effects on TF binding. Unlike previous methods for prioritizing putative non-coding drivers, our method does not rely on the recurrence of mutations across tumor samples. Instead, we consider mutations to be potentially 'significant' if they lead to larger changes in TF binding affinity than expected by chance in that particular genomic region. Thus, the magnitude of the mutations' effects, rather than their recurrence, is the basis for prioritizing mutations and regulatory regions for further studies.

To predict the quantitative effects of non-coding variants on TF binding we use QBiC-Pred [19], a computational method based on regression models of TF-DNA binding specificity trained on high-throughput *in vitro* data [20]. We focus on single-nucleotide mutations, since they are the dominant type of somatic mutation identified in cancer genomes [2]. Importantly, our method links enhancers and promoters to the genes they are likely to regulate, and it combines evidence from all regulatory regions of each gene in order to infer whether a gene is potentially dysregulated due to non-coding mutations. Finally, we use gene expression data from donors with versus without mutations in promoters and enhancers in order to validate that our method prioritizes biologically relevant mutations and regulatory regions.

2 Data and Methods

2.1 ICGC Simple Somatic Mutations and Gene Expression Data

To develop and test our new method for prioritizing putative regulatory driver mutations, we used the Liver Cancer-RIKEN, Japan (LIRI-JP) project from the International Cancer Genome Consortium (ICGC) [1]. We chose this project because it has a large number of donors with whole-genome simple somatic mutation (SSM) data (258 donors), as well as gene expression data (RNA-seq) for 230 out of the 258 donors with SSM data. The study reported a total of ~3.8 million mutations, most of which are single nucleotide mutations (~3.5 million).

2.2 Promoter and Enhancer Data

We focused our analyses on mutations within promoter and enhancer regions, as TF binding sites are located in these regions. We defined promoters as the genomic sequences within ±1000 bp of each RefSeq [21] transcription start site (TSS), excluding any RefSeq exon sequences. We focused on promoters of protein-coding genes, using only TSSs that map to genes within the HUGO gene nomenclature (HGNC) [22]. These criteria resulted in a set of 21,543 promoters.

For enhancers, we used the experimentally determined enhancers from the FANTOM5 project [23,24], which are frequently used in studies of non-coding mutations (e.g. [2,7]). Importantly, the FANTOM consortium provides information about the linkage between enhancers and associated TSSs, which is critical for being able to connect our enhancer results to gene expression data. After removing enhancers that overlapped with promoter regions, we obtained a total of 41,254 enhancers.

We further filtered the 21,543 promoters and the 41,254 enhancers to keep only those that contain mutations. Since most enhancers are hundreds of base pairs long (median = 254bp) and promoters are ∼2,000 bp, the majority of enhancers and a fair number of promoters do not contain any mutations (Figure S1). Thus, after removing these regions without mutations, we obtained a set of 12,612 promoters and 9,018 enhancers with mutations in the LIRI-JP study.

2.3 Defining the Effects of Mutations on TF Binding, and the Significance of These Effects

We use TF binding changes to prioritize mutations that might act as drivers within each regulatory region (enhancer or promoter). For each region, we ask whether the mutations detected in that region lead to *larger* TF binding changes than expected by chance, i.e. according to a background model of random mutations (Fig. 1A). To assess the effect of non-coding mutations on TF-DNA binding we use QBiC-Pred [19], a method we recently developed to quantify TF binding changes based on regression models trained on high-throughput *in vitro* binding data. While other binding specificity models can be used to predict the effects of mutations on TF binding, here we use QBiC-Pred because it performed better than methods based on position weight matrix or deep learning models of specificity [19,20].

For each mutation of interest, QBiC-Pred reports its predicted effect on the binding specificity of 582 human TFs, based on models derived from 667 universal protein binding microarray (PBM) data sets. The effect of a mutation m on TF T is reported in terms of the difference (Δ_m) in the logarithm of the PBM binding intensity signal for the mutated sequence relative to the wild-type sequence according to the binding model for TF T [19]. Positive values represent increased TF binding, while negative values represent decreased binding. Although here we focus on binding changes predicted with QBiC-Pred, our approach can directly use other binding specificity models, as long as they accurately reflect the quantitative TF binding changes induced by DNA mutations.

For a transcription factor T and a regulatory region R that has one or more mutations in the data set of interest, we compute the largest effect on TF binding (either positive or negative) over all mutations in R ($\Delta_{R,T}$, Fig. 1A). Next, to determine if this effect is significant, we compare it against the distribution of effects expected by chance, according to a background model that takes into account: 1) the mutation spectra in that particular cohort, and 2) the particular DNA sequence of regulatory region R. Conceptually, if there are k total mutations in region R ($k = 3$ in Fig. 1A), the full distribution of possible binding effects will be computed taking into account all possible sets of k mutations across the region. Each set i of k mutations will have a particular effect on TF binding ($\mathbf{D_i}$) and will occur with a particular probability (P_i) (Fig. 1B). The effect $\mathbf{D_i}$ is computed by taking the largest effect over the k mutations, similarly to the case of the real mutations. The probability P_i of a particular set of k mutations is computed as described below.

Using all single-nucleotide somatic mutations reported in the LIRI-JP study, we computed the mutation spectra for this cohort, analyzing mutations in their trinucleotide contexts [25]. As in previous studies [25,26], we consider 6 mutation

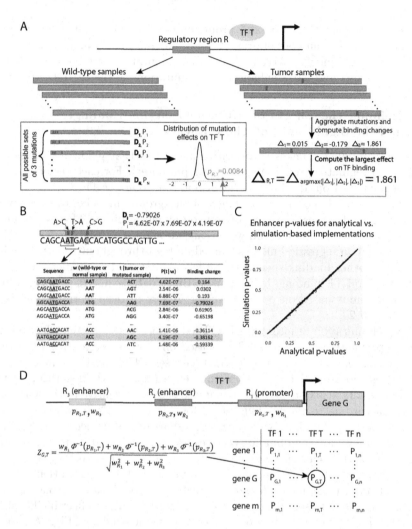

Fig. 1. TF-centric approach to prioritize genes based on mutations in their regulatory regions. (A) For each regulatory region R and TF T, we aggregate the mutations across all patients in the cohort of interest, and compute the largest effect (either positive or negative) across all mutations. This effect, $\mathbf{\Delta}_{R,T}$, is then compared to the distribution of effects computed for all possible sets of k mutations, where k is the number of actual mutations observed in region R (here, $k = 3$). (B) For any set i of k mutations, we compute the binding change with the largest magnitude among these k mutations ($\mathbf{D}_i$), and the probability of that set of k mutations (P_i), as described in Data and Methods. (C) Comparison of p-values computed for mutation effects on MYC binding, for 9,018 enhancers. Plot shows the high correlation between p-values calculated using the analytical versus the simulation-based approach. (D) For genes with multiple regulatory regions, we compute the combined significance of the TF binding changes for TF T across all regions R_i by combining their p-values $p_{R_i,T}$ using Liptak's method, also known as weighted Stouffer's method, as described in Data and Methods. We then use the combined p-values, adjusted for multiple hypothesis testing, to rank genes according to the smallest p-value across all TFs.

types (C > A, C > G, C > T, T > A, T > C, T > G) each in 16 possible contexts, based on the nucleotide right before and right after the mutated position. Mutations and their reverse complements (e.g. **ATG** > **ACG** and **CAT** > **CGT**) are counted together. For each trinucleotide, we estimate the probability of mutating the central base (i.e. $P(t \neq w | w = ATG)$, where t is the trinucleotide in the tumor sample and w is the trinucleotide in the corresponding normal sample) by taking the ratio between how many times that base was mutated in that context and how many times the wild-type trinucleotide occurs in regulatory regions (enhancers or promoters) across all normal samples. We calculate mutation probabilities separately for enhancers versus promoters, since they may be affected differently by mutagenic processes; indeed, we saw significant differences between the mutation spectra at enhancers versus promoters (Figure S2).

When the central nucleotide of a trimer is mutated, i.e. $t \neq w$, there are three possible mutations, e.g. **ATG** to **AAG**, **ACG**, or **AGG**. We estimated the probability of each mutation type (e.g. **ATG** to **AAG**) given that a mutation exists at the central nucleotide, i.e. $P(t = AAG | w = ATG, w \neq t)$, as the number of times we observed that particular mutation type divided by the total number of times the central nucleotide was mutated in that context, in the regulatory regions of interest. Next, to calculate the probability of a particular mutation in a particular trinucleotide context, we multiply the probability that the trinucleotide is mutated with the probability of the specific mutation, e.g.:

$$P(t = AAG | w = ATG) = P(w \neq t | w = ATG) \times P(t = AAG | w = ATG, w \neq t) \tag{1}$$

which can be simplified to the number of **ATG** to **AAG** mutations in enhancers divided by the total number of **ATG** in enhancers across all normal samples.

We note that the k mutations aggregated over region R are typically from different samples in our cohort, and can thus be considered independent. Therefore, for a set i of k mutations we compute the overall probability P_i of that particular set by multiplying the individual probabilities $P(t|w)$ for each of the k mutations, as illustrated in Fig. 1A, B. Finally, to assess the significance of $\Delta_{R,T}$, we compare this value against the distribution of effects for random sets of k mutations in region R, with the p-values being computed efficiently from P_i and $\mathbf{D}_i$ values using either an analytical or a simulation-based approach, as described below.

2.4 Analytical and Simulation-Based Approaches to Compute the Significance of Mutation Effects on TF Binding

Our simulation-based approach uses the mutation probabilities described in Sect. 2.3 to repeatedly sample mutations in the regulatory region of interest. Let us consider a regulatory region R of length L that contains k mutations across all patients in our cohort. Since there are 3 possible mutations for each position in R, we have a total of $3L$ mutations to consider in this region. At each iteration of our simulation-based approach, we randomly sample k out of the $3L$ possible mutations, with replacement. We do the sampling with replacement because the exact same mutation can occur in two or more patients. For this sampling process, the probability of selecting a particular set i of k mutations (P_i) is computed as described in Sect. 2.3. The TF binding change for this set of mutations ($\mathbf{D}_i$) is computed by taking the maximum effect over the k randomly chosen mutations, as described in Sect. 2.3 and illustrated in Fig. 1B. By repeatedly sampling sets of k mutations using this procedure, we can approximate the distribution of mutation effects on TF binding (Fig. 1A), and use it to compute empirical p-values for $\mathbf{\Delta}_{R,T}$, taking the sign of the effect into account. This simulation-based approach is simple to understand and implement. However, simulations are time consuming and unfeasible for generating background distributions of mutations effects for all regulatory regions (totalling 12,612 promoters and 9,018 enhancers) and all TFs (totalling 582 TFs with 667 binding models available).

Alternatively, we can use an analytical approach to directly compute the p-value for the effect $\mathbf{\Delta}_{R,T}$. Conceptually, the p-value of interest is the probability of obtaining an effect on TF binding at least as large as $\mathbf{\Delta}_{R,T}$ when we randomly choose k of the $3L$ possible mutations in the regulatory region R. For simplicity, let us consider these effects in absolute value. For a set i of k mutations, if at least one of the mutations leads to an absolute change in binding of TF T that is $\geq |\mathbf{\Delta}_{R,T}|$, then $|\mathbf{D}_i| \geq |\mathbf{\Delta}_{R,T}|$. On the other hand, if all the k mutations lead to absolute binding changes $< |\mathbf{\Delta}_{R,T}|$, then we have $|\mathbf{D}_i| < |\mathbf{\Delta}_{R,T}|$. Thus, focusing on the absolute values of the effects of mutations on TF binding, we can compute our p-value of interest as:

$$P(|\text{effect of random set of } k \text{ mutations}| \geq |\mathbf{\Delta}_{R,T}|) = \tag{2}$$

$$1 - P(|\text{effect of random set of } k \text{ mutations}| < |\mathbf{\Delta}_{R,T}|) = 1 - \sum_{\substack{\text{Sets } i \text{ of } k \text{ mutations} \\ s.t. |\mathbf{D_i}| < |\mathbf{\Delta}_{R,T}|}} P_i$$

The total number of possible sets i of k mutations in regulatory region R is $\binom{3L+k-1}{k}$, which is the number of possible unordered outcomes when sampling k out of $3L$ mutations with replacement. Even when choosing only the sets for which all k mutations have absolute binding changes $< |\mathbf{\Delta}_{R,T}|$, the number of possibilities can be very large and not feasible to compute explicitly. To overcome this problem, we compute a vector $\pi = (\pi_1, \pi_2, \ldots, \pi_l)$, $0 \leq l < 3L$, with the probabilities of all individual mutations m in region R for which $|\Delta_m| < |\mathbf{\Delta}_{R,T}|$. The sum in Eq. 2 can then be written in terms of the vector π, as the sum of the all elements in the outer product of π with itself, taken k times. For example, for $k = 3$, each element P_i in Eq. 2 is an element of $\pi \otimes \pi \otimes \pi$. Importantly, we do not need to compute this outer product explicitly, as we are only interested in the sum of all elements in the product, which can be written as $(\pi_1 + \pi_2 + \ldots + \pi_l)^k$. Thus, our p-value of interest can be calculated as:

$$P(|\text{effect of random set of } k \text{ mutations}| \geq |\mathbf{\Delta}_{R,T}|) = 1 - (\pi_1 + \pi_2 + \ldots + \pi_l)^k \tag{3}$$

Finally, since we want to take into account the directionality of the effect of mutations on TF binding, i.e. decreased binding ($\Delta_{R,T} < 0$) or increased binding ($\Delta_{R,T} > 0$), we calculate the one-sided p-value as $p_{R,T} = (1 - (\sum_{j=1}^{l} \pi_j)^k)/2$.

We confirmed that the results from our analytical and simulation-based approaches agree with each other (Fig. 1C). For this test, we randomly picked one of the TFs, MYC, and we calculated the p-values for changes in MYC binding specificity for all 9018 mutated enhancers, using a simulation process with one million iterations. Since the simulation-based approach is more time consuming and has limited precision in estimating the p-values of interest (due to the limited iterations), we used the analytical approach for all subsequent analyses.

2.5   Integrating Results Across All Regulatory Regions of a Gene

Genes encoded in the human genome typically have multiple regulatory regions (enhancers and promoters); mutations in either of these regions could affect a gene's expression. Thus, it is of interest to integrate the effects of mutations across all regulatory regions of each gene. As detailed in Sect. 2.2, we define gene promoters based on TSS coordinates in the RefSeq database [21], and we leverage TSS-enhancer links from the FANTOM consortium [24], considering all the cell types and tissues with available data. In other words, if a genomic region has been identified as an enhancer for gene G in one tissue, then we consider that region as part of the regulatory landscape of gene G, in order to be as inclusive as possible. On average, each TSS is associated with 4.9 enhancers according to the FANTOM data. For genes that have multiple TSSs in RefSeq, we consider each TSS separately. Thus, some genes may appear multiple times in our final results, with different p-values that correspond to its different TSSs.

Given a transcription factor T and a gene G with r regulatory regions containing at least one mutation in our cohort of interest, we can use the approach in Sect. 2.4 to calculate the significance $p_{R_i,T}$ of the effects ($\Delta_{R_i,T}$) of mutations in each regulatory region R_i on the binding specificity of TF T. Next, we want to integrate these effects over all r regulatory regions by combining their p-values. Intuitively, our null hypothesis (H_0) is that the effects for all regulatory regions ($\Delta_{R_i,T}$) come simply from the background distribution of effects due to random mutations. The alternative hypothesis (H_1) is that at least one regulatory region of gene G has an effect $\Delta_{R_i,T}$ significantly larger than expected by chance according to the background model of mutations in regulatory regions. Importantly, as the promoter and enhancer regions used in our analysis do not overlap, we can consider the p-values $p_{R_i,T}$ as coming from independent tests.

One approach to combine the r p-values computed for gene G is Fisher's method [27], which is often used in meta-analyses, including analyses of non-coding mutations in cancer [9,28,29]. However, Fisher's method would not take into account the fact that different regulatory regions have different lengths and different probabilities of harboring mutations. If a regulatory region R_i is very long, then the number of possible mutations, and thus the number of possible effects on TF T, is also large. In comparison, a shorter regulatory region, R_j, will have fewer possible mutations and fewer possible effects on TF T. If the two regions have the same p-value, i.e. $p_{R_i,T} = p_{R_j,T}$, we might still consider, intuitively, that the mutation effect in R_i is more significant because for R_i it would be easier to achieve a large effect on TF binding, since more mutations can occur in R_i than R_j. This is similar to meta-analyses of studies with very different sample sizes, where a weighted version of Stouffer's method, developed by Liptak [30] and also known as the weighted Z-method or weighted Z-test, was found to be superior to Fisher's method when combining p-values from independent tests [31,32].

Here, we use Liptak's method [30] to combine the p-values of all regulatory regions of a gene, with the weights computed based on the mutations probabilities in each region, combined using Shannon's entropy. Specifically, for a regulatory region R of length L we compute its weight as $w_R = -\sum_{m=1}^{3L} p_m log(p_m)$, where p_m is the probability of the mth possible mutation, computed according to the trinucleotide centered at that position (see Fig. 1B). We note that weights computed in this manner are overall correlated with the length of the regulatory regions, but avoid giving an out-sized importance to very long regions (Figure S3). Thus, for TF T and gene G with r regulatory regions, we compute the weighted test statistic $\sum_{i=1}^{r} w_{R_i} z_{R_i,T}$, where $z_{R_i,T} = \Phi^{-1}(p_{R_i,T})$ and Φ^{-1} is the inverse of the standard normal cumulative distribution function, as initially proposed by Liptak [30,33]. Under the null hypothesis (H_0), this test statistic follows a normal distribution $N(0,\sum_{i=1}^{r} w_{R_i})$, for any choice of weights [33,34]. This allows us to compute the p-value of the combined test, $P_{G,T}$, as follows:

$$Z_{G,T} = \frac{\sum_{i=1}^{r} w_{R_i} z_{R_i,T}}{\sqrt{\sum_{i=1}^{r} w_{R_i}^2}} \text{ and } P_{G,T} = 2 \times (1 - \Phi(|Z_{G,T}|)) \tag{4}$$

where Φ is standard normal cumulative distribution function. Finally, we used the Benjamini-Hochberg correction [35] to adjust for multiple hypothesis testing across all genes and all TFs. We then ranked genes according to the smallest p-value across all TFs, and we analyzed the top genes for differences in gene expression. In total, we analyzed 5,336 genes with mutations in enhancers, 11,721 genes with mutations in promoters, and 13,982 genes with mutations in at least one regulatory region (enhancer or promoter).

3 Results

3.1 Integrated Analysis Across Regulatory Regions Identifies 54 Genes with Significant TF Binding Changes Due to Mutations in Regulatory DNA

Using the single nucleotide mutation data from the LIRI-JP study, we identified 13,982 genes with mutations in either the promoter or the enhancer regions. For each of these genes, we analyzed all 582 human TFs for which QBiC-Pred models are available [19], and we computed the smallest p-value $P_{G,T}$ across all

TFs, adjusted for multiple testing, as described in Data and Methods (Fig. 1). We chose to take the minimum p-value, rather than integrate the p-values across factors, because TF binding specificities are oftentimes highly correlated, especially for closely-related paralogous TFs. Next, we ranked genes according to the smallest adjusted p-value across all TFs, and we identified 54 genes for which this p-value was < 0.05 (Fig. 2, Table S1). In other words, for each of these 54 genes, our analysis revealed at least one TF for which the mutations in the regulatory regions of the gene have larger effects on TF binding specificity than expected by chance based on the mutation spectra in our cohort.

To determine if these 54 prioritized genes are potentially relevant in cancer, we first asked whether this set is enriched for cancer prognostic genes. Using pathology data from The Human Protein Atlas [36], we found that 33 out of the 54 prioritized genes are indeed prognostic markers in at least one cancer type (Fig. 2B). This represents a significant enrichment ($p = 0.095$, Fisher's exact test) when compared to genes ranked in the bottom half of our ranked list.

We performed similar analyses focusing only on promoters or only on enhancer regions. For the enhancer-only analysis, we identified 5 significant genes (at a minimum adjusted p-value cutoff of 0.05) among the 5,336 genes with enhancer mutations (Table S1). For the promoter-only analysis, we identified 73 significant genes among the 11,721 genes with promoter mutations (Table S1), none of which were also prioritized in the enhancer-only analysis. These results are not surprising, given that some genes only have either enhancer or promoter mutations, but not both. In addition, here we use a stringent set of enhancers, as reported by the FANTOM consortium, in order to limit the number of false positive enhancer calls; however, there are likely a large number of false negatives, i.e. enhancers that are missing from our data. Among the 54 genes identified in the combined analysis of promoters and enhancers, 4 of them are also prioritized in the enhancer-only analysis, and 47 of them are prioritized in the promoter-only analysis (Figure S4). Three genes, ETS1, CELF6, and PALT1 were identified only in the combined analysis (Fig. 2B).

We also found a significant enrichment of cancer prognostic genes in the set of 73 genes prioritized in the promoter-only analysis (44 of the 73 prioritized genes are prognostic markers, Fisher's exact test $p = 0.084$). For the enhancer-only analysis, we found that 3 of the 5 prioritized genes are prognostic markers. However, given the small number of genes prioritized in this analysis, the enrichment in prognostic markers was not significant.

3.2 Genes with Significant Mutations in Their Regulatory Regions Show Large Expression Differences in Mutated Versus Non-mutated Samples

For the 54 genes prioritized based on mutations in enhancers and promoters, we also asked whether the mutations are likely to affect gene expression. To test this, we leveraged the gene expression data (EXP-S) available in ICGC for our cohort of interest (Sect. 2.1). For each gene, we compared its expression level (i.e. normalized read counts, or normalized TPM values) for donors with versus without

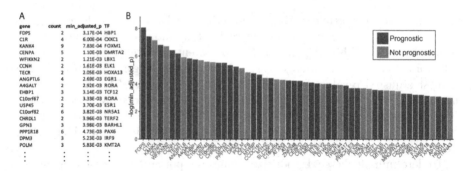

Fig. 2. Genes prioritized based on mutations in their regulatory regions. (A) Top genes with the smallest adjusted p-values. The table shows each gene's name, the number of mutations in all its regulatory regions, the minimum p-value after adjusting for multiple testing, and the TF with the most significant binding changes for that gene. (B) Barplot shows the top 54 prioritized genes from our combined promoter-enhancer analysis. Y-axis shows the negative logarithm of the minimum adjusted p-value. Cancer prognostic markers, as reported in The Human Protein Atlas, are shown in yellow. (Color figure online)

mutations in the regulatory regions of that gene, and we used a Wilcoxon rank-sum test to assess the significance of the observed gene expression differences. Our analysis revealed that the difference in gene expression between donors with vs. without mutations in regulatory regions are much more significant for the set of 54 prioritized gene (i.e. those with minimum adjusted Liptak's test $p < 0.05$) compared to a control set of genes (i.e. those with $p \geq 0.1$) (Fig. 3A). Gene with intermediate p-values ($0.05 \leq p < 0.1$) also showed significant gene expression differences (Fig. 3A).

We note that gene expression analyses could not be performed for all genes, as for some genes with mutations in regulatory regions there was no expression data available from the donors where the mutations were observed. Among the 54 prioritized genes, 43 genes had expression data for one or more donors with mutations in enhancers or promoters, and in most of those cases the number of such donors was one, making it difficult to reach statistical significance. Nevertheless, we found significant gene expression differences for 8 of the 43 prioritized genes with available data (Fig. 3B–H).

Among these genes, TM4SF18 and CENPA (Fig. 3B, C) are prognostic markers in liver cancers according to The Human Protein Atlas [36]. In addition, CENPA has been shown to be aberrantly expressed in hepatocellular carcinomas (HCCs) compared to non-tumor tissues [37]. Another one of the significant genes, CTNNA3, is a tumor suppressor in HCCs [38]; according to our analysis, its expression is very low in the only donor with mutations in the regulatory regions of CTNNA3 (Fig. 3D), which is consistent with the gene's role as a tumor suppressor. Gene DPM3 is part of the DPM family, whose members were found to be significantly correlated with shorter overall survival in liver cancer patients [39]; in our analysis, the only donor with mutations in the regulatory

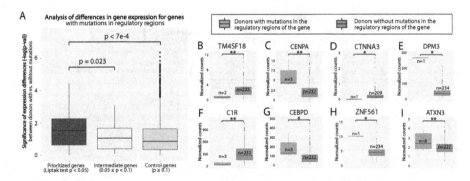

Fig. 3. Analysis of differences in gene expression for genes with mutations in enhancer and promoter regions, comparing donors with vs. without mutations in regulatory DNA. (A) Genes prioritized by our analysis (minimum adjusted Liptak's test $p < 0.05$) have larger expression differences due to mutations in regulatory regions compared to intermediate ($0.05 \leq p < 0.1$) and control ($p \geq 0.1$) genes. To represent the expression differences (y-axis), we show the negative logarithm of the Wilcoxon rank-sum test p-values between the gene's expression level in the two donor groups, i.e. with vs. without mutations in regulatory regions. (B-H) Boxplots showing the 8 prioritized genes with significant gene expression differences. The y-axes show the gene expression levels, as normalized read counts, as reported in ICGC. P-values are calculated using a two-sided Wilcoxon rank-sum test, with ** denoting $p < 0.05$ and * denoting $0.05 \leq p < 0.1$.

regions of DPM3 had a very high DPM3 expression level (Fig. 3E). Genes C1R, CEBPD, and ZNF561 (Fig. 3F, G, H) are prognostic markers in renal cancers, which have been shown to metastasize to liver [40].

The remaining prioritized gene with significant expression differences, ATXN3 (Fig. 3I) was interestingly not among the genes characterized as prognostic markers in cancer. ATXN3 plays important roles in several tumours, e.g. by deubiquitinating PTEN in lung cancer [41] and KLF4 in breast cancer [42], although a role for ATXN3 in liver cancer has not yet been reported. According to our analysis, mutations in the ATXN3 enhancers result in significant gain-of-binding mutations for RUNX1 (Fig. 4), a TF involved in tumour initiation and development in hematopoietic cells and several tissues [43]. Among the eight donors with mutations in ATXN3 enhancers, five have mutations that lead to significant changes in RUNX1 binding specificity, and for these five donors the expression level of ATXN3 is significantly higher than for all other donors ($p = 0.011$, Wilcoxon rank-sum test), despite the small sample size. Overall, these results show that our method is able to identify and prioritize relevant genes that are likely to be significantly affected by mutations in their regulatory regions.

We also analyzed the expression levels of genes prioritized in the promoter- and enhancer-only analyses. For the 5 genes prioritized according to mutations in enhancers, we found more significant expression differences (between donors with vs. without regulatory mutations) than in the control gene set ($p = 0.03$,

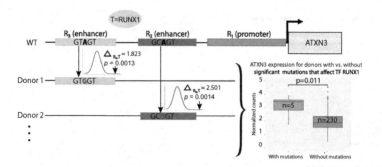

Fig. 4. Mutations in the ATXN3 enhancers that alter the binding specificity of TF RUNX1. Out of 8 donors with ATXN3 regulatory mutations, 5 have significant ($p < 0.1$, one-sided Wilcoxon rank-sum test) mutations in enhancers which result in the gain of binding of TF RUNX1. Two enhancers where RUNX1 has the most significant binding changes among all TFs are shown. Donors with significant RUNX1 mutations (orange boxplot) have higher ATXN3 expression compared to donors without ATXN3 regulatory mutations (grey boxplot), $p < 0.011$ two-sided Wilcoxon rank-sum test. (Color figure online)

one-sided Wilcoxon rank-sum test; Figure S6B). Interestingly, the set of 73 genes prioritized in the promoter-only analysis (Figure S5) did not differ significantly from a control set ($p = 0.231$; Figure S6A). Of these 73 genes, 47 were also prioritized in the combined analysis of promoters and enhancers, while the remaining 26 genes appeared only in the promoter-only analysis, suggesting that the promoter-only analysis may be more prone to false positive, i.e. prone to prioritizing genes for which promoter mutations do not correspond to changes in gene expression. To further investigate this, we focused on the 26 genes, and found that only one of them (DDX21) has significant expression differences between donors with vs. without mutations (Figure S6C). This suggests that integrating regulatory mutations over enhancers and promoters, rather than promoters alone, has the best potential to prioritize genes that are dys-regulated in cancer.

4   Discussion

In summary, we developed a new approach to prioritize putative regulatory driver mutations in cancer, based on quantitative predictions of the effects of single nucleotide variants on TF binding [19]. Our method is orthogonal to existing tools (e.g. [4–7]), in that it does not require the putative driver mutations to be highly recurrent; instead, we assess the significance of the mutations by testing whether they cause *larger* TF binding changes than expected in the case of completely random mutations. Using this approach, we identified 54 potentially dysregulated genes (Fig. 2) by prioritizing genes for which mutations in their enhancers or promoters lead to significant changes in TF binding specificity. Our analyses show that these genes are enriched for cancer prognostic markers, and they have higher differences in expression levels between donors with vs.

without mutations in regulatory regions, compared to control gene sets (Fig. 3). We also linked the potentially dysregulated genes with the TFs whose binding events are altered the most by putative regulatory mutations (Fig. 4).

Our approach can be applied to any somatic mutation data. Here, we used single nucleotide mutations; however, our method can also be applied, with minor modifications, to small indels. In addition, our method can be adapted for patient-level analyses. Current driver identification approaches combine mutations from different patients to gain more statistical power to identify significant regulatory mutations. However, a disadvantage of such approaches is that they may not work well on small cohorts or on cohorts that are highly heterogeneous. For approaches that identify drivers based on recurrence, it is impossible to run the analysis for individual patients. However, since our method does not require the driver mutations to be recurrent, it can be applied to identify potentially dysregulated genes for each patient, using our simulation-based approach, with minor modifications to the resampling process. Briefly, instead of using sampling with replacement, we would sample without replacement for patient-level analyses because a patient cannot have the same mutation more than once. The patient-level analysis would need to be further refined so that it has more coherent evidence to prioritize genes for follow-up experimental validations; however, it would provide a very different perspective than the cohort-level approaches.

Overall, the TF-centric approach described here uses a distinctive pipeline to prioritize putative regulatory driver mutations in cancer, by focusing on the magnitude of the effect of the mutations. Our approach is orthogonal to existing methods and thus serves to complement existing tools and resources for analyzing and prioritizing putative non-coding drivers (e.g. [44]). Our results show that most of the potentially dysregulated genes prioritized by our method either have large expression differences in donors with vs. without mutations, or are cancer prognostic genes, or both. While experimental validations are needed to determine whether these genes actually contribute to cancer development, our results suggest that regulatory mutations should be investigated further, not just based on their recurrence, but also based on their functional effects, in order to uncover dysregulated genes that may drive tumorigenesis.

5 Acknowledgements, Code Availability, and Supplemental Materials

This work was supported by National Science Foundation grant MCB-1715589 (R.G.) and National Institutes of Health grant R01-GM135658 (R.G.). All R code used in this work, as well as all supplemental tables and figures are available in GitHub at https://github.com/jz132/cancer-mutations.

References

1. ICGC/TCGA Pan-Cancer Analysis of Whole Genomes Consortium: Pan-cancer analysis of whole genomes. Nature **578**(7793), 82–93 (2020)

2. Khurana, E., Fu, Y., Chakravarty, D., Demichelis, F., Rubin, M., Gerstein, M.: Role of non-coding sequence variants in cancer. Nat. Rev. Genet. **17**(2), 93–108 (2016)
3. Elliott, K., Larsson, E.: Non-coding driver mutations in human cancer. Nat. Rev. Cancer **21**(8), 500–509 (2021)
4. Lochovsky, L., Zhang, J., Fu, Y., Khurana, E., Gerstein, M.: LARVA: an integrative framework for large-scale analysis of recurrent variants in noncoding annotations. Nucleic Acids Res. **43**(17), 8123–8134 (2015)
5. Lochovsky, L., Zhang, J., Gerstein, M.: MOAT: efficient detection of highly mutated regions with the mutations overburdening annotations tool. Bioinformatics **34**(6), 1031–1033 (2018)
6. Rheinbay, E., et al.: Recurrent and functional regulatory mutations in breast cancer. Nature **547**(7661), 55–60 (2017)
7. Weinhold, N., Jacobsen, A., Schultz, N., Sander, C., Lee, W.: Genome-wide analysis of noncoding regulatory mutations in cancer. Nat. Genet. **46**(11), 1160–1165 (2014)
8. Lawrence, M.S., et al.: Mutational heterogeneity in cancer and the search for new cancer-associated genes. Nature **499**(7457), 214–218 (2013)
9. Rheinbay, E., et al.: Analyses of non-coding somatic drivers in 2,658 cancer whole genomes. Nature **578**(7793), 102–111 (2020)
10. Heinz, S., et al.: Simple combinations of lineage-determining transcription factors prime cis-regulatory elements required for macrophage and B cell identities. Mol. Cell **38**(4), 576–589 (2010)
11. Link, V.M., Romanoski, C.E., Metzler, D., Glass, C.K.: MMARGE: motif mutation analysis for regulatory genomic elements. Nucleic Acids Res. **46**(14), 7006–7021 (2018)
12. Shen, Z., Hoeksema, M.A., Ouyang, Z., Benner, C., Glass, C.K.: MAGGIE: leveraging genetic variation to identify DNA sequence motifs mediating transcription factor binding and function. Bioinformatics **36**(Suppl_1), i84–i92 (2020)
13. Horn, S., et al.: TERT promoter mutations in familial and sporadic melanoma. Science **339**(6122), 959–961 (2013)
14. Huang, F.W., Hodis, E., Xu, M.J., Kryukov, G.V., Chin, L., Garraway, L.A.: Highly recurrent TERT promoter mutations in human melanoma. Science **339**(6122), 957–959 (2013)
15. Buisson, R., et al.: Passenger hotspot mutations in cancer driven by APOBEC3A and mesoscale genomic features. Science **364**(6447), 06 (2019)
16. Mas-Ponte, D., Supek, F.: DNA mismatch repair promotes APOBEC3-mediated diffuse hypermutation in human cancers. Nat. Genet. **52**(9), 958–968 (2020)
17. Perera, D., Poulos, R.C., Shah, A., Beck, D., Pimanda, J.E., Wong, J.W.: Differential DNA repair underlies mutation hotspots at active promoters in cancer genomes. Nature **532**(7598), 259–263 (2016)
18. Kim, E., et al.: Systematic functional interrogation of rare cancer variants identifies oncogenic alleles. Cancer Discov. **6**(7), 714–726 (2016)
19. Martin, V., Zhao, J., Afek, A., Mielko, Z., Gordân, R.: QBiC-Pred: quantitative predictions of transcription factor binding changes due to sequence variants. Nucleic Acids Res. **47**(W1), W127–W135 (2019)
20. Zhao, J., Li, D., Seo, J., Allen, A.S., Gordân, R.: Quantifying the impact of noncoding variants on transcription factor-DNA binding. Res. Comput. Mol. Biol. **10229**, 336–352 (2017)
21. O'Leary, N.A., et al.: Reference sequence (RefSeq) database at NCBI: current status, taxonomic expansion, and functional annotation. Nucleic Acids Res. **44**(D1), D733-745 (2016)

22. Tweedie, S., et al.: Genenames.org: the HGNC and VGNC resources in 2021. Nucleic Acids Res. **49**(D1), D939–D946 (2021)
23. Andersson, R., et al.: An atlas of active enhancers across human cell types and tissues. Nature **507**(7493), 455–461 (2014)
24. Lizio, M., et al.: Gateways to the FANTOM5 promoter level mammalian expression atlas. Genome Biol. **16**, 22 (2015)
25. Alexandrov, L.B., et al.: The repertoire of mutational signatures in human cancer. Nature **578**(7793), 94–101 (2020)
26. Jusakul, A., et al.: Whole-genome and epigenomic landscapes of etiologically distinct subtypes of cholangiocarcinoma. Cancer Discov. **7**(10), 1116–1135 (2017)
27. Fisher, R.A.: Statistical Methods for Research Workers, 4th edn. Oliver & Boyd, Edinburgh (1934)
28. Lawrence, M.S., et al.: Discovery and saturation analysis of cancer genes across 21 tumour types. Nature **505**(7484), 495–501 (2014)
29. Araya, C.L., et al.: Identification of significantly mutated regions across cancer types highlights a rich landscape of functional molecular alterations. Nat. Genet. **48**(2), 117–125 (2016)
30. Lipták, T.: On the combination of independent tests. Magyar Tud Akad Mat Kutato Int Kozl **3**, 171–197 (1958)
31. Whitlock, M.C.: Combining probability from independent tests: the weighted Z-method is superior to Fisher's approach. J. Evol. Biol. **18**(5), 1368–1373 (2005)
32. Zaykin, D.V.: Optimally weighted Z-test is a powerful method for combining probabilities in meta-analysis. J. Evol. Biol. **24**(8), 1836–1841 (2011)
33. van Zwet, W.R., Oosterhoff, J.: On the combination of independent test statistics. Ann. Math. Stat. **38**(3), 659–680 (1967)
34. Heard, N.A., Rubin-Delanchy, P.: Choosing between methods of combining p-values. Biometrika **105**(1), 239–246 (2018)
35. Hochberg, Y.: A sharper Bonferroni procedure for multiple tests of significance. Biometrika **75**(4), 800–802 (1988)
36. Uhlen, M., et al.: A pathology atlas of the human cancer transcriptome. Science **357**(6352), 08 (2017)
37. Li, Y., et al.: ShRNA-targeted centromere protein A inhibits hepatocellular carcinoma growth. PLoS ONE **6**(3), e17794 (2011)
38. He, B., et al.: CTNNA3 is a tumor suppressor in hepatocellular carcinomas and is inhibited by miR-425. Oncotarget **7**(7), 8078–8089 (2016)
39. Li, M., Xia, S., Shi, P.: DPM1 expression as a potential prognostic tumor marker in hepatocellular carcinoma. PeerJ **8**, e10307 (2020)
40. Bianchi, M., et al.: Distribution of metastatic sites in renal cell carcinoma: a population-based analysis. Ann. Oncol. **23**(4), 973–980 (2012)
41. Sacco, J.J., et al.: The deubiquitylase Ataxin-3 restricts PTEN transcription in lung cancer cells. Oncogene **33**(33), 4265–4272 (2014)
42. Zou, H., Chen, H., Zhou, Z., Wan, Y., Liu, Z.: ATXN3 promotes breast cancer metastasis by deubiquitinating KLF4. Cancer Lett. **467**, 19–28 (2019)
43. Otálora-Otálora, B.A., Henríquez, B., López-Kleine, L., Rojas, A.: RUNX family: oncogenes or tumor suppressors (review). Oncol. Rep. **42**(1), 3–19 (2019)
44. Liu, E.M., Martinez-Fundichely, A., Bollapragada, R., Spiewack, M., Khurana, E.: CNCDatabase: a database of non-coding cancer drivers. NAR **49**(D1), D1094–D1101 (2021)

DeepMinimizer: A Differentiable Framework for Optimizing Sequence-Specific Minimizer Schemes

Minh Hoang[1]([envelope]) [ID], Hongyu Zheng[2] [ID], and Carl Kingsford[2] [ID]

[1] Computer Science Department, Carnegie Mellon Univeristy,
Pittsburgh, PA 15213, USA
qhoang@andrew.cmu.edu
[2] Computational Biology Department, Carnegie Mellon Univeristy,
Pittsburgh, PA 15213, USA
hongyuz1@andrew.cmu.edu, carlk@cs.cmu.edu

Abstract. Minimizers are k-mer sampling schemes designed to generate sketches for large sequences that preserve sufficiently long matches between sequences. Despite their widespread application, learning an effective minimizer scheme with optimal sketch size is still an open question. Most work in this direction focuses on designing schemes that work well on expectation over random sequences, which have limited applicability to many practical tools. On the other hand, several methods have been proposed to construct minimizer schemes for a specific target sequence. These methods, however, require greedy approximations to solve an intractable discrete optimization problem on the permutation space of k-mer orderings. To address this challenge, we propose: (a) a reformulation of the combinatorial solution space using a deep neural network re-parameterization; and (b) a fully differentiable approximation of the discrete objective. We demonstrate that our framework, DEEP-MINIMIZER, discovers minimizer schemes that significantly outperform state-of-the-art constructions on genomic sequences.

Keywords: Sequence sketching · Optimization · Deep learning

1 Introduction

Minimizers [15,16] are deterministic methods to sample k-mers from a sequence at approximately regular intervals such that sufficient information about the identity of the sequence is preserved. Sequence sketching with minimizers is widely used to reduce memory consumption and processing time in bioinformatics programs such as read mappers [7,10], k-mer counters [3,5] and genome assemblers [17]. Given a choice of k-mer length k and window length w, a minimizer selects the lowest priority k-mer from every overlapping window in the target sequence according to some total ordering π over all k-mers. Minimizer performance is measured by its density [12] on a target sequence, which is proportional to the induced sketch size.

© The Author(s), under exclusive license to Springer Nature Switzerland AG 2022
I. Pe'er (Ed.): RECOMB 2022, LNBI 13278, pp. 52–69, 2022.
https://doi.org/10.1007/978-3-031-04749-7_4

Depending on the choice of π, the resulting density can significantly vary. The theoretical lower-bound of density achievable by any minimizer scheme is given by $1/w$ [12]. On the other hand, a random initialization of π will yield an expected density of approximately $2/w$ [16], which is frequently used as a baseline for comparing minimizer performance. This motivates the question: How do we effectively optimize π to improve the performance of minimizers?

An exhaustive search over the combinatorial space of π suffices for very small k, but quickly becomes intractable for values of k used in practice (i.e., $k \geq 7$) (Sect. 3.1). To work around this, many existing approaches focus on constructing minimizer schemes from mathematical objects with appealing properties such as universal hitting sets (UHS) [4,11,12,14,19]. While these schemes provide upper-bound guarantees for expected densities on random sequences, they only obtain modest improvements over a random minimizer when used to sketch a specific sequence [19].

The idea of learning minimizer schemes tailored towards a target sequence has been previously explored, although to a lesser extent. Current approaches include heuristic designs [1,8], greedy pruning [2] and construction of k-mer sets that are well-spread on the target sequence [20]. However, these methods only learn crude approximations of π by dividing k-mers into disjoint subsets with different priorities to be selected. Within each subset, the relative ordering is arbitrarily assigned to recover a valid minimizer, hence they are not necessarily optimal. We give a detailed overview of these methods in Sect. 2.

This paper instead tackles the problem of directly learning a total order π. The hardness of solving such a task comes from two factors, which we will review in detail in Sect. 3.1: (1) the search space of k-mer orderings is very large; and (2) the density minimizing objective is discrete. To overcome the above challenges, we propose to reformulate the original problem as parameter optimization of a deep learning system. This results in the first fully-differentiable minimizer selection framework that can be efficiently optimized using gradient-based learning techniques. Our contributions are:

- We define a more well-behaved search space that is suitable for gradient-based optimization. This is achieved by implicitly representing k-mer orderings as continuous score assignments. The space of these assignments is parameterized by a neural network called PRIORITYNET, whose architecture guarantees that every output assignment is *consistent* (i.e., corresponding to valid minimizer schemes). The modelling capacity of PRIORITYNET can be controlled via increasing its architecture depth, which implies a mild restriction on the candidate space in practice (Sect. 3.2).
- We approximate the discrete learning objective by a pair of simpler tasks. First, we design a complementary neural network called TEMPLATENET, which outputs potentially *inconsistent* assignments (i.e., template) with guaranteed low densities on the target sequence (Sect. 3.4). We then search for *consistent* assignments (i.e., valid minimizers) around these templates, which potentially will yield similar densities. This is achieved via a fully differentiable proxy objective (Sect. 3.3) that minimizes a novel divergence (Sect. 3.5) between these networks.

– We compare our framework, DEEPMINIMIZER, against various state-of-the-art benchmarks and observe that DEEPMINIMIZER yields sketches with significantly lower densities on genomic sequences (Sect. 4).

2   Related Work

UHS-Based Methods. Most existing minimizer selection schemes with performance guarantees over random sequences are based on the theory of universal hitting sets (UHS) [11,14]. Particularly, a (w, k)-UHS is defined as a set of k-mers such that every window of length w (from any possible sequence) contains at least one of its elements. Every UHS subsequently defines a family of corresponding minimizer schemes whose expected densities on random sequences can be upper-bounded in terms of the UHS size [12]. As such, to obtain minimizers with provably low density, it suffices to construct small UHS, which is often the common learning objective of many existing approaches [4,12,19].

In the context of *sequence-specific* minimizers, there are several concerns with this approach. First, the requirement of UHS to "hit" all windows of *every possible* sequence is often too strong with respect to the need of sketching a specific string and results in sub-optimal universal hitting sets [20]. Additionally, since real sequences rarely follow a uniform distribution [18], there tends to be little correspondence between the provable upper-bound on expected density and the actual density measured on a target sequence. In practice, the latter is usually more pessimistic on sequences of interest, such as the human reference genome [19,20], which drives the development of various *sequence-specific* minimizer selection methods.

Heuristic Methods. Several minimizer construction schemes rank k-mers based on their frequencies in the target sequence [1,8], such that rare k-mers are more likely to be chosen as minimizers. These constructions nonetheless rely on the assumption that rare k-mers are spread apart and ideally correspond to a sparse sampling. Another greedy approach is to sequentially remove k-mers from an arbitrarily constructed UHS, as long as the resulting set still hits every w-long window on the target sequence [2]. Though this helps to fine-tune a given UHS with respect to the sequence of interest, there is no guarantee that such an initial set will yield the optimal solution after pruning.

Polar Set Construction. Recently, a novel class of minimizer constructions was proposed based on polar sets of k-mers, whose elements are sufficiently far apart on the target sequence [20]. The sketch size induced by such a polar set is shown to be tightly bounded with respect to its cardinality. This reveals an alternate route to low-density minimizer schemes through searching for the minimal polar set. Unfortunately, this proxy objective is NP-hard and currently approximated by a greedy construction [20].

Remark. In all of the above methods, the common objective to be optimized can be seen as a partition of the set of all k-mers into disjoint subsets. For example, frequency values are used to denote different buckets of k-mers [1,8]. Others [2,4,19,20] employ a more fine-grained partitioning scheme defined by the constructed UHS/polar set. Each subset has an assigned priority value, such that k-mers from higher priority subsets are always chosen over k-mers from lower priority subsets. However, it remains inconclusive how k-mers from within the same subset can be optimally selected to recover a total ordering π. Practically, these methods resort to using a pre-determined arbitrary ordering to resolve such situations. In contrast, our work investigates a novel approach to directly learn this ordering.

3 Methods

3.1 Background

Let Σ be an alphabet of size $|\Sigma| = \sigma$ and S be a sequence containing exactly l overlapping k-mers defined on this alphabet, i.e., $S \in \Sigma^{l+k-1}$. For some $w \in \mathbb{N}^+$ such that $l \geq w$, we define a (w, k)-window as a substring in S of length $w+k-1$, which contains exactly w overlapping k-mers. For ease of notation, we further let $l_w \triangleq l - w + 1$ denote the number of (w, k)-windows in S. For the rest of this paper, we assume that w and k are fixed and given as application-specific parameters.

Definition 1 (Minimizer). A minimizer scheme $m : \Sigma^{w+k-1} \to [1..w]$ is uniquely specified by a total ordering π on Σ^k. Here, we encode π as a function $\rho : \Sigma^k \to \mathbb{N}^+$ that maps k-mers to its position in π. Given a (w, k)-window ω, m then returns the smallest k-mer in ω according to ρ:

$$m(\omega; \pi) \triangleq \operatorname*{argmin}_{i \in [1..w]} \rho(\omega[i]; \pi) \equiv \operatorname*{argmin}_{i \in [1..w]} \sum_{s \in \Sigma^k} \mathbb{I}(s <_\pi \omega[i]) , \qquad (1)$$

where $\mathbb{I}$ denotes the indicator function, $\omega[i]$ denotes the i-th k-mer in ω, and $s <_\pi \omega[i]$ implies s precedes $\omega[i]$ in π. We break ties by prioritizing k-mers that occur earlier in (i.e., to the left of) the window.

When applied to a sequence S, the above scheme selects one k-mer position from every overlapping window to construct the sequence sketch $\mathcal{L}(S; m) = \{t+m(\omega_t) \mid t \in [1, l_w]\}$, with ω_t denoting the tth window in S. Naturally, a smaller sketch leads to more space and cost savings. As such, we measure minimizer performance by the density factor metric $\mathcal{D}(S; m) \triangleq |\mathcal{L}(S; m)| \times (w+1)/l_w$, which approximates the number of k-mers selected per window [12]. The minimizer selection problem is then formalized as density minimization with respect to π:

$$\pi_* = \operatorname*{argmin}_{\pi} \mathcal{D}(S; m(\cdot; \pi)) \equiv \operatorname*{argmin}_{\pi} |\mathcal{L}(S; m(\cdot; \pi))| . \qquad (2)$$

This objective, however, is intractable to optimize for two reasons. First, the number of all k-mer permutations scales super-exponentially with k and σ (i.e.,

$\sigma^k!$), thus renders any form of exhaustive search on this space impossible under most practical settings. Furthermore, the set counting operation $|\mathcal{L}(S; m(\cdot; \pi))|$ is non-differentiable even if the solution space is continuous, which makes efficient gradient-based optimizers inaccessible. The remainder of this section therefore proposes a deep-learning strategy to address both these challenges, and is organized as follows.

Section 3.2 describes a unifying view of existing methods as reparameterizations of ρ (Definition 1). We then propose a novel deep parameterization called PRIORITYNET, which relaxes the permutation search space of Eq. 2 into a well-behaved weight space of a neural network.

Section 3.3 shows that density optimization with respect to PRIORITYNET can be approximated by two sub-tasks via introducing another complementary network, called TEMPLATENET. This approximation can be formalized as a fully-differentiable proxy objective that minimizes divergence between TEMPLATENET and PRIORITYNET.

Section 3.4 and Sect. 3.5 then respectively discuss the parameterization of TEMPLATENET and the divergence measure in our proxy objective, thus completing the specification of our framework, DEEPMINIMIZER. An overview of our framework is given in Fig. 1.

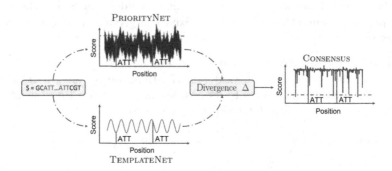

Fig. 1. Our DEEPMINIMIZER framework employs a twin network architecture. PRIORITYNET generates valid minimizers, but has no guarantee on density. In contrast, TEMPLATENET generates low-density templates that might not correspond to valid minimizers. We minimize the divergence between these networks to arrive at consensus minimizers with low densities on the target sequence.

3.2 Search Space Reparameterization

We first remark that many existing methods can be seen as different reparameterizations of ρ in Definition 1. For example, ρ can be parameterized with frequency information from the target sequence [1,8], i.e., $\rho(\omega_i; S) \propto \sum_{j=1}^{l_w} \mathbb{I}(\omega_j = \omega_i)$; or instantiated with a UHS v [4,19], i.e., $\rho(\omega_i; v) = \mathbb{I}(\omega_i \notin v)$. Similar set-ups have been explored in the context of sequence-specific minimizers using a pruned UHS $v(S)$ [2] and a polar set $\zeta(S)$ [20] constructed for the target sequence. We note that there are fewer discrete values potentially assigned by ρ than the total number of k-mers in all these re-parameterizations. As such, these methods still rely on a pre-determined arbitrary ordering to break ties in windows with two or more similarly scored k-mers. When collisions occur frequently, this could have unexpected impact on the final density.

DEEPMINIMIZER instead employs a continuous parameterization of ρ using a feed-forward neural network parameterized by weights α, which takes as input the multi-hot encoding of a k-mer (i.e., a concatenation of its character one-hot encodings) and returns a real-valued score in $[0, 1]$. This continuous scheme practically eliminates the chance for scoring collisions. Furthermore, the solution space of this re-parameterization is only restricted by the modelling capacity encoded by our architecture weight space. This limitation quickly diminishes as we employ sufficiently large number of hidden layers in the network. We can subsequently rewrite Eq. 2 as optimizing a neural network with density as its loss function:

$$\alpha_* = \operatorname*{argmin}_{\alpha} \mathcal{D}(S; \rho(\cdot; \alpha)) . \tag{3}$$

Applying this network on every k-mer along S can be compactly written as a convolutional neural network, denoted by f, which maps the entire sequence S to a *score assignment* vector. We require this score assignment to be *consistent* across different windows in order to recover a valid ordering π from such implicitly encoded ρ. Specifically, one k-mer can not be assigned different scores at different locations in S. To enforce this, we let the first convolution layer of our architecture, PRIORITYNET, have kernel size k, and all subsequent layers to have kernel size 1. An illustration for PRIORITYNET when $k = 2$ is given in Fig. 2.

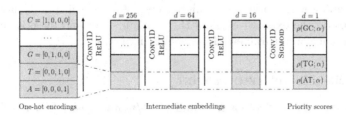

Fig. 2. Our PRIORITYNET architecture for $k = 2$, parameterized by weights α, maps sequence multi-hot encoding to priority scores through a series of 3 convolution layers with kernel size $[k, 1, 1]$ and $[256, 64, 16]$ embedding channels respectively. Fixing network weights α, the computation of assigned priority score to any k-mer is deterministic given its character one-hot encodings.

3.3 Proxy Objective

The density computation in Eq. 3, however, is not differentiable with respect to the network weights. As such, α cannot be readily optimized with established gradient back-propagation techniques used in most deep learning methods. To work around this, we introduce a proxy optimization objective that approximates Eq. 3 via coupling PRIORITYNET with another function called TEMPLATENET. Unlike the former, TEMPLATENET relaxes the *consistency* requirement and generates *template* score assignments that might not correspond to valid minimizer schemes. In exchange, such *templates* are guaranteed to yield low densities by design.

Intuitively, the goals of these networks are complementary: PRIORITYNET generates valid minimizer schemes in the form of *consistent* priority score assignments, whereas TEMPLATENET pinpoints neighborhoods of low-density score assignments situated around its output templates. This reveals an alternative optimization route where these networks negotiate towards a consensus solution that (a) satisfies the constraint enforced by PRIORITYNET; and (b) resembles a template in the output space of TEMPLATENET, thus potentially yielding low density. Let f and g denote our proposed PRIORITYNET and TEMPLATENET, respectively parameterized by weights α and β, we formalize this objective as minimizing some divergence measure Δ between their outputs:

$$(\alpha_*, \beta_*) = \underset{\alpha, \beta}{\text{argmin}} \; \Delta \left(f(S; \alpha), g(S; \beta) \right) . \tag{4}$$

In the remainder of this paper, we detail the full specification of our proxy objective, which requires two other ingredients. First, Sect. 3.4 discusses the parameterization of our TEMPLATENET g to consistently generate templates that achieve the theoretical lower-bound density [12] on the target sequence. Furthermore, we note that the proxy objective in Eq. 4 will perform best when the divergence measure Δ reflects the difference in densities of two score assignments. Section 3.5 then discusses a practical choice of Δ to accurately capture high-performing neighborhoods of minimizers. These specifications have strong implications on the expressiveness of the solution space and directly influences the performance of our framework, as shown in Sect. 4.

3.4   Specification of TEMPLATENET

The well-known theoretical lower bound $1 + 1/w$ for density factor [12] implies that the optimal minimizer, if it exists, samples k-mers exactly w positions apart. As a result, we want to guarantee that the output of TEMPLATENET approximates this scenario given any weights initialization. Without loss of generality, we impose that TEMPLATENET is given by a continuous function $g : \mathbb{R} \to [0, 1]$, such that its output template $\mathbf{v} = [g(i)]_{i \in [l]}$ consists of evaluations of g restricted to integer inputs (i.e., k-mer positions). Then, Proposition 1 below shows a sufficient construction for g that approximately yields the optimal density.

Proposition 1. *Let $g : \mathbb{R} \to [0, 1]$ be a periodic function with minimal period w, such that g has a unique minimum value on every w-long interval. Formally, g satisfies:*

$$(1): \; \forall t \in \mathbb{R} : g(t) = g(t + w)$$

$$(2): \ \forall i,j \in \operatorname*{arginf}_{t} g(t), \ \exists u \in \mathbb{N} : |i - j| = uw \ .$$

Then, the template generated by g induces a sketch with density factor $1+1/w+$ *$o(1)$ on S when S is sufficiently long (i.e., $l_w \gg w^2$).*

Proof. We give a detailed proof of Proposition 1 in Appendix A.

Note that the resulting sketch induced by g does not necessarily correspond to a valid minimizer. While this sketch has low density, it does not preserve the sequence identity like a minimizer sketch, hence is not useful for downstream applications. However, it is sufficient as a guiding template to help PRIORITYNET navigate the space of orderings.

Proposition 1 leaves us with infinitely many candidate functions to choose from. In fact, TEMPLATENET can be as simple as $g(t) = \sin(2\pi t/w)$ to generate a near-optimal score assignment. This naïve specification, however, encodes exactly one template (i.e., one that picks k-mers from the set of interval positions $\{w, 2w, \dots\}$), whose proximal neighborhood might not contain any valid minimizer scheme. For example, consider a sequence S in which some particular k-mer occurs exactly at positions $t \in \{\frac{1}{2}w, \frac{3}{2}w, \dots\}$. Ideally, we would want to align the *template minima* with these locations, which is not possible given the above choice of g. As such, it is necessary that the specification of TEMPLATENET is sufficiently expressive for Eq. 4 to find an optimal solution.

In particular, we want to construct a parameterized function such that every k-mer position can be sampled by at least one sketch encoded in its parameter space. Furthermore, we note that the periodic property is only a sufficient condition to obtain low-density sketches. In practice, we only want the template minima to periodically occur at fixed intervals. Enforcing the scores assigned at all positions to exactly follow a sinusoidal pattern is restrictive and might lead to overlooking good templates. To address these design goals, we propose the following ensemble parameterization:

$$g(t) = \sigma \left(\sum_{\phi=0}^{w-1} \beta_\phi \sin\left(\frac{2\pi}{w}(t + \phi) \right) \right) , \tag{5}$$

where σ denotes a sigmoid activation function, which ensures that $g(t)$ appropriately maps to $[0,1]$; $\beta = \{\beta_\phi\}_{\phi=0}^{w-1}$ are optimizable parameters such that $\beta_\phi \geq 0$ and $\sum_{\phi=1}^{w} \beta_\phi = 1$.

Optimizing β has two implications. First, by adjusting the dominant phase shift $\phi_{\max} = \operatorname{argmax}_\phi \beta_\phi$, we can control the offset of the periodic template minima, which leads to good coverage on the target sequence. Second, by adjusting the magnitudes of the remaining phase shifts $\{\beta_\phi\}_{\phi \neq \phi_{\max}}$, we can have more degrees of freedom to assign scores outside the template minima. Lastly, the non-negative and sum-to-one constraints help to avoid the trivial assignment of squashing all magnitudes to 0 and are easily guaranteed by letting β be the output of a softmax layer.

3.5   Specification of the Divergence Measure Δ

As standard practice, we first consider instantiating Δ with the squared ℓ^2-distance. Specifically, let $\mathbf{v}_f = f(S; \alpha)$ and $\mathbf{v}_g = g(S; \beta)$ denote the score assignments respectively output by PRIORITYNET and TEMPLATENET given S, then $\Delta_{\ell^2}(\mathbf{v}_f, \mathbf{v}_g) \triangleq \sum_{i=1}^{l}(\mathbf{v}_f[i] - \mathbf{v}_g[i])^2$. This divergence measure, however, places an excessively strict matching objective at all locations along $\mathbf{v}_f$ and $\mathbf{v}_g$. Such a perfect match is unnecessary as long as the k-mers outside sampled locations are assigned higher scores, and will take away the degrees of freedom needed for the proxy objective to satisfy the constraints implied by PRIORITYNET.

Consequently, we are interested in constructing a divergence that: (a) strategically prioritizes matching $\mathbf{v}_f$ to the minima of the template $\mathbf{v}_g$; and (b) enables flexible assignment at other positions to admit more solutions that meet the consistency requirement. To accomplish these design goals, we propose the following asymmetrical divergence:

$$\Delta(\mathbf{v}_f, \mathbf{v}_g) \triangleq \sum_{i=1}^{l} \left[(1 - \mathbf{v}_g[i]) \cdot (\mathbf{v}_f[i] - \mathbf{v}_g[i])^2 + \lambda \cdot \mathbf{v}_g[i] \cdot (1 - \mathbf{v}_f[i])^2 \right]. \quad (6)$$

Specifically, the idea behind the first component $(1 - \mathbf{v}_g[i]) \cdot (\mathbf{v}_f[i] - \mathbf{v}_g[i])^2$ in the summation is to weight each position-wise matching term $(\mathbf{v}_f[i] - \mathbf{v}_g[i])^2$ by its corresponding template score: the weight term $(1 - \mathbf{v}_g[i])$ implies stronger matching preference around the minima of $\mathbf{v}_g$ where the template scores $\mathbf{v}_g[i]$ are low, and vice versa weaker preference at other locations. Furthermore, to ensure that f properly assigns higher scores to the locations outside the minima of $\mathbf{v}_g$, the second component $\mathbf{v}_g[i] \cdot (1 - \mathbf{v}_f[i])^2$ subsequently encourages f to maximize its assigned scores wherever possible, again weighted by the relative relevance of each location. The trade-off between these components is controlled by the hyper-parameter λ. Finally, we confirm that this divergence measure is fully differentiable with respect to α and β, hence can be efficiently optimized using gradient-based techniques. Particularly, the parameter gradients of both networks are given by:

$$\frac{\partial}{\partial \alpha} \Delta(\mathbf{v}_f, \mathbf{v}_g) = \sum_{i=1}^{l} a_i \cdot \frac{\partial}{\partial \alpha} \mathbf{v}_f[i]$$

$$\frac{\partial}{\partial \beta} \Delta(\mathbf{v}_f, \mathbf{v}_g) = \sum_{i=1}^{l} b_i \cdot \frac{\partial}{\partial \beta} \mathbf{v}_g[i] , \quad (7)$$

where the gradients of network outputs are obtained via back-propagation and their respective constants are given by $a_i = 2 \cdot (1 - \mathbf{v}_g[i]) \cdot (\mathbf{v}_f[i] - \mathbf{v}_g[i]) + 2\lambda \cdot \mathbf{v}_g[i] \cdot (\mathbf{v}_f[i] - 1)$ and $b_i = 2 \cdot (\mathbf{v}_g[i] - 1) \cdot (\mathbf{v}_f[i] - \mathbf{v}_g[i]) - (\mathbf{v}_f[i] - \mathbf{v}_g[i])^2 + \lambda \cdot \mathbf{v}_g[i] \cdot (\mathbf{v}_f[i] - 1)$.

4   Results

Implementation Details. We implement our method using PyTorch and deploy all experiments on a RTX-2060 GPU. Due to limited GPU memory, each training epoch only computes a batch divergence which averages over $N = 10$ randomly sampled subsequences of length $l = 500 \times (w + k)$. We set $\lambda = 1$ and use architectures of PRIORITYNET and TEMPLATENET as given in Fig. 2 and Sect. 3.4 respectively. Network weights are optimized using the ADAM optimizer [9] with learning rate $\eta = 5 \times 10^{-3}$. Our implementation is available at https://github.com/Kingsford-Group/deepminimizer.

Comparison Baselines. We compare DEEPMINIMIZER with the following benchmarks: (a) random minimizer baseline; (b) Miniception [19]; (c) PASHA [4]; and (d) PolarSet Minimizer [20]. Among these methods, (d) is a sequence-specific minimizer scheme. For each method, we measure the density factor $\mathcal{D}$ obtained on different segments of the human reference genome: (a) chromosome 1 (CHR1); (b) chromosome X (CHRX); (c) the centromere region of chromosome X [13] (which we denote by CHRXC); and (d) the full genome (HG38). We used lexicographic ordering for PASHA as suggested by [19]. Random ordering is used to rank k-mers within the UHS for Miniception, and outside the layered sets for PolarSet.

Visualizing the Mechanism of DEEPMINIMIZER. First, we show the transformation of the priority scores assigned by SCORENET and TEMPLATENET over 600 training epochs. Figure 3 plots the outputs of these networks evaluated on positions 500 to 1000 of CHRXC, and their corresponding locations of sampled k-mers. Initially, the PRIORITYNET assignment resembles that of a random minimizer and expectedly yields $\mathcal{D} = 2.05$. After training, the final

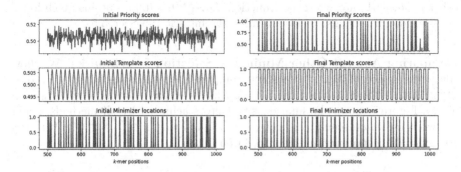

Fig. 3. Visualization of PRIORITYNET and TEMPLATENET score assignments on positions 500–1000 of CHRXC with $w = 13$, $k = 8$. Left: Initial assignments ($\mathcal{D} = 2.05$); Right: Final assignments after 600 training epochs ($\mathcal{D} = 1.39$). The bottom plots show corresponding locations of sampled k-mers: a value of 1 means selected, and 0 otherwise.

TEMPLATENET assignment converges with a different phase shift than its initial assignment, but its period remains the same. Simultaneously, the PRIORITYNET assignment learns to match this template, hence induces a visibly sparser sketch with $\mathcal{D} = 1.39$. This result clearly demonstrates the negotiating behaviour of our twin architecture to find optimal neighborhood of score assignments.

Convergence of Our Proxy Objective. We further demonstrate that our proxy objective meaningfully improves minimizer performance as it is optimized. The first two columns of Fig. 4 show the best density factors achieved by our method over 600 epochs on two scenarios: (a) varying k with fixed w; and (b) varying w with fixed k. The experiment is repeated on CHRXC and HG38. In every scenario, DEEPMINIMIZER starts with $\mathcal{D} \simeq 2.0$, which is only comparable to a random minimizer. We observe steady decrease of $\mathcal{D}$ over the first 300 epochs before reaching convergence, where total reduction ranges from 11–23%.

Generally, larger k values lead to better performance improvement at convergence. This is expected since longer k-mers are more likely to occur uniquely in the target sequence, which makes it easier for a minimizer to achieve sparse sampling. In fact, previous results have shown that when k is much smaller than $\log w$, no minimizer will be able to achieve the theoretical lower-bound $\mathcal{D}$ [19]. On the other hand, larger w values lead to smaller improvements and generally slower convergence. This is because our ensemble parameterization of TEMPLATENET scales with the window size w and becomes more complicated to optimize as w increases.

Evaluating Our Proposed Divergence Measure. The last column of Fig. 4 shows the density factors achieved by our DEEPMINIMIZER method, respectively specified by the proposed divergence function in Eq. 6 and ℓ^2-divergence. Here, we fix $w = 14$ and vary $k \in \{6, 8, 10, 12, 14\}$ and observe that with the ℓ^2-divergence, we only obtain performance similar to a random minimizer. On the other hand, with our divergence function, DEEPMINIMIZER obtains much lower densities on all settings, thus confirming the intuition in Sect. 3.5.

Comparing Against Other Minimizer Selection Benchmarks. We show the performance of DEEPMINIMIZER compared to other benchmark methods. DEEPMINIMIZER is trained for 600 epochs to ensure convergence, as shown above. Figure 5 shows the final density factors achieved by all methods, again on two comparison scenarios: (a) fix $w = 13$, and vary $k \in \{6, 8, 10, 12, 14\}$; and (b) fix $k = 14$, and vary $w \in \{10, 25, 40, 55, 70, 85\}$.

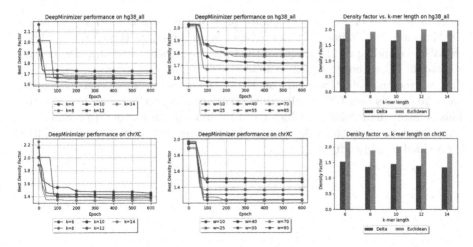

Fig. 4. Best density factors obtained by DEEPMINIMIZER on HG38 (above) and CHRXC (below) over 600 training epochs. Left: fix $w = 13$, and vary $k \in \{6, 8, 10, 12, 14\}$; Center: fix $k = 14$, and vary $w \in \{10, 25, 40, 55, 70, 85\}$; Right: Comparing proposed Δ-divergence and ℓ^2-divergence.

DEEPMINIMIZER consistently achieves better performance compared to *non-sequence-specific* minimizers (i.e., PASHA, Miniception) on all settings. We observe up to 40% reduction of density factor (e.g., on CHRXC, $w = 70$, $k = 14$), which clearly demonstrates the ability of DEEPMINIMIZER to exploit *sequence-specific* information. Furthermore, we also observe that DEEPMINIMIZER outperforms our *sequence-specific* competitor, PolarSet, in a majority of settings. The improvements over PolarSet are especially pronounced for smaller k values, which are known harder tasks for minimizers [19]. On larger w values, our method performs slightly worse than PolarSet in some settings. This is likely due to the added complexity of optimizing TEMPLATENET, as described in convergence ablation study of our method.

In addition, we also conduct investigation on the centromere region of chromosome X (i.e., CHRXC), which contains highly repetitive subsequences [6] and has been shown to hamper performance of PolarSet [20]. Figure 5 shows that PolarSet and the UHS-based methods perform similarly to a random minimizer, whereas our method is consistently better. Moreover, we observe that DEEPMINIMIZER obtains near-optimal densities with CHRXC on several settings. For example, we achieved $\mathcal{D} = 1.22$ when $k = 14$, $w \in \{40, 70\}$, which is significantly better than the results on CHR1 and CHRX. This suggests that CHRXC is not necessarily more difficult to sketch, but rather good sketches have been excluded by the UHS and polar set reparameterizations, which is not the case with our framework.

Runtime Performance. DEEPMINIMIZER runs efficiently with GPU computing. In all of our experiments, each training epoch takes approximately 30 seconds to 2 minutes, depending on the choice of k and w, which controls the batch size. Performance evaluation takes between several minutes (CHRXC) to 1 hour (HG38), depending on the length of the target sequence. Generally, our method is cost-efficient without frequent evaluations. Our most cost-intensive experiment (i.e., convergence ablation study on HG38) requires a full-sequence evaluation every 20 epochs over 600 epochs, thus takes approximately 2 days to complete. This is faster than PolarSet, which has a theoretical runtime of $\mathcal{O}(n^2)$ and takes several days to run with HG38. A more detailed runtime ablation study on CHR1 is provided in Appendix B.

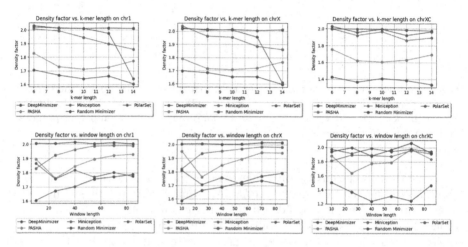

Fig. 5. Density factors obtained by DEEPMINIMIZER (600 training epochs), Random Minimizer, PASHA, Miniception and PolarSet on CHR1, CHRX and CHRXC. Above: fix $w = 13$, and vary $k \in \{6, 8, 10, 12, 14\}$; Below: fix $k = 14$, and vary $w \in \{10, 25, 40, 55, 70, 85\}$.

5   Conclusion

We introduce a novel framework called DEEPMINIMIZER for learning *sequence-specific* minimizers. This is achieved via casting minimizer selection as optimizing a k-mer scoring function ρ. We propose a more well-behaved search space for minimizers, given by a neural network parameterization of ρ, called PRIORITYNET. Then, we introduce a complementary network, called TEMPLATENET which pinpoints optimal scoring templates and guides PRIORITYNET to the neighborhood of low-density assignments around them. Coupling these networks leads to a fully differentiable proxy objective that can effectively leverage gradient-based learning techniques. DEEPMINIMIZER obtains better performance than state-of-the-art sequence-agnostic and sequence-aware minimizer selection schemes, especially on known hard tasks such as sketching the repetitive centromere region of Chromosome X. However, we also observe mild limitations in several settings with large window length w, which hampers the performance of DEEPMINIMIZER. This is likely due to the heuristic construction of our TEMPLATENET component, which we will investigate in our future work.

Acknowledgements. This work was supported in part by the Gordon and Betty Moore Foundation's Data-Driven Discovery Initiative [GBMF4554 to C.K.], by the US National Institutes of Health [R01GM122935], and the US National Science Foundation [DBI-1937540]. Conflict of Interest: C.K. is a co-founder of Ocean Genomics, Inc.

A   Proof of Proposition 1

We first re-express the density factor of S in terms of a priority score assignment $\mathbf{v} \in [0,1]^l$. Note that this expression will hold regardless of whether $\mathbf{v}$ satisfies the consistency constraint (Sect. 3.2). Particularly, let $\gamma_1 = 1$ and $\gamma_t = \mathbb{I}\left(\underset{j\in\omega_t}{\operatorname{argmin}}\ \mathbf{v}[j] \neq \underset{j'\in\omega_{t-1}}{\operatorname{argmin}}\ \mathbf{v}[j']\right)$ indicate the event the t-th window picks a different k-mer than the $(t-1)$-th window, we have:

$$\mathcal{D}(S;\mathbf{v}) \;=\; \frac{w+1}{l_w} \times |\mathcal{L}(S;\mathbf{v})| = \frac{w+1}{l_w}\sum_{t=1}^{l_w}\gamma_t \;. \tag{8}$$

Without loss of generality, we assume $0 \in \underset{t}{\operatorname{arginf}}\ g(t)$ since this can always be achieved via adding a constant phase shift to g. As g has a fundamental period of w, this implies $\{uw \mid u \in \mathbb{N}\} \subseteq \operatorname{arginf}_t g(t)$, which further reduces to $\{uw \mid k \in \mathbb{N}\} = \operatorname{arginf}_t g(t)$ when condition (2) holds.

Let us now derive the values of γ_t for $t \in \mathcal{I}_u \triangleq [(u-1)w+1, uw]$, $u \in \mathbb{N}^+$. We have:

- $uw \in \operatorname{arginf}\ g(t)$,
- $\forall t \in \mathcal{I}_u$ such that $t \neq uw$, we have $t \notin \underset{t}{\operatorname{arginf}}\ g(t)$, and
- $\forall t \in \mathcal{I}(u) : uw \in \omega_t$, which follows from the above argument and the definition of window ω_t.

Together, these observations imply that $\forall t \in \mathcal{I}_u : \operatorname{argmin}_{j\in\omega_t}\mathbf{v}[j] = uw$ and consequently $\gamma_t = 0$ for all values of $t \in \mathcal{I}_u$ except $t = (u-1)w+1$. For $u = 1$, we trivially have $\gamma_{(u-1)w+1} = 1$ by definition of γ_1. For $u > 1$, we have $\operatorname{argmin}_{j\in\omega_{(u-1)w}}\mathbf{v}[j] = (u-1)w \neq \operatorname{argmin}_{j\in\omega_{(u-1)w+1}}$, which also implies $\gamma_{(u-1)w+1} = 1$. Following the above derivations, we have:

$$D(S;\mathbf{v}) \;=\; \frac{w+1}{l_w}\sum_{t=1}^{l_w}\gamma_t = \frac{w+1}{l_w}\left(c + \sum_{u=1}^{\lfloor\frac{l_w}{w}\rfloor}\sum_{t\in\mathcal{I}_u}\gamma_t\right) = \frac{w+1}{l_w}\left(c + \left\lfloor\frac{l_w}{w}\right\rfloor\right),$$

$$\tag{9}$$

where the third equality follows from the derived values γ_t for $t \in \mathcal{I}_u$. Finally, using the fact that $c = \sum_{t=\lfloor\frac{l_w}{w}\rfloor w+1}^{l_w}\gamma_t < w$ and the sufficient length assumption

$l_w \gg w^2$, we have:

$$\frac{w+1}{l_w}\left(c + \left\lfloor \frac{l_w}{w} \right\rfloor\right) \le \frac{w^2}{l_w} + \frac{w+1}{w} \;=\; 1 + \frac{1}{w} + o(1)\,, \tag{10}$$

which concludes our proof. □

B   Other Empirical Results

This section contains extra experiments that showcase various aspects of our DEEPMINIMIZER framework. For all experiments, we use the same implementation, benchmarks and settings as detailed in Sect. 4.

Density Performance of DEEPMINIMIZER on More Sequence Baselines.
We deploy DEEPMINIMIZER on CHR1 and CHRX. For both sequences, we observe the best density factor obtained over 600 training epochs for various values of k and w. Figure 6 shows that DEEPMINIMIZER consistently improves density factors until convergence, which tends to happen between 200–300 training epochs for all experiments.

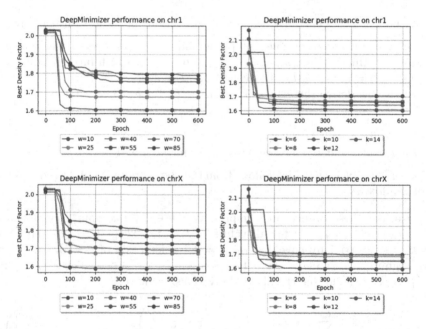

Fig. 6. Demonstrating convergence of DEEPMINIMIZER on CHR1 (left) and CHRX (right) with different w, k values.

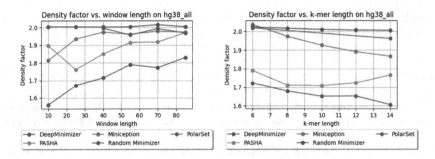

Fig. 7. Comparing performance of DEEPMINIMIZER with other benchmarks on HG38 for different values of w, k.

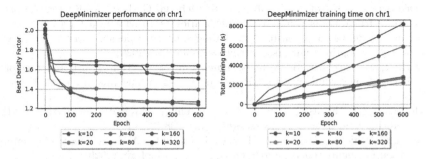

Fig. 8. Best density obtained (left) and runtime (right) of DEEPMINIMIZER for $k \in \{10, 20, 40, 80, 160, 320\}$ on CHR1.

DEEPMINIMIZER Outperforms Other Baselines on Large Sequences. Figure 7 compares the performance of DEEPMINIMIZER and various comparison baselines on the entire human genome HG38. We measure the best density factor obtained over 600 training epochs for various values of k and w and observe that DEEPMINIMIZER consistently achieves the best performance among comparison baselines.

Density Performance of DEEPMINIMIZER on Large Values of k. Figure 8 (left) showcases the performance of DEEPMINIMIZER on CHR1 with large values of k. We fix $w = 13$ and observe the best density factor obtained over 600 training epochs for various values of k up to 320. We show that DEEPMINIMIZER behaves similarly for large k, and achieves the best density $\mathcal{D} = 1.22$ with $k = 160$.

Runtime Performance of DEEPMINIMIZER on Large Values of k. Figure 8 (right) measures runtime (in seconds) of DEEPMINIMIZER on CHR1 over 600 epochs. Larger k values require PRIORITYNET to have more parameters. Expectedly, we observe runtime for $k = 40, 80, 160, 320$ to increase in the same order. For $k = 10$ and 20, however, the runtimes are approximately the same as $k = 80$. We note that a smaller k value means there are more k-mers in the same sequence.

As such, even though PRIORITYNET is more compact for these values of k, we will incur some overhead from querying it more often.

References

1. Chikhi, R., Limasset, A., Medvedev, P.: Compacting de Bruijn graphs from sequencing data quickly and in low memory. Bioinformatics **32**(12), i201–i208 (2016)
2. DeBlasio, D., Gbosibo, F., Kingsford, C., Marçais, G.: Practical universal k-mer sets for minimizer schemes. In: Proceedings of the 10th ACM International Conference on Bioinformatics, Computational Biology and Health Informatics, BCB 2019, pp. 167–176. Association for Computing Machinery, New York (2019)
3. Deorowicz, S., Kokot, M., Grabowski, S., Debudaj-Grabysz, A.: KMC 2: fast and resource-frugal k-mer counting. Bioinformatics **31**(10), 1569–1576 (2015)
4. Ekim, B., Berger, B., Orenstein, Y.: A randomized parallel algorithm for efficiently finding near-optimal universal hitting sets. bioRxiv (2020)
5. Erbert, M., Rechner, S., Müller-Hannemann, M.: Gerbil: a fast and memory-efficient k-mer counter with GPU-support. Algorithms Mol. Biol. **12**(1), 1–12 (2017)
6. Fukagawa, T., Earnshaw, W.C.: The centromere: chromatin foundation for the kinetochore machinery. Dev. Cell **30**(5), 496–508 (2014)
7. Jain, C., Rhie, A., Hansen, N., Koren, S., Phillippy, A.M.: A long read mapping method for highly repetitive reference sequences. bioRxiv (2020)
8. Jain, C., Rhie, A., Zhang, H., Chu, C., Walenz, B.P., Koren, S., Phillippy, A.M.: Weighted minimizer sampling improves long read mapping. Bioinformatics **36**(Suppl._1), i111–i118 (2020)
9. Kingma, D.P., Ba, J.: ADAM: a method for stochastic optimization. arXiv preprint arXiv:1412.6980 (2014)
10. Li, H.: Minimap2: pairwise alignment for nucleotide sequences. Bioinformatics **34**(18), 3094–3100 (2018)
11. Marçais, G., DeBlasio, D., Kingsford, C.: Asymptotically optimal minimizers schemes. Bioinformatics **34**(13), i13–i22 (2018)
12. Marçais, G., Pellow, D., Bork, D., Orenstein, Y., Shamir, R., Kingsford, C.: Improving the performance of minimizers and winnowing schemes. Bioinformatics **33**(14), i110–i117 (2017)
13. Miga, K.H., et al.: Telomere-to-telomere assembly of a complete human X chromosome. Nature **585**(7823), 79–84 (2020)
14. Orenstein, Y., Pellow, D., Marcais, G., Shamir, R., Kingsford, C.: Designing small universal k-mer hitting sets for improved analysis of high-throughput sequencing. PLOS Comput. Biol. **13**, e1005777 (2017)
15. Roberts, M., Hayes, W., Hunt, B., Mount, S., Yorke, J.: Reducing storage requirements for biological sequence comparison. Bioinformatics **20**, 3363–9 (2005)
16. Schleimer, S., Wilkerson, D., Aiken, A.: Winnowing: local algorithms for document fingerprinting. In: Proceedings of the ACM SIGMOD International Conference on Management of Data 10 (2003)
17. Ye, C., Ma, Z.S., Cannon, C.H., Pop, M., Douglas, W.Y.: Exploiting sparseness in de novo genome assembly. In: BMC Bioinformatics, vol. 13, pp. 1–8. BioMed Central (2012)

18. Zhang, Z.D., et al.: Statistical analysis of the genomic distribution and correlation of regulatory elements in the encode regions. Genome Res. **17**(6), 787–797 (2007)

19. Zheng, H., Kingsford, C., Marçais, G.: Improved design and analysis of practical minimizers. Bioinformatics **36**(Suppl._1), i119–i127 (2020)

20. Zheng, H., Kingsford, C., Marçais, G.: Sequence-specific minimizers via polar sets. Bioinformatics **37**, i187–i195 (2021)

MetaCoAG: Binning Metagenomic Contigs via Composition, Coverage and Assembly Graphs

Vijini Mallawaarachchi⬤ and Yu Lin$^{(\boxtimes)}$⬤

School of Computing, Australian National University, Canberra, Australia
{vijini.mallawaarachchi,yu.lin}@anu.edu.au

Abstract. Metagenomics has allowed us to obtain various genetic material from different species and gain valuable insights into microbial communities. Binning plays an important role in the early stages of metagenomic analysis pipelines. A typical pipeline in metagenomics binning is to assemble short reads into longer contigs and then bin into groups representing different species in the metagenomic sample. While existing binning tools bin metagenomic contigs, they do not make use of the assembly graphs that produce such assemblies. Here we propose MetaCoAG, a tool that utilizes assembly graphs with the composition and coverage information to bin metagenomic contigs. MetaCoAG uses single-copy marker genes to estimate the number of initial bins, assigns contigs into bins iteratively and adjusts the number of bins dynamically throughout the binning process. Experimental results on simulated and real datasets demonstrate that MetaCoAG significantly outperforms state-of-the-art binning tools, producing similar or more high-quality bins than the second-best tool. To the best of our knowledge, MetaCoAG is the first stand-alone contig-binning tool to make direct use of the assembly graph information.

Availability: MetaCoAG is freely available at https://github.com/Vini2/MetaCoAG.

Keywords: Metagenomics · Binning · Contigs · Assembly graphs

1   Introduction

The development of high-throughput sequencing technologies has paved the way for metagenomics studies to analyze microbial communities without the need for culturing, especially in large scale metagenomics studies such as the Human Microbiome Project [39]. These microbial communities consist of a large number of micro-organisms including bacteria. Samples obtained directly from the environment can be sequenced to obtain large amounts of sequencing reads. In order to characterize the composition of a sample and the functions of the microbes present, we perform *metagenomics binning* where we cluster sequences into bins that represent different taxonomic groups [36].

I. Pe'er (Ed.): RECOMB 2022, LNBI 13278, pp. 70–85, 2022.
https://doi.org/10.1007/978-3-031-04749-7_5

Next-generation sequencing (NGS) technologies such as Illumina allow us to sequence microbial communities and obtain highly accurate short sequences called *reads*. These reads can be binned [1,8,12,22,29,34,41] prior to assembly, but results can be less reliable due to their short lengths [47]. Hence, a widely used pipeline for metagenomics analysis is to first assemble reads into longer sequences called *contigs* and then bin these assembled contigs into groups that belong to different taxonomic groups [36]. Current contig-binning approaches fall into two broad categories [48]: (1) *reference-based* binning approaches [16, 25,29,43] which classify contigs into known taxonomic groups by comparing against a reference database and (2) *reference-free* binning approaches which cluster contigs into unlabeled bins based on genomic features of these contigs. Reference-free binning approaches [3,14,42,44,45] have become more popular as they enable the identification of new species that are not available in the current databases. Reference-free contig-binning tools mainly make use of two features to perform binning: (1) *composition*, obtained as normalized frequencies of oligonucleotides (also known as k-mers) of a given length and (2) *coverage*, considered as the average number of reads that map to each base of the contig. These tools achieve improved performance by combining both the composition and the coverage information. However, it still remains challenging for these binning tools to accurately reconstruct microbial genomes of species with similar composition and coverage profiles.

Another challenge in metagenomics binning is to estimate the number of species present in a given sample. Recent binning tools have made use of *single-copy marker genes* (appear only once in the genome and are conserved in the majority of bacterial genomes [2,10,45]) to estimate the number of species. In tools such as MaxBin/MaxBin2 [44,45], only one marker gene is utilized to estimate the number of initial bins which may lead to an underestimation of the number of species. Hence, it is worth investigating how to make use of multiple single-copy marker genes together to obtain a better estimate for the number of bins and to explore more features of contigs that can improve the binning result.

Contigs are obtained by assembling reads into longer sequences, and there are many tools to perform assembly. Most existing metagenomic assemblers [19, 28,31] use *assembly graphs* as the key data structure (*e.g.*, simplified de Bruijn graph [32]) to assemble reads into contigs. Previous studies indicated that contigs connected to each other in the assembly graph are more likely to belong to the same taxonomic group [5,23]. Although popular metagenomic assemblers such as metaSPAdes [28] output contigs along with their connection information in the assembly graph, most existing binning tools ignore the valuable connection information between contigs. More recently, bin-refinement tools such as Graph-Bin [23], GraphBin2 [24] and METAMVGL [49] have been developed to refine existing binning results using assembly graphs. These tools rely upon the bins produced by an existing binning tool and cannot dynamically adjust the number of bins. Moreover, metabinners such as DAS tools [38] and MetaWRAP[40] have been introduced to integrate and optimize the results of multiple binning approaches. Although these tools achieve improved binning performance, they still require initial binning results obtained from other existing binning tools and some tools cannot dynamically adjust the number of bins. Hence, it is worth exploring methods to develop a stand-alone contig-binning tool that makes use of the assembly graph information.

In this paper, we introduce MetaCoAG, a reference-free stand-alone approach for binning metagenomic contigs. In addition to composition and abundance information, MetaCoAG makes use of the connectivity information from assembly graphs for binning. To the best of our knowledge, MetaCoAG is the first contig-binning tool to make direct use of the assembly graph information. We benchmark MetaCoAG against state-of-the-art contig-binning tools using simulated and real datasets. The experimental results show that MetaCoAG significantly outperforms other contig-binning tools, *e.g.*, improving the completeness of bins while maintaining high purity levels and producing more high-quality bins.

2 Methods

Figure 1 shows the overall workflow of MetaCoAG. Each step of MetaCoAG is explained in detail in the following sections.

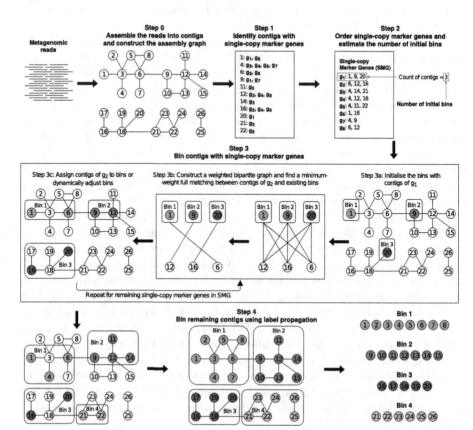

Fig. 1. MetaCoAG workflow

2.1   Step 0: Assemble Reads into Contigs and Construct the Assembly Graph

This preprocessing step is carried out to assemble the reads into contigs and obtain the assembly graph. Metagenomic assemblers first use graph models to connect overlapping reads or k-mers and to infer contigs as non-branching paths. After graph simplification, the vertices represent contigs and edges represent connections between contigs in the assembly graph. Here we use the popular metagenomic assembler metaSPAdes [28] to derive input contigs and assembly graphs. Note that the assembly graphs can also be obtained similarly using other metagenomic assemblers such as MEGAHIT [19] and metaFlye [17].

2.2   Step 1: Identify Contigs with Single-Copy Marker Genes

Single-copy marker genes appear only once in a bacterial genome and are conserved in the majority of bacterial genomes [2,10,45]. For each single-copy marker gene, we use FragGeneScan [33] and HMMER [11] to identify the contigs which contain this marker gene. A single-copy marker gene is considered to be contained in a contig if more than 50% of the gene length is aligned to this contig. Similar to approaches such as MaxBin [45] and MaxBin2 [44], Meta-CoAG uses single-copy marker genes to distinguish contigs belonging to different species (*i.e.*, if these contigs contain the same single-copy marker gene).

2.3   Step 2: Order Single-Copy Marker Genes and Estimate the Number of Initial Bins

For a given single-copy marker gene, the contigs containing this marker gene should come from different species (*e.g.*, if two contigs contain the same marker gene, then the two contigs should belong to two different species). In the ideal case, if we have a near-perfect assembly, the number of contigs that contain the same single-copy marker gene should be equal to the number of species present in the sample. However, in reality, assemblies can be fragmented and erroneous, which may make it challenging to recover all single-copy marker genes and hence, lowering the counts of contigs containing each single-copy marker gene.

To get a better estimation of the number of species, we obtain the counts of contigs containing each single-copy marker gene. We also record the single-copy marker genes found in each contig. For a single-copy marker gene, the number of contigs that it can distinguish is the number of contigs containing this gene. Therefore, we order all the single-copy marker genes according to the descending order of the number of contigs containing them. We refer to this list of ordered marker genes as SMG where a single-copy marker gene g_i has a set of contigs $C(g_i)$ containing g_i. The number of initial bins is empirically set to be the number of contigs that contain the first gene in SMG, in order to recover the maximum number of species possible from the marker gene information.

2.4   Step 3: Bin Contigs with Single-copy Marker Genes

Step 3a: Initialize Bins. We initialize the bins using the contigs of the first single-copy marker gene g_1 in SMG; *i.e.*, we initialize a new bin B for each contig in $C(g_1)$ (as shown in Step 3a of Fig. 1). We define the initialized set of bins as $BINS$. Please note that the number of bins $|BINS|$ may change during the binning process.

Calculating Composition and Coverage Similarities. Previous studies on metagenomics binning have used genomic signatures as they follow species-specific patterns [9,45]. The most commonly used genomic signatures to characterize composition information are *tetranucleotide frequencies* (136 canonical 4-mers, also known as *tetramers*) [3,14,27,42,44,45]. For each contig c, we normalize the tetranucleotide frequencies using its total number of tetranucleotides to obtain the normalized tetranucleotide frequency vector, $tetra(c)$. We obtain the tetranucleotide composition distance between contigs c and c' as $d_{tetra}(c, c') = dist_E(tetra(c), tetra(c'))$ where $dist_E$ is the Euclidean distance function.

We use the same formula proposed by Wu *et al.* [45] to estimate how similar c and c' are (*i.e.* belonging to the same species) based on their composition, $S_{comp}(c, c')$ as shown in Eq. 1.

$$S_{comp}(c, c') = \frac{N_{intra}\left(d_{tetra}(c,c')|\mu_{intra},\sigma^2_{intra}\right)}{N_{intra}\left(d_{tetra}(c,c')|\mu_{intra},\sigma^2_{intra}\right) + N_{inter}\left(d_{tetra}(c,c')|\mu_{inter},\sigma^2_{inter}\right)} \quad (1)$$

N_{intra} and N_{inter} are Gaussian distributions with μ_{intra}, σ_{intra}, μ_{inter} and σ_{inter} set according to the latest values of MaxBin 2.2.7 [44] which have been calculated by analysing the Euclidean distance between the tetranucleotide frequencies of pairs of sequences sampled from the same genome (*intra*) and different genomes (*inter*). If the distance is lower between two sequences, they are more similar, and are more likely to belong to the same genome.

We use the coverage information of the contigs as coverage carries important information about the abundance of species and has been used in previous metagenomics binning studies [2,14,27,42,45]. Shotgun sequencing has shown to follow the Lander-Waterman model [18] and the Poisson distribution has been used to obtain the sequencing coverage of nucleotides and applied in metagenomics binning [45,46]. Modifying the definition found in Wu *et al.* [45], we estimate how similar c and c' are in terms of their coverage values in each sample, $S_{cov}(c, c')$ as shown in Eq. 2.

$$S_{cov}(c, c') = \min\left(\prod_{n=1}^{M} Poisson(cov_n(c)|cov_n(c')), \prod_{n=1}^{M} Poisson(cov_n(c')|cov_n(c))\right) \quad (2)$$

Here $cov_n(c)$ and $cov_n(c')$ refer to the coverage values of the contigs c and c' respectively in the sample n where M is the number of samples. *Poisson* is the Poisson probability mass function.

Step 3b: Construct a Weighted Bipartite Graph and Find a Minimum-Weight Full Matching. In the previous steps, we have used single-copy marker genes to identify pairs of contigs that belong to different species. Remind that contigs in different bins in $BINS$ are expected to belong to different species and contigs in $C(g_i)$ are also expected to belong to different species. However, there is no measurement to measure how likely a contig c in $C(g_i)$ belongs to an existing bin B in $BINS$. Therefore, we introduce a bipartite graph between $C(g_i)$ and $BINS$ and propose a weight $w_{c2B}(c, B)$ between a contig c in $C(g_i)$ and an existing bin B in $BINS$ as shown in Eq. 3.

$$w_{c2B}(c, B) = \frac{\sum_{c' \in B} w_{c2c}(c, c')}{|B|} \tag{3}$$

In Eq. 3, $w_{c2c}(c, c')$ is the weight that measures how likely a pair of contigs c and c' belong to the same species and is computed using Eq. 4. $S_{comp}(c, c')$ and $S_{cov}(c, c')$ are calculated according to Eqs. 1 and 2 resp.

$$w_{c2c}(c, c') = -\big(log(S_{comp}(c, c')) + log(S_{cov}(c, c'))\big) \tag{4}$$

Now we find a minimum-weight full matching (minimum-cost assignment) [15] for the above bipartite graph between $C(g_i)$ and $BINS$ where every contig c in $C(g_i)$ will get paired with exactly one bin B in $BINS$. For this purpose, we use the minimum-weight full matching algorithm implemented in the *NetworkX* python library which is based on the algorithm proposed by Karp [15] and the time complexity is $O(|C(g_i)| \times |BINS| \times log(|BINS|))$.

In the next step, we will see how we can assign the contigs to existing bins based on the minimum-weight full matching we have obtained.

Step 3c: Assign Contigs to Existing Bins and Dynamically Adjust Bins. Previous studies have observed that contigs connected to each other in the assembly graph are more likely to belong to the same taxonomic group [5,23]. While $w_{c2B}(c, B)$ considers both composition and coverage information, the assembly graph has not yet been incorporated into the binning process. Therefore, we introduce $d_{graph}(c, B)$ to measure how well contig c is connected to contigs in bin B within the assembly graph. Specifically, $d_{graph}(c, B)$ is defined as the average length of the shortest-path distances between contig c and all the contigs in bin B in the assembly graph. Note that both $w_{c2B}(c, B)$ and $d_{graph}(c, B)$ will be used to assign contigs to existing bins or dynamically adjust the bins.

We define the thresholds intra-species weight $w_{intra} = -\big(log(p_{intra})\big) \times M$ and inter-species weight $w_{inter} = -\big(log(p_{inter})\big) \times M$ where M is the number of samples in the dataset. Each candidate pair (c, B) obtained from the minimum-weight full matching falls under one of the following three cases (refer Fig. 2).

– **Case 1:** If the weight of the candidate pair $w_{c2B}(c, B)$ is less than or equal to w_{intra} and the average distance $d_{graph}(c, B)$ is less than or equal to d_{limit}, then contig c will be assigned to bin B, i.e., $B \leftarrow B \cup \{c\}$ (e.g., contig 4 and Bin 1 in Fig. 2).

- **Case 2:** If the weight of the candidate pair $w_{c2B}(c, B)$ is greater than w_{inter} and the average distance $d_{graph}(c, B)$ is greater than d_{limit}, then a new bin B' is created and contig c is assigned to that new bin, i.e., $B' = \{c\}$ and $BINS \leftarrow BINS \cup \{B'\}$. (e.g., contig 21 in Fig. 2).
- **Case 3:** If $w_{c2B}(c, B)$ and $d_{graph}(c, B)$ satisfy neither Case 1 nor Case 2, then contig c will not be assigned to any bin (e.g., contig 14 in Fig. 2).

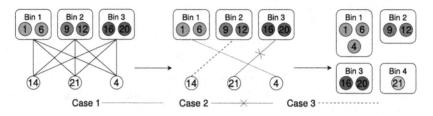

Fig. 2. Cases 1, 2 and 3 in assigning contigs to existing bins or adjusting bins

The default values for parameters p_{intra}, p_{inter}, d_{limit} were chosen empirically and set to 0.1, 0.01 and 20 respectively. Now we iteratively perform Steps 3b and 3c to process all the contigs containing single-copy marker genes. The remaining challenge is to bin the contigs which do not contain single-copy marker genes which will be addressed in Step 4.

2.5   Step 4: Bin Remaining Contigs Using Label Propagation

After we bin the contigs with single-copy marker genes, each such contig receives a label corresponding to its bin. Now we will propagate labels from these contigs to other unlabeled contigs within the same connected component.

Step 4a: Propagate Labels Within Connected Components. MetaCoAG uses composition, coverage and distance information from the assembly graph to propagate labels from labeled contigs to the unlabeled contigs located within the same connected components. More specifically, for each unlabeled long contig c (at least 1,000 bp long because short contigs result in unreliable composition and coverage information) directly connected or connected via short contigs to a labeled contig c', MetaCoAG computes a candidate propagation action $(c', c, d(c, c'), w_{c2B}(c, B'))$ where $d(c, c')$ is the shortest distance between c and c' using only unlabeled vertices and $w_{c2B}(c, B')$ is computed according to Eq. 3 where B' is the bin to which contig c' is assigned. Given two candidate propagation actions (a, b, d, w) and (a', b', d', w'), (a, b, d, w) has a higher priority than (a', b', d', w') if $d < d'$ or $(w < w'$ and $d = d')$. MetaCoAG iteratively selects the candidate propagation action with the highest priority and executes the corresponding label propagation. If a contig to be labeled contains single-copy marker genes, the relevant candidate propagation action is executed if the single-copy marker genes of the contig are not present in the intended bin. We restrict the depth of the search for labeled contigs in this step to 10.

Step 4b: Propagate Labels Across Different Components. Note that some components in the assembly graph may not have any labeled contigs and we need to propagate labels from labeled bins to unlabeled contigs across components. Calculating pair-wise weights $w_{c2c}(c, c')$ for all the remaining contigs becomes time consuming. Hence, for each bin B we create a representative contig $c(B)$ which has a composition profile and a coverage profile calculated by averaging the normalized tetranucleotide frequency vectors and coverage vectors of all the contigs in bin B, respectively. These profiles will provide a better representation of the composition and coverage of the bins. Then, for each unlabeled contig c, MetaCoAG identifies a bin B that minimizes $w_{c2c}(c, c(B))$ which is calculated according to Eq. 4, and assigns contig c into that bin B. This propagation is limited to long contigs (at least 1,000 bp long by default). If an unlabeled contig contains single-copy marker genes, it is assigned to bin B that minimizes $w_{c2c}(c, c(B))$ if the single-copy marker genes of the contig are not present in bin B. Then, Step 4a is performed again to further propagate labels.

Step 4e: Postprocessing. In this step, we will make final adjustments on the current bins. Two bins B and B' are *mergeable* if they have no common marker genes and $w_{c2c}(c(B), c(B'))$ (calculated by Eq. 4) is upper bounded by w_{intra} (defined in Step 3c). Then, MetaCoAG creates a graph where vertices denote current bins and edges between two vertices denote that the corresponding two bins are mergeable. Now we use the implementation of python-igraph library to find maximal cliques (igraph_maximal_cliques) in this graph and merge the bins found in each maximal clique. After merging bins, we also remove the bins which contain less than one third (set by default) of the single-copy marker genes. Finally, MetaCoAG outputs the bins along with their corresponding contigs.

3 Experimental Setup

3.1 Datasets and Tools

Simulated Datasets. We evaluated the binning performance on the simulated **simHC+** dataset [45] which consists of 100 bacterial species. Paired-end reads were simulated using InSilicoSeq [13] with the predefined MiSeq error model.

Real Datasets. We used three real datasets to evaluate the binning performance on real-world metagenomic data: (1) Preborn infant gut metagenome [37] with 18 samples (NCBI accession number *SRA052203*), referred as **Sharon**, (2) Metagenomics of the Chronic Obstructive Pulmonary Disease (COPD) Lung Microbiome [6] with 18 samples (NCBI BioProject number *PRJEB9034*), referred as **COPD** and (3) Human metagenome sample from tongue dorsum of a participant from the Deep WGS HMP clinical samples [21] with 8 samples (NCBI accession number *SRX378791*), referred as **Deep HMP TD**.

Tools Used. We used the popular metagenomic assembler metaSPAdes [28] (from SPAdes version 3.15.2 [4]) to assemble reads into contigs and obtain the assembly graphs. For the datasets containing multiple samples, the contigs and assembly graph were obtained by co-assembling the reads from all the samples together. The mean coverage of each contig in each sample was calculated using CoverM (available at https://github.com/wwood/CoverM).

MetaCoAG was benchmarked against the binning tools MaxBin2 (version 2.2.7) [44] in its default settings, MetaBAT2 (version 2.12.1) [14] with -m 1500 and Vamb (version 3.0.1) [27] in co-assembly mode (for a fair comparison with other tools) with the parameter --minfasta 200000 as per authors.

The binning results were evaluated using the tools AMBER [26] (version 2.0.2), CheckM [30] (version 1.1.3) and GTDB-Tk [7] (version 1.5.0).

3.2   Evaluation Metrics

Since the ground truth species for the simHC+ dataset were available, we used Minimap2 [20] to map the contigs to the reference genomes and determine the ground truth. With this ground truth annotation of contigs, we used AMBER [26] to assess the binning results of the simHC+ dataset. We set the recall as AMBER completeness and precision as AMBER purity and calculated the F1-score as $2 \times$ (precision $\times$ recall)/(precision + recall) for each bin/species.

For all the datasets, we determined the completeness and contamination of all the bins using CheckM [30]. We define purity as $1/(1 +$ CheckM contamination). Then, we set the recall as CheckM completeness and precision as purity, and calculate F1-score as $2 \times$ (precision$\times$recall)/(precision+recall) for each bin. Furthermore, we counted the number of high-quality bins (bins with >80% completeness and >90% purity), medium-quality bins (bins with >50% completeness and >80% purity) and low-quality bins (bins which are not considered as high-quality or medium-quality).

4   Results and Discussion

4.1   Benchmarks Using SimHC+ Dataset

We first benchmarked MetaCoAG against two popular contig-binning tools, MaxBin2 [44] and MetaBAT2 [14] on the simulated dataset **simHC+** [45][1]. We evaluated the binning results of the simHC+ dataset produced by all the tools using the two popular evaluation tools AMBER [26] and CheckM [30]. AMBER assesses the quality of bins based on the ground truth annotations provided and CheckM assesses the quality of bins based on sets of single-copy marker genes. We analyzed the purity, completeness and F1-score of the binning results calculated by AMBER (at the nucleotide level) and CheckM (refer to Appendix Table 2).

---

[1] Please note that the recently published tool Vamb [27] was not used to evaluate the simHC+ dataset as the number of contigs was less than the number recommended by the authors (https://github.com/RasmussenLab/vamb#recommended-workflow).

MetaCoAG has recovered bins with a better trade-off between purity and completeness when compared to other binning tools (Fig. 3(a)). This better trade-off is demonstrated from the best F1-score results produced by MetaCoAG with a median F1-score of 95.69% from AMBER and a median F1-score of 98.48% from CheckM (Fig. 3(b) and (c) respectively, where each point denotes a bin) when compared with other binning tools. Furthermore, MetaCoAG has recovered the highest number of high-quality bins (69 bins) and the lowest number of low-quality bins (13 bins) (Refer to Appendix Table 3).

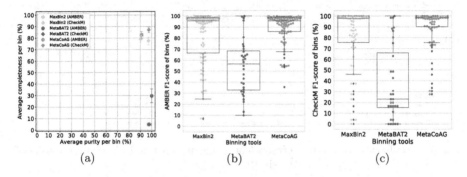

(a)   (b)   (c)

Fig. 3. Binning results of the simHC+ dataset: (a) Average completeness per bin vs. average purity per bin from AMBER and CheckM results of each binning tool, (b) Swarm plot with overlaid box plot for the AMBER F1-score of the bins produced by each binning tool and (c) Swarm plot with overlaid box plot for the CheckM F1-score of the bins produced by each binning tool.

A major challenge faced by most existing binning tools is how to accurately separate contigs of species belonging to the same genus, where such species tend to have similar oligonucleotide composition and appear in similar abundances. For example, the three species, *S. pneumoniae*, *S. thermophilus* and *S. suis* from simHC+ belong to the *Streptococcus* genus, and they have very similar oligonucleotide composition and coverage values (Refer to Fig. 4(a)). Not surprisingly, contigs from these three species were incorrectly binned by MaxBin2 and even ignored by MataBAT2 because they share similar composition and coverage profiles (Refer to Fig. 4(b)). On the contrary, MetaCoAG was able to accurately bin most of the contigs from these three species because they naturally form three subgraphs in the assembly graph (Refer to Fig. 4(b)), thus improving the F1-scores of *S. pneumoniae* from 46.51% to 93.40%, *S. thermophilus* from 49.97% to 95.67% and *S. suis* from 72.39% to 95.95%. Figure 4(b) demonstrates that the use of assembly graph in MetaCoAG can assist in separating species, despite the high similarity in oligonucleotide composition and coverage of certain species. Furthermore, we observed that the assembly graphs help MetaCoAG to bin species with high variance of intra-species oligonucleotide composition and coverage profiles while most existing tools suffer from the assumption that the oligonucleotide composition and coverage are conserved within the same species.

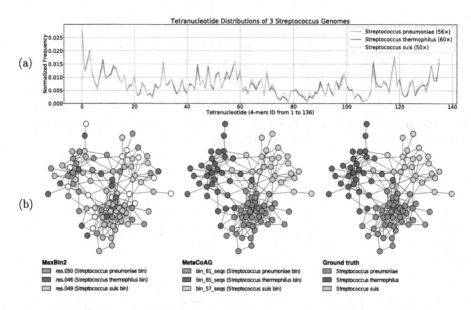

Fig. 4. Visualization of the (a) tetranucleotide composition and (b) binning results of three *Streptococcus* genomes in the simHC+ dataset

4.2 Benchmarks Using Real Datasets

We benchmarked MetaCoAG against MaxBin2 [44], MetaBAT2 [14] and Vamb [27] on three real metagenomic datasets; **Sharon** [37], **COPD** [6] and **Deep HMP TD** [21]. Similar to the simHC+ dataset, we again use CheckM [30] to evaluate the bins produced by all the binning tools and identify high-quality bins. Figure 5 shows that MetaCoAG has also achieved the best binning result in terms of the median F1-score for the real datasets. For the Sharon dataset, MetaCoAG records a median F1-score of 99.24% while the second-best tool (Vamb) has a median F1-score of 83.88%. For the COPD dataset, MetaCoAG records a median F1-score of 75.68% while the second-best tool (MaxBin2) has a median F1-score of 25.13%. For the Deep HMP TD dataset, MetaCoAG records a median F1-score of 76.34% while the second-best tool (MaxBin2) has a median F1-score of 37.40%. Furthermore, MetaCoAG has produced the highest number of high-quality bins for all the real datasets (Please refer to Appendix Table 3).

We used GTDB-Tk [7] to annotate all high-quality bins produced by the three best-performing tools; MetaCoAG, MaxBin2 and Vamb for the real datasets. Then we compared such taxonomic annotations (up to the species level) with the results in original publications for these datasets as shown in Table 1. The comparisons show that MetaCoAG achieves the best consistency with the comprehensive analysis results in original publications. These results demonstrate that MetaCoAG has been able to recover species in real metagenomics samples that are ignored by other binning tools, as well as recover more species correctly with respect to the original analysis of these real datasets.

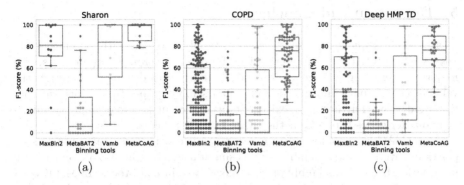

Fig. 5. Swarm plots with overlaid box plots for the F1-score from CheckM results of the real datasets. Each point denotes a bin.

Table 1. GTDB-Tk annotations of high-quality species for the real datasets

Dataset	Species	MaxBin2 [44]	Vamb [27]	MetaCoAG	Present in original analysis
Sharon [37]	Cutibacterium avidum	✓	✓	✓	✓
	Enterococcus faecalis	✓	✓	✓	✓
	Peptoniphilus lacydonensis	✓	✓	✓	✓
	Staphylococcus aureus	✓	✓	✓	✓
	Staphylococcus epidermidis	✓	✓	✓	✓
	Staphylococcus hominis	✓	✗	✓	✓
	Leuconostoc citreum	✗	✗	✓	✓
COPD [6]*	Peptostreptococcus sp	✓	✓	✓	✓
	SR1 bacterium human oral taxon HOT-345	✓	✓	✓	✗†
	Prevotella pallens	✗	✓	✓	✓
	Haemophilus sputorum	✗	✓	✓	✓
	Herbaspirillum huttiense	✗	✓	✓	✓
	Capnocytophaga gingivalis	✓	✗	✓	✓
	Capnocytophaga leadbetteri	✓	✗	✓	✓
	Lancefieldella sp000564995	✓	✗	✓	✓
	Actinomyces graevenitzii	✓	✗	✗	✓
	Actinomyces oris	✓	✗	✗	✓
	Anaeroglobus micronuciformis	✓	✗	✗	✗
	Eubacterium sulci	✗	✗	✓	✓
	Prevotella shahii	✗	✗	✓	✓
	Prevotella histicola	✗	✗	✓	✓
	Lachnospiraceae bacterium oral taxon 096	✗	✗	✓	✗†
Deep HMP TD [21]*	Actinomyces graevenitzii	✓	✓	✓	✓
	Saccharimonadaceae TM7x sp900557595	✓	✓	✓	✗†
	Neisseria subflava_C	✓	✗	✓	✓
	Prevotella pallens	✓	✗	✓	✓
	Anaeroglobus micronuciformis	✓	✗	✓	✗
	Actinomyces sp. ICM47	✓	✗	✓	✓
	Lancefieldella sp000564995	✓	✗	✗	✗
	Eubacterium_B sulci	✗	✗	✓	✓

✓ denotes that the species is present and ✗ denotes that the species is absent in the result/analysis. Green colored items match the original analysis whereas the red colored items do not match the original analysis.

* The species were determined based on the most abundant genera presented.

† These species were added to NCBI taxonomy in year 2020 [35] which is after the relevant analysis [6,21].

5   Discussion and Conclusion

Metagenomic sequencing and *de novo* assembly, coupled with binning methods have facilitated the characterization of different microbial communities. The majority of existing metagenomic contig-binning tools do not make use of the valuable connectivity information found in assembly graphs from which the contigs are derived. Furthermore, existing tools do not make use of multiple single-copy marker genes throughout the entire binning process.

MetaCoAG is a tool for binning metagenomic contigs that makes use of composition, coverage and assembly graphs simultaneously. The use of connectivity information from the assembly graphs makes the binning process of MetaCoAG robust against similar inter-species oligonucleotide composition and coverage (among species within the same genus) as well as high variance of intra-species oligonucleotide composition and coverage (within the same species). Experimental results on both simulated and real datasets show that MetaCoAG achieves the best binning results compared to state-of-the-art tools, especially in terms of bin quality. However, assembly issues such as misassemblies can be challenging to handle and requires further investigation.

MetaCoAG can be easily extended to work with other metagenomics assemblers. In the future, we plan to extend MetaCoAG to support overlapped binning [24] (*i.e.*, contigs may belong to multiple species) and multi-sampled binning [27] (*i.e.*, integration across multiple samples instead of co-assembly). Furthermore, we plan to incorporate MetaCoAG with assembly pipelines that may lead to more efficient and accurate analysis for metagenomic datasets.

Acknowledgments. This research was undertaken with the assistance of resources and services from the National Computational Infrastructure (NCI), which is supported by the Australian Government.

A   Appendix

Table 2. AMBER and CheckM evaluation results for the simHC+ dataset.

Evaluation criteria	MaxBin score (%)	MetaBAT2 score (%)	MetaCoAG score (%)
Average purity per bin (AMBER)	90.36	**98.30**	91.07
Average purity per bin (CheckM)	97.25	**100.0**	97.55
Average completeness per bin (AMBER)	79.34	13.02	**82.73**
Average completeness per bin (CheckM)	77.51	29.59	**87.17**
F1-score per bin (AMBER)	84.50	23.00	**86.70**
F1-score per bin (CheckM)	80.64	37.25	**89.44**
Accuracy (AMBER)	77.07	14.38	**84.46**
Binned fraction (AMBER)	84.90	14.79	**92.04**

Table 3. The number of high-quality, medium-quality and low-quality bins.

Dataset	Binning tool	No. of bins detected	High-quality	Medium-quality	Low-quality
simHC+ [45]	MaxBin2	95	59	**11**	25
	MetaBAT2	32	4	4	24
	MetaCoAG	90	**69**	8	**13**
Sharon [37]	MaxBin2	14	6	2	6
	MetaBAT2	24	2	2	20
	Vamb	10	5	1	4
	MetaCoAG	10	**7**	**3**	**0**
COPD [6]	MaxBin2	156	9	24	123
	MetaBAT2	76	0	2	74
	Vamb	61	6	7	48
	MetaCoAG	68	**17**	**25**	**26**
Deep HMP TD [21]	MaxBin2	69	**8**	**15**	46
	MetaBAT2	61	0	1	60
	Vamb	29	2	3	13
	MetaCoAG	27	**8**	9	**10**

References

1. Alanko, J., Cunial, F., Belazzougui, D., et al.: A framework for space-efficient read clustering in metagenomic samples. BMC Bioinform. **18**(3), 59 (2017)
2. Albertsen, M., Hugenholtz, P., Skarshewski, A., et al.: Genome sequences of rare, uncultured bacteria obtained by differential coverage binning of multiple metagenomes. Nat. Biotechnol. **31**(6), 533–538 (2013)
3. Alneberg, J., Bjarnason, B.S., de Bruijn, I., et al.: Binning metagenomic contigs by coverage and composition. Nat. Methods **11**, 1144–1146 (2014)
4. Bankevich, A., Nurk, S., Antipov, D., et al.: SPAdes: a new genome assembly algorithm and its applications to single-cell sequencing. J. Comput. Biol. **19**(5), 455–477 (2012). PMID: 22506599
5. Barnum, T.P., Figueroa, I.A., Carlström, C.I., et al.: Genome-resolved metagenomics identifies genetic mobility, metabolic interactions, and unexpected diversity in perchlorate-reducing communities. ISME J. **12**(6), 1568–1581 (2018)
6. Cameron, S.J.S., Lewis, K.E., Huws, S.A., et al.: Metagenomic sequencing of the chronic obstructive pulmonary disease upper bronchial tract microbiome reveals functional changes associated with disease severity. PLoS ONE **11**(2), 1–16 (2016)
7. Chaumeil, P.A., Mussig, A.J., Hugenholtz, P., et al.: GTDB-Tk: a toolkit to classify genomes with the genome taxonomy database. Bioinformatics **36**(6), 1925–1927 (2019)
8. Cleary, B., Brito, I.L., Huang, K., et al.: Detection of low-abundance bacterial strains in metagenomic datasets by eigengenome partitioning. Nat. Biotechnol. **33**, 1053 (2015)
9. Deschavanne, P.J., Giron, A., Vilain, J., et al.: Genomic signature: characterization and classification of species assessed by chaos game representation of sequences. Mol. Biol. Evol. **16**(10), 1391–1399 (1999)
10. Dupont, C.L., Rusch, D.B., Yooseph, S., et al.: Genomic insights to SAR86, an abundant and uncultivated marine bacterial lineage. ISME J. **6**(6), 1186–1199 (2012)

11. Eddy, S.R.: Accelerated profile HMM searches. PLoS Comput. Biol. **7**(10), 1–16 (2011)
12. Girotto, S., Pizzi, C., Comin, M.: MetaProb: accurate metagenomic reads binning based on probabilistic sequence signatures. Bioinformatics **32**(17), i567–i575 (2016)
13. Gourlé, H., Karlsson-Lindsjö, O., Hayer, J., et al.: Simulating Illumina metagenomic data with InSilicoSeq. Bioinformatics **35**(3), 521–522 (2018)
14. Kang, D., Li, F., Kirton, E.S., et al.: MetaBAT 2: an adaptive binning algorithm for robust and efficient genome reconstruction from metagenome assemblies. PeerJ **7**, e27522v1 (2019)
15. Karp, R.M.: An algorithm to solve the m × n assignment problem in expected time o(mn log n). Networks **10**(2), 143–152 (1980)
16. Kim, D., Song, L., Breitwieser, F.P., et al.: Centrifuge: rapid and sensitive classification of metagenomic sequences. Genome Res. **26**(12), 1721–1729 (2016)
17. Kolmogorov, M., Bickhart, D.M., Behsaz, B., et al.: metaFlye: scalable long-read metagenome assembly using repeat graphs. Nat. Methods **17**(11), 1103–1110 (2020)
18. Lander, E.S., Waterman, M.S.: Genomic mapping by fingerprinting random clones: a mathematical analysis. Genomics **2**(3), 231–239 (1988)
19. Li, D., Liu, C.M., Luo, R., et al.: MEGAHIT: an ultra-fast single-node solution for large and complex metagenomics assembly via succinct de Bruijn graph. Bioinformatics **31**(10), 1674–1676 (2015)
20. Li, H.: Minimap2: pairwise alignment for nucleotide sequences. Bioinformatics **34**(18), 3094–3100 (2018)
21. Lloyd-Price, J., Mahurkar, A., Rahnavard, G., et al.: Strains, functions and dynamics in the expanded human microbiome project. Nature **550**(7674), 61–66 (2017)
22. Luo, Y., Yu, Y.W., Zeng, J., et al.: Metagenomic binning through low-density hashing. Bioinformatics **35**(2), 219–226 (2018)
23. Mallawaarachchi, V., Wickramarachchi, A., Lin, Y.: GraphBin: refined binning of metagenomic contigs using assembly graphs. Bioinformatics **36**(11), 3307–3313 (2020)
24. Mallawaarachchi, V.G., Wickramarachchi, A.S., Lin, Y.: GraphBin2: refined and overlapped binning of metagenomic contigs using assembly graphs. In: Kingsford, C., Pisanti, N. (eds.) 20th International Workshop on Algorithms in Bioinformatics (WABI 2020). Leibniz International Proceedings in Informatics (LIPIcs), vol. 172, pp. 8:1–8:21. Schloss Dagstuhl-Leibniz-Zentrum für Informatik, Dagstuhl, Germany (2020)
25. Menzel, P., Ng, K.L., Krogh, A.: Fast and sensitive taxonomic classification for metagenomics with Kaiju. Nat. Commun. **7**, 11257 (2016)
26. Meyer, F., Hofmann, P., Belmann, P., et al.: AMBER: assessment of metagenome BinnERs. GigaScience **7**(6), giy069 (2018)
27. Nissen, J.N., Johansen, J., Allesøe, R.L., et al.: Improved metagenome binning and assembly using deep variational autoencoders. Nat. Biotechnol. **39**(5), 555–560 (2021)
28. Nurk, S., Meleshko, D., Korobeynikov, A., et al.: metaSPAdes: a new versatile metagenomic assembler. Genome Res. **27**(5), 824–834 (2017)
29. Ounit, R., Wanamaker, S., Close, T.J., et al.: CLARK: fast and accurate classification of metagenomic and genomic sequences using discriminative k-mers. BMC Genomics **16**(1), 236 (2015)
30. Parks, D.H., Imelfort, M., Skennerton, C.T., et al.: CheckM: assessing the quality of microbial genomes recovered from isolates, single cells, and metagenomes. Genome Res. **25**(7), 1043–1055 (2015)

31. Peng, Y., Leung, H.C.M., Yiu, S.M., et al.: IDBA-UD: a de novo assembler for single-cell and metagenomic sequencing data with highly uneven depth. Bioinformatics **28**(11), 1420–1428 (2012)

32. Pevzner, P.A., Tang, H., Waterman, M.S.: An Eulerian path approach to DNA fragment assembly. Proc. Natl. Acad. Sci. **98**(17), 9748–9753 (2001)

33. Rho, M., Tang, H., Ye, Y.: FragGeneScan: predicting genes in short and error-prone reads. Nucleic Acids Res. **38**(20), e191–e191 (2010)

34. Schaeffer, L., Pimentel, H., Bray, N., et al.: Pseudoalignment for metagenomic read assignment. Bioinformatics **33**(14), 2082–2088 (2017)

35. Schoch, C.L., Ciufo, S., Domrachev, M., et al.: NCBI taxonomy: a comprehensive update on curation, resources and tools. Database **2020** (2020)

36. Sedlar, K., Kupkova, K., Provaznik, I.: Bioinformatics strategies for taxonomy independent binning and visualization of sequences in shotgun metagenomics. Comput. Struct. Biotechnol. J. **15**, 48–55 (2017)

37. Sharon, I., Morowitz, M.J., Thomas, B.C., et al.: Time series community genomics analysis reveals rapid shifts in bacterial species, strains, and phage during infant gut colonization. Genome Res. **23**(1), 111–120 (2013)

38. Sieber, C.M., Probst, A.J., Sharrar, A., et al.: Recovery of genomes from metagenomes via a dereplication, aggregation and scoring strategy. Nat. Microbiol. **3**(7), 836–843 (2018)

39. Turnbaugh, P.J., Ley, R.E., Hamady, M., et al.: The human microbiome project. Nature **449**(7164), 804–810 (2007)

40. Uritskiy, G.V., DiRuggiero, J., Taylor, J.: MetaWRAP—a flexible pipeline for genome-resolved metagenomic data analysis. Microbiome **6**(1), 1–13 (2018)

41. Vinh, L.V., Lang, T.V., Binh, L.T., et al.: A two-phase binning algorithm using l-mer frequency on groups of non-overlapping reads. Algorithms Mol. Biol. **10**(1), 2 (2015)

42. Wang, Z., Wang, Z., Lu, Y.Y., et al.: SolidBin: improving metagenome binning with semi-supervised normalized cut. Bioinformatics **35**(21), 4229–4238 (2019)

43. Wood, D.E., Salzberg, S.L.: Kraken: ultrafast metagenomic sequence classification using exact alignments. Genome Biol. **15**(3), R46 (2014)

44. Wu, Y.W., Simmons, B.A., Singer, S.W.: MaxBin 2.0: an automated binning algorithm to recover genomes from multiple metagenomic datasets. Bioinformatics **32**(4), 605–607 (2015)

45. Wu, Y.W., Tang, Y.H., Tringe, S.G., et al.: MaxBin: an automated binning method to recover individual genomes from metagenomes using an expectation-maximization algorithm. Microbiome **2**(1), 26 (2014)

46. Wu, Y.W., Ye, Y.: A novel abundance-based algorithm for binning metagenomic sequences using l-tuples. J. Comput. Biol. **18**(3), 523–534 (2011). PMID: 21385052

47. Yu, G., Jiang, Y., Wang, J., Zhang, H., et al.: BMC3C: binning metagenomic contigs using codon usage, sequence composition and read coverage. Bioinformatics **34**(24), 4172–4179 (2018)

48. Yue, Y., Huang, H., Qi, Z., et al.: Evaluating metagenomics tools for genome binning with real metagenomic datasets and CAMI datasets. BMC Bioinform. **21**(1), 334 (2020)

49. Zhang, Z., Zhang, L.: METAMVGL: a multi-view graph-based metagenomic contig binning algorithm by integrating assembly and paired-end graphs. bioRxiv (2020)

A Fast, Provably Accurate Approximation Algorithm for Sparse Principal Component Analysis Reveals Human Genetic Variation Across the World

Agniva Chowdhury[1], Aritra Bose[2] (ID), Samson Zhou[3], David P. Woodruff[3], and Petros Drineas[4](✉) (ID)

[1] Computer Science and Mathematics Division, Oak Ridge National Laboratory, Oak Ridge, TN, USA
chowdhurya@ornl.gov

[2] Computational Genomics, IBM T.J. Watson Research Center, Yorktown Heights, NY, USA
a.bose@ibm.com

[3] School of Computer Science, Carnegie Mellon University, Pittsburgh, PA, USA
dwoodruf@cs.cmu.edu

[4] Department of Computer Science, Purdue University, West Lafayette, IN, USA
pdrineas@purdue.edu

Abstract. Principal component analysis (PCA) is a widely used dimensionality reduction technique in machine learning and multivariate statistics. To improve the interpretability of PCA, various approaches to obtain sparse principal direction loadings have been proposed, which are termed Sparse Principal Component Analysis (SPCA). In this paper, we present `ThreSPCA`, a provably accurate algorithm based on thresholding the Singular Value Decomposition for the SPCA problem, without imposing any restrictive assumptions on the input covariance matrix. Our thresholding algorithm is conceptually simple; much faster than current state-of-the-art; and performs well in practice. When applied to genotype data from the 1000 Genomes Project, `ThreSPCA` is faster than previous benchmarks, at least as accurate, and leads to a set of interpretable biomarkers, revealing genetic diversity across the world.

Keywords: Sparse PCA · Population stratification · Principal Component Analysis · Population structure

1 Introduction

Principal Component Analysis (PCA) and the related Singular Value Decomposition (SVD) are fundamental data analysis and dimensionality reduction tools

A. Chowdhury, A. Bose and S. Zhou—Equal Contribution.
This work was done when the author was a student in the Department of Statistics, Purdue University, IN.

I. Pe'er (Ed.): RECOMB 2022, LNBI 13278, pp. 86–106, 2022.
https://doi.org/10.1007/978-3-031-04749-7_6

that are used across a wide range of areas including machine learning, multivariate statistics, and many others. These tools return a set of orthogonal vectors of decreasing importance that are often interpreted as fundamental latent factors that underlie the observed data. Even though the vectors returned by PCA and SVD have strong optimality properties, they are notoriously difficult to interpret in terms of the underlying processes generating the data [18], since they are linear combinations of *all* available data points or *all* available features. The concept of Sparse Principal Components Analysis (SPCA) was introduced in the seminal work of [11], where sparsity constraints were enforced on the singular vectors in order to improve interpretability; see for example, document analysis applications in [11,18,22].

Formally, given a positive semidefinite (PSD) matrix $\mathbf{A} \in \mathbb{R}^{n \times n}$, SPCA can be defined as the constrained maximization problem:[1]

$$\mathcal{Z}^* = \max_{\mathbf{x} \in \mathbb{R}^n,\ \|\mathbf{x}\|_2 \leq 1} \mathbf{x}^\top \mathbf{A} \mathbf{x}, \qquad \text{subject to } \|\mathbf{x}\|_0 \leq k. \tag{1}$$

In the above formulation, $\mathbf{A}$ is a covariance matrix representing, for example, all pairwise feature or object similarities for an underlying data matrix. Therefore, SPCA can be applied to either the object or feature space of the data matrix, while the parameter k controls the sparsity of the resulting vector and is part of the input. Let $\mathbf{x}^*$ denote a vector that achieves the optimal value $\mathcal{Z}^*$ in the above formulation. Intuitively, the optimization problem of Eq. (1) seeks a *sparse*, unit norm vector $\mathbf{x}^*$ that maximizes the data variance. It is well-known that solving the above optimization problem is NP-hard [20] and that its hardness is due to the sparsity constraint. Indeed, if the sparsity constraint were removed, then the resulting optimization problem can be easily solved by computing the top left or right singular vector of $\mathbf{A}$ and its maximal value $\mathcal{Z}^*$ is equal to the top singular value of $\mathbf{A}$.

In this work, we explore the potential of SPCA in the analysis of genetics data leveraging a *provably accurate* thresholding algorithm for SPCA. In genetics, PCA is a tool of paramount importance and is ubiquitously used to estimate population structure and extract ancestry information [23]. It is well-known that genome-wide association studies (GWAS) that attempt to identify genetic markers that are associated with complex traits in a typical case/control setting can be grossly confounded by the underlying population structure, due to the presence of subgroups in the population that belong to different ancestries in both the case and control groups [24]. To account for such population stratification and to minimize the underlying spurious associations, researchers typically use the top few principal components as covariates in the underlying model. However, the principal components are linear combinations of all available genetic markers and, therefore, are not interpretable. SPCA is an obvious remedy towards that end, since one can use it to identify Single Nucleotide Polymorphisms (SNPs)

---

[1] Recall that the p-th power of the ℓ_p norm of a vector $\mathbf{x} \in \mathbb{R}^n$ is defined as $\|\mathbf{x}\|_p^p = \sum_{i=1}^n |\mathbf{x}_i|^p$ for $0 < p < \infty$. For $p = 0$, $\|x\|_0$ is a semi-norm denoting the number of non-zero entries of x.

or genetic markers carrying information about the underlying genetic ancestry. See also [12,13,16] for prior work motivating and using SPCA in the context of human genetics data analysis.

1.1   Our Contributions

Thresholding is a simple algorithmic concept, where each coordinate of, say, a vector is retained if its value is sufficiently high; otherwise, it is set to zero. Thresholding naturally preserves entries that have large magnitude while creating sparsity by eliminating small entries. Therefore, it seems like a logical strategy for SPCA: after computing a dense vector that approximately solves a PCA problem, perhaps with additional constraints, thresholding can be used to sparsify it.

We present a simple, provably accurate, thresholding algorithm (ThreSPCA, Sect. 2.1) for SPCA that leverages the fact that the top singular vector is an optimal solution for the SPCA problem without the sparsity constraint. Our algorithm actually uses a thresholding scheme that leverages the top few singular vectors of the underlying covariance matrix; it is simple and intuitive, yet offers tradeoffs in running time vs. accuracy, the first of its kind. Our algorithm returns a vector that is provably sparse and, when applied to the input covariance matrix $\mathbf{A}$, provably captures the optimal solution $\mathcal{Z}^*$ up to a small additive error. Indeed, our output vector has a sparsity that depends on k (the target sparsity of the original SPCA problem of Eq. (1)) and ε (an accuracy parameter between zero and one). Our analysis provides unconditional guarantees for the accuracy of the solution of the proposed thresholding scheme. To the best of our knowledge, no such analyses have appeared in prior work (see Sect. 1.2 for details). We emphasize that our approach only requires an approximate SVD and, as a result, ThreSPCA runs very quickly. In practice, ThreSPCA is much faster than current state-of-the-art and at least as accurate in the analysis of human genetics datasets. An additional contribution of our work is that, unlike prior work, our algorithm has a clear trade-off between quality of approximation and output sparsity. Indeed, by increasing the density of the final SPCA vector, one can improve the amount of variance that is captured by our SPCA output. See Theorem 1 for details on this sparsity vs. accuracy trade-off for ThreSPCA.

Importantly, we evaluate ThreSPCA on the genotype dataset from 1000 Genomes (1KG) Project [10] and on simulated genotype data in order to practically assess its performance. ThreSPCA identifies functionally relevant, interpretable SNPs from the 1KG data and, from an accuracy perspective, it performs comparably to current state-of-the-art SPCA algorithms while being much faster than its competitors.

1.2   Prior Work

SPCA was formally introduced by [11]; however, previously studied PCA approaches based on rotating [14] or thresholding [7] the top singular vector of the input matrix seemed to work well, at least in practice, given sparsity constraints. Following [11], there has been an abundance of interest in SPCA, with extensions based on LASSO (ScoTLASS) on an ℓ_1 relaxation of the problem [15] or a non-convex regression-type approximation, penalized similar to LASSO [28].

Prior work that offers provable guarantees, typically given *some assumptions about the input matrix*, includes [22], which analyzed a specific set of vectors in a low-dimensional eigenspace of the input matrix and presented relative error guarantees for the optimal objective, given the assumption that the input covariance matrix has a decaying spectrum. The time complexity of the algorithm of [22] is given by $\mathcal{O}(n^{d+1} \log n)$ (due to solving an exact SVD), where d is the low rank parameter that affects the accuracy of the output. Even for $d = 1$, the theoretical time complexity boils down to $\mathcal{O}(n^2 \log n)$ and it is not clear how to make use of an approximate SVD algorithm to improve this running time without affecting its theoretical bound. Furthermore, for a high precision output, one generally needs d to be larger than one, in which case the practical running time also increases drastically. [1] gave a polynomial-time algorithm that solves sparse PCA *exactly* for input matrices of constant rank. [8] showed that sparse PCA can be approximated in polynomial time within a factor of $n^{-1/3}$ and also highlighted an additive PTAS of [2] based on the idea of finding multiple disjoint components and solving bipartite maximum weight matching problems. This PTAS needs time $n^{\text{poly}(1/\varepsilon)}$, whereas ThreSPCA has running time that depends on the sparsity of the input data.

SPCA has been applied in the context of human genetics before, in the form of sparse factor analysis (SFA) [12] and with a penalty term in LASSO (L-PCA) or Adaptive LASSO (AL-PCA) [16]. However, there are a number of aspects that our work improves compared to prior studies. First, unlike ThreSPCA, the SFA method used some prior assumptions on the genotype matrix and none of these previous studies come with a theoretical guarantee showing a clear sparsity vs. accuracy trade-off.

Second, prior work has to tune the penalty parameter in [16] several times in order to achieve a specific sparsity value in practice, which increases the running time of the method. Third, the convergence of the SPCA algorithm proposed by [16] depends on an initial PC score, which typically relies on the top right singular vector of the data and necessitates the computation of an exact SVD, which is expensive. It is not clear whether replacing the exact SVD with a fast approximate SVD would affect the results of [16].

2   Materials and Methods

2.1   The THRESPCA Algorithm

Notation. We use bold letters to denote matrices and vectors. For a matrix $\mathbf{A} \in \mathbb{R}^{n \times n}$, we denote its (i,j)-th entry by $A_{i,j}$; its i-th row by $\mathbf{A}_{i*}$, and its j-th column by $\mathbf{A}_{*j}$; its 2-norm by $\|\mathbf{A}\|_2 = \max_{\mathbf{x} \in \mathbb{R}^n,\ \|\mathbf{x}\|_2 = 1} \|\mathbf{A}\mathbf{x}\|_2$; and its (squared) Frobenius norm by $\|\mathbf{A}\|_F^2 = \sum_{i,j} A_{i,j}^2$. We use the notation $\mathbf{A} \succeq 0$ to denote that the matrix $\mathbf{A}$ is symmetric positive semidefinite (PSD) and $\text{Tr}(\mathbf{A}) = \sum_i A_{i,i}$ to denote its trace, which is also equal to the sum of its singular values. Given a PSD matrix $\mathbf{A} \in \mathbb{R}^{n \times n}$, its Singular Value Decomposition is given by $\mathbf{A} = \mathbf{U}\mathbf{\Sigma}\mathbf{U}^T$, where $\mathbf{U}$ is the matrix of left/right singular vectors and $\mathbf{\Sigma}$ is the diagonal matrix of singular values.

Our Approach: SPCA via SVD Thresholding. To achieve nearly input sparsity runtime, our thresholding algorithm is based upon using the top ℓ right (or left) singular vectors of the PSD matrix $\mathbf{A}$. Given $\mathbf{A}$ and an accuracy parameter ε, our approach first computes $\mathbf{\Sigma}_\ell \in \mathbb{R}^{\ell \times \ell}$ (the diagonal matrix of the top ℓ singular values of $\mathbf{A}$) and $\mathbf{U}_\ell \in \mathbb{R}^{n \times \ell}$ (the matrix of the top ℓ left singular vectors of $\mathbf{A}$), for $\ell = 1/\varepsilon$. Then, it *deterministically* selects a subset of $\mathcal{O}\left(k/\varepsilon^3\right)$ rows of $\mathbf{U}_\ell$ using a simple thresholding scheme based on their squared row norms (recall that k is the sparsity parameter of the SPCA problem). In the last step, it returns the *top right singular vector* of the matrix consisting of the columns of $\mathbf{\Sigma}_\ell^{1/2}\mathbf{U}_\ell^\top$ that correspond to the row indices of $\mathbf{U}_\ell$ chosen in the thresholding step. Notice that this right singular vector is an $\mathcal{O}\left(k/\varepsilon^3\right)$-dimensional vector, which is finally expanded to a vector in $\mathbb{R}^n$ by appropriate padding with zeros. This sparse vector is our approximate solution to the SPCA problem of Eq. (1).

This simple algorithm is somewhat reminiscent of prior thresholding approaches for SPCA. However, to the best of our knowledge, no provable a priori bounds were known for such algorithms without strong assumptions on the input matrix. This might be due to the fact that prior approaches focused on thresholding only the top right singular vector of $\mathbf{A}$, whereas our approach thresholds the top $\ell = 1/\varepsilon$ right singular vectors of $\mathbf{A}$. This slight relaxation allows us to present provable bounds for the proposed algorithm.

In more detail, let the SVD of $\mathbf{A}$ be $\mathbf{A} = \mathbf{U}\mathbf{\Sigma}\mathbf{U}^T$. Let $\mathbf{\Sigma}_\ell \in \mathbb{R}^{\ell \times \ell}$ be the diagonal matrix of the top ℓ singular values and let $\mathbf{U}_\ell \in \mathbb{R}^{n \times \ell}$ be the matrix of the top ℓ right (or left) singular vectors. Let $R = \{i_1, \ldots, i_{|R|}\}$ be the set of indices of rows of $\mathbf{U}_\ell$ that have squared norm at least ε^2/k and let $\bar{R}$ be its complement. Here $|R|$ denotes the cardinality of the set R and $R \cup \bar{R} = \{1, \ldots, n\}$. Let $\mathbf{R} \in \mathbb{R}^{n \times |R|}$ be a sampling matrix that selects[2] the rows of $\mathbf{U}_\ell$ whose indices are in the set R. Given this notation, we are now ready to state Algorithm 1. Notice that $\mathbf{R}\mathbf{y}$ satisfies $\|\mathbf{R}\mathbf{y}\|_2 = \|\mathbf{y}\|_2 = 1$ (since $\mathbf{R}$ has orthogonal columns) and $\|\mathbf{R}\mathbf{y}\|_0 = |R|$. Since R is the set of rows of $\mathbf{U}_\ell$ with squared norm at least ε^2/k and $\|\mathbf{U}_\ell\|_F^2 = \ell = 1/\varepsilon$, it follows that $|R| \le k/\varepsilon^3$. Thus, the vector returned by Algorithm 1 has k/ε^3 sparsity and unit norm. (See the Appendix for more details.)

Theorem 1. *Let k be the sparsity parameter and $\varepsilon \in (0, 1]$ be the accuracy parameter. Then, the vector $\mathbf{z} \in \mathbb{R}^n$ (the output of Algorithm 1) has sparsity k/ε^3, unit norm, and satisfies*

$$\mathbf{z}^\top \mathbf{A}\mathbf{z} \ge \mathcal{Z}^* - 3\varepsilon\text{Tr}(\mathbf{A}).$$

---

[2] Each column of $\mathbf{R}$ has a single non-zero entry (set to one), corresponding to one of the $|R|$ selected columns. Formally, $\mathbf{R}_{i_t,t} = 1$ for $t = 1, \ldots, |R|$; all other entries of $\mathbf{R}$ are set to zero.

Algorithm 1. ThreSPCA: fast thresholding SPCA via SVD

Input: $\mathbf{A} \in \mathbb{R}^{n \times n}$, sparsity k, error parameter $\varepsilon > 0$.
Output: $\mathbf{y} \in \mathbb{R}^n$ such that $\|y\|_2 = 1$ and $\|\mathbf{y}\|_0 = k/\varepsilon^2$.

1: $\ell \leftarrow 1/\varepsilon$;
2: Compute $\mathbf{U}_\ell \in \mathbb{R}^{n \times \ell}$ (top ℓ left singular vectors of $\mathbf{A}$) and $\mathbf{\Sigma}_\ell \in \mathbb{R}^{\ell \times \ell}$ (the top ℓ singular values of $\mathbf{A}$);
3: Let $R = \{i_1, \ldots, i_{|R|}\}$ be the set of rows of $\mathbf{U}_\ell$ with squared norm at least ε^2/k and let $\mathbf{R} \in \mathbb{R}^{n \times |R|}$ be the associated sampling matrix (see text for details);
4: $\mathbf{y} \in \mathbb{R}^{|R|} \leftarrow \mathrm{argmax}_{\|\mathbf{x}\|_2=1} \left\| \mathbf{\Sigma}_\ell^{1/2} \mathbf{U}_\ell^\top \mathbf{R} \mathbf{x} \right\|_2^2$;
5: **return** $\mathbf{z} = \mathbf{R}\mathbf{y} \in \mathbb{R}^n$;

The optimality gap of Theorem 1 depends on $\mathrm{Tr}(\mathbf{A})$, which is the sum of the eigenvalues of $\mathbf{A}$ and can also be viewed as the total variance of the data. Therefore, if we divide both sides of the bound in Theorem 1 by $\mathrm{Tr}(\mathbf{A})$, the resulting bound is given by $(\mathsf{prop}^* - \widetilde{\mathsf{prop}}) \leq 3\varepsilon$, where for a given k, $\widetilde{\mathsf{prop}}$ is the proportion of the total variance explained by the output of ThreSPCA and prop^* is the proportion of the total variance explained by the optimal Sparse PC. Now, trivially, we have $(\mathsf{prop}^* - \widetilde{\mathsf{prop}}) \geq 0$, since prop^* is the maximum variance explained by Sparse PC for a given sparsity value. Thus, combining these two yields $0 \leq (\mathsf{prop}^* - \widetilde{\mathsf{prop}}) \leq 3\varepsilon$, which can be interpreted as the quality-of-approximation in terms of the proportion of total variance explained by ThreSPCA.

The proof of Theorem 1 is deferred to the appendix. See Sect. 1.A for the proof of Theorem 1 as well as an intermediate result (Lemma 1) that leads to the final bound in Theorem 1. The running time of Algorithm 1 is dominated by the computation of the top ℓ singular vectors and singular values of the matrix $\mathbf{A}$. One could always use the SVD of the full matrix $\mathbf{A}$ ($\mathcal{O}\left(n^3\right)$ time) to compute the top ℓ singular vectors and singular values of $\mathbf{A}$. In practice, any iterative method, such as subspace iteration using a random initial subspace or the Krylov subspace of the matrix, can be used towards this end. We now address the inevitable approximation error incurred by such approximate SVD methods below.

Using Approximate SVD Algorithms. Although the guarantees of Theorem 1 in Algorithm 1 use an exact SVD computation, which could take time $\mathcal{O}\left(n^3\right)$, we can further improve the running time by using an approximate SVD algorithm such as the randomized block Krylov method of [21], which runs in nearly input sparsity running time. Our analysis uses the relationships $\|\mathbf{\Sigma}_{\ell,\perp}^{1/2}\|_2^2 \leq \mathrm{Tr}(\mathbf{A})/\ell$ and $\sigma_1(\Sigma_\ell) \leq \mathrm{Tr}(\mathbf{A})$. The randomized block Krylov method of [21] recovers these guarantees up to a multiplicative $(1 + \varepsilon)$ factor, in $\mathcal{O}\left(\log n/\varepsilon^{1/2} \cdot \mathrm{nnz}(\mathbf{A})\right)$ time. Here $\mathrm{nnz}(\mathbf{A})$ denotes the number of non-zero entries of the matrix $\mathbf{A}$, which is $\mathcal{O}\left(n^2\right)$ for dense matrices.

Extracting Additional Sparse PCs. To get multiple sparse PCs using Algorithm 1, we remove the top principal component from the data and run `ThreSPCA` on the residual dataset. In other words, let $\mathbf{X} \in \mathbb{R}^{m \times n}$ be the *mean-centered* data matrix corresponding to $\mathbf{A}$, *i.e.*, $\mathbf{A} = \mathbf{X}^\top \mathbf{X}$. Let $\mathbf{v} \in \mathbb{R}^n$ be the top right singular vector of $\mathbf{X}$; then, in order to get the second sparse PC, we run `ThreSPCA` on the covariance matrix $\mathbf{A}_1 = \mathbf{X}_1^\top \mathbf{X}_1$, where $\mathbf{X}_1 = \mathbf{X} - \mathbf{X}\mathbf{v}\mathbf{v}^\top$.

2.2   Data

1000 Genome Data. In order to evaluate the speed and accuracy of `ThreSPCA` as well as to interpret its output, we first analyzed data from the 1000 Genome Project (1KG) [10], which contained genotype data from 2,503 individuals with 39,517,397 SNPs sampled from 26 different populations across all continents. After performing Quality Control (QC) with minor allele frequency below 5% and, subsequently, pruning related genotypes for Linkage Disequilibrium (LD) using a window size of 50 kb and $r^2 > 0.2$, we finally retained 360,498 variants.

Simulated Data. We generated simulated data emulating real-world populations to evaluate whether `ThreSPCA` can correctly identify markers which contribute to the genetic differences between and within the populations. Based on previous work [4], we simulated two datasets varying $m = \{5000, 10000\}$ SNPs genotyped across $n = \{500, 1000\}$ individuals based on the Pritchard-Stephens-Donelly (PSD) model [25] with the mixing parameter between populations, $\alpha = 0.01$. The allele frequencies were simulated based on real-world data from three divergent populations, namely CEU (Utah residents with Northern and Western European ancestry), ASW (African ancestry in Southwestern US), and MXL (Mexican ancestry in California) from the HapMap Phase 3 data [17]. We selected a threshold t and varied it across the range $t = \{100, 250, 500\}$, representing the number of SNPs which contribute to population structure between the populations (true positives); the remaining $m - t$ genotypes were simulated such as they had minimal genetic differences between populations (false positives). We simulated 200 data sets (100 each for values of m and n) and applied `ThreSPCA`, `L-PCA` and `AL-PCA` for comparative analyses to evaluate their efficacy.

2.3   Experiments

We performed QC on the 1KG data, including LD pruning using PLINK2.0 [9]. PCA was performed using TeraPCA [5]. Annotation of `ThreSPCA` derived variants were performed in Ensembl Variant Effect Predictor (VEP) [19]. We performed Gene Ontology (GO) pathway analyses using `clusterProfiler` [27] in R. We ran `ThreSPCA`, with the threshold parameter ℓ, fixed to one.

3   Results

3.1   `ThreSPCA` Reveals Genetic Diversity Across the World

We applied `ThreSPCA` with a sparsity threshold of $k = 500$ on the 1KG data after quality control and pruning for correlated SNPs. We obtained sets of informative markers of cardinality k from each of the PCs. We restricted our analysis to the top three PCs, resulting in a total of 1,500 SNPs, which explained approximately 83% of the variance. Thus, we performed PCA on a reduced 1KG data with 2,503 individuals and 1,500 SNPs. We observed that both the PCA plot and the allele frequency bar plot, grouped by populations across the world, are almost identical. The squared Pearson correlation coefficient (r^2) between the top two PCs from the original 1KG data and ThreSPCA informed variants are very high, equal to 0.98, 0.97 and 0.94 for PCs 1, 2 and 3 respectively. Thus, the PCA plot of the informative markers clearly preserves the clusters of each subgroup (Fig. 1) and reveals fine-scale population structure among the groups.

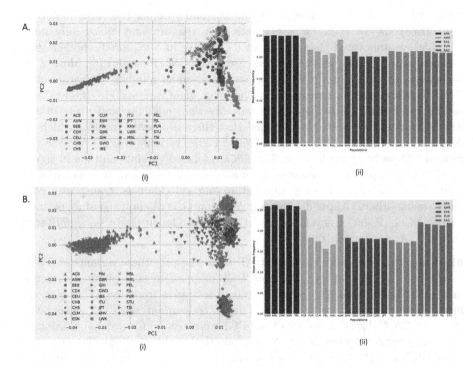

Fig. 1. Population structure of world populations from: A. pruned 1KG data with 360,498 SNPs, and B. 1KG data with 1,500 ThreSPCA derived variants corresponding to the top three PCs, captured by (i) PCA plot and (ii) mean allele frequency bar plots colored by continental populations arranged in order from Africa (AFR), Americas (AMR), East Asia (EAS), Europe (EUR) and South Asia (SAS).

Examining each of the three PCs closely shows that the mean allele frequency distribution (Appendix Figs. 5) from PC1 is skewed towards the African populations and also from the mixed ancestry populations of ASW (Africans in Southwestern US) and ACB (African Caribbeans from Barbados). SNPs obtained from PC2 were almost equally distributed across the continental populations with a slightly higher frequencies in East Asians. PC3 shows a skewness towards South

Asian populations. To make an informed choice of the sparsity threshold k, we computed the *PC scores* from the top two PCs by projecting the sparse vectors obtained from ThreSPCA on the original pruned 1KG data for a range of values of $k = \{500, 1000, 5000\}$.

We computed r^2 between the *PC scores* obtained from each PC for each value of k and the original PC obtained from the pruned 1KG data. We observe high correlation values for the top two PCs, cumulatively reaching their peak when the sparsity parameter k is set to approximately 500 (Appendix Fig. 7 (left)).

3.2   Interpretability of ThreSPCA Informed Variants

Annotating the Selected Variants. To understand whether the variants derived from ThreSPCA for each PC are functionally relevant and biologically interpretable, we annotated them using VEP [19]. We also explored whether these variants were mapping to a trait or disease in the GWAS catalog [6]. Most of the variants were introns with some intergenic and small number of Transcription Factor binding sites, upstream and downstream gene variants, etc. Interestingly, among the coding consequences, 58 variants were missense and likely disease causing and further statistics revealed that there are seven variants which are deleterious and nine *probably* or *possibly* damaging variants (Fig. 2). We also performed GO pathway analyses on ThreSPCA informed variants and found significantly enriched pathways common to humans across the world, such as pathways related to synapses and potassium, cation and ion channels, transporter complex, among others (Appendix Fig. 6a). We found the calcium signaling pathway from KEGG (Kyoto Encyclopedia of Genes and Genomes) to be significantly enriched (Appendix Fig. 6b).

Mapping the Selected Variants to Traits. Mapping these variants in GWAS catalog, we found that variants from PC1 mapped to skin pigment measurement (Appendix Table 2), justifying our observation from the PCA plot and mean allele frequency distribution. This is concordant with our observation that ThreSPCA observed variants from PC1 were skewed towards populations of African ancestry (Appendix Fig. 5), who are darker skinned than the rest of the world. PC2 and PC3 on the other hand mapped to various traits which are commonly found to be varying in populations across the world such as body height, BMI, hip and waist circumference, circadian rhythm, gut microbiome, smoking status, cardiovascular diseases, calcium channel blocker use (concordant with calcium signaling pathway found in GO analyses), blood measurements, among others.

3.3   Comparing ThreSPCA to State-of-the-Art

Simulation Studies. We designed a simulation study to evaluate the correctness of ThreSPCA and compare it with the state-of-the-art SPCA methods in genetics, namely, L-PCA and AL-PCA from [16]. The population structure of the

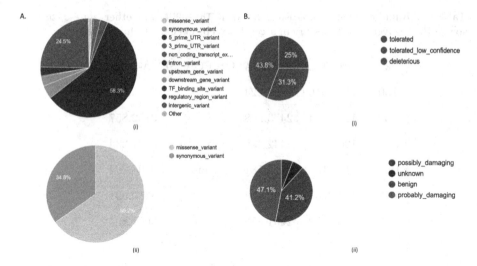

Fig. 2. Pie charts showing the percentage of variants from A. (i) most severe consequences and (ii) coding consequences obtained from VEP. B. Deleterious and probably damaging from (i) SIFT and (ii) PolyPhen.

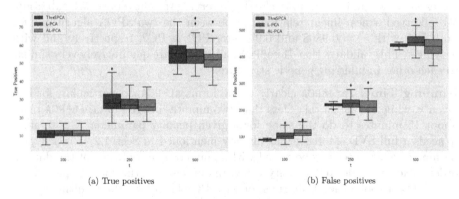

(a) True positives   (b) False positives

Fig. 3. Box and whisker plots comparing between ThreSPCA, L-PCA and AL-PCA for true and false positives obtained from the simulated dataset of $m = 10,000$ and $n = 1,000$ and varying values of t, i.e., the number of SNPs which contribute to population structure.

simulation shows three distinct clusters for each population with signs of admixture between them (Appendix Fig. 4). Applying ThreSPCA on the simulated dataset with 10,000 markers and 1,000 individuals, we observed that ThreSPCA identified similar numbers (mean) of true positives, i.e., markers contributing to the genetic diversity between and within the populations when compared to its counterparts L-PCA and AL-PCA, while identifying a significantly smaller number of false positives, i.e., noisy markers which have no difference in allele frequencies between populations (Fig. 3b).

Table 1. Running time comparisons between ThreSPCA and other *state-of-the-art* sparse PCA solvers. All times are in seconds except CWPCA, which is in hours.

k	ThreSPCA	AL-PCA	CWPCA	SPCA-Lowrank
150	117.3016	2287.224		3253.057
800	126.8674	2473.908		3152.857
1000	120.9341	2442.435		3121.234
1500	119.6183	2715.581	> 5hrs	3408.294
6000	123.2763	2440.104		3319.691
12000	126.3872	2451.353		3071.864

Real Data. We applied both ThreSPCA and AL-PCA[3] on the 1KG data with $k = 500$ and compared the *PC1 scores vs. PC2 scores* generated from the outputs of the aforementioned methods. ThreSPCA and AL-PCA are almost identical to the corresponding standard PC plot, clearly preserving the clusters of each subgroup. We observed a near-linear relationship between the two SPCA algorithms for both PCs with $r^2 = 0.9808$ and 0.9426 for PC1 & PC2, respectively and with varying k. This validates that ThreSPCA and AL-PCA are qualitatively very similar to each other in inferring genetic structure.

Running Time. ThreSPCA clearly outperforms AL-PCA. In particular, for any given k, while ThreSPCA takes less than two minutes in 1KG data, AL-PCA takes about 15 minutes to do the same for a given penalty parameter $\lambda > 0$, since it needs a full SVD. Moreover, as already mentioned in Sect. 1.2, λ is a hyper-parameter which needs to be tuned with many cross-nested runs of the data in order to achieve a desired sparsity value. In our case, for the sparsity parameter set to 500, it took at least six runs for each PC. Therefore, the resulting speed-up achieved by ThreSPCA is more than 45x for real data set and around 80x for simulated data.

Finally, we also compare the output of our algorithm against other state-of-the-art SPCA approaches, including the coordinate-wise optimization algorithm of [3] (cwpca), and the spannogram-based algorithm of [22] (spca-lowrank). To measure the accuracy of the of various SPCA algorithms, we first looked at the term $\mathbf{z}^\top \mathbf{A} \mathbf{z}$ (for varying k), which is nothing but the total variability explained (VE) by the sparse output $\mathbf{z}$. In terms of VE, we noticed that ThreSPCA matches that of the other *state-of-the-art* SPCA solvers for all the sparsity values observed, which are much larger than that of AL-PCA (Appendix Fig. 7 (right)). In addition, we also found that ThreSPCA is not only among the most accurate algorithms, but also is the fastest (Table 1) among all (takes about 100 s to 120 s

---

[3] Results from L-PCA are qualitatively very similar to AL-PCA and we only report results for the latter.

to run for each k, while other solvers including `AL-PCA` run in time at least 2,200 s for each k. See details in Appendix Sect. 1.B.3).

4 Discussion

We present `ThreSPCA`, a simple and intuitive approximation algorithm for SPCA, based on a deterministic thresholding scheme, without imposing any restrictive assumption on the input covariance matrix. `ThreSPCA` comes with a provable accuracy guarantee and provides a clear sparsity vs. accuracy trade-off. In practice, it is much faster than the other *state-of-the-art* SPCA methods and indeed, can be implemented in nearly input sparsity time.

Applying `ThreSPCA` on the 1KG data, we observed that the set of derived SNPs accurately approximates the genetic diversity across world populations. For each PC, the derived set of k SNPs (we used $k = 500$ throughout the analyses) captured genetic structure within different continental populations. Together, the top three PCs which explain most of the variance in the 1KG data, we observed that `ThreSPCA` selected 1500 meaningful, ancestry information preserving SNPs which leads to similar inference of population structure across the world as the original 1KG data with 360,498 SNPs. Annotating `ThreSPCA` derived variants further showed that they are interpretable and mostly missense in nature, thus likely disease causing. To interpret this, we mapped these variants to various traits in GWAS catalog and found that indeed these variants were mapped to different common traits such as body height, BMI, etc. which vary within and between populations across the world, sometimes leading to spurious associations due to population structure among populations [26]. These variants also mapped to various diseases, which vary across populations such as cardiovascular diseases. Although the scale of the data used in this analysis is small when compared to large-scale genomic data, we observe that `ThreSPCA` is designed to handle biobank-scale datasets since it only need to run a randomized SVD/PCA analysis, which can be implemented efficiently in out-of-core settings [5]. `ThreSPCA` can also be used in GWAS as a population stratification correction step by identifying informative markers which highlight the ancestry stratification of cases/controls with fine-grained details which is often overlooked by a standard PCA.

In summary, `ThreSPCA` provides a fast and provably accurate approximate method for computing SPCA. It provides a method to find interpretable markers in population genetics, which can immensely help understand population stratification, a major cause of spurious associations in GWAS. Also, it highlights the genetic sub-structure among different populations and the `ThreSPCA` derived variants are likely disease causing, often mapped to potential diseases and traits.

Acknowledgements. PD and AC were partially supported by National Science Foundation (NSF) 10001390, NSF III-10001674, NSF III-10001225, and an IBM Faculty Award to PD. AB was supported by IBM. DPW and SZ would like to thank partial support from NSF grant No. CCF-181584, Office of Naval Research (ONR) grant N00014-18-1-2562, National Institute of Health (NIH) grant 5401 HG 10798-2, and a Simons Investigator Award.

Code Availability. A Python implementation of `ThreSPCA` can be found at: https://github.com/aritra90/ThreSPCA.

Appendix 1.A   SPCA via Thresholding: Discussions and Proofs

The intuition behind Theorem 1 is that we can decompose the value of the optimal solution into the value contributed by the coordinates in R, the value contributed by the coordinates outside of R, and a cross term. The first term we can upper bound by the output of the algorithm, which maximizes with respect to the coordinates in R. For the latter two terms, we can upper bound the contribution due to the upper bound on the squared row norms of indices outside of R and due to the largest singular value of $\mathbf{U}$ being at most the trace of $\mathbf{A}$.

We highlight that, as an intermediate step in the proof of Theorem 1, we need to prove the following Lemma 1, which is very much at the heart of our proof of Theorem 1 and, unlike prior work, allows us to provide provably accurate bounds for the thresholding Algorithm 1. At a high level, the proof of Lemma 1 first decomposes a basis for the columns spanned by $\mathbf{U}$ into those spanned by the top ℓ singular vectors and the remaining $n - \ell$ singular vectors. We then lower bound the contribution of the top ℓ singular vectors by upper bounding the contribution of the remaining $n - \ell$ singular vectors after noting that the largest remaining singular value is at most a $1/\ell$-fraction of the trace. We look at the detailed proof of Lemma 1 below where we use the notation of Sect. 2.1. For notational convenience, let $\sigma_1, \ldots, \sigma_n$ be the diagonal entries of the matrix $\mathbf{\Sigma} \in \mathbb{R}^{n \times n}$, i.e., the singular values of $\mathbf{A}$.

Lemma 1. *Let $\mathbf{A} \in \mathbb{R}^{n \times n}$ be a PSD matrix and $\mathbf{\Sigma} \in \mathbb{R}^{n \times n}$ (respectively, $\mathbf{\Sigma}_\ell \in \mathbb{R}^{\ell \times \ell}$) be the diagonal matrix of all (respectively, top ℓ) singular values and let $\mathbf{U} \in \mathbb{R}^{n \times n}$ (respectively, $\mathbf{U}_\ell \in \mathbb{R}^{n \times \ell}$) be the matrix of all (respectively, top ℓ) singular vectors. Then, for all unit vectors $\mathbf{x} \in \mathbb{R}^n$,*

$$\left\| \mathbf{\Sigma}_\ell^{1/2} \mathbf{U}_\ell^\top \mathbf{x} \right\|_2^2 \geq \left\| \mathbf{\Sigma}^{1/2} \mathbf{U}^\top \mathbf{x} \right\|_2^2 - \varepsilon \mathrm{Tr}(\mathbf{A}).$$

Proof. Let $\mathbf{U}_{\ell,\perp} \in \mathbb{R}^{n \times (n-\ell)}$ be a matrix whose columns form a basis for the subspace perpendicular to the subspace spanned by the columns of $\mathbf{U}_\ell$. Similarly, let $\mathbf{\Sigma}_{\ell,\perp} \in \mathbb{R}^{(n-\ell) \times (n-\ell)}$ be the diagonal matrix of the bottom $n - \ell$ singular values of $\mathbf{A}$. Notice that $\mathbf{U} = [\mathbf{U}_\ell \ \ \mathbf{U}_{\ell,\perp}]$ and $\mathbf{\Sigma} = [\mathbf{\Sigma}_\ell \ \mathbf{0}; \ \mathbf{0} \ \ \mathbf{\Sigma}_{\ell,\perp}]$; thus,

$$\mathbf{U}\mathbf{\Sigma}^{1/2}\mathbf{U}^\top = \mathbf{U}_\ell \mathbf{\Sigma}_\ell^{1/2} \mathbf{U}_\ell^\top + \mathbf{U}_{\ell,\perp} \mathbf{\Sigma}_{\ell,\perp}^{1/2} \mathbf{U}_{\ell,\perp}^\top.$$

By the Pythagorean theorem,

$$\left\|\mathbf{U}\mathbf{\Sigma}^{1/2}\mathbf{U}^\top\mathbf{x}\right\|_2^2 = \left\|\mathbf{U}_\ell\mathbf{\Sigma}_\ell^{1/2}\mathbf{U}_\ell^\top\mathbf{x}\right\|_2^2 + \left\|\mathbf{U}_{\ell,\perp}\mathbf{\Sigma}_{\ell,\perp}^{1/2}\mathbf{U}_{\ell,\perp}^\top\mathbf{x}\right\|_2^2.$$

Using invariance properties of the vector two-norm and sub-multiplicativity, we get

$$\left\|\mathbf{\Sigma}_\ell^{1/2}\mathbf{U}_\ell^\top\mathbf{x}\right\|_2^2 \geq \left\|\mathbf{\Sigma}^{1/2}\mathbf{U}^\top\mathbf{x}\right\|_2^2 - \left\|\mathbf{\Sigma}_{\ell,\perp}^{1/2}\right\|_2^2 \left\|\mathbf{U}_{\ell,\perp}^\top\mathbf{x}\right\|_2^2.$$

We conclude the proof by noting that $\left\|\mathbf{\Sigma}^{1/2}\mathbf{U}^\top\mathbf{x}\right\|_2^2 = \mathbf{x}^\top\mathbf{U}\mathbf{\Sigma}\mathbf{U}^\top\mathbf{x} = \mathbf{x}^\top\mathbf{A}\mathbf{x}$ and

$$\left\|\mathbf{\Sigma}_{\ell,\perp}^{1/2}\right\|_2^2 = \sigma_{\ell+1} \leq \frac{1}{\ell}\sum_{i=1}^n \sigma_i = \frac{\mathrm{Tr}(\mathbf{A})}{\ell}.$$

The inequality above follows since $\sigma_1 \geq \sigma_2 \geq \ldots \sigma_\ell \geq \sigma_{\ell+1} \geq \ldots \geq \sigma_n$. We conclude the proof by setting $\ell = 1/\varepsilon$.

Theorem 1. *Let k be the sparsity parameter and $\varepsilon \in (0,1]$ be the accuracy parameter. Then, the vector $\mathbf{z} \in \mathbb{R}^n$ (the output of Algorithm 1) has sparsity k/ε^3, unit norm, and satisfies*

$$\mathbf{z}^\top\mathbf{A}\mathbf{z} \geq \mathcal{Z}^* - 3\varepsilon\mathrm{Tr}(\mathbf{A}).$$

Proof. Let $R = \{i_1, \ldots, i_{|R|}\}$ be the set of indices of rows of $\mathbf{U}_\ell$ (columns of $\mathbf{U}_\ell^\top$) that have squared norm at least ε^2/k and let $\bar{R}$ be its complement. Here $|R|$ denotes the cardinality of the set R and $R \cup \bar{R} = \{1, \ldots, n\}$. Let $\mathbf{R} \in \mathbb{R}^{n \times |R|}$ be the sampling matrix that selects the columns of $\mathbf{U}_\ell$ whose indices are in the set R and let $\mathbf{R}_\perp \in \mathbb{R}^{n \times (n-|R|)}$ be the sampling matrix that selects the columns of $\mathbf{U}_\ell$ whose indices are in the set $\bar{R}$. Thus, each column of $\mathbf{R}$ (respectively $\mathbf{R}_\perp$) has a single non-zero entry, equal to one, corresponding to one of the $|R|$ (respectively $|\bar{R}|$) selected columns. Formally, $\mathbf{R}_{i_t,t} = 1$ for all $t = 1, \ldots, |R|$, while all other entries of $\mathbf{R}$ (respectively $\mathbf{R}_\perp$) are set to zero; $\mathbf{R}_\perp$ can be defined analogously. The following properties are easy to prove: $\mathbf{R}\mathbf{R}^\top + \mathbf{R}_\perp\mathbf{R}_\perp^\top = \mathbf{I}_n$; $\mathbf{R}^\top\mathbf{R} = \mathbf{I}$; $\mathbf{R}_\perp^\top\mathbf{R}_\perp = \mathbf{I}$; $\mathbf{R}_\perp^\top\mathbf{R} = \mathbf{0}$. Recall that $\mathbf{x}^*$ is the optimal solution to the SPCA problem from Eq. (1). We proceed as follows:

$$\begin{aligned}
\left\|\mathbf{\Sigma}_\ell^{1/2}\mathbf{U}_\ell^\top\mathbf{x}^*\right\|_2^2 &= \left\|\mathbf{\Sigma}_\ell^{1/2}\mathbf{U}_\ell^\top(\mathbf{R}\mathbf{R}^\top + \mathbf{R}_\perp\mathbf{R}_\perp^\top)\mathbf{x}^*\right\|_2^2 \\
&\leq \left\|\mathbf{\Sigma}_\ell^{1/2}\mathbf{U}_\ell^\top\mathbf{R}\mathbf{R}^\top\mathbf{x}^*\right\|_2^2 + \left\|\mathbf{\Sigma}_\ell^{1/2}\mathbf{U}_\ell^\top\mathbf{R}_\perp\mathbf{R}_\perp^\top\mathbf{x}^*\right\|_2^2 \\
&\quad + 2\left\|\mathbf{\Sigma}_\ell^{1/2}\mathbf{U}_\ell^\top\mathbf{R}\mathbf{R}^\top\mathbf{x}^*\right\|_2 \left\|\mathbf{\Sigma}_\ell^{1/2}\mathbf{U}_\ell^\top\mathbf{R}_\perp\mathbf{R}_\perp^\top\mathbf{x}^*\right\|_2 \\
&\leq \left\|\mathbf{\Sigma}_\ell^{1/2}\mathbf{U}_\ell^\top\mathbf{R}\mathbf{R}^\top\mathbf{x}^*\right\|_2^2 + \sigma_1\left\|\mathbf{U}_\ell^\top\mathbf{R}_\perp\mathbf{R}_\perp^\top\mathbf{x}^*\right\|_2^2 \\
&\quad + 2\sigma_1\left\|\mathbf{U}_\ell^\top\mathbf{R}\mathbf{R}^\top\mathbf{x}^*\right\|_2 \left\|\mathbf{U}_\ell^\top\mathbf{R}_\perp\mathbf{R}_\perp^\top\mathbf{x}^*\right\|_2.
\end{aligned} \qquad (2)$$

The above inequalities follow from the Pythagorean theorem and sub-multiplicativity. We now bound the second term in the right-hand side of the above inequality.

$$\left\|\mathbf{U}_\ell^\top \mathbf{R}_\perp \mathbf{R}_{\bar{\perp}}^\top \mathbf{x}^*\right\|_2 = \|\sum_{i=1}^n (\mathbf{U}_\ell^\top \mathbf{R}_\perp)_{*i}(\mathbf{R}_{\bar{\perp}}^\top \mathbf{x}^*)_i\|_2$$

$$\leq \sum_{i=1}^n \|(\mathbf{U}_\ell^\top \mathbf{R}_\perp)_{*i}\|_2 \cdot |(\mathbf{R}_{\bar{\perp}}^\top \mathbf{x}^*)_i| \leq \sqrt{\frac{\varepsilon^2}{k}} \sum_{i=1}^n |(\mathbf{R}_{\bar{\perp}}^\top \mathbf{x}^*)_i|$$

$$\leq \sqrt{\frac{\varepsilon^2}{k}}\|\mathbf{R}_{\bar{\perp}}^\top \mathbf{x}^*\|_1 \leq \sqrt{\frac{\varepsilon}{k}}\sqrt{k} = \varepsilon. \tag{3}$$

In the above derivations we use standard properties of norms and the fact that the columns of $\mathbf{U}_\ell^\top$ that have indices in the set $\bar{R}$ have squared norm at most ε^2/k. The last inequality follows from $\|\mathbf{R}_{\bar{\perp}}^\top \mathbf{x}^*\|_1 \leq \|\mathbf{x}^*\|_1 \leq \sqrt{k}$, since $\mathbf{x}^*$ has at most k non-zero entries and Euclidean norm at most one.

Recall that the vector $\mathbf{y}$ of Algorithm 1 maximizes $\|\mathbf{\Sigma}_\ell^{1/2}\mathbf{U}_\ell^\top \mathbf{R}\mathbf{x}\|_2$ over all vectors $\mathbf{x}$ of appropriate dimensions (including $\mathbf{R}\mathbf{x}^*$) and thus

$$\|\mathbf{\Sigma}_\ell^{1/2}\mathbf{U}_\ell^\top \mathbf{R}\mathbf{y}\|_2 \geq \left\|\mathbf{\Sigma}_\ell^{1/2}\mathbf{U}_\ell^\top \mathbf{R}\mathbf{R}^\top \mathbf{x}^*\right\|_2. \tag{4}$$

Combining Eqs. (2), (3), and (4), we get that for sufficiently small ε,

$$\left\|\mathbf{\Sigma}_\ell^{1/2}\mathbf{U}_\ell^\top \mathbf{x}^*\right\|_2^2 \leq \|\mathbf{\Sigma}_\ell^{1/2}\mathbf{U}_\ell^\top \mathbf{z}\|_2^2 + 2\varepsilon \mathrm{Tr}(\mathbf{A}). \tag{5}$$

In the above we used $\mathbf{z} = \mathbf{R}\mathbf{y}$ (as in Algorithm 1) and $\sigma_1 \leq \mathrm{Tr}(\mathbf{A})$. Notice that

$$\mathbf{U}_\ell \mathbf{\Sigma}_\ell^{1/2}\mathbf{U}_\ell^\top \mathbf{z} + \mathbf{U}_{\ell,\perp}\mathbf{\Sigma}_{\ell,\perp}^{1/2}\mathbf{U}_{\ell,\perp}^\top \mathbf{z} = \mathbf{U}\mathbf{\Sigma}^{1/2}\mathbf{U}^\top \mathbf{z},$$

and using the Pythagorean theorem we get

$$\|\mathbf{U}_\ell \mathbf{\Sigma}_\ell^{1/2}\mathbf{U}_\ell^\top \mathbf{z}\|_2^2 + \|\mathbf{U}_{\ell,\perp}\mathbf{\Sigma}_{\ell,\perp}^{1/2}\mathbf{U}_{\ell,\perp}^\top \mathbf{z}\|_2^2 = \|\mathbf{U}\mathbf{\Sigma}^{1/2}\mathbf{U}^\top \mathbf{z}\|_2^2.$$

Using the unitary invariance of the two norm and dropping a non-negative term, we get the bound

$$\|\mathbf{\Sigma}_\ell^{1/2}\mathbf{U}_\ell^\top \mathbf{z}\|_2^2 \leq \|\mathbf{\Sigma}^{1/2}\mathbf{U}^\top \mathbf{z}\|_2^2. \tag{6}$$

Combining Eqs. (5) and (6), we conclude

$$\left\|\mathbf{\Sigma}_\ell^{1/2}\mathbf{U}_\ell^\top \mathbf{x}^*\right\|_2^2 \leq \|\mathbf{\Sigma}^{1/2}\mathbf{U}^\top \mathbf{z}\|_2^2 + 2\varepsilon \mathrm{Tr}(\mathbf{A}). \tag{7}$$

We now apply Lemma 1 to the optimal vector $\mathbf{x}^*$ to get

$$\left\|\mathbf{\Sigma}^{1/2}\mathbf{U}^\top \mathbf{x}^*\right\|_2^2 - \varepsilon \mathrm{Tr}(\mathbf{A}) \leq \left\|\mathbf{\Sigma}_\ell^{1/2}\mathbf{U}_\ell^\top \mathbf{x}^*\right\|_2^2.$$

Combining with Eq. (7) we get

$$\mathbf{z}^\top \mathbf{A}\mathbf{z} \geq \mathcal{Z}^* - 3\varepsilon \mathrm{Tr}(\mathbf{A}).$$

In the above we used $\|\mathbf{\Sigma}^{1/2}\mathbf{U}^\top \mathbf{z}\|_2^2 = \mathbf{z}^\top \mathbf{A}\mathbf{z}$ and $\left\|\mathbf{\Sigma}^{1/2}\mathbf{U}^\top \mathbf{x}^*\right\|_2^2 = (\mathbf{x}^*)^\top \mathbf{A}\mathbf{x}^* = \mathcal{Z}^*$. The result then follows from rescaling ε.

Appendix 1.B   Additional Experiments

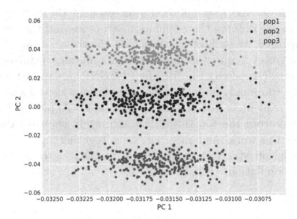

Fig. 4. PCA plot of the simulated data with three distinct populations simulated from the PSD model with an $\alpha = 0.01$, $n = 1,000$, $m = 10,000$ and $t = 100$

Appendix 1.B.1   Simulated Studies

The genotype matrix $\mathbf{X} \in \mathbb{R}^{m \times n}$ consisting of the simulated allele frequencies was generated using the algorithms of [25]. More specifically, we set $\mathbf{F} = \mathbf{TS}$, where $\mathbf{T} \in \mathbb{R}^{m \times d}$ and $\mathbf{S} \in \mathbb{R}^{d \times n}$, where $d \leq n$ is the number of population groups. $\mathbf{S}$ is the indicator matrix that encapsulates structure with n individuals and contained in d populations. On the other hand, $\mathbf{T}$ characterizes how the structure is manifested in the allele frequencies of each SNP. Finally, projecting $\mathbf{S}$ onto the column space of $\mathbf{T}$, we obtain the allele frequency matrix $\mathbf{F}$. We sample $\mathbf{X}$ as a special case of $\mathbf{F}$ for the Pritchard-Stephens-Donelly (PSD) model. We simulate $\mathbf{S}$ using i.i.d draws from the Dirichlet distribution with varying values of α, which denotes the parameter influencing the relatedness between the individuals and is directly proportional to the admixture of populations. Appendix Fig. 4 shows the population structure observed in this simulated data.

As it is difficult to establish notions of statistical significance in ThreSPCA capturing the ancestry informative markers from the original data, we simulated data sets with varying numbers of individuals (n) and SNPs (m) and allowed t true SNPs that contribute to genetic ancestry. For the random markers that do not contribute to the genetic differentiation, we sampled the Fst distances between the individuals from a uniform distribution in the range $\{0, 0.005\}$, which indicates minimum difference in populations. Thus, with this step we achieve the "true" markers contributing to genetic difference are the t SNPs and the remaining $m - t$ SNPs, we conclude, are noise.

Appendix 1.B.2   Experiments on 1KG data

Population Structure Captured by PCA Plots. We filtered the original 1KG data for the `ThreSPCA` derived $k = 500$ SNPs for each of the first three PCs and in the PCA plots we observe the population structure and the allele frequency distribution captured by each of the PCs. We clearly see that the SNPs from PC1 loadings are most frequent in the African populations or mixed populations of African ancestry (Appendix Fig. 5). The PC2 SNPs are most frequent in East Asians, although commonly found in other populations as well and the third PC SNPs are most frequent in South Asian populations (Appendix Fig. 5)). Thus, the SNP loadings from the top three PCs accurately captures the population structure across the world and merging them together, we not only capture the entire population structure in the PCA plot but also discover fine-grain substructure of populations (Fig. 1).

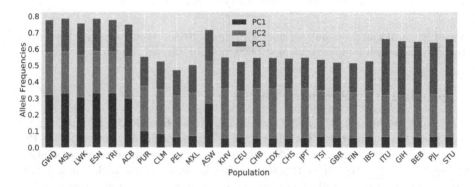

Fig. 5. Mean allele frequencies obtained from the first three PCs from `ThreSPCA` with $k = 500$.

Tuning Input Sparsity k. We tried a range of k's varying it from 50 to 1500 and observed the r^2 between the PCs derived from the original 1KG data and the 1500 SNPs derived from `ThreSPCA`. We observed that for the top two PCs the r^2 is high from 0.96 to 0.99 wit the peak for both the PCs reaching around $k = 500$. PC1 continues to increase by two decimal points before saturing at $k = 1000$. Thus, we selected $k = 500$ for all the experiments as both the PCs reached their respective peaks (Fig. 6).

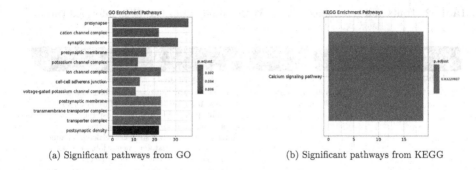

(a) Significant pathways from GO   (b) Significant pathways from KEGG

Fig. 6. GO pathway analyses of the `ThreSPCA` informed variants, colored by p-values.

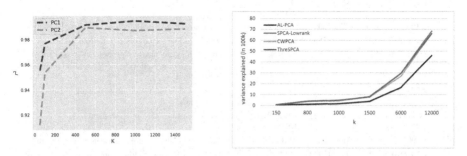

Fig. 7. Left: Line plot between the r^2 between the PC scores of each PC obtained from `ThreSPCA` and the original PC from 1KG data with varying values of sparsity, k. Right: Variance explained by `ThreSPCA`, `AL-PCA`, and other state-of-the-art SPCA solvers for varying k.

Appendix 1.B.3   Comparing `ThreSPCA` with the *State-of-the-Art*

Simulated Data. We observed that increasing the threshold of true positives (markers that contribute to genetic structure) t led to an increase of the number of true positives observed in `ThreSPCA`.

Real Data. For $k = 500$, on 1KG data we found perfect correlation with `ThreSPCA` and `AL-PCA` for PC1 and PC2 with $r^2 = 0.97$ and 0.94 respectively. We also observed similar trends for $k = 1000$ and $k = 1500$ (squared correlations larger than 0.9 for both PC1 and PC2).

Comparing the output of `ThresPCA` against other state-of-the-art SPCA approaches, we compare the *greedy coordinate-wise* (GCW) method of `cwpca` and set the low-rank parameter d of `spca-lowrank` to one. We performed these evaluations on an Intel Xeon Gold 6126 processor running at 2.6 GHz with 96 GB of RAM and a 64-bit CentOS Linux 7 OS.

Table 2. Traits and genes mapped in GWAS catalog from **ThreSPCA** informed variants.

PCs	SNP	CHR	POS	MAPPED GENE	MAPPED TRAITS
PC1	rs35399673	5	104529307	RAB9BP1	skin pigmentation measurement
PC2	rs11556924	7	129663496	ZC3HC1	coronary artery disease, diastolic and systolic blood pressure, myocardial infarction, platelet count, parental longevity, testosterone measurement, Agents acting on the renin-angiotensin system use, Calcium channel blocker use, hematocrit, hemoglobin count, myeloid white cell count, body height, leukocyte count, cardiovascular disease age at menarche
	rs12525051	6	151913710	CCDC170	heel bone mineral density
	rs1938679	11	69272096	MYEOV - LINC02747	body height
	rs196052	6	22057200	CASC15	Corneal astigmatism
	rs2069235	22	39747780	SYNGR1	primary biliary cirrhosis rheumatoid arthritis
	rs4714599	6	42285815	TRERF1	eosinophil percentage of granulocytes, neutrophil percentage of granulocytes, eosinophil percentage of leukocytes
	rs5747035	22	17718606	ADA2	word list delayed recall measurement, memory performance
	rs7714191	5	131341541	ACSL6-AS1, ACSL6	cortical surface area measurement
	rs7901883	10	103186838	BTRC	smoking behavior smoking status measurement
	rs7976816	12	124315343	DNAH10	BMI-adjusted waist circumference waist circumference
	rs8002164	13	58248732	PCDH17	upper aerodigestive tract neoplasm
	rs847888	12	112151742	ACAD10	diastolic blood pressure
	rs907183	8	8729761	MFHAS1, MFHAS1	Calcium channel blocker use measurement
PC3	rs10164546	2	106141004	FHL2	pursuit maintenance gain measurement
	rs1020410	2	176784138	EXTL2P1 - LNPK	physical activity
	rs10896109	11	66080023	TMEM151A - CD248	circadian rhythm
	rs1264423	6	30571471	PPP1R10	mean corpuscular volume
	rs12679528	8	15566164	TUSC3	body mass index
	rs16942383	15	89405052	ACAN	BMI-adjusted hip circumference
	rs2988114	13	80870878	SPRY2	gut microbiome measurement
	rs34672598	20	7884260	HAO1	QT interval
	rs3828919	6	31466057	MICB	platelet count
	rs41492548	9	130607359	ENG	monocyte count
	rs4679760	3	155855418	KCNAB1	birth weight, parental genotype effect measurement
	rs744680	10	131741695	EBF3	visual perception measurement
	rs76496105	2	110447667	BMS1P19 - SRSF3P6	platelet count platelet crit

References

1. Asteris, M., Papailiopoulos, D., Karystinos, G.N.: Sparse principal component of a rank-deficient matrix. In: 2011 IEEE International Symposium on Information Theory Proceedings, pp. 673–677 (2011)
2. Asteris, M., Papailiopoulos, D., Kyrillidis, A., Dimakis, A.G.: Sparse PCA via bipartite matchings. In: Advances in Neural Information Processing Systems, pp. 766–774 (2015)
3. Beck, A., Vaisbourd, Y.: The sparse principal component analysis problem: optimality conditions and algorithms. J. Optim. Theory Appl. **170**(1), 119–143 (2016). https://doi.org/10.1007/s10957-016-0934-x
4. Bose, A., Burch, M.C., Chowdhury, A., Paschou, P., Drineas, P.: Clustrat: a structure informed clustering strategy for population stratification. bioRxiv (2020)
5. Bose, A., Kalantzis, V., Kontopoulou, E.M., Elkady, M., Paschou, P., Drineas, P.: TeraPCA: a fast and scalable software package to study genetic variation in tera-scale genotypes. Bioinformatics **35**(19), 3679–3683 (2019)
6. Buniello, A., et al.: The NHGRI-EBI GWAS catalog of published genome-wide association studies, targeted arrays and summary statistics 2019. Nucleic Acids Res. **47**(D1), D1005–D1012 (2019)
7. Cadima, J., Jolliffe, I.T.: Loading and correlations in the interpretation of principal components. J. Appl. Stat. **22**(2), 203–214 (1995)
8. Chan, S.O., Papailliopoulos, D., Rubinstein, A.: On the approximability of sparse PCA. In: Proceedings of the 29th Conference on Learning Theory, pp. 623–646 (2016)
9. Chang, C.C., Chow, C.C., Tellier, L.C., Vattikuti, S., Purcell, S.M., Lee, J.J.: Second-generation plink: rising to the challenge of larger and richer datasets. Gigascience **4**(1), s13742-015 (2015)
10. Consortium, G.P., et al.: A global reference for human genetic variation. Nature **526**(7571), 68 (2015)
11. d'Aspremont, A., Ghaoui, L.E., Jordan, M.I., Lanckriet, G.R.G.: A direct formulation for sparse PCA using semidefinite programming. SIAM Rev. **49**(3), 434–448 (2007)
12. Engelhardt, B.E., Stephens, M.: Analysis of population structure: a unifying framework and novel methods based on sparse factor analysis. PLoS Genet. **6**(9), e1001117 (2010)
13. Hsu, Y.L., Huang, P.Y., Chen, D.T.: Sparse principal component analysis in cancer research. Transl. Cancer Res. **3**(3), 182 (2014)
14. Jolliffe, I.T.: Rotation of principal components: choice of normalization constraints. J. Appl. Stat. **22**(1), 29–35 (1995)
15. Jolliffe, I.T., Trendafilov, N.T., Uddin, M.: A modified principal component technique based on the LASSO. J. Comput. Graph. Stat. **12**(3), 531–547 (2003)
16. Lee, S., et al.: Sparse principal component analysis for identifying ancestry-informative markers in genome-wide association studies. Genet. Epidemiol. **36**(4), 293–302 (2012)
17. Li, J.Z., et al.: Worldwide human relationships inferred from genome-wide patterns of variation. Science **319**(5866), 1100–1104 (2008)
18. Mahoney, M.W., Drineas, P.: CUR matrix decompositions for improved data analysis. Proc. Natl. Acad. Sci. **106**(3), 697–702 (2009)
19. McLaren, W., et al.: The ensembl variant effect predictor. Genome Biol. **17**(1), 1–14 (2016)

20. Moghaddam, B., Weiss, Y., Avidan, S.: Generalized spectral bounds for sparse LDA. In: Proceedings of the 23rd International Conference on Machine Learning, pp. 641–648 (2006)
21. Musco, C., Musco, C.: Randomized block krylov methods for stronger and faster approximate singular value decomposition. In: Advances in Neural Information Processing Systems 28: Annual Conference on Neural Information Processing Systems, pp. 1396–1404 (2015)
22. Papailiopoulos, D., Dimakis, A., Korokythakis, S.: Sparse PCA through low-rank approximations. In: Proceedings of the 30th International Conference on Machine Learning, pp. 747–755 (2013)
23. Patterson, N., Price, A.L., Reich, D.: Population structure and eigenanalysis. PLoS Genet. **2**(12), e190 (2006)
24. Price, A.L., Patterson, N.J., Plenge, R.M., Weinblatt, M.E., Shadick, N.A., Reich, D.: Principal components analysis corrects for stratification in genome-wide association studies. Nat. Genet. **38**(8), 904–909 (2006)
25. Pritchard, J.K., Stephens, M., Donnelly, P.: Inference of population structure using multilocus genotype data. Genetics **155**(2), 945–959 (2000)
26. Sohail, M., et al.: Polygenic adaptation on height is overestimated due to uncorrected stratification in genome-wide association studies. Elife **8**, e39702 (2019)
27. Yu, G., Wang, L.G., Han, Y., He, Q.Y.: clusterProfiler: an R package for comparing biological themes among gene clusters. OMICS J. Integr. Biol. **16**(5), 284–287 (2012)
28. Zou, H., Hastie, T.: Regularization and variable selection via the elastic net. J. Roy. Stat. Soc. B **67**(2), 301–320 (2005)

Gene Set Priorization Guided by Regulatory Networks with p-values through Kernel Mixed Model

Haohan Wang[1], Oscar L. Lopez[2], Wei Wu[3(✉)], and Eric P. Xing[4(✉)]

[1] Language Technologies Institute, School of Computer Science,
Carnegie Mellon University, Pittsburgh, USA
[2] Alzheimer's Disease Research Center, University of Pittsburgh Medical Center,
Pittsburgh, USA
[3] Computational Biology Department, School of Computer Science,
Carnegie Mellon University, Pittsburgh, USA
weiwu2@cs.cmu.edu
[4] Machine Learning Department, School of Computer Science,
Carnegie Mellon University, Pittsburgh, USA
epxing@cs.cmu.edu

Abstract. The transcriptome association study has helped prioritize many causal genes for detailed study and thus further helped the development of many therapeutic strategies for multiple diseases. How- ever, prioritizing the causal gene only does not seem always to be able to offer sufficient guidance to the downstream analysis. Thus, in this paper, we propose to perform the association studies from another perspective: we aim to prioritize genes with a tradeoff between the pursuit of the causality evidence and the interest of the genes in the pathway. We introduce a new method for transcriptome association study by incorporating the information of gene regulatory networks. In addition to directly building the regularization into variable selection methods, we also expect the method to report p-values of the associated genes so that these p-values have been empirically proved trustworthy by geneticists. Thus, we introduce a high-dimension variable selection method with the following two merits: it has a flexible modeling power that allows the domain experts to consider the structure of covariates so that prior knowledge, such as the gene regulatory network, can be integrated; it also calculates the p-value, with a practical manner widely accepted by geneticists, so that the identified covariates can be directly assessed with statistical guarantees. With simulations, we demonstrate the empirical strength of our method against other high-dimension variable selection methods. We further apply our method to Alzheimer's disease, and our method identifies interesting sets of genes.

1 Introduction

While genome-wide association studies (GWAS) have identified many genetic variants significantly associated with diseases, they face the challenge of the

I. Pe'er (Ed.): RECOMB 2022, LNBI 13278, pp. 107–125, 2022.
https://doi.org/10.1007/978-3-031-04749-7_7

result interpretation due to linkage disequilibrium. Transcriptome-wide association studies (TWAS) have been introduced as a response to this challenge (Gamazon et al. 2015; Gusev et al. 2016; Barbeira et al. 2018). TWAS typically involves three steps: the training of expression-prediction model from SNPs based on reference data, the prediction of expression from SNPs in the GWAS data, and the association mapping between the predicted gene expression and the traits (Wainberg et al. 2019).

In this paper, we only focus on the last step, *i.e.*, the association between the gene expression and the trait. Usually, this last step is completed using conventional statistical tools for association study, such as univariate heritability estimation (Gusev et al. 2016) or multivariate regression (Gamazon et al. 2015), where conventional Wald test sometimes plays the central role in testing for the significantly associated variant (Feng et al. 2021a, b). In this project, to facilitate our method development specially designed for the last step, we directly work with collected gene expression data, instead of the ones that are predicted from models trained on reference panels. Therefore, it is worth noting that the association between the transcriptomes and the traits has been studied for a long time with collected gene expressions, especially on the statistics side of the community (*e.g.*, Zou and Hastie 2005; Ding and Peng 2005; Meinshausen and Bühlmann 2010).

From GWAS to TWAS, statistical methods have helped prioritize multiple genetic factors that leads to significant achievements in understanding the underlying mechanism of certain diseases and the follow-up development of therapies (Visscher et al. 2017). However, in other times, the pinpoint of the one causal gene may not always be helpful in the downstream study. For example, multiple statistical methods have repeatedly identify the gene *APOE* for sporadic Alzheimer's disease (*e.g.*, Bertram and Tanzi 2009; Tost and Reitz 2013), leading to a reasonably well understanding of the disease mechanisms at different levels (*e.g.*, Zetterberg and Mattsson 2014; Masters et al. 2015; Fan et al. 2020), yet the community is still in an active search of effective treatment of this disease (*e.g.*, Yiannopoulou and Papageorgiou 2020).

This disparity encourages us to conjecture that prioritizing only the genes with the most causality evidence may not always be the best strategy. Instead, prioritizing a set of genes with a tradeoff between the emphasis on the causality evidence and the interests in the regulatory structure may help the downstream analysis further. The interests of the association study of gene sets has been explored in the context of GWAS previously (de Leeuw et al. 2015), sometimes with the concept of epistasis (Crawford et al. 2017; Wang et al. 2019), validating that our conjecture is worth further exploration.

Therefore, motivated by the lack of statistical methods designed to take into account characteristics of the gene expression data, *e.g.*, the co-regulatory network structure of genes (Wainberg et al. 2019) and the belief that accounting for this structure can potentially improve the power of the association study at a significant scale (Gamazon et al. 2015; Lonsdale et al. 2013), we aim to introduce a statistical method that can incorporate the regulatory network and can prioritize genes together with ones in its regulatory network.

A number of statistical methods have been developed to allow one to incorporate various structural knowledge of the data in the regression analysis. For example, (Li and Li 2008) introduced the network-constrained regularization that incorporates the graph of covariates while performing regression; (Kim and Xing 2010) proposed the tree-structured regularization that integrates the graph of responses while performing multi-response regression; (Puniyani et al. 2010) introduced a multi-population regularization that accounts for the heterogeneity of samples; (Wang et al. 2018) leveraged the precision matrix of covariates to account for the dependencies between genes. While these regularization-based methods have the flexibility of incorporating all kinds of structural knowledge, they are limited by the lack of the ability to provide p-values, therefore, limited due to the shortage of direct measurement in assessing the reliability of the findings.

On the other hand, regarding the calculation of the p-value, the statistics community has pushed the frontier with high-dimension models using a family of methods (*e.g.,* Bühlmann 2013; Zhang and Zhang 2014; Lockhart et al. 2014; Javanmard and Montanari 2014). While these methods have delivered the promises of assembling the advantages of regularization and p-value calculation, they have not entered the phase of incorporating all kinds of structural knowledge as the previous methods incorporated yet. Also, these methods typically involve the calculation of the precision matrix of covariates, which can be forbiddingly expensive in terms of the computational power for whole-genome studies.

Thus, we will introduce a new statistical tool that can test for the association between gene expression and the trait in a multivariate regression manner, while incorporating the regulatory network structure as prior knowledge, and reporting p-values that can be used by geneticists as a reference to analyze their results. Our method is built upon the success of how the linear mixed model calculates p-values, which is proven trustworthy empirically (*e.g.,* Kang et al. 2008, 2010; Lippert et al. 2011; Yang et al. 2014). The central idea of our method relies on the resemblance between the linear mixed model and Ridge regression (Wang et al. 2022). Following the simulation setup in (Li and Li 2008), we test our method against multiple competing methods that can either incorporate the graph of covariates as prior knowledge or perform high-dimension variable selection with p-values. The simulations endorse the strength of our method empirically. We further apply it to Alzheimer's disease and report the findings. Our results suggest that there are also interesting connections between genes in the cerebellum and Alzheimer's disease, aligning with previous questions raised (Jacobs et al. 2018).

Finally, to summarize the three major contributions of our work:

- We propose that the association study of gene sets is also important in TWAS, and we propose to prioritize the genes considering its regulatory networks.
- We introduce a statistical method for this goal, which can incorporate the network structure as prior knowledge as flexible as regularized regression methods, while reporting p-values empirically trustworthy to geneticists.

- We apply our methods to study Alzheimer's disease and notice that: although the cerebellum is less believed to play a role in Alzheimer's disease, our results suggest interesting mechanisms in this tissue.

2   Method

To introduce our method, we will first introduce the connections between linear mixed model and Ridge regression as background. Then, we will introduce our design rationales building upon this connection, and the design rationales will lead to the discussion of our main model. Finally, we will introduce the overall algorithm for the method.

Notations. We use $\mathbf{X} \in \mathcal{R}^{n \times p}$ to denote the data of n samples and p covariates, $\mathbf{y} \in \mathcal{R}^{n \times 1}$ to denote the response variable, $\mathbf{C} \in \mathcal{R}^{p \times p}$ to denote the *closeness* of covariates (*e.g.* one *closeness* measure can be the gene network), $\mathbf{x} \in \mathcal{R}^{n \times 1}$ to denote one gene (*e.g.* $\mathbf{x}$ can be the ith column of $\mathbf{X}$, denoted as $\mathbf{x} = \mathbf{X}_i$), and β_i to denote the corresponding effect size to be estimated.

2.1   Background

We consider the setup of a high-dimension linear regression

$$\mathbf{y} = \mathbf{X}\beta + \epsilon,$$

where $\epsilon \sim N(0, \mathbf{I}\sigma_\epsilon)$ and denotes the noises. We are given $\mathbf{X}$ and $\mathbf{y}$ and we aim to estimate β.

We first consider using Ridge regression with the optimization goal of

$$\beta_{rr} = \arg\min_{\beta} \|\mathbf{y} - \mathbf{X}\beta\|_2^2 + \lambda\|\beta\|_2^2, \tag{1}$$

where λ is the regularization weight.

Without loss of generality, we pay a particular attention to i^{th} column of $\mathbf{X}$ and we use $\mathbf{x}$ to denote that column, and we use $\mathbf{X}_{-i}$ to denote the remaining columns. Correspondingly, we use $\alpha = \beta_i$ and β_{-i} to denote the effect sizes. Thus, we can rewrite (1) to

$$\beta_{rr} = \arg\min_{\beta} \|\mathbf{y} - \mathbf{x}\alpha - \mathbf{X}_{-i}\beta_{-i}\|_2^2 + \lambda_1\|\alpha\|_2^2 + \lambda_2\|\beta_{-i}\|_2^2, \tag{2}$$

where we can allow λ_1 and λ_2 to be different.

Conveniently, we can have the closed form solution of (2) as

$$\widehat{\alpha}_{rr} = \frac{\mathbf{x}^T(\mathbf{I} - \mathbf{X}_{-i}(\mathbf{X}_{-i}^T\mathbf{X}_{-i} + \lambda_2\mathbf{I})^{-1}\mathbf{X}_{-i}^T)\mathbf{y}}{\mathbf{x}^T(\mathbf{I} - \mathbf{X}_{-i}(\mathbf{X}_{-i}^T\mathbf{X}_{-i} + \lambda_2\mathbf{I})^{-1}\mathbf{X}_{-i}^T)\mathbf{x} + \lambda_1}, \tag{3}$$

On the other hand, if we are only interested in $\mathbf{x}$ and its effect size, we can consider estimating it through linear mixed model, which assumes the generation of data as

$$\mathbf{y} = \mathbf{x}\alpha + \mathbf{X}_{-i}\mathbf{u} + \epsilon, \tag{4}$$

where $\mathbf{u} \sim N(0, \mathbf{I}\sigma_u^2)$ and denotes the random effects, and the dimension of $\mathbf{I}$ can be inferred from context.

If we aim to estimate the parameters by maximizing the likelihood of (4), a commonly used estimator (Lippert et al. 2011) will lead us to

$$\widehat{\alpha}_{lmm} = \frac{\mathbf{x}^T(\delta\mathbf{I} + \mathbf{K})^{-1}\mathbf{y}}{\mathbf{x}^T(\delta\mathbf{I} + \mathbf{K})^{-1}\mathbf{x}},$$

where $\mathbf{K}$ is the kinship matrix used to estimate the parameters. In practice, $\mathbf{K}$ is often set to be $\mathbf{K} = \mathbf{X}_{-i}\mathbf{X}_{-i}^T$, thus, together with the Woodbury identity, we will have:

$$\widehat{\alpha}_{lmm} = \frac{\mathbf{x}^T(\mathbf{I} - \mathbf{X}_{-i}(\mathbf{X}_{-i}^T\mathbf{X}_{-i} + \delta\mathbf{I})^{-1}\mathbf{X}_{-i}^T)\mathbf{y}}{\mathbf{x}^T(\mathbf{I} - \mathbf{X}_{-i}(\mathbf{X}_{-i}^T\mathbf{X}_{-i} + \delta\mathbf{I})^{-1}\mathbf{X}_{-i}^T)\mathbf{x}}, \tag{5}$$

where δ is the REML estimation of $\sigma_\epsilon^2/\sigma_u^2$ (Thompson et al. 1962).

By comparing (3) and (5), we can see that linear mixed model (4) will lead to the same solution of Ridge regression (2) if we are only interested in the effect size of $\mathbf{x}$ and we set $\lambda_1 = 0$ (Maldonado 2009; Heckerman 2018; Wang et al. 2022).

2.2   Method

With the background above, we start to introduce our method: we consider an extension of Ridge regression by leveraging the closeness information $\mathbf{C}$ to regularize the effect sizes. For example, to regularize that the genes nearby shall have similar effect sizes, we can solve the following problem:

$$\beta = \arg\min_{\beta} \|\mathbf{y} - \mathbf{X}\beta\|_2^2 + \lambda \sum_{i,j} \mathbf{C}_{i,j}(\beta_i - \beta_j)^2 \tag{6}$$

where λ is again the regularization weight.

(6) can be equivalently re-written into (Li and Li 2008):

$$\beta = \arg\min_{\beta} \|\mathbf{y} - \mathbf{X}\beta\|_2^2 + \frac{\lambda}{2}\beta^T\mathbf{L}\beta \tag{7}$$

where $\mathbf{L}$ is the Laplacian matrix, defined as $\mathbf{L} = \mathbf{D} - \mathbf{C}$, where $\mathbf{D}$ is a diagonal matrix and $\mathbf{D}_{i,i} = \sum_j^p \mathbf{C}_{i,j}$.

Further, the connections between LMM and Ridge regression offer us a convenient way of extending (7) and formalize the extended equation into the solution of LMM. Specifically, we first extend (7) to be:

$$\beta = \arg\min_{\beta} \|\mathbf{y} - \mathbf{x}\alpha - \mathbf{X}_{-i}\beta_{-i}\|_2^2 + \lambda\beta_{-i}^T\mathbf{L}_{-i}\beta_{-i} \tag{8}$$

where we drop the $\frac{1}{2}$ because it only affects the scale of the regularization weight. Then we write out the closed form solution of α:

$$\widehat{\alpha} = \frac{\mathbf{x}^T(\mathbf{I} - \mathbf{X}_{-i}(\lambda\mathbf{L}_{-i} + \mathbf{X}_{-i}^T\mathbf{X}_{-i})^{-1}\mathbf{X}_{-i}^T)\mathbf{y}}{\mathbf{x}^T(\mathbf{I} - \mathbf{X}_{-i}(\lambda\mathbf{L}_{-i} + \mathbf{X}_{-i}^T\mathbf{X}_{-i})^{-1}\mathbf{X}_{-i}^T)\mathbf{x}} \tag{9}$$

Notice that the (7) is not written into (8) because they are equivalent. It is only written into (8) as how LMM resembles ridge regression. Fortunately, the estimated coefficients with (8) are close to those with (7) with bounded differences, as shown by (Wang et al. 2022).

Now, if we consider the below model,

$$\mathbf{y} \sim N(\mathbf{x}\beta_i, \mathbf{X}(\mathbf{I} + \mathbf{H} + \mathbf{H}^2)\mathbf{X}^T\sigma_u^2 + \mathbf{I}\sigma_\epsilon^2) \tag{10}$$

where $\mathbf{H}$ denotes the normalized network, calculated by:

$$\mathbf{H} = \mathbf{D}^{-1/2}\mathbf{C}\mathbf{D}^{-1/2}$$

We use $\mathbf{J}$ to denote $\mathbf{I} + \mathbf{H} + \mathbf{H}^2$ for simplicity. We use the P3D method (Zhang et al. 2010) and the re-parametrization trick (Lippert et al. 2011) for estimation of parameters, which leads to the following solution for α:

$$\widehat{\alpha} = \frac{\mathbf{x}^T(\widehat{\delta}\mathbf{I} + \mathbf{X}\mathbf{J}\mathbf{X}^T)^{-1}\mathbf{y}}{\mathbf{x}^T(\widehat{\delta}\mathbf{I} + \mathbf{X}\mathbf{J}\mathbf{X}^T)^{-1}\mathbf{x}} \tag{11}$$

which is the solution of a linear mixed model whose kinship matrix is set to be $\mathbf{K} = \mathbf{X}\mathbf{J}\mathbf{X}^T$, and the $\widehat{\delta}$ is estimated through maximum likelihood estimation.

Notice that with Woodbury identity we can show that (9) and (11) are the same if $\mathbf{J} \propto \mathbf{L}^{-1}$ and λ is chosen accordingly.

Therefore, we have demonstrated how we can simply use a linear mixed model with a designed kernel matrix to replace the network regularized multivariate regression method (Li and Li 2008). Our method also have two additional merits: it can calculate p-values in a manner that is accepted by geneticists; it is free of the hyperparameter λ as the regularization weight. However, these merits come at a price, through some approximations that may not be considered rigorous by statisticians. Fortunately, these approximations have been widely used in the study of GWAS and have been proven efficient in identifying associated covariates.

For example, after we estimate α, p-values can be calculated following standard hypothesis procedure, such as Wald Test, as conducted by a variety of genetics studies (*e.g.*, Kang et al. 2008, 2010; Lippert et al. 2011; Yang et al. 2014).

Another challenge is that $\mathbf{L}$ is likely to be singular so the inverse does not exist. Despite that there are many methods to approximate the Moore-Penrose inverse of a Laplacian matrix (Bozzo and Franceschet 2012; Bozzo 2013; Van Mieghem et al. 2017), we notice a simple solution that does not involve the decomposition of $\mathbf{L}$.

Algorithm 1: Algorithm of the proposed method KMM

Input: gene expressions $\mathbf{X}$, traits $\mathbf{y}$, the network of gene closeness (e.g., regulary network) $\mathbf{C}$, correlation threshold τ;
Output: test statistics of genes $\mathbf{p}$;
Calculate the $\mathbf{D}_{i,i} = \sum_j^p \mathbf{C}_{i,j}$ and $\mathbf{H} = \mathbf{D}^{-1/2}\mathbf{C}\mathbf{D}^{-1/2}$;
Calculate $\mathbf{K} = \mathbf{X}(\mathbf{I} + \mathbf{H} + \mathbf{H}^2)\mathbf{X}^T$;
for *every gene* $\mathbf{x}$ *in* $\mathbf{X}$ **do**

> Calculate its correlation with every other gene in $\mathbf{X}$, denote the correlation as $\mathbf{t_x}$;
> Index the genes $c(\mathbf{x}) = \{i | \mathbf{t_x}(i) \geq \tau\}$;
> $\mathbf{K}' = \mathbf{K} - \mathbf{X}_{c(\mathbf{x})}(\mathbf{I} + \mathbf{H} + \mathbf{H}^2)_{c(\mathbf{x})}\mathbf{X}_{c(\mathbf{x})}^T$;
> Estimate $\delta_{\mathbf{x}}$ with $\mathbf{K}'$, $\mathbf{X}$, and $\mathbf{y}$ following (Thompson et al., 1962);
> Estimate $\beta_{\mathbf{x}}$ with (11);
> Calculate $\mathbf{p_x}$ as the p-value for $\mathbf{x}$ through Wald test.

end
Report $\mathbf{p}$ as the p-value for all the genes;

We consider the normalized Laplacian matrix defined as $\mathbf{L}' = \mathbf{I} - \mathbf{H}$, where $\mathbf{H}$ is defined in preceding texts. We notice that the normalization leads to a convenient property about the eigenvalues e of $\mathbf{H}$ that $|e_i(\mathbf{H})| < 1$. Therefore, we can approximate the inverse of $\mathbf{L}'$ with (Petersen et al. 2008):

$$\mathbf{L}'^{-1} = (\mathbf{I} - \mathbf{H})^{-1} \approx \mathbf{I} + \mathbf{H} + \mathbf{H}^2.$$

Finally, while the kinship matrix is ideally constructed with $\mathbf{K} = \mathbf{X}_{-i}\mathbf{X}_{-i}^T$ when we study $\mathbf{x}$ (*i.e.*, $\mathbf{X}_i$), which means that for every covariate we study, we need to construct the kinship matrix and estimate δ once, and this repeated procedure will introduce additional computational load. A convenient way to circumvent the computational load is to construct $\mathbf{K}$ once only with $\mathbf{K} = \mathbf{X}\mathbf{X}^T$ which is used by multiple genetic studies (*e.g.*, Kang et al. 2008, 2010; Lippert et al. 2011; Yang et al. 2014). Regarding this approximation, (Yang et al. 2014) empirically showed that the statistical power in terms of p-value will be reduced if the covariate of interest $\mathbf{x}$ is included in the kinship matrix over SNP data, while, on the other hand, (Wang et al. 2022) showed an statistical argument that the approximation barely degrades the performance for high-dimension data when the covariates are independent.

However, due to the correlated expressions of the genes, the inclusion of the gene of interest ($\mathbf{x}$) and its correlated genes will significantly degrade its performances in identifying associated genes (as we tested empirically). To counter this, we introduce to construct $\mathbf{K}$ with $\mathbf{K} = \mathbf{X}_{-c(\mathbf{x})}\mathbf{X}_{-c(\mathbf{x})}^T$, where $-c(\mathbf{x})$ denotes all the genes there are not correlated with $\mathbf{x}$. We define the *not correlated* as the Spearman correlation smaller than a threshold. This threshold is introduced as a hyperparameter of our method. Intuitively, the higher this threshold is set, the less effect it has in filtering genes, thus the less power the method has in countering the effects introduced by correlated genes during proritization. On the other hand, the lower this threshold is set, the less genes will remain in the construction of kinship matrix, thus the more similar the method will be with vanilla marginal regression.

In summary, we formally introduce our method in Algorithm 1.

3   Simulation Experiments

3.1   Competing Methods

We consider the following competing methods:

- KMM: our proposed method, which we name kernel mixed model to describe the fact that the essential idea is to add a kernel function on the kinship matrix of linear mixed model.
- LMM: linear mixed model (*e.g.*, Lippert et al. 2011), which can be seen as our method without the network structure.
- NCL (Network-constrained Lasso) (Li and Li 2008): an extension of Lasso that regularizes the coefficients to follow network structure.
- Wald Test: Standard Wald test with the standard FDR control using the Benjamini-Hochberg (BH) procedure (Benjamini and Hochberg 1995).
- Lasso: Linear regression with ℓ_1 norm regularizations.
- AL (Adaptive Lasso): an extension of Lasso that weighs the regularization term (Zou 2006) (enabled by the method introduced in (Huang et al. 2008) for high-dimensional data).

For models that do not report p-value, we consider the variables with non-zero estimated coefficients as identified ones. For regularized regression methods, the hyperparameters are chosen with a prior knowledge of the number of associated genes following (Wang et al. 2018). Due to the statistical limitation of selecting hyperparameters using cross-validation and information criteria in high dimensional data (*e.g.*, see discussions in (Wang et al. 2018)), in practice, it is arguably more reliable for the practitioners to provide the expected number of variables to be selected.

3.2   General Data Generation Process

We follow the simulation set-up in (Li and Li 2008) to simulate data and to test the performance of our method KMM. (Li and Li 2008) assumed that genes that are regulated by the same Transcription Factor (TF) will either have or not have causal effects consistent with the causality of their regulating TFs.

Following (Li and Li 2008), we defined the parameters that will be used in our simulation studies and introduced how they were simulated as follows:

- First, we denoted the sample size as n , the number of TFs as t, the number of total covariates (i.e. including all genes and TFs) as p. We assumed that each TF regulates a fixed number of genes. Thus, there will be $p_g = \frac{p}{t} - 1$ genes regulated by one same TF. Also, we assumed that in all t TF groups, there will be exactly c groups of causal TFs and genes.

- X is an $n \times p$ feature matrix, where we assumed that each row $X_{i:}$ (i.e. each sample) is independently distributed. We assumed that any pairs of TFs will follow an identical independent normal distribution as $X_{TF} \sim N(0,1)$. Each gene was assumed to be jointly distributed with its corresponding TF, which was indicated by a bivariate normal with a correlation of r. Such a bivariate normal distribution can be converted into a conditional distribution as $N(X_{TF} \times r, 1 - r^2)$. Thus, we simulated TFs first, based on which we simulated the rest of all genes using above conditional distribution.
- β is a $p \times 1$ vector such that,

$$\beta_i = \begin{cases} w_j, & \text{if the } i^{th} \text{ covariate is the } j^{th} \text{ TF} \\ \frac{w_j}{\sqrt{p_g}}, & \text{if the } i^{th} \text{ covariate is regulated by the } j^{th} \text{ TF} \end{cases}$$

where w is a vector of length t and w_j is none-zero only when the jth TF and genes regulated by it all have causal effects.
- The input regulatory network N will be a $p \times p$ symmetric matrix, where N_{ij} = 1 only when either the ith covariate is a TF that regulates the jth covariate, or the jth covariate is a TF that regulates the ith covariate. Otherwise, N_{ij} = 0. The gene regulatory networks are known prior to the methods.
- The response vector y is given as $y = X\beta + \epsilon$, where the random noise ϵ was simulated following a normal distribution that guarantees the signal to noise ratio to always be 30.

Further, we extend the detailed configurations of (Li and Li 2008) to more challenging and diverse scenarios, with which, we can better understand the strength and limitations of our methods. For example, we consider a more challenging configuration where $n = 500$ and $p = 8800$ (800 TFs) as we believe this configuration is closer to the configurations of the real data we have. We set $r = 0.7$ so that regulated genes are correlated with the TF strongly (we will also explore different choices of r and how it affects our model).

In (Li and Li 2008), the values of w are either 3 or 5 with sign flips. Since ϵ will be calculated according to the effect size w to maintain a fixed signal to noise ratio of 30, the effects of choices of values of w seems negligible. However, we notice that different choices of w will lead to a different level of diversity of the effect sizes (as the standard deviation of w will change), so we also explore the different choices of w and the effects on the performances. Further, we also explore how different choices of r will affect the performances. Finally, as our method will rely on a good choice of the network known in prior, we also test the different scenarios where the network is not always correct.

3.3   Results

To introduce a diverse of magnitudes of w, we configure the data generation process with six non-zero values of the vector w (effect sizes of the TF), with the first three of them set to be $3, 3, -3$, and the remaining three of them set to be $v, v, -v$. The rest of the data generation process are the same as the general data generation above.

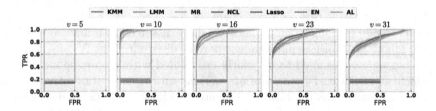

Fig. 1. ROC curve of competing methods over different magnitude choices of the effect sizes

We repeat the experiments three times and plot the ROC curve with standard deviation plotted as the shady areas in Fig. 1. The greater the v is, the larger the deviation of the effect sizes will be (because we fix three effect sizes to be 3 and allow the remaining ones scaled with v.)

As Fig. 1 shows, when v is small (i.e., when the deviation of the effect sizes are small), most hypothesis testing method can solve the problem well enough. However, as v increases, our methods start to show the advantages over previous hypothesis testing methods. We conjecture this is because the previous method tend to identify the genes with larger effect sizes and our regularization helps deemphasize these TFs and force the model to focus more on the regulated genes.

This experiment is mainly designed to test the methods for varying magnitudes of effect sizes in the simulation data. For further empirical evidence over other different configurations of the simulation of the data, we direct the readers to the appendix. In summary, our proposed methods outperform the competing methods in different setups, even when the network structure as prior knowledge is misspecified.

4   Study of Transcriptome Association of Alzheimer's Disease

In the real data application, we compare our method KMM with the Wald test. We do not show the results of the LMM method because it does not report any significant genes, and we do not compare the other methods we assessed in the simulation experiment due to their lack of ability to report p-values for identified associations.

Data and Preprocessing. We apply our method and the Wald test to the late-onset Alzheimer's disease (AD) dataset provided by Harvard Brain Tissue Resource Center and Merck Research Laboratories (Zhang et al. 2013). The gene expression profiling data are collected over three different brain regions, namely dorsolateral prefrontal cortex (PFC), visual cortex (VC), and cerebellum (CB). There are 230 samples, 101 of which are diagnosed with AD, and the rest are normal controls. We first filter the data and consider only the genes usually expressed in brain, as collected in the Human Protein Atlas database

(Pontén et al. 2008). Then we use the protein-protein interaction networks from BioGRID (Oughtred et al. 2018) as the networks of genes to be incorporated into the model as prior knowledge.

Also, we adjust the gene expression data for age and gender. We demonstrate the effectiveness of correcting the age and gender confounding effects by comparing and showing the differences of correlations between expression levels of genes and these covariates before and after the regression in the appendix.

Gene Sets of Interest. The definition of gene sets can vary dependent on the context (Uffelmann et al. 2021). Here, as one of the strengths of our method is the incorporation of the network structure of the genes in the association model, we are interested in finding out whether this would facilitate us to identify disease-associated genes within an interacting network. In order to do so, we first identify around 5000 gene interaction groups, which we will name after the hub gene in each group, in the BioGrid Protein-protein interaction database (Oughtred et al. 2018). Since we would like to focus on those gene interaction groups most likely implicated in Alzheimer's disease, we design a scoring system to rank the gene interaction groups in terms of their relevancy to AD. With the help of GeneCards (Safran et al. 2010), which can inform us how strongly the community believes each gene is associated with Alzheimer's disease, we retrieve the ranks of the genes in each group believed to be associated with Alzheimer's disease. Then by comparing the average rank for all the genes in each group, we are able to identify the top 10 gene interaction groups implicated in AD, which are known to play key roles in AD, and thus we will focus on these groups in the downstream analysis.

Table 1. Hypergeometric test of the overlap of genes identified by the methods and the interacting genes denoted by the hub gene.

hub gene		APOE	APP	BACE1	CASP1	CASP4	CASP6	MAPT	NCSTN	PSEN1	SNCA
PFC	WALD	0.382	**0.011**	0.921	**0.021**	0.962	0.392	0.068	0.445	0.195	0.075
	KMM	0.064	**0.012**	0.194	0.144	0.552	**0.003**	0.084	0.552	**0.043**	**0.033**
CR	WALD	0.896	0.204	0.59	0.311	0.947	0.557	0.193	0.993	0.782	0.275
	KMM	0.378	0.231	**0.02**	0.105	**0.038**	0.717	0.545	**0.008**	**0.029**	0.706

In order to test how well the compared methods can identify the AD-associated genes in a gene interaction group, we use the following procedure. We first apply each method to test whether each gene in a gene interaction group is associated with the AD phenotype with a p-value. Then a gene with a p-value smaller than the threshold of 0.05 is considered as significantly associated with AD. Finally, for each gene interaction group, we conducted a hypergeometric test to assess whether the identified AD-associated genes are significantly enriched in this group (i.e., p-value from the hypergeometric test < 0.05). We applied our method KMM and the conventional Wald test to the gene expression data collected from all the three available compartments: PFC, CR,

Table 2. Enrichment analysis for the identified genes in the network with hub *CASP6*.

	Significant results	*p*-value	genes
GO	Glial cell development	4.89E−02	*APP, VIM, MAPT*
	Intermediate filament cytoskeleton organization	1.44E−02	*PPL, KRT18, VIM*
	Positive regulation of neuron death	4.47E−02	*APP, EIF2S1, MAPT*
	Positive regulation of NF-kappaB transcription factor activity	1.31E−02	*CFLAR, APP, PRKCZ, TRAF1*
	Memory	4.85E−02	*APP, MAPT, PRKCZ*
Pathway	Alzheimer disease-amyloid secretase pathway	2.06E−03	*APP, PSEN2, PRKCZ*

Table 3. Enrichment analysis for the identified genes in the *NCSTN* subnetwork.

	Significant results	*p*-value	Genes
GO	amyloid-beta formation	3.15E−08	*PSEN1, BACE1, APH1B, NCSTN*
	T cell activation	1.93E−02	*PSEN1, ALB1, NCSTN*
	regulation of long-term synaptic potentiation	3.74E−02	*ALB1, NCSTN*
	regulation of synaptic plasticity	1.10E−02	*PSEN1, ALB1, NCSTN*
	glial cell differentiation	9.63E−03	*PSEN1, ALB1, NCSTN*
	cerebellum development	3.53E−03	*PSEN1, ALB1, NCSTN*
Pathway	Notch signaling pathway	1.23E−05	*PSEN1, APH1B, NCSTN*
	Alzheimer disease-amyloid secretase pathway	3.23E−07	*PSEN1, BACE1, APH1B, NCSTN*
	Alzheimer disease-presenilin pathway	1.94E−06	*PSEN1, BACE1, APH1B, NCSTN*

and VC. Table 1 reports the *p*-values of the hypergeometric tests for the top-10 AD-implicated gene interaction groups. We notice that our method can identify more gene interaction groups significantly enriched with the AD-associated genes than what the Wald test can identify. In particular, KMM reported four significant networks in PFC and CR each, while Wald testing reported two significant networks only in PFC. Neither of these methods report any networks in VC. Next, we will examine our findings in the PFC and CR more closely.

Findings in the PFC Region. PFC is known to play key roles in Alzheimer's disease (Salat et al. 2001). We first investigate the gene interaction group with the most significant *p*-value in PFC (Table 1): the *CASP6* subnetwork. Our method identifies 18 AD-associated genes in this subnetwork. Functional GO and pathway enrichment analysis of these genes suggests that they are clearly implicated in Alzheimer's disease, as shown in Table 2. For example, recent studies have found AD-associated loci in or near genes that are expressed in microglia (Hemonnot et al. 2019; Efthymiou and Goate 2017), suggesting that the biological process of glial cell development may play a role in the development of

Table 4. Enrichment analysis for the identified genes in the *PSEN1* subnetwork.

	Significant results	*p*-value	Genes
GO	amyloid-beta formation	3.19E−06	*PSEN1, BACE1, APH1B, NCSTN*
	Notch receptor processing	9.17E−05	*PSEN1, ALB1, NCSTN*
	positive regulation of response to stimulus	3.53E−03	*DLL1, PRKACA, CASP1, CTNNA1, CTNNB1, CASP4, PSEN1, CIB1, TCF7L2*
	positive regulation of skeletal muscle tissue development	1.62E−02	*DLL1, CTNNB1*
	astrocyte development	1.23E−03	*DLL1, GFAP, PSEN1*
	glial cell differentiation	2.05E−04	*DLL1, GFAP, PSEN1, CTNNB1, NCSTN*
	cellular response to amyloid-beta	1.35E−03	*CASP4, PSEN1, BACE1*
	cell fate determination	1.65E−03	*DLL1, CTNNB1, NOTCH4*
	angiogenesis	1.43E−03	*DLL1, CTNNB1, NOTCH4, PRKACA, CIB4*
Pathway	neuron projection regeneration	2.81E−02	*GFAP, CTNNA1*
	neuron death	3.57E−03	*PSEN1, CASP7, NCSTN*
	cerebellum development	1.15E−02	*DLL1, NCSTN, PSEN1*
	regulation of synaptic plasticity	4.43E−02	*GFAP, NCSTN, PSEN1*
	lymphocyte activation	2.94E−02	*DLL1, PSEN1, CTNNB1, NCSTN*
	positive regulation of MAPK cascade	4.87E−02	*PRKACA, CTNNB1, PSEN1, CIB1*

Alzheimer's disease. With evidence suggesting the linkage between neurodegeneration the progressive accumulation of abnormal filamentous protein (Cairns et al. 2004), the biological process intermediate filament cytoskeleton organization is probably implicated with Alzheimer's disease. Neural cell death and memory are clearly linked to Alzheimer's disease (*e.g.*, Niikura et al. 2006). Recent evidence also suggests that the transcription factor nuclear factor-kappa B as a major risk factor in cellular, invertebrate and vertebrate models of AD (Jones and Kounatidis 2017).

Findings in the CR Region. Interestingly, we notice that our method also reports genes in four networks in Table 1. Even though the understanding of function relevance in the cerebellum is still in the early stage (Jacobs et al. 2018), our results lead us to inspect further in this region. We examine the identified AD-associated genes in two of the subnetworks, *NCSTN* and *PSEN1*, and the results are reported in Tables 3 and 4, respectively. Similarly, these biological processes and pathways are clearly implicated in the development of Alzheimer's disease.

The functional enrichment analysis of the identified AD-associated genes in the *NCSTN* subnetwork is reported in Table 3. The role of amyloid-beta formation plays in Alzheimer's disease has been widely studied with a large body of evidence (e.g., Murpy and LeVine III 2010; Sadigh-Eteghad et al. 2015). The activities of T-cells have been a putative factor for Alzheimer's disease for a long time (Town et al. 2005), although recent evidence starts to raise alternative thoughts (Dhanwani et al. 2020). There is also plenty of evidence suggesting that the dynamic gain and loss of synapses is linked to Alzheimer's disease (e.g., Subramanian et al. 2020)

The enrichment analysis of the identified AD-associated genes in the *PSEN1* network is reported in Table 4. Evidence suggests that the dysfunction of Notch signaling could play a critical role in the development of Alzheimer's disease (Kapoor and Nation 2020). Astrocyte development is clearly linked with the development of Alzheimer's disease (González-Reyes et al. 2017; Perez-Nievas and Serrano-Pozo 2018). Angiogenesis is also hypothesized to be implicated with Alzheimer's disease (Vagnucci Jr and Li 2003).

5   Conclusion

In this paper, we aim to improve the existing methods used for understanding the association between transcriptome and phenotype. By allowing the incorporation of gene network structures in the association model, our method can facilitate the identification of the disease-associated genes with implications for the disease mechanism. Our proposed method KMM combines the merits of several previous methods such as being multivariate, regularized by network structures, as well as reporting *p*-values in a manner accepted by geneticists. In particular, KMM is built upon the connections between the linear mixed model and ridge regression to further improve the linear mixed model's modeling power with the capability to incorporate of the network structures. After conducting simulation experiments to verify the effectiveness of the proposed method, we apply our method to real data collected from patients with Alzheimer's disease. In our application, we notice that the AD-associated genes reported by our method in the *CASP6, NCSTN, PSEN1* subnetworks are particularly interesting: our enrichment analysis suggests these genes are implicated in Alzheimer's disease. Thus, we demonstrate the efficacy of our method, and our released software can help the community identify the disease-associated genes with implications for the disease mechanism in the future.

A   Additional Simulation Experiments

Different Strengths of the Regulation. Further, we study how the strength of regulation will affect the performances of our methods, and we model this shift of strength with variations of the parameter r in the data generation process, while the rest of the configurations remain the same as the data generation process. Also, we continue to focus on the intermediate level of the previous example where we set $v = 16$.

Similarly, we repeat the experiments three times and plot the ROC curve with standard deviation plotted as the shady areas in Fig. 2.

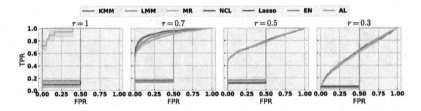

Fig. 2. ROC curve of competing methods over different regulation strength of the TF.

As Fig. 2 shows, our method is on par with previous hypothesis testing methods over most correlation levels. When $r = 1$, the regulated genes are distributed in the same way as the TF, although are associated with smaller effect sizes. Both LMM and KMM are good enough to uncover the associated genes in this case. When r is smaller (0.5 or 0.3), the regulated genes are less dependent on the TF, the hypothesis testing methods all perform similarly, probably because that when the regulated genes are more independent from the TF, the network structure does not introduce advantages. However, when $r = 0.7$, the KMM method starts to show a clear advantage over other methods. In summary, our proposed method can outperform other methods when there is a strong correlation between the TF and regulated genes (but not too strong when the regulated genes and TF are identically distributed). We believe this is the most frequently seen scenarios in real-world data. In addition, in other scenarios, our method does not perform worse than other methods, so there is no loss in using our method in general. In fact, if one calculates the area under ROC curve for Fig. 2, our method performs the best in all these four tested scenarios, although the advantages of our method in the other three scenarios are marginal.

Misspecified Network Structure. Finally, as our method is built upon the knowledge of network structure, we are interested in knowing what if the network structure is misspecified since in practice, we may not always be able to obtain a network structure faithful to the underlying regulatory mechanism. To simulate this, we introduce another hyperparameter q in the data generation process. When we generate the network structure N, we drop the edges in the network structure with the probability $1 - q$. The rest configuration of data generation is the same as the general one introduced in the preceding texts.

Again, we repeat the experiments three times and plot the ROC curve with standard deviation plotted as the shady areas in Fig. 3.

As Fig. 3 shows, our method is surpringly robust to the misspefication of the prior network structure. When $q = 1$, the input network is faithful to the underlying regulatory network, and the KMM method certainly outperforms the competing methods. Interestingly, the advantages of the KMM method maintain even when half of the edges of the input network are missing ($q = 0.5$). When $q = 0.3$, which means that 70% of the edges of the underlying regulatory network are missing in the input network for the model, the proposed method start to

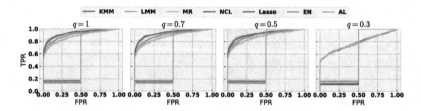

Fig. 3. ROC curve of competing methods when the prior network is misspecified (the edges of a network is dropped with probability $1 - q$).

perform similarly to the previous hypothesis testing methods. Even this case, the calculated area under ROC score of KMM will be higher than those competing methods, although this advantage cannot be observed in the ROC curves.

B   Covaraite Regressing

To demonstrate the success correction of these factors, we compared the Spearman's correction between the expressions and the covariates before and after the correction. Figure 4 shows the comparison of the Spearman's correlation between the gene expressions and the covariates before and after the regressing across the three different compartments studied in this work, and we can see that the correlation between each genes and the age covaraites drops significantly after the regression.

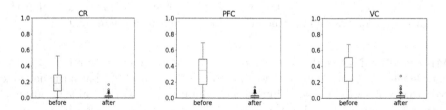

Fig. 4. The comparison of the Spearman's correlation between the gene expressions and the covariates before and after the regressing.

References

Barbeira, A.N., et al.: Exploring the phenotypic consequences of tissue specific gene expression variation inferred from GWAS summary statistics. Nat. Commun. **9**(1), 1–20 (2018)

Benjamini, Y., Hochberg, Y.: Controlling the false discovery rate: a practical and powerful approach to multiple testing. J. Roy. Stat. Soc. Ser. B (Methodological) **57**, 289–300 (1995)

Bertram, L., Tanzi, R.E.: Genome-wide association studies in alzheimer's disease. Hum. Mol. Genet. **18**(R2), R137–R145 (2009)

Bozzo, E.: The moore-penrose inverse of the normalized graph laplacian. Linear Algebra Appl. **439**(10), 3038–3043 (2013)

Bozzo, E., Franceschet, M.: Approximations of the generalized inverse of the graph laplacian matrix. Internet Math. **8**(4), 456–481 (2012)

Bühlmann, P.: Statistical significance in high-dimensional linear models. Bernoulli **19**(4), 1212–1242 (2013)

Cairns, N.J., Lee, V.M.-Y., Trojanowski, J.Q.: The cytoskeleton in neurodegenerative diseases. J. Pathol. J. Pathol. Soc. Great Britain Ireland **204**(4), 438–449 (2004)

Crawford, L., Zeng, P., Mukherjee, S., Zhou, X.: Detecting epistasis with the marginal epistasis test in genetic mapping studies of quantitative traits. PLoS Genet. **13**(7), e1006869 (2017)

de Leeuw, C.A., Mooij, J.M., Heskes, T., Posthuma, D.: Magma: generalized gene-set analysis of GWAS data. PLoS Comput. Biol. **11**(4), e1004219 (2015)

Dhanwani, R., et al.: T cell responses to neural autoantigens are similar in alzheimer's disease patients and age-matched healthy controls. Front. Neurosci. **14**, 874 (2020)

Ding, C., Peng, H.: Minimum redundancy feature selection from microarray gene expression data. J. Bioinform. Comput. Biol. **3**(02), 185–205 (2005)

Efthymiou, A.G., Goate, A.M.: Late onset alzheimer's disease genetics implicates microglial pathways in disease risk. Mol. Neurodegener. **12**(1), 1–12 (2017)

Fan, L., et al.: New insights into the pathogenesis of alzheimer's disease. Front. Neurol. **10**, 1312 (2020)

Feng, H., et al.: Leveraging expression from multiple tissues using sparse canonical correlation analysis and aggregate tests improves the power of transcriptome-wide association studies. PLoS Genet. **17**(4), e1008973 (2021)

Feng, H., Mancuso, N., Pasaniuc, B., Kraft, P.: Multitrait transcriptome-wide association study (TWAS) tests. Genetic Epidemiol. **108**, 240–256 (2021b)

Gamazon, E.R., et al.: A gene-based association method for mapping traits using reference transcriptome data. Nat. Genet. **47**(9), 1091–1098 (2015)

González-Reyes, R.E., Nava-Mesa, M.O., Vargas-Sánchez, K., Ariza-Salamanca, D., Mora-Muñoz, L.: Involvement of astrocytes in alzheimer's disease from a neuroinflammatory and oxidative stress perspective. Front. Mol. Neurosci. **10**, 427 (2017)

Gusev, A., et al.: Integrative approaches for large-scale transcriptome-wide association studies. Nat. Genet. **48**(3), 245–252 (2016)

Heckerman, D.: Accounting for hidden common causes when inferring cause and effect from observational data. arXiv:1801.00727 (2018)

Hemonnot, A.-L., Hua, J., Ulmann, L., Hirbec, H.: Microglia in alzheimer disease: well-known targets and new opportunities. Front. Aging Neurosci. **11**, 233, e1004219 (2019)

Huang, J., Ma, S., Zhang,C.-H.: Adaptive lasso for sparse high-dimensional regression models. Statistica Sinica **18**, 1603–1618 (2008)

Jacobs, H.I., et al.: The cerebellum in alzheimer's disease: evaluating its role in cognitive decline. Brain **141**(1), 37–47 (2018)

Javanmard, A., Montanari, A.: Hypothesis testing in high-dimensional regression under the gaussian random design model: asymptotic theory. IEEE Trans. Inf. Theory **60**(10), 6522–6554, e1004219 (2014)

Jones, S.V., Kounatidis, I.: Nuclear factor-kappa B and alzheimer disease, unifying genetic and environmental risk factors from cell to humans. Front. Immunol. **8**, 1805 (2017)

Kang, H.M., et al.: Efficient control of population structure in model organism association mapping. Genetics **178**(3), 1709–1723 (2008)

Kang, H.M., et al.: Variance component model to account for sample structure in genome-wide association studies. Nat. Genet. **42**(4), 348–354 (2010)

Kapoor, A., Nation, D.A.: Role of notch signaling in neurovascular aging and alzheimer's disease. In: Seminars in Cell and Developmental Biology. Elsevier (2020)

Kim, S., Xing, E.P.: Tree-guided group lasso for multi-task regression with structured sparsity (2010)

Li, C., Li, H.: Network-constrained regularization and variable selection for analysis of genomic data. Bioinformatics **24**(9), 1175–1182 (2008). ISSN: 1367–4803. https://doi.org/10.1093/bioinformatics/btn081

Lippert, C., Listgarten, J., Liu, Y., Kadie, C.M., Davidson, R.I., Heckerman, D.: Fast linear mixed models for genome-wide association studies. Nat. Methods **8**(10), 833–835 (2011)

Lockhart, R., Taylor, J., Tibshirani, R.J., Tibshirani, R.: A significance test for the lasso. Ann. Stat. **42**(2), 413 (2014)

Lonsdale, J., et al.: The genotype-tissue expression (GTEX) project. Nat. Genet. **45**(6), 580–585 (2013)

Maldonado, Y.M.: Mixed models, posterior means and penalized least-squares. Lecture Notes-Monograph Series, pp. 216–236 (2009)

Masters, C.L., Bateman, R., Blennow, K., Rowe, C.C., Sperling, R.A., Jeffrey, L.: Cummings 2015. "alzheimer's disease". Nature Reviews Disease Primers (2015). https://doi.org/10.1038/nrdp

Meinshausen, N., Bühlmann, P.: Stability selection. J. Roy. Stat. Soc. Ser. B (Stat. Methodol.) **72**(4), 417–473, e1004219 (2010)

Murpy, M., LeVine III, H.,: Alzheimer's disease and the β-amyloid peptide. J. Alzheimers Dis. **19**(1), 311–323 (2010)

Niikura, T., Tajima, H., Kita, Y.: Neuronal cell death in alzheimer's disease and a neuroprotective factor, humanin. Curr. Neuropharmacol. **4**(2), 139–147 (2006)

Oughtred, R., et al.: The biogrid interaction database: 2019 update. Nucleic Acids Res. **47**(D1), D529–D541 (2018)

Perez-Nievas, B.G., Serrano-Pozo, A.: Deciphering the astrocyte reaction in alzheimer's disease. Front. Aging Neurosci. **10**, 114, e1004219 (2018)

Petersen, K.B., Pedersen, M.S., et al.: The matrix cookbook. Tech. Univ. Denmark **7**(15), 510, e1004219 (2008)

Pontén, F., Jirström, K., Uhlén, M.: The human protein atlas-a tool for pathology. J. Pathol. J. Pathol. Soc. Great Britain Ireland **216**(4), 387–393, e1004219 (2008)

Puniyani, K., Kim, S., Xing, E.P.: Multi-population GWA mapping via multi-task regularized regression. Bioinformatics **26**(12), i208–i216, e1004219 (2010)

Sadigh-Eteghad, S., Sabermarouf, B., Majdi, A., Talebi, M., Farhoudi, M., Mahmoudi, J.: Amyloid-beta: a crucial factor in alzheimer's disease. Med. Princ. Pract. **24**(1), 1–10 (2015)

Safran, M., et al.: Genecards version 3: the human gene integrator. Database 2010 (2010)

Salat, D.H., Kaye, J.A., Janowsky, J.S.: Selective preservation and degeneration within the prefrontal cortex in aging and alzheimer disease. Arch. Neurol. **58**(9), 1403–1408 (2001)

Subramanian, J., Savage, J.C., Tremblay, M.È.: Synaptic loss in alzheimer's disease: mechanistic insights provided by two-photon in vivo imaging of transgenic mouse models. Front. Cell. Neurosci. **14**, 445 (2020)

Thompson, W.A., et al.: The problem of negative estimates of variance components. Ann. Math. Stat. **33**(1), 273–289 (1962)

Tosto, G., Reitz, C.: Genome-wide association studies in alzheimer's disease: a review. Curr. Neurol. Neurosci. Rep. **13**(10), 381 (2013)

Town, T., Tan, J., Flavell, R.A., Mullan, M.: T-cells in alzheimer's disease. NeuroMol. Med. **7**(3), 255–264 (2005)

Uffelmann, E., et al.: Genome-wide association studies. Nat. Rev. Methods Primers **1**(1), 1–21 (2021)

Vagnucci, A.H., Jr., Li, W.W.: Alzheimer's disease and angiogenesis. Lancet **361**(9357), 605–608, e1004219 (2003)

Van Mieghem, P., Devriendt, K., Cetinay, H.: Pseudoinverse of the Laplacian and best spreader node in a network. Phys. Rev. E **96**(3), 032311 (2017)

Visscher, P.M., et al.: 10 years of gwas discovery: biology, function, and translation. Am. J. Hum. Genet. **101**(1), 5–22, e1004219 (2017)

Wainberg, M., et al.: Opportunities and challenges for transcriptome-wide association studies. Nat. Genet. **51**(4), 592–599 (2019)

Wang, H., Lengerich, B.J., Aragam, B., Xing, E.P.: Precision lasso: accounting for correlations and linear dependencies in high-dimensional genomic data. Bioinformatics **35**(7), 1181–1187 (2018)

Wang, H., Yue, T., Yang, J., Wu, W., Xing, E.P.: Deep mixed model for marginal epistasis detection and population stratification correction in genome-wide association studies. BMC Bioinf. **20**(23), 1–11, e1004219 (2019)

Wang, H., Aragam, B., Xing, E.P.: Tradeoffs of linear mixed models in genome-wide association studies. J. Comput. Biol. (2022). (to appear)

Yang, J., Zaitlen, N.A., Goddard, M.E., Visscher, P.M., Price, A.L.: Advantages and pitfalls in the application of mixed-model association methods. Nat. Genet. **46**(2), 100–106 (2014)

Yiannopoulou, K.G., Papageorgiou, S.G.: Current and future treatments in alzheimer disease: an update. J. Central Nerv. Syst. Dis. **12**, 1179573520907397, e1004219 (2020)

Zetterberg, H., Mattsson, N.: Understanding the cause of sporadic alzheimer's disease. Expert Rev. Neurother. **14**(6), 621–630 (2014)

Zhang, B., et al.: Integrated systems approach identifies genetic nodes and networks in late-onset alzheimer's disease. Cell **153**(3), 707–720 (2013)

Zhang, C.-H., Zhang, S.S.: Confidence intervals for low dimensional parameters in high dimensional linear models. J. Roy. Stat. Soc. Ser. B (Stat. Methodol.) **76**(1):217–242 (2014). https://doi.org/10.2307/24772752

Zhang, Z., et al.: Mixed linear model approach adapted for genome-wide association studies. Nat. Genet. **42**(4), 355–360 (2010)

Zou, H.: The adaptive lasso and its oracle properties. J. Am. Stat. Assoc. **101**(476), 1418–1429 (2006)

Zou, H., Hastie, T.: Regularization and variable selection via the elastic net. J. Roy. Stat. Soc. Ser. B (Stat. Methodol.) **67**(2), 301–320, e1004219 (2005)

Real-Valued Group Testing
for Quantitative Molecular Assays

Seyran Saeedi[1,5], Myrna Serrano[2,6], Dennis G. Yang[3], J. Paul Brooks[4],
Gregory A. Buck[2,6], and Tomasz Arodz[1,6(✉)]

[1] Department of Computer Science, College of Engineering, Virginia Commonwealth
University, Richmond, VA, USA
`tarodz@vcu.edu`
[2] Department of Microbiology and Immunology, School of Medicine,
Virginia Commonwealth University, Richmond, VA, USA
[3] Department of Mathematics, College of Arts and Sciences, Drexel University,
Philadelphia, PA, USA
[4] Department of Information Systems, School of Business, Virginia Commonwealth
University, Richmond, VA, USA
[5] Department of Electrical and Computer Engineering, University of California,
Santa Barbara, CA, USA
[6] Center for Microbiome Engineering and Data Analysis, Virginia Commonwealth
University, Richmond, VA, USA

Abstract. Combinatorial group testing and compressed sensing both
focus on recovering a sparse vector of dimensionality n from a much
smaller number $m < n$ of measurements. In the first approach, the problem is defined over the Boolean field – the goal is to recover a Boolean vector and measurements are Boolean; in the second approach, the unknown
vector and the measurements are over the reals. Here, we focus on real-
valued group testing setting that more closely fits modern testing protocols relying on quantitative measurements, such as qPCR, where the
goal is recovery of a sparse, Boolean vector and the pooling matrix needs
to be Boolean and sparse, but the unknown input signal vector and the
measurement outcomes are nonnegative reals, and the matrix algebra
implied in the test protocol is over the reals. With the recent renewed
interest in group testing, focus has been on quantitative measurements
resulting from qPCR, but the method proposed for sample pooling were
based on matrices designed with Boolean measurements in mind. Here,
we investigate constructing pooling matrices dedicated for the real-valued
group testing. We provide conditions for pooling matrices to guarantee
unambiguous decoding of positives in this setting. We also show a deterministic algorithm for constructing matrices meeting the proposed condition, for small matrix sizes that can be implemented using a laboratory
robot. Using simulated data, we show that the proposed approach leads
to matrices that can be applied for higher positivity rates than combinatorial group testing matrices considered for viral testing previously.
We also validate the approach through wet lab experiments involving
SARS-CoV-2 nasopharyngeal swab samples.

© The Author(s), under exclusive license to Springer Nature Switzerland AG 2022
I. Pe'er (Ed.): RECOMB 2022, LNBI 13278, pp. 126–142, 2022.
https://doi.org/10.1007/978-3-031-04749-7_8

Keywords: Group testing · Compressed sensing · qPCR · SARS-CoV-2 testing

1   Introduction

Widely-available, fast-turnover molecular testing for the presence of highly contagious infectious diseases is considered a key tool in limiting their spread [36]. For newly emerging viral diseases, the gold-standard detection approach involves molecular assays based on polymerase chain reaction (PCR) [10], which can be quickly designed once the genetic sequence of the virus becomes available. However, rapid scaling of testing to cover the affected communities may face obstacles, leading to interest in pooling strategies that allow for using m tests to screen many more than m samples, out of which up to k are expected to be positive [13,29,30,33,34]. Traditional, adaptive pooling approaches, such as Dorfman pooling [1,13] and its improved variants [21,30], combine biological material from multiple individuals into pools, each tested using one test. This allows for quickly eliminating a large fraction of virus-negative pools of samples. However, follow-up testing is required to confirm which individual samples in the pools that tested positive are positive, introducing delays and requiring protocols for storing and retrieving previously tested samples for re-testing.

Non-adaptive, single-step protocols in which tests do not depend on each other and can be done in parallel have been studied under the umbrella of combinatorial group testing [2,14] or Boolean compressed sensing [3,27], and also in information retrieval [23]. The key challenge in non-adaptive group testing protocols is the design of a binary measurement matrix A (see Fig. 1), which prescribes that sample j should be assigned to pool i if $A_{ij} = 1$. In many applications, the matrix should be sparse, and several authors considered combinatorial group testing with sparse matrices [17,18]. The matrix needs to guarantee that the identity of the positive samples can be decoded from measurements of the sample pools. Probabilistic group testing relaxes that requirement to allow the decoding guarantee to fail with some low probability [7]. These approaches all share the underlying Boolean algebra – the measurement result for each pool is binary, and only provides information whether the tested pool is all-negative or whether it contains at least one positive sample. Combinatorial quantitative group testing [37] extends this approach to measurements that provide the number of positive samples in the pool. The focus on binary or integer measurements puts a limit on how small the number of pools, m, can be for a given number of tests, n.

Contemporary molecular assays often provide more than a binary readout – for example, cycle threshold (C_t) values in qPCR can be used to provide an estimate of the quantity of the measured molecule – but the availability of this quantitative information is underutilized in pooling matrix design. Compressed sensing [4,6,9,11] approaches generalize group testing to quantitative, real-valued measurements resulting from real-valued linear algebra involving A and the unknown vector x^*, similar to how qPCR would provide an estimate of the combined abundance of the molecule in the pool. However, matrices designed for compressed sensing typically involves real-valued elements, for example sampled from a normal distribution, and thus are not feasible to implement for laboratory pooling of samples. Compressed sensing also focuses on the quality of approximating the real-valued unknown signal vector, instead of just its pattern of nonzeros that indicates which samples are positive, and on applications with high-dimensional signals, such as 3D imaging [26]. Nonnegative compressed sensing [5, 12, 24, 25] is more aligned with molecular testing by focusing on unknown vectors that, like molecular abundance vectors, involve nonnegative reals, but is still optimizing the matrix design towards approximating the vector x^* instead of recovering its pattern of nonzeros.

Combination group testing and compressed sensing have gained renewed interest recently in the context of SARS-CoV-2 testing. It has been argued that combinatorial, nonadaptive pooling designs become advantageous compared to multi-step, adaptive pooling and to testing individual samples as the positivity rate, the fraction of samples in the tested group, increases [8]. One recent nonadaptive pooling method, P-BEST, achieves 8-fold reduction in the number of tests for groups of samples with positivity rate of around 1% or less. Another method, Tapestry [19], achieves 2.3-fold reduction for 1.9% positivity rates, extending to 10-fold reduction for 0.2% positivity rates. Both P-BEST and Tapestry utilize quantitative measurements and employ decoding techniques from compressed sensing domain, but both use matrix construction strategies designed for Boolean-measurement combinatorial group testing: P-BEST relies on Reed-Solomon codes, and Tapestry utilizes Kirkman Triple Systems resulting in 2-disjunct matrices. Neither of these approaches considered whether the matrix design can be improved if measurements are assumed to be real-valued.

1.1   Problem Statement and Contribution

We introduce real-valued group testing: an approach that exploits the quantitative nature of molecular assays and aims at recovering the nonzero patterns in sparse vector. Consider an unknown nonnegative vector $x^* \in \mathbb{R}_+^n$ of molecular abundances in samples from n individuals, with up to k positives; we call such vectors k-sparse. Equal amounts of biological sample from individual i are placed by a laboratory robot into d_i distinct testing pools, with pool j having material from p_j samples. The initial amount of sample, and the time it takes for the laboratory robot to perform the work, puts limits on d_i and p_j. The assignment of samples to pools is given by a binary matrix A, where $A_{ji} = 1$ indicates that a portion of sample i is placed in pool j. The total abundances in the pools are then equal to $y^* = Ax^*$. In practice, to ensure that each sample has the same total contribution to the measurement pools, we normalize each column of the binary matrix by dividing it by the number of ones in it. The abundances are quantified by a molecular assay, such as qPCR, leading to observed measurement vector $\hat{y} \in \mathbb{R}^m$. Readouts are estimated, from C_t values, on a logarithmic scale, leading to noise that is approximately multiplicative; here, we assume that $|\hat{y}_j - y_j^*| \propto y_j^*$. The goal is to design matrix A that will allow us to uncover supp(x^*), the support of x^*, from $\hat{y}$, that is, find which samples are positive. Matrices designed

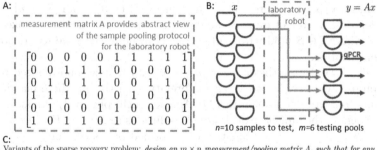

Fig. 1. Conceptual illustration of the proposed real-valued group testing. **A:** According to matrix A, genetic material from sample no. 1 (left-most column) goes into testing pool 4 and 6, and testing pool no. 1 (top row) will contain genetic material from samples 6–10. **B:** A robot programmed according to matrix A will distribute samples to pools, then qPCR assays will provide quantitative readout of the amount of viral genetic material in each pool. A decoding algorithm will resolve which cases are positive. **C:** Comparison of real-valued group testing with existing approaches. Note that in our approach, we construct a binary matrix but we column-normalize it to have unit column norms prior to use in testing; this ensures samples have equal contribution to the measurement vector.

specifically for real-valued group testing allows for reducing the number of pools for a given number of samples compared to matrices designed for Boolean combinatorial group testing. For example, for the matrix in Fig. 1, quantitative measurements allow for distinguishing between the scenario with one positive at the third column and a scenario with two positives, at first and third columns. The matrix would not be appropriate for Boolean measurements, which would not be able to distinguish between these two scenarios.

We show a new necessary and sufficient condition for normalized binary matrices to guarantee unambiguous recovery of support of k-sparse nonnegative signals. We also provide a deterministic method for constructing matrices meeting the proposed condition, for small values of k, m, n that are relevant for the viral testing setting. The approach is validated using simulated data as well as limited laboratory experiments.

2 Methods

2.1 Notation

Let $[n] = \{1, ..., n\}$. Define support of a vector x by $\operatorname{supp}(x) = \{i \in [n] : x_i \neq 0\}$. The L_0 pseudo-norm is $||x||_0 = |\operatorname{supp}(x)|$. A vector is k-sparse if $||x||_0 \leq k$, that is, if it has up to k non-zeros. For n-dimensional vector x, for $S \subset [n]$, we use x_S to denote a vector of dimensionality equal to x, with entries x_i taken from vector x where $i \in S$, and with null entries elsewhere. By $\bar{S}$ we will denote the complement of S in $[n]$.

By $\mathbb{B}^{m \times n}$ we denote the space of $m \times n$ binary matrices normalized to have unit sum of each column; we use $\mathbb{B}$ for brevity even though these matrices are not binary, but instead have entries $1/\|A_i\|_0$ for each column A_i of matrix A. By $\ker A$ we denote the nullspace of A, a set of solutions to $Ax = 0$. For $m \times n$ matrix A, given $S \subset [n]$, A_S is a submatrix of A formed by its columns $A_i = A(\cdot, i)$, where $i \in S$. Similarly, for $T \subset [m]$, A_T is a submatrix of A formed by its rows $A_j = A(j, \cdot)$, where $j \in T$. It will be clear from the context whether we are considering a rows or columns.

2.2   Overview of the Matrix Design and Decoding Algorithms

Our approach to constructing pooling matrices is based on three observations. First, in Sects. 2.3 and 2.3, we prove conditions that a matrix must meet to allow for unambiguous decoding of the positive samples. Next, in Sect. 2.3, we show that a matrix meeting the conditions can be obtained by starting with a wide initial matrix and removing some columns. Finally, in Sect. 2.3, we show that the computational cost of finding which columns to remove can be substantially reduced by focusing on a series of submatrices instead of on the initial wide matrix, and that all the submatrices are essentially the same up to permutation of rows and columns, so in fact only one submatrix needs to be considered.

These observations together allow us to formulate a deterministic matrix design method (Algorithm 1) and a corresponding algorithm for decoding positive samples from noisy measurements (Algorithm 2).

Algorithm 1. Pooling Matrix Design Algorithm

Input: m - number of rows; k - sparsity; d - maximum number of nonzeros per matrix column

Output: A - pooling matrix
1: POOLINGMATRIX(m, k, d)
2:      $D_m = $ WidestBinaryMatrix(m, d)
3:      $D_{dk} = $ WidestBinaryMatrix(dk, d)
4:      $V_{dk} = $ ViolatingColumnSubsets(D_{dk})
5:      $V_m = \emptyset$
6:      **for all** $S : dk$-row subset of D_m **do**
7:          $V'_{dk} = $ MapColumnIDs$_{D_{dk} \to D_m}(V_{dk}, S)$
8:          $V_m = V_m \cup V'_{dk}$
9:      $H = $ HittingSet(columns of D_m, sets V_m)
10:     $A = $ remove columns H from D_m
11:     **return** column-normalized A

Algorithm 2. COMP-NNLS Decoding Algorithm

Input: $\hat{y}$ - measurements; A - pooling matrix

Output: P - set of positives
1: COMP-NNLS$(\hat{y}, A)$
2:      **for all** i **do**
3:          **if** $\hat{y}_i = 0$ **then**
4:              remove columns j from A if $A_{ij} = 1$
5:              remove row i from $A, \hat{y}$
6:      $\hat{x} = \arg\min_x \|Ax - \hat{y}\|_2$ s.t. $x \geq 0$
7:      $P = \text{supp}(\hat{x})$
8:      **return** P

Pooling Matrix Design Algorithm. The pooling matrix construction proceeds in three broad steps. First, (ln. 2 in Algorithm 1), an initial wide binary m-row matrix is constructed by concatenating all possible binary columns containing between 2 and d nonzeros. This matrix will lead to many ambiguities if used in group testing. Second (ln. 3–8), we analyze which sets of columns lead to ambiguities – we perform the analysis on a smaller, dk-row binary matrix (ln. 3–4), and translate the results to the initial matrix (ln. 5–8). These steps rely on our main technical contribution, Theorems 2 and 4 and Lemmas 1 and 2 described below in Sects. 2.3–2.3. Each of these column sets needs to be broken apart by removing at least one column from each set from the initial matrix, which can be achieved (ln. 9) by solving an instance of the hitting set problem, as described in Sect. 2.3.

Decoding Algorithm. The decoding algorithm (Algorithm 2) has two steps. First, in a manner similar to combinatorial orthogonal matching pursuit (COMP) [7], to simplify computations, we eliminate from A all columns that have nonzero entry in rows with null measured y_j, and we also eliminate these rows from A and y (Algorithm 2, ln. 2–6). This leaves us with a smaller set of possible positives. For pooling matrices constructed by the proposed approach, in the noise-free case Theorem 2 shows that the set of positives can be unambiguously decoded from measurements y^* by solving $y^* = Ax$ s.t. $x \geq 0$ and taking the support of the solution vector. When noisy measurements $\hat{y}$ are available instead of y^*, this constrained linear system may have empty set of sparse solutions. Indeed, the space of possible measurement noise vectors $\hat{y} - y^*$ is m-dimensional, while the set of k-sparse nonnegative solutions to $y = Ax$ s.t. ≥ 0 is a union of a finite number of k-dimensional sets, with m typically much higher than k. Thus, to find a sparse x we use (Algorithm 2, ln. 7) nonnegative least squares (NNLS), that is, we find the signal with the smallest, in the L_2 sense, change in $\hat{y}$ needed to find a nonnegative solution. This approach is aligned with the decoding guarantee for the noisy case provided by Theorem 4.

2.3   Constructing Matrices for Real-Valued Group Testing

Necessary and Sufficient Condition in Noise-Free Case. We will show that any matrix with the following property allows for recovering the pattern of non-zeros of any unknown nonnegative signal with sparsity k.

Definition 1. *k-balanced nullspace property. Matrix $A \in \mathbb{B}^{m \times n}$ has k-balanced nullspace property if for all $\eta \in \ker A \setminus \{0\}$, at least $k+1$ entries η_i are positive and at least $k+1$ entries η_j are negative.*

Intuitively (see Fig. 2), the property precludes having $\eta = x' - x''$ with a k-sparse, nonnegative x' and another nonnegative, nonzero x'' with $\text{supp}(x') \neq \text{supp}(x'')$ that could serve as an alternative set of positive cases. If we recover a set of up to k positive cases from the measurement vector y^*, we are guaranteed that there is no other set of positive cases, not just with up to k cases but of any cardinality, consistent with y^*. Thus, the guarantee is stronger than in group testing based on k-disjunct matrices, where the guarantee is limited to alternative sets of positives with cardinality up to k. The intuition is formalized as follows.

Theorem 2. *A matrix $A \in \mathbb{B}^{m \times n}$ allows for decoding* supp(x^*) *from y^* given by $y^* = Ax^*$ if and only if it meets the k-balanced nullspace property.*

Proof. ($\Longleftarrow$): Let x^* be a k-sparse vector with support S, let $y^* = Ax^*$, and let $x \neq x^*$ be another nonnegative vector with different pattern of non-zeros (i.e., supp$(x^*) \neq$ supp(x)) such that $y^* = Ax$. Take $\eta = x^* - x$; then $\eta \in \ker A \setminus \{0\}$. Vector η constructed this way will have at most k positive entries, since $-x$ only contributes negative or null entries and x^* is k-sparse nonnegative; k-balanced nullspace property will not hold. Thus, the property implies no such x exists. The unknown pattern of non-zeros in x^* can be recovered from the unique solution to the constrained linear system of equations $y^* = Ax$, $x \geq 0$.

($\Longrightarrow$): Assume k-balanced nullspace property is violated. We will show that recovering supp(x^*) unambiguously is not possible for some nonnegative x^*, that is, k-balanced nullspace property is necessary for real-valued group testing. For A with only positive entries, any $\eta \in \ker A \setminus \{0\}$ must have at least one positive and one negative entry. Let S be the support of the positive entries in η, and $-S$ the support of the negative entries; both are nonempty. Both η_S and $-\eta_{-S}$ are nonnegative, at least one of them is k-sparse, and supp$(\eta_S) \neq$ supp$(-\eta_{-S})$. We have $A\eta = A\eta_S + A\eta_{-S} = 0$. Both η_S and $-\eta_{-S}$ lead to the same measurement vector $y = A\eta_S = A(-\eta_{-S})$; given y and A, S and $-S$ resulting from η provide two different, equally possible sets of positive cases. $\qquad\blacksquare$

Sufficient Condition for the Multiplicative Noise Case. The measurements of $y = Ax$ are expected to be noisy, with the observed $\hat{y} \neq y$. For nonnegative x and A, the true y is also nonnegative, and in qPCR and similar settings, $\hat{y}$ is also constrained to be nonnegative by the nature of the measurement process. Often, with high probability, null y_j leads to null $\hat{y}_j$. To fit these characteristics, we focus on multiplicative noise.

Consider a k-sparse solution $x \in \mathbb{R}^n_+$ resulting in noise-free measurement $y \in \mathbb{R}^m$. The observed results of the measurements is $\hat{y}$, with noise of magnitude limited by $\delta > 0$ in the sense $|\hat{y}_i - y_i| \leq \delta y_i$. This noise bound is not symmetric, that is, bound on $\hat{y}_i/y_i$ is $1 + \delta$ while the bound on $y_i/\hat{y}_i$ is not $1/(1 + \delta)$; for small values of δ it approximates the symmetric multiplicative noise model that captures qPCR noise well.

The bound on the magnitude of noise can help us establish how much the measurement noise can impact the decoding of the positive cases from $\hat{y}$.

Fig. 2. Illustration of k-balanced nullspace property for $k = 2$. A: Vector η that violates it by having only $k = 2$ positive entries. B: The violation allows for two alternative solutions, $x^* = \eta_S$ and $x = -\eta_{-S}$, with the same measurement outcome.

Definition 3. k, l-**sparse δ-robustness property.** *Matrix $A \in \mathbb{B}^{m \times n}$ has k, l-sparse δ-robustness property if for all k-sparse $x' \in \mathbb{R}_+^n$ with $||x'||_1 = 1$, for all l-sparse $x'' \in \mathbb{R}_+^n$ with $\mathrm{supp}(x') \neq \mathrm{supp}(x'')$, we have $||Ax' - Ax''||_2 > 2\delta$.*

The δ-robustness property is an extension of the k-balanced property, which is a special case for $\delta = 0$ and $l = n - 1$. The property guarantees, via Theorem 4, that for two different sets of positives, the corresponding noise-free measurements are so different that even in the presence of measurement noise, the two sets of positives can be distinguished by the decoding algorithm (see Fig. 3).

Theorem 4. *Consider $A \in \mathbb{B}^{m \times n}$, and noisy measurements such that for every $x \in \mathbb{R}_+^n$ we observe $\hat{y}$ instead of $y = Ax$, with multiplicative error $|\hat{y}_i - y_i| \leq \delta y_i$ of magnitude limited by a constant $\delta \in \mathbb{R}_+$. Consider a k-sparse $x^* \in \mathbb{R}_+^n$ and a set $X \subset \mathbb{R}_+^n$ of alternative, l-sparse solutions with different support than x^*. If A has k, l-sparse δ-robustness property, then $\mathrm{supp}(\arg \min_x ||Ax - \hat{y}||_2 \text{ s.t. } x \geq 0)$ will correctly decode $\mathrm{supp}(x^*)$.*

Proof. For any $A \in \mathbb{B}^{m \times n}$, we have $||y||_1 = ||Ax||_1 = ||x||_1$, since columns of $A \in \mathbb{B}^{m \times n}$ add up to one. Noise of the form $|\hat{y}_i - y_i| \leq \delta y_i$ guarantees that $||\hat{y} - y||_2 \leq \delta ||y||_2 \leq \delta ||y||_1 = \delta ||x||_1$. Consider any $x' \in X$ and let $y' = Ax'$; also, let $y^* = Ax^*$. For x' to possibly be a solution $x' = \arg \min_x ||Ax - \hat{y}||_2 \text{ s.t. } x \geq 0$ instead of x^*, we need $||y' - \hat{y}||_2 \leq ||y^* - \hat{y}||_2$. This implies $||y' - \hat{y}||_2 \leq \delta ||x||_1$. From triangle inequality, it further implies $||y' - y^*||_2 \leq ||y^* - \hat{y}||_2 + ||y' - \hat{y}||_2 \leq 2\delta ||x^*||_1$, that is, $||Ax' - Ax^*||_2 \leq 2\delta ||x^*||_1$. If A has k, l-sparse δ-robustness property, we have a contradiction.

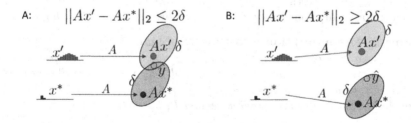

Fig. 3. Illustration of k, l-sparse δ-robustness property. A: Violation of the property may lead to ambiguity in decoding noisy measurement $\hat{y}$. B: If the property holds, x^* can be decoded from $\hat{y}$.

Property-Violating Column Sets. To arrive with a matrix meeting the k-balanced nullspace property and the k, l-sparse δ-robustness property, we can start with a matrix that does not meet the property, and then trim the matrix, as shown by the following result.

Definition 5. Violating set. *A violating set of a matrix $A \in \mathbb{B}^{m \times n}$ is a subset V of columns of A such that submatrix A_V does not meet k-balanced nullspace property, or the k, l-order δ-robustness property. A* minimal violating set *is a violating set that contains no smaller violating sets.*

Lemma 1. *For a matrix D, a submatrix $A \in \mathbb{B}^{m \times n}$, formed by eliminating columns from D, meets the k-balanced nullspace property and the k, l-order δ-robustness property if one column has been removed from all minimal violating sets of D.*

Proof. Removing columns from a matrix does not introduce new violating sets, but removing even one column from a minimal violating set removes the violating set. Thus, removing one column from each minimal violating set eliminates all violating sets.

All minimal violating sets for the k-balanced nullspace property can be obtained by enumerating all minimal linearly dependent sets of the matrix and filtering them according to the number of positive and negative linear weights. Finding minimal dependent sets is a special case of enumerating all circuits in a matroid [22]. For our experiments, we use the implementation in the function `circuits` of 4ti2 [35]. For the δ-robustness property, quadratic programming iterated over sets of up to $k + l$ columns can produce all minimal violating sets.

Reducing the Search for Property-Violating Sets to a Small Base Problem. Detecting all minimal dependent sets for a large $m \times n$ matrix in $\mathbb{B}^{m \times n}$ has combinatorial complexity and is computationally infeasible. For matrices $A \in \mathbb{B}^{m \times n}$ with up to d non-zeros in each column, $y = Ax$ will be sparse for nonnegative, k-sparse x, and we have the following decomposition that can make the search for testing matrices with m rows more efficient, by allowing us to focus on a small submatrices with dk rows, irrespective of how large m is.

Lemma 2. Decomposition into Base Problem. *Consider $A \in \mathbb{B}^{m \times n}$ with $\|A_i\|_0 \leq d$ for each column i. The matrix meets the k-balanced nullspace property or the k, l-sparse δ-robustness property if and only if for every dk-sized subset $T \in [m]$, the submatrix A_T formed by taking only rows in T, and then eliminating columns that have support extending beyond T, has those properties.*

Proof. Whether a set of matrix columns violates either of the properties is not affected by removing rows that have all-null values in these columns. If the property is violated for A_T, it is violated by some specific columns of A_T. Extending these columns by bringing back the remaining rows of A, which have all-null values, does not remove the violation involving the columns, and A does not meet the property. Conversely, if A does not meet the property, then some k-sparse vector x that leads to a measurement with at most dk nonzeros has an alternative x' that also has to lead to a measurement with up to dk nonzeros, in the same set T, and the property that guaranties lack of ambiguity must be violated for submatrix A_T.

It is not necessary to detect all minimal dependent sets in all the submatrices A_T, we only need to do it once, for a small matrix. Up to permutation of columns and rows, all these submatrices are identical to, or a submatrix of, of a single small matrix D_{dk} with dk rows and $\sum_{i=2}^{d} \binom{dk}{i}$ columns that results from concatenating all possible binary columns with two ones, with three ones, and up to dk ones. Given all minimal violating column sets in D_{dk}, a simple, efficient procedure can compute column indices in A from indices in D_{dk}. We can iteratively translate minimal violating sets of D_{dk} to minimal violating sets of each kd-row subset of A, which together form all the minimal violating sets of A.

Constructing the Pooling Matrix Based on Property-Violating Column Sets. For a given k and d, finding matrix A with the largest n for a given m by starting from some initial matrix D_m and then removing smallest number of columns that break up all minimal violating sets corresponds to the hitting set problem. The hitting set problem involves a universe U, and a set Σ of subsets of U. The goal is to select the smallest number of items from U such that each set from Σ contains at least one of those items. Here, the initial matrix D_m is an m-row matrix that has all $N = \binom{m}{2} + ... + \binom{m}{d}$ possible binary vectors with $2, ..., d$ ones as columns. Universe U is the set of columns of the initial matrix D_m, and Σ are all the minimal violating sets of D_m. We aim to break each minimal violating set by removing as few columns from D_m as possible. The problem, through equivalence with set cover problem, is NP-hard. The trivial approach of checking every subset of columns of D_m matrix has complexity of $|\Sigma|2^N$, and algorithms that require instead $N|\Sigma|2^{|\Sigma|}$ also have been formulated [16], but in our case $|\Sigma| > N$. While an approximation algorithm can be used for large set cover problems, for small values of m, k, and d considered here, integer programming is effective. We used GUROBI [20] solver with the number of threads limited to 32, limiting the time of calculations for each m to 500,000 s. All calculations were performed on a machine with 4 Intel Xeon E7-8894V4 2.4 GHz CPUs, with 24 cores per CPU, and 6 TB RAM.

3   Results

3.1   Comparison of Matrix Properties with Existing Approaches

We used the proposed approach to construct two $k = 2$, $d = 3$ testing matrices, for $m = 12$ and $m = 16$. These two numbers of pools correspond well to the standard dimensions of a $384 = 16 \times 24$ well plate. A robot with a multichannel pipette can work on multiple columns (for $m = 16$) or half-rows (for $m = 12$) in parallel. We used the noise limit of $\delta = 0.125$ for $m = 12$ and $\delta = 0.15$ for $m = 16$. The properties of the two pooling matrices are presented in Table 1. For comparison, we also summarized the properties of matrices of similar sizes resulting from existing approaches for combinatorial group testing or nonnegative compressed sensing.

In combinatorial group testing, unambiguous decoding of k-sparse signals from Boolean measurements is guaranteed for a binary pooling matrix if the matrix is k-disjunct [15], that is, if for any set of k columns, every other column has at least one non-zero element in a row where all the k columns are null. Reed-Solomon codes [32] with Kautz-Singleton construction [23] are a popular way of constructing k-disjunct matrices. The three RS/KS matrices with smallest m, comparable to o m in our matrices, are 12×16 (RS code: $[3,2]_4$), 15×25 ($[3,2]_5$), and 21×49 ($[3,2]_7$); these have lower compression rates than our matrices.

RS/KS construction is not optimal [31]. To find optimal codes for a given m, we used integer programming to find 2-disjunct matrices with highest possible n for values of $m = 12, 16$ used in our matrices. The resulting widest-possible 2-disjunct matrices are 12×20 and 16×37. These have much lower compression rates n/m than matrices constructed using our approach that exploits the fact that the matrices will be used with real-valued measurements instead of Boolean measurements.

Table 1. Properties of pooling matrices resulting from our approach for $k = 2$, $d = 3$, compared with existing methods for matrix construction. For each $m \times n$ matrix, we provide the maximum number of non-zeros per column (d_{max}) and per row (p_{max}), the maximum sample dilution dil_{max} resulting from d_{max} and p_{max}, maximum sparsity assumed in matrix design, k_{max}, the maximum positivity rate, k_{max}/n the matrix is designed for, and the testing compression rate, n/m.

Matrix Design method	m	n	d_{max}	p_{max}	dil_{max}	k_{max}	k_{max}/n	n/m
Proposed method	16	66	3	14	42	2	3.03%	4.13:1
Proposed method	12	36	3	10	30	2	5.55%	3.00:1
Combinatorial Group Testing Matrices								
Tapestry [19]	45	105	3	8	24	2	1.90%	2.33:1
P-BEST [33]	48	384	6	48	288	4	1.04%	8.00:1
$[3,2]_7$ Reed-Solomon/Kautz-Singleton	21	49	3	7	21	2	4.08%	2.33:1
$[3,2]_5$ RS/KS [23,32]	15	25	3	5	15	2	8.00%	1.67:1
$[3,2]_4$ RS/KS [23,32]	12	16	3	4	12	2	12.50%	1.33:1
Optimal 2-disjunct [15]	16	37	3	7	21	2	5.40%	2.32:1
Optimal 2-disjunct [15]	12	20	3	5	15	2	10.00%	1.67:1
Nonnegative Compressed Sensing Matrices								
Random-permutation [24]	16	66	3	15	45	–	–	4.13:1
Random-permutation [24]	12	36	3	12	36	–	–	3.00:1
Random-binomial [25]	16	66	7	17	119	–	–	4.13:1
Random-binomial [25]	12	36	5	13	65	–	–	3.00:1
One-sided coherence [5]	16	20	4	5	20	2	10.00%	1.25:1
One-sided coherence [5]	12	9	4	3	12	2	22.20%	0.75:1

RS codes are better suited for larger m and n, and have been used recently in SARS-CoV-2 testing for settings that involve lower positivity rates. P-BEST [33], a method based on RS codes, involves a 48×384 matrix designed for positivity rates up to 1%, much lower rate than our matrices; it also has much higher sample dilution rate. However, it offers two-fold higher compression rate than our matrices, and is a better choice in low-positivity scenarios.

Another method proposed for SARS-CoV-2 testing is Tapestry [19], which relies on Kirkman Triple Systems [28] for constructing 2-disjunct matrices. The 45×105 Tapestry matrix, while having lower compression rate than our 16×66 matrix, also only has unambiguous decoding guarantees for positivity of up to 1.9%.

In addition to matrices used in combinatorial group testing, we also evaluated matrix design approaches proposed in the nonnegative compressed sensing domain which, unlike our matrices, are designed with the goal of recovering the vector, not only its support. One-side coherence [5] has been proposed as a computationally-efficient sufficient condition for sparse recovery of nonnegative signals. It uses a condition $\rho/(1 + \rho) < 1/2k$ with $\rho = \max_{i \neq j} |A_i^T A_j|/\|A_i\|_2^2$; for $k = 2$, one needs $\rho < 1/3$. The condition was initially used for random $[0, 1]$-uniform matrices, which have low $A_i^T A_j$, but which are not feasible to use in laboratory setting. For binary matrices, $A_i^T A_j$ reduces to overlap and $\|A_i\|_2^2$ to $\|A_i\|_0$; to have a chance of finding A with $\rho < 1/3$ we need $\|A_i\|_0 \geq 4$, that is, columns with at least four non-zeros. Maximum independent set on the graph of all possible $\|A_i\|_0 \geq 4$ columns, with edge if $A_i^T A_j > 1$, is tractable for small m, but highest-n-for-m matrices are very poor: 12×9, and 16×20.

Other approaches [24,25] for creating matrices for nonnegative compressed sensing use random binary matrices. We tested matrices with entries sampled from Bernoulli distribution, an approach that has been used previously to construct sensing matrices that meet sparse recovery guarantees with high probability, including in the nonnegative case [25]. In order to increase the quality of the random matrices we used as comparison to our approach, instead of single random matrix, we sampled 100 Bernoulli matrices, and we picked the matrix that has highest mean of sensitivity and specificity when tested on 100 random 2-sparse inputs x. Bernoulli matrices may have highly varying number of ones between columns; as an alternative, we used a permutation approach that guarantees that the number of elements in each column is between 2 and 3, as in matrices resulting from our method. A similar approach was used previously in nonnegative sparse recovery [24]. Again, we picked the best of 100 random matrices for comparison with our approach. Random matrices only provide probability guarantees that hold with high probability for large n, large m. The small random matrices used here do not come with guarantees for unambiguous recovery for any sparsity value k, which is indicated by a dash in Table 1.

3.2   Effectiveness on Simulated Data

We have used synthetic, simulated data to evaluate the sensitivity and specificity of decoding the positive cases from real-valued measurements using the matrices

Table 2. Sensitivity and specificity of pooling matrices resulting from our approach for varying positivity rates in simulated noise-free and noisy measurements scenarios, for 10,000 simulated experiments. For each matrix, we provide its dimensions and its compression rate. Maximum positivity rate for which the matrix was designed is marked with an asterisk.

MATRIX	POSITIVITY RATE	NOISE-FREE MEASUREMENTS		MEASUREMENTS WITH SIMULATED NOISE	
		SENSIT. [%]	SPECIF. [%]	SENSIT. [%]	SPECIF. [%]
PROPOSED 16 × 66, 4.13 : 1	1.52%	100.0	100.0	100.00	99.99
PROPOSED 16 × 66, 4.13 : 1	3.03% *	100.0	100.0	99.48	99.18
PROPOSED 16 × 66, 4.13 : 1	4.55%	100.0	99.25	98.06	96.50
PROPOSED 12 × 36, 3 : 1	2.78%	100.0	100.0	100.00	99.91
PROPOSED 12 × 36, 3 : 1	5.56% *	100.0	100.0	99.26	97.88
PROPOSED 12 × 36, 3 : 1	8.33%	100.0	97.39	97.72	92.36

in Table 1. For each $m \times n$ matrix, we simulated 10,000 random input vectors of dimensionality n and with given number of nonzeros. The nonzero elements of each vector were sampled uniformly at random from $[1.0, 1000.0]$ range, that is, from a range spanning around 10 qPCR C_t cycles. True, noise-free measurements were obtained by setting $y = Ax$. To simulate noisy qPCR measurements $\hat{y}$, we used a realistic model of qPCR noise [19], of the form $\hat{y}_i = (1 + q)^{\mathcal{N}(0,\sigma^2)} y_i$, with the hyperparameters set to $q = 0.95$ and $\sigma = 0.1$. Given A and $\hat{y}$, we used COMP-NNLS decoding algorithm (algorithm2) to identify the positive samples.

First, we focused on the matrices resulting from the proposed method, and analyzed how they behave when the set of samples have varying positivity rates. We used three positivity rates: two that are within the rate range for which the matrix is designed for, and one that exceeds the rate by half. Results in Table 2 show that within the designed range of positivity rates, the matrices have high sensitivity, above 99%, and high specificity, above 99% for the 16 × 66 matrix and above 97% for the 14 × 36 matrix. When the maximum number of positives that the matrix was designed for is exceeded by half, the sensitivity drops by about 1.5% and remains above 97%, while the specificity drops to between 92% and 96%, indicating a small increase in the number of false negatives and a moderate increase in the number of false positives.

Next, we compared the proposed matrices with matrices from existing approaches in experiments with $k = 2$ positives, that is, the highest number of positives the matrices are designed for (only P-BEST is designed for higher number of positives, $k_{max} = 4$). As Table 3 show, in the scenario with noise-free measurements, all the matrices, with the exception of some of the random matrices, exhibit perfect accuracy.

As expected, matrices originating from Boolean combinatorial group testing, which are designed to work without quantitative information available for the decoding process, are not significantly affected by the presence of multiplicative measurement noise. This comes at the cost of having relatively low compression rates, (small m, n matrices) or being limited to low positivity rates (large m, n P-BEST and Tapestry).

Random matrices originating from nonnegative compressed sensing approaches are by design of the same shape as matrices from our approach, with the same compression rates. Permutation-based matrices exhibit slightly lower sensitivity than our deterministically-designed matrices, and show specificity lower by 0.78% for the 16×66 matrix, and by 1.44% for the 12×36 matrix – in both cases, this drop translates to almost doubling of the number of false positives. Binomial random matrices have lower specificity and sensitivity. Matrices based on one-sides coherence, while having high sensitivity and specificity, achieve that by offering very low compression rate.

Table 3. Sensitivity and specificity pooling matrices resulting from our approach compared to existing approaches in simulated noise-free and noisy measurements scenarios, for 10,000 simulated experiments with $k = 2$ positive samples. For each matrix, we provide its dimensions and its compression rate.

MATRIX	NOISE-FREE MEASUREMENTS		MEASUREMENTS WITH SIMUL. NOISE	
	SENSIT. [%]	SPECIF. [%]	SENSIT. [%]	SPECIF. [%]
PROPOSED METHOD 16×66, $4.13 : 1$	100.0	100.0	99.48	99.18
PROPOSED METHOD 12×36, $3 : 1$	100.0	100.0	99.26	97.88
TAPESTRY 45×105, $2.33 : 1$	100.0	100.0	99.93	100.0
P-BEST 48×384, $8 : 1$	100.0	100.0	99.92	100.0
$[3, 2]_7$ RS/KS 21×49, $2.33 : 1$	100.0	100.0	99.87	100.0
$[3, 2]_5$ RS/KS 15×25, $1.67 : 1$	100.0	100.0	99.81	100.0
$[3, 2]_4$ RS/KS 12×16, $1.33 : 1$	100.0	100.0	99.84	100.0
OPT. 2-DISJ. 16×37, $2.31 : 1$	100.0	100.0	99.85	100.0
OPT. 2-DISJ. 12×20, $1.67 : 1$	100.0	100.0	99.79	100.0
RAND-PERM 16×66, $4.13 : 1$	100.0	99.77	99.10	98.40
RAND-PERM 12×36, $3 : 1$	100.0	99.19	98.87	96.44
RAND-BINOM 16×66, $4.13 : 1$	95.39	98.18	94.50	95.30
RAND-BINOM 12×36, $3 : 1$	100.0	97.57	99.00	93.09
1-SIDED COH. 16×20, $1.25 : 1$	100.0	100.0	99.77	100.0
1-SIDED COH. 12×9, $0.75 : 1$	100.0	100.0	99.76	100.0

3.3   Effectiveness in Wet Lab

To validate the proposed approach in a laboratory setting, we performed limited testing of biological samples using qPCR assay. We focused on the 12×36 pooling matrix that offers higher maximum positivity rate, $k/n = 5.5\%$, than the 16×66 matrix. We used one set of 36 human samples: two samples were previously confirmed to be positive for SARS-CoV-2, and the remaining 34 were confirmed negatives. Briefly, nasopharyngeal swabs were collected and immediately transferred to 3 ml of PrimeStore MTM medium to deactivate the virus. Nucleic acids were purified using ThermoFisher MagMAX Viral/Pathogen Nucleic Acid Isolation Kit. Nucleic Acid-extracts were stored at −80 °C. Virus detection and viral loads were performed by qPCR using IDT 2019-nCoV CDC qPCR Probe Assay targeting the SARS-CoV-2 nucleocapsid gene (N1 and N2). The human RNase P gene was used as a control for sample integrity. Amplification was performed in our ViiA7 Real-Time PCR System using the TaqMan Fast Virus 1-Step Master Mix.

We randomly assigned the two positive samples to two columns of the pooling matrix. Samples were thawed in ice and spun down for 5 s. We then prepared 12 master wells. For each $A_{ij} = 1$ in the sensing matrix A, we pipetted 3 µl of biological material from sample j into master well i. We then pipetted 5 µl of volume from each master well into a separate testing well. The qPCR assay was performed in a 20 µl volume containing 5 µl of 4×TaqMan Fast Virus 1-Step Master Mix (Thermo Fisher Scientific), 1.5 µl of primers/probe set, and 8.5 µl DEPC-treated water. The qPCR was performed using ViiA7 Real-Time PCR System (Thermo Fisher Scientific) with the following cycling conditions: reverse transcription at 50 °C for 15 min. and 95 °C for 2 min., followed by 40 cycles of PCR at 95 °C for 3 s. and 55 °C for 30 s. We ran 12 individual qPCR assays, one per testing well. Total viral load in each testing well was estimated from the qPCR C_t value. Briefly, a standard curve was obtained by amplification of known amounts of SARS-CoV-2 (IDT 2019-CoV Plasmid Controls). Five consecutive dilutions (dilution factor 1:10) were prepared containing from 104 to 1 copies/reaction. The amounts of SARS-CoV-2 in samples were obtained by plotting C_t values onto the standard curve.

A second round of 12 qPCR assays was carried out using 5 µl volume from each master wells, leading to a second set of measurements, with the same underlying viral loads but differing due to technical variability of the qPCR assay. Finally, we performed the same experiment on the same pooled master wells, by diluting the master well 1:5, then pipetting 5 µl of volume into a testing well, and performing qPCR. This set of tests was again performed in duplicate, resulting in two additional data sets differing due to technical variability. For all four data sets, we used NNLS to recover the viral loads and, subsequently, the sparsity pattern of the 36 samples. In all four data sets, the method correctly identified both positive samples, and correctly labeled the remaining 34 samples as negative.

4   Conclusion

We provided a theoretical and empirical exploration of real-valued group testing, a setting that is relevant for improving efficiency of community testing for viral diseases. The proposed approach is focused on small-to-medium sized matrices that are convenient in a laboratory setting. The resulting matrices are useful for positivity rates below 5%, where they offer higher reduction in the number of tests than matrices designed for Boolean combinatorial group testing while maintaining high sensitivity and specificity. For higher positivity rates, the approach is not recommended. For much lower positivity rates, up to about 1%, existing approaches such as P-BEST offer higher reduction in the number of tests.

Acknowledgement. T.A., G.A.B, and M.S. are funded by NSF grant CBET-2034995.

References

1. Abdalhamid, B., Bilder, C.R., McCutchen, E.L., Hinrichs, S.H., Koepsell, S.A., Iwen, P.C.: Assessment of specimen pooling to conserve SARS CoV-2 testing resources. Am. J. Clin. Pathol. **153**(6), 715–718 (2020)
2. Aldridge, M., Johnson, O., Scarlett, J., et al.: Group testing: an information theory perspective. Found. Trends Commun. Inf. Theory **15**(3–4), 196–392 (2019)
3. Atia, G.K., Saligrama, V.: Boolean compressed sensing and noisy group testing. IEEE Trans. Inf. Theory **58**(3), 1880–1901 (2012)
4. Berinde, R., Gilbert, A.C., Indyk, P., Karloff, H., Strauss, M.J.: Combining geometry and combinatorics: a unified approach to sparse signal recovery. In: 46th Annual Allerton Conference on Communication, Control, and Computing, pp. 798–805. IEEE (2008)
5. Bruckstein, A.M., Elad, M., Zibulevsky, M.: On the uniqueness of non-negative sparse & redundant representations. In: 2008 IEEE International Conference on Acoustics, Speech and Signal Processing, pp. 5145–5148. IEEE (2008)
6. Candes, E.J., Romberg, J.K., Tao, T.: Stable signal recovery from incomplete and inaccurate measurements. Commun. Pure Appl. Math. **59**(8), 1207–1223 (2006)
7. Chan, C.L., Che, P.H., Jaggi, S., Saligrama, V.: Non-adaptive probabilistic group testing with noisy measurements: near-optimal bounds with efficient algorithms. In: 2011 49th Annual Allerton Conference on Communication, Control, and Computing (Allerton), pp. 1832–1839. IEEE (2011)
8. Cleary, B., et al.: Using viral load and epidemic dynamics to optimize pooled testing in resource-constrained settings. Sci. Transl. Med. **13**(589) (2021)
9. Cohen, A., Dahmen, W., DeVore, R.: Compressed sensing and best k-term approximation. J. Am. Math. Soc. **22**(1), 211–231 (2009)
10. Corman, V.M., et al.: Detection of 2019 novel coronavirus (2019-nCoV) by real-time RT-PCR. Eurosurveillance **25**(3), 2000045 (2020)
11. Donoho, D.L.: Compressed sensing. IEEE Trans. Inf. Theory **52**(4), 1289–1306 (2006)
12. Donoho, D.L., Tanner, J.: Sparse nonnegative solution of underdetermined linear equations by linear programming. Proc. Natl. Acad. Sci. **102**(27), 9446–9451 (2005)
13. Dorfman, R.: The detection of defective members of large populations. Ann. Math. Stat. **14**(4), 436–440 (1943)
14. Du, D., Hwang, F.K., Hwang, F.: Combinatorial Group Testing and Its Applications. World Scientific, River Edge (2000)
15. D'yachkov, A.G., Rykov, V.V.: Bounds on the length of disjunctive codes. Problemy Peredachi Informatsii **18**(3), 7–13 (1982)

16. Fomin, F.V., Kratsch, D.: Exact Exponential Algorithms. Springer-Verlag, Heidelberg (2010). https://doi.org/10.1007/978-3-642-16533-7
17. Gandikota, V., Grigorescu, E., Jaggi, S., Zhou, S.: Nearly optimal sparse group testing. IEEE Trans. Inf. Theory **65**(5), 2760–2773 (2019)
18. Gebhard, O., Hahn-Klimroth, M., Parczyk, O., Penschuck, M., Rolvien, M.: Near optimal sparsity-constrained group testing: improved bounds and algorithms. arXiv:2004.11860 (2020)
19. Ghosh, S., et al.: A compressed sensing approach to pooled RT-PCR testing for Covid-19 detection. IEEE Open J. Signal Process. **2**, 248–264 (2021)
20. Gurobi Optimization, L.: Gurobi Optimizer Reference Manual (2020). http://www.gurobi.com
21. Heidarzadeh, A., Narayanan, K.R.: Two-stage adaptive pooling with RT-qPCR for COVID-19 screening. arXiv:2007.02695 (2020)
22. Hemmecke, R.: On the positive sum property and the computation of Graver test sets. Math. Program. Ser. B **96**, 247–269 (2003)
23. Kautz, W., Singleton, R.: Nonrandom binary superimposed codes. IEEE Trans. Inf. Theory **10**(4), 363–377 (1964)
24. Khajehnejad, M.A., Dimakis, A.G., Xu, W., Hassibi, B.: Sparse recovery of nonnegative signals with minimal expansion. IEEE Trans. Sig. Process. **59**(1), 196–208 (2010)
25. Kueng, R., Jung, P.: Robust nonnegative sparse recovery and the nullspace property of 0/1 measurements. IEEE Trans. Inf. Theory **64**(2), 689–703 (2017)
26. Lustig, M., Donoho, D.L., Santos, J.M., Pauly, J.M.: Compressed sensing MRI. IEEE Sig. Process. Mag. **25**(2), 72–82 (2008)
27. Malioutov, D., Malyutov, M.: Boolean compressed sensing: LP relaxation for group testing. In: 2012 IEEE International Conference on Acoustics, Speech and Signal Processing (ICASSP), pp. 3305–3308. IEEE (2012)
28. Ngo, H.Q., Du, D.Z.: A survey on combinatorial group testing algorithms with applications to DNA library screening. Discrete Math. Probl. Med. Appl. **55**, 171–182 (2000)
29. Perchetti, G.A., et al.: Pooling of SARS-CoV-2 samples to increase molecular testing throughput. J. Clin. Virol. 104570 (2020)
30. Pilcher, C.D., Westreich, D., Hudgens, M.G.: Group testing for severe acute respiratory syndrome-coronavirus 2 to enable rapid scale-up of testing and real-time surveillance of incidence. J. Infect. Dis. **222**(6), 903–909 (2020)
31. Porat, E., Rothschild, A.: Explicit nonadaptive combinatorial group testing schemes. IEEE Trans. Inf. Theory **57**(12), 7982–7989 (2011)
32. Reed, I.S., Solomon, G.: Polynomial codes over certain finite fields. J. Soc. Ind. Appl. Math. **8**(2), 300–304 (1960)
33. Shental, N., et al.: Efficient high-throughput SARS-CoV-2 testing to detect asymptomatic carriers. Sci. Adv. eabc5961 (2020)
34. Täufer, M.: Rapid, large-scale, and effective detection of COVID-19 via nonadaptive testing. J. Theor. Biol. **506**, 110450 (2020)
35. 4ti2 team: 4ti2–a software package for algebraic, geometric and combinatorial problems on linear spaces (2020). https://4ti2.github.io
36. Walensky, R.P., Del Rio, C.: From mitigation to containment of the COVID-19 pandemic: Putting the SARS-CoV-2 genie back in the bottle. JAMA-J. Am. Med. Assoc. **323**, 1889–1890 (2020)
37. Wang, C., Zhao, Q., Chuah, C.N.: Optimal nested test plan for combinatorial quantitative group testing. IEEE Trans. Sig. Process. **66**(4), 992–1006 (2017)

On the Effect of Intralocus Recombination on Triplet-Based Species Tree Estimation

Max Hill[(✉)][iD] and Sebastien Roch[iD]

Department of Mathematics, University of Wisconsin–Madison, Madison, USA
bacharach@wisc.edu

Abstract. We consider species tree estimation from multiple loci subject to intralocus recombination. We focus on R^*, a summary coalescent-based method using rooted triplets. We demonstrate analytically that intralocus recombination gives rise to an inconsistency zone, in which correct inference is not assured even in the limit of infinite amount of data. In addition, we validate and characterize this inconsistency zone through a simulation study that suggests that differential rates of recombination between closely related taxa can amplify the effect of incomplete lineage sorting and contribute to inconsistency.

Keywords: Phylogenetics · Species tree estimation · Intralocus recombination

1 Introduction

Species tree estimation from genomic data is complicated by various biological phenomena which generate phylogenetic conflict, among them hybridization, horizontal gene transfer, gene duplication and loss, and incomplete lineage sorting (ILS) [23]. In particular, ILS may cause phylogenetic conflict in which a gene tree exhibits a different topology from that of the species tree, and is of greatest concern for species trees with short internal branches [23]. Of some interest is the existence of an anomaly zone for species trees, in which the most probable topology in the gene tree distribution differs from the topology of the species tree [5,7,8] (see also [2,21] for a more recent discussion of these and other relevant issues).

The existence of an anomaly zone has served as an impetus for the development of summary coalescent-based methods, quartets, such as R^*, MP-EST, BUCKy, ASTRAL, and others [6,15,16,20]. Some of these methods are based on the fact that rooted triples and unrooted quartets are special cases in which no anomaly zone exists [5,13] and also provide sufficient information to reconstruct

MH was supported by NSF grants DMS-1902892 (to SR) and DMS-2023239 (TRIPODS Phase II). SR was supported by NSF grants DMS-1902892 and DMS-2023239 (TRIPODS Phase II).

I. Pe'er (Ed.): RECOMB 2022, LNBI 13278, pp. 143–158, 2022.
https://doi.org/10.1007/978-3-031-04749-7_9

the full phylogeny [24,27]. Provided that the gene trees are estimated without error, such methods can provide statistically consistent methods of estimating species tree topology [30].

A common assumption of coalescent-based models based on the multispecies coalescent (MSC) [21,22] is that recombination occurs between genes (or loci)— so that gene trees may be assumed unlinked or statistically independent— but that *intralocus recombination* (i.e., recombination occurring *within* gene sequences), does not occur [2,9]. The significance of the latter assumption— that is, the impact of intralocus recombination on phylogenetic inference—is a matter of present interest [2,32] and much debate about its significance when unaccounted for [9,14,26]. One justification for assuming no intralocus recombination is that within-gene recombination may break gene function [23].

An influential simulation study argued that even high levels of intralocus recombination do not present a significant challenge for species tree estimation relative to other biological phenomena [14]. On the other hand, the authors of [25] suggest the absence of intralocus recombination may be an unreasonable assumption in real data, such as protein-coding genes in eukaryotes [2,19], and particularly in the case of species phylogenies with many taxa [26]. In particular, the potential for intralocus recombination to distort gene tree frequencies has been recognized as a challenge to summary coalescent-based methods, and [14] has been critiqued for its focus on shallow divergences and limitation to a low number of loci and taxa [26].

In this paper we take an analytical approach to investigate the effect of intralocus recombination. We prove that intralocus recombination has the potential to confound R^*, a summary coalescent-based methods based on inferring rooted triples. That is, we show that correct inference of rooted triplets cannot be guaranteed in the presence of intralocus recombination, assuming a distance-based approach is used for gene tree reconstruction. We then present a simulation study which characterizes the "inconsistency zone", i.e. the regime of parameters for S in which rooted triple inference does not converge to S as $m \to \infty$. We find that the effect arises when differential rates of recombination are exhibited between closely-related taxa.

1.1 Key Definitions

A *species phylogeny* $S = (V_S, E_S; r, \bar{\rho}, \bar{\tau}, \bar{\theta})$ is a directed binary tree with vertex set V_S, edge set E_S, root $r \in V_S$, and n labeled leaves $L_S = [n]$, such that each edge $e \in E_S$ is associated with a length $\tau_e \in (0, \infty)$, expressed in coalescent units, a recombination rate $\rho_e \in [0, \infty)$, and a mutation rate $\theta_e \in [0, \infty)$. It is assumed that there exists an ancestral population common to all leaves of S, i.e., a population above the root, with respective mutation and recombination parameters. Mutation rates are assumed to be per site per coalescent unit (a coalescent unit being $2N_e$ generations for diploid organisms, where N_e is the effective population size); recombination rates are per locus per coalescent unit.

The general question considered here is how to reconstruct the topology of the species phylogeny from gene sequence data sampled from its leaves.

This sequence data takes the form of multiple sequence alignments; a *multiple sequence alignment* (MSA) is an $n \times k$ matrix M whose entries are letters in the nucleotide alphabet $\{A, T, C, G\}$ such that entries in the same column are assumed to share a common ancestor. The phylogenetic reconstruction problem in this paper is to recover the topology of S from m independent samples of M.

We define a *rooted triple* to be a rooted binary phylogenetic tree with label set of size three; we use the notation $XY|Z$ (or equivalently $YX|Z$) to denote a rooted triple with leaves X, Y, Z having the property that the path from X to Y does not intersect the path from Z to the root [24]. The term *species triplet* refers to a restriction of S to three of its leaves. A rooted triple $XY|Z$ is said to be *uniquely favored* if it appears in more gene samples than either of the other two rooted triples $XZ|Y$ or $YZ|X$.

1.2   Inference Methods

This paper considers *Majority-Rule Rooted Triple*, or R^*, a consensus-based pipeline for species tree estimation. R^* utilizes the fact that the full topology of S is uniquely determined by, and hence can be recovered from, its rooted triples [27]. The R^* pipeline has three steps: first, for each gene, infer a rooted triple for each triplet of leaves $X, Y, Z \in L_S$. Second, make a list of uniquely favored triples from the m sampled genes. Finally, construct the most-resolved topology containing only uniquely favored triples. When gene trees are drawn independently according to the MSC, it holds that for every set of three taxa, the most probable rooted triple in the gene tree distribution matches the rooted triple obtained by restricting the species tree S to that set of three taxa; for this reason, the topology of the R^* consensus tree converges to that of S [6].

Since we are interested in the inference of the species-tree topology from *sequence data*, we consider a distance-based approach in which a species triplet with leaves X, Y, Z is inferred to have topology $XY|Z$ if

$$\delta_{XY} < \delta_{XZ} \wedge \delta_{YZ}. \tag{1}$$

where $\delta_{XY} = \delta_{XY}(M_k)$ is the number of mismatching nucleotides between sequences $\mathbf{s}_X$ and $\mathbf{s}_Y$ $(X, Y \in L_S)$. We refer to this inference procedure as R^* **with sequence distances**.

1.3   Multispecies Coalescent with Recombination

The model considered here, which we term the *Multispecies Coalescent with Intralocus Recombination*, or MSCR, uses the ancestral recombination graph (ARG) model from [10] (see also [1]) within the framework of the multi-species coalescent (MSC) [8,21,22]. In the single-population ARG [10], ancestors are represented by edges in the graph (see Fig. 1a), and the number N of ancestors, or *gene lineages*, at time t is a bottom-up birth-death process in which births (recombination events) occur at rate ρN and deaths (coalescent events) occur at rate $N(N-1)/2$. When a coalescent event happens, two edges are chosen at

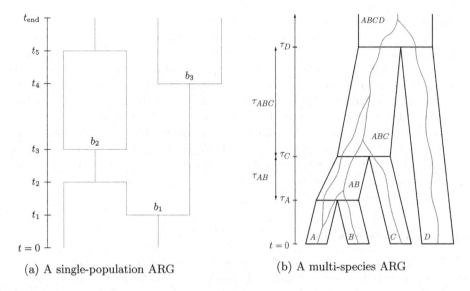

(a) A single-population ARG

(b) A multi-species ARG

Fig. 1. Two depictions of an ARG, in a single population (left) and in the multispecies case (right). In Fig. 1a, two lineages enter the population at time 0 and three exit at time t_{end}. Coalescent events occurred at times t_2 and t_5. Recombinations with breakpoints b_1, b_2, b_2 occurred at times t_1, t_3, and t_4. In Fig. 1b, the lineages of a multispecies ARG are shown in blue within a 4-taxa species tree S (the thick tree) with fixed edge lengths $\tau_A, \tau_B, \ldots, \tau_{ABC}$. (Color figure online)

random and merged into one. When recombination occurs, a randomly chosen lineage splits into two parent lineages. Each recombination vertex is labeled by a number b, chosen uniformly on [0,1]; this number is the *breakpoint* of the recombination.

The single-population ARG can be extended to multiple species in a manner similar to the MSC: at time $t = 0$, each leaf of S begins with a single lineage, and these lineages evolve in a bottom-up manner according to the ARG process along each edge of a fixed species tree (see Fig. 1b). If $\mathcal{G}$ is a rooted directed graph with edge lengths and leaf and breakpoint labels obtained in this manner, then we say that $\mathcal{G}$ is **generated according to the MSCR process on** S. In this scheme, the locus is modeled by the unit interval, and for each site $x \in [0, 1]$, a *marginal gene tree* $\mathcal{T}(x)$ can be obtained by tracing upward along the edges of $\mathcal{G}$ starting from the leaves; if a recombination vertex is reached with breakpoint b, take the left path if $x \leq b$ and the right path if $x > b$. This yields a collection of rooted edge-weighted binary trees; a simple example is shown in Fig. 2. The set of marginal gene trees $\mathcal{M} := \{\mathcal{T}(x) : 0 \leq x \leq 1\}$ is almost surely finite [10]. For each $T_g \in \mathcal{M}$, define $I(T_g) = \{x \in [0, 1] : \mathcal{T}(x) = T_g\}$, and define $w_g = |I(T_g)|$, where $|\cdot|$ denotes Lebesgue measure. In words, $I(T_g)$ is the identical-by-descent segment of the locus having genealogy T_g, and w_g is the proportion of sites with genealogy T_g.

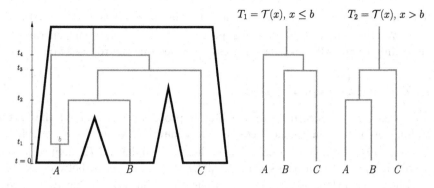

Fig. 2. On the left, an ancestral recombination graph (in blue) is shown within a 3-taxa tree S (in black). The times of coalescence and recombination events are labeled $t_1, \ldots t_4$ on the time axis, and the breakpoint associated with the recombination event is labeled $b \in [0,1]$. On the right, the corresponding marginal gene trees T_1 and T_2 are shown. This particular example also illustrates how intralocus recombination may contribute to phylogenetic conflict by allowing for 'partial' ILS, whereby one or more of the marginal gene trees (in this case T_1) exhibits a topology different from that of S.

Measuring time in coalescent units, this paper assumes that the per-site mutation rate is given by a fixed number $\theta > 0$ which does not vary on S. For each $x \in [0,1]$, site x evolves independently according to the Jukes-Cantor process [12,27] on the tree $\mathcal{T}(x)$. A somewhat more general description of this algorithm can be found in [4].

Thus, to model the evolution of a genetic locus consisting of k sites in which recombination breakpoints are distributed uniformly between them, a two-step process is followed. First, a multispecies ARG $\mathcal{G}$ is generated according to the MSCR process on S, from which a marginal gene tree $\mathcal{T}(x)$ is obtained for each $x \in [0,1]$. Second, for each $x \in [0,1]$ the Jukes-Cantor process is run with input tree $\mathcal{T}(x)$ in order to generate a nucleotide $\mathcal{N}(i,x) \in \{A,T,C,G\}$ for each $i \in L_S$. The MSA M_k is then defined as the $n \times k$ random matrix with rows $\mathbf{s}_1, \ldots, \mathbf{s}_n$ where for each $X \in [n]$, $\mathbf{s}_X = (s_X(1), \ldots, s_X(k))$ where $s_X(j) = \mathcal{N}(X, \frac{j}{k-1})$, $j = 0, 1, \ldots, k-1$. In this case, we say that M_k is **generated according to the MSCR-JC(k) process on** S.

In words, the MSCR-JC(k) process models the evolution of n homologous genes situated at a common genetic locus consisting of k sites, and which may have experienced intralocus recombination; these homologous genes are assumed to have been drawn from n distinct species whose true species phylogeny is represented by S. The resulting homologous aligned DNA sequences are the rows of the $n \times k$ matrix M_k. The phylogenetic reconstruction problem considered here pertains to whether the topology of S can be recovered from sequence data generated in this manner, or more precisely:

Problem: Let S be a species phylogeny with leaf labels $L_S = [n]$. Fix $k \geq 2$. Given m independent samples $M_k^{(1)}, \ldots, M_k^{(m)}$, each generated according to the MSCR-JC(k) process on S, recover the topology of S.

1.4   Estimating Sequence Distances

Let $\mathcal{G}$ be generated according to the MSCR process on S, and $\mathcal{M}$ the corresponding set of marginal gene trees. Given a marginal gene tree $T_g \in \mathcal{M}$, let $d_{XY}^{T_g}$ be the *evolutionary distance* between leaves X and Y on T_g, defined as the expected number of mutations per site along the unique path between X and Y. It follows from the assumptions about the mutation process that $d_{XY}^{T_g} = 2\theta t$, where t is the time of the most recent common ancestor of X and Y on T_g. For example in Fig. 2, $d_{AB}^{T_1} = 2\theta t_4$ and $d_{AB}^{T_2} = 2\theta t_2$. Define the *breakpoint-weighted uncorrected distance* by

$$\Delta_{XY} := \frac{3}{4} \sum_{T_g \in \mathcal{M}} w_g \left(1 - e^{-\frac{4}{3} d_{XY}^{T_g}} \right). \tag{2}$$

This formula, due to [28], generalizes the uncorrected Jukes-Cantor distance to the setting of intralocus recombination; if no intralocus recombination occurs, then the right-hand side has only a single summand and reduces to the inverse of the Jukes-Cantor distance correction formula for a single non-recombining locus.

Our first lemma shows that δ_{XY} can be approximated by $k\Delta_{XY}$ when k is large.

Lemma 1. *If M_k is generated according to the MSCR-JC(k) process on S then for all $X, Y \in L_S$, conditioned on $\mathcal{G}$, $\delta_{XY}(M_k) = k\Delta_{XY} + o(k)$ almost surely as $k \to \infty$.*

2   Inconsistency of R*

2.1   Statement and Overview

The main result is the following:

Theorem 1. *For k sufficiently large, R^* using sequence distances is not statistically consistent under the MSCR-JC(k) model. That is, there exists a species phylogeny S such that the topology of the output of R^* using sequence distances does not converge in probability to the topology of the species tree.*

To prove Theorem 1, it suffices to consider a species tree S with $L_S = \{A, B, C\}$ and topology $AB|C$. Denote edges of S, or *populations*, by the letters A, B, C, AB, and ABC as depicted in Fig. 3 where A, B, C correspond to the leaf populations, AB is the parent edge of A and B, and ABC is edge extending above the root. The key idea is to allow recombination only in population A. In order to keep the analysis tractable, the recombination rate and length of edge A

are chosen so that with high probability the number of recombinations is 0 or 1, so that the number of lineages on the ARG exiting population A (backwards-in-time) is either one or two. By choosing the internal branch length τ_{AB} sufficiently small, ILS occurs along that edge with high probability, so that all coalescent events on the ancestral recombination graph occur in the root population ABC. In that case, as long as the mutation rate is not too large, we show that, on the event $R_1 C_0$ (see Fig. 3), taxa B and C are more likely to be inferred as more closely related than taxa A and B, so that R^* converges to the wrong topology $BC|A$ as the number m of samples grows.

The mutation rate θ is assumed to be the same in all populations. The vector of recombination rates $\bar{\rho}$ is defined by setting $\rho_A = \rho > 0$ and $\rho_X = 0$ for all $X \neq A$. Assume S to be ultrametric. The populations A and B have length $\tau_A = \tau_B > 0$, the internal population AB has length τ_{AB}, the age of the root t_{root} is given by $t_{\text{root}} = \tau_A + \tau_{AB} = \tau_C$. For now assume that $\tau_{AB} > 0$ and $\tau_A > 0$; their precise values will be determined later in the proof.

Let M_k be generated according to the MSCR-JC(k) process on S, and let $E_{XY|Z}$ be the event that the rooted triple inferred from M_k using (1) is $XY|Z$. The following lemma implies that to prove Theorem 1, it will suffice to prove

$$\mathbb{P}[E_{YZ|X}] > \mathbb{P}[E_{XY|Z}]. \tag{3}$$

The *consistency zone* for R^* with sequence distances under the MSCR-JC(k) model is the set of species phylogenies S such that the topology of the R^* consensus tree converges in probability to the topology of S as $m \to \infty$.

Lemma 2. *A necessary and sufficient condition for S to lie in the consistency zone for R^* with sequence distances under the MSCR-JC(k) model is that for all $XY|Z \in \mathcal{R}(S)$,*

$$\mathbb{P}[E_{XY|Z}] > \mathbb{P}[E_{XZ|Y}] \vee \mathbb{P}[E_{YZ|X}] \tag{4}$$

Here $\mathcal{R}(S) = \{S|J : J \subseteq L_S, |J| = 3, \text{ and } S|J \text{ is binary}\}$ is the set of restricted rooted triples of S (see [24]).

By Lemma 1, with probability one, an ancestral recombination graph $\mathcal{G}$ generated according to the MSCR process has the property that sequences of increasing length k generated on it by the Jukes-Cantor process satisfy the almost sure limit $\frac{1}{k}\delta_{XY}(M_k) \to \Delta_{XY}$ as $k \to \infty$. Since almost sure convergence implies convergence in distribution, it holds that under the joint process which combines both genealogical and mutational processes, $\frac{1}{k}\delta_{XY}(M_k) \Rightarrow \Delta_{XY}$ as $k \to \infty$ for all $X, Y \in L_S$. Therefore, since the distribution function of Δ_{XY} is continuous, $\mathbb{P}[E_{XY|Z}] \to \mathbb{P}[E]$ and $\mathbb{P}[E_{YZ|X}] \to \mathbb{P}[F]$ as $k \to \infty$, where $E := [\Delta_{AB} < \Delta_{AC} \wedge \Delta_{BC}]$ and $F := [\Delta_{BC} < \Delta_{AB} \wedge \Delta_{AC}]$. Therefore inequality (3) will hold for sufficiently large k provided that

$$\mathbb{P}[F] > \mathbb{P}[E]. \tag{5}$$

We detail the proof next.

2.2 Key Lemmas

In what follows, set intersection is denoted with product notation (i.e. so that $XY = X \cap Y$ for events X, Y) and the important events to be considered are

$$R_i = [\text{exactly } i \text{ recombinations occur in the time interval } (0, \tau_A)]$$
$$C_i = [\text{exactly } i \text{ coalescences occur during the time interval } (0, t_{\text{root}})]$$
$$C_{0,X} = [\text{no coalescence occurs in population } X].$$

Since recombination occurs only in population A, the number of recombination events is governed by the recombination rate ρ and the duration τ_A of population A. The following lemma shows that τ_A can be chosen sufficiently small that with high probability, zero or one recombination occurs.

Lemma 3 (Recombination Probabilities). *For all* $\rho, \tau_A \geq 0$, $\mathbb{P}[R_0] = e^{-\rho\tau_A}$ *and* $\mathbb{P}[R_1] \geq \mathbb{P}[R_1 C_{0,A}] \geq \rho\tau_A e^{-(1+2\rho)\tau_A}$. *As* $\tau_A \to 0^+$, $\mathbb{P}[\cup_{k\geq 2} R_k] = O(\rho^2\tau_A^2)$.

For the case where no recombination occurs, the probabilities of E and F are estimated in the following lemma using elementary MSC calculations.

Lemma 4 (No Recombination Case). $\mathbb{P}[E|R_0] - \mathbb{P}[F|R_0] \leq \tau_{AB}$.

For the case where *exactly one* recombination occurs, the following lemma characterizes the behavior of coalescent events occurring below the root of S. Intuitively, it says that coalescence in population AB is rare when τ_{AB} is small.

Lemma 5 (Effect of Small Internal Edge). *As* $\tau_{AB} \to 0^+$, $\mathbb{P}[C_0|R_1] = K + O(\tau_{AB})$, $\mathbb{P}[C_{0,A}|R_1 C_1] = O(\tau_{AB})$, *and* $\mathbb{P}[C_2|R_1] = O(\tau_{AB})$, *where* $K = \mathbb{P}[C_{0,A}|R_1] \in (0, 1)$ *depends only on* τ_A *and* ρ, *and satisfies* $\lim_{\tau_A \to 0} K = 1$ *for any fixed* $\rho > 0$.

Next we apply Lemma 5 to show that $\mathbb{P}[E|R_1 C_1] - \mathbb{P}[F|R_1 C_1]$ is small, tending to zero as $\tau_{AB} \to 0^+$.

Lemma 6. $\mathbb{P}[E|R_1 C_1] - \mathbb{P}[F|R_1 C_1] = O(\tau_{AB})$ *as* $\tau_{AB} \to 0^+$,

We now come to a key part of the calculation: the event $R_1 C_0$, depicted in Fig. 3. The next lemma demonstrates that as long as θ is not too large, conditional on $R_1 C_0$, the event F is more likely than E.

Lemma 7. *The quantity* $\bar{\alpha} := \mathbb{P}[F|R_1 C_0] - \mathbb{P}[E|R_1 C_0]$ *depends only on* θ *and is positive if* $\theta \in (0, 3/4)$.

Proof Sketch. We sketch the proof idea here. Conditional on $R_1 C_0$, four distinct lineages enter population ABC at time t_{root}. Denote these lineages by A_1, A_2, B, and C, as shown in Fig. 3. Since no recombination occurs in population ABC, the order in which the lineages coalesce determines a *labeled history* (an ultrametric rooted binary tree with labeled tips and internal nodes rank-ordered according

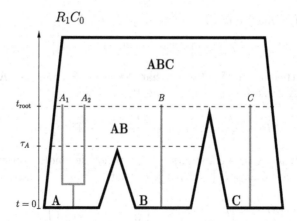

Fig. 3. A depiction of the event R_1C_0. The portion of the ancestral recombination graph more ancient than t_{root} is not shown.

to age [21]), whose tips are taken to be the lineages A_1, A_2, B and C at time t_{root}. There are 18 such labeled histories $\gamma_1, \ldots, \gamma_{18}$. Since pairs of lineages coalesce uniformly at random under the coalescent, $\mathbb{P}[\gamma_j|R_1C_0] = \frac{1}{18}$ for all j, and hence

$$\mathbb{P}[F|R_1C_0] - \mathbb{P}[E|R_1C_0] = \frac{1}{18}\sum_{j=1}^{18}\left(\mathbb{P}[F|R_1C_0\gamma_j] - \mathbb{P}[E|R_1C_0\gamma_j]\right). \qquad (6)$$

Having conditioned a particular labeled history, the probabilities $\mathbb{P}[E|R_1C_0\gamma_j] - \mathbb{P}[F|R_1C_0\gamma_j]$ for $j = 1, \ldots, 18$ are computed in a straightforward manner, so that the right hand side of (6) is positive provided that not too much signal is lost by a high mutation rate. In particular, since there are *two* lineages from A and only one from each of B and C, at least one of the A lineages is more likely to be included in the final coalescing pair, favoring greater pairwise distances between A and the other two taxa than those between B and C. $\qquad \square$

The next lemma applies Lemmas 5, 6, and 7 to show that $P[F|R_1] > \mathbb{P}[E|R_1]$ when the internal branch length τ_{AB} is small and the mutation rate θ is not too large.

Lemma 8. *If $\theta \in (0, 3/4)$, then $\mathbb{P}[E|R_1] - \mathbb{P}[F|R_1] = -\bar{a}K + O(\tau_{AB})$ as $\tau_{AB} \to 0^+$ (where the term $-\bar{a}K$ does not depend on τ_{AB}).*

2.3 Proof of Theorem 1

Proof of Theorem 1. It suffices to prove (5) for some choice of parameters ρ, θ, τ_A, and τ_{AB}. Let $\rho > 0$ and $\theta \in (0, 3/4)$ be arbitrary; we will show that τ_A, and τ_{AB} can be chosen sufficiently small that (5) holds. Conditioning on the number of recombination events in population A,

$$\mathbb{P}[F] - \mathbb{P}[E] >$$
$$(\mathbb{P}[F|R_0] - \mathbb{P}[E|R_0])\,\mathbb{P}[R_0] + (\mathbb{P}[F|R_1] - \mathbb{P}[E|R_1])\,\mathbb{P}[R_1] - \mathbb{P}[\cup_{k\geq 2}R_k].$$

Therefore by Lemma 4 and the trivial inequality $\mathbb{P}[R_0] \leq 1$,

$$\mathbb{P}[F] - \mathbb{P}[E] > -\tau_{AB} + (\mathbb{P}[F|R_1] - \mathbb{P}[E|R_1])\,\mathbb{P}[R_1] - \mathbb{P}[\cup_{k \geq 2} R_k].$$

By Lemma 8, there exists $\delta > 0$ such that $\mathbb{P}[F|R_1] - \mathbb{P}[E|R_1] > \bar{\alpha}K/2$ whenever $0 < \tau_{AB} < \delta$. Assume further that $\tau_{AB} \in (0, \delta)$. Then

$$\mathbb{P}[F] - \mathbb{P}[E] > -\tau_{AB} + \frac{\bar{\alpha}K}{2}\mathbb{P}[R_1] - \mathbb{P}[\cup_{k \geq 2} R_k].$$

By Lemma 3, there exists constants $C, D > 0$ not depending on τ_{AB} such that $\mathbb{P}[R_1] \geq C\rho\tau_A$ and $\mathbb{P}[\cup_{i \geq 2} R_i] \leq D\rho^2\tau_A^2$, so that

$$\mathbb{P}[F] - \mathbb{P}[E] > -\tau_{AB} + \left(\frac{1}{2}\bar{\alpha}KC - D\rho\tau_A\right)\rho\tau_A.$$

Since K does not depend on τ_{AB} and $K \to 1$ as $\tau_A \to 0$ by Lemma 5, there exists $\tau_A > 0$ sufficiently small that both $K > 1/2$ and $\epsilon := \bar{\alpha}C/4 - D\rho\tau_A > 0$. It follows that $\mathbb{P}[F] - \mathbb{P}[E] > -\tau_{AB} + \epsilon\rho\tau_A$. Since ϵ does not depend on τ_{AB}, it follows that $\mathbb{P}[F] - \mathbb{P}[E] > 0$ for τ_{AB} sufficiently small.   □

3   Simulation Study

We performed a simulation study to characterize the inconsistency zone established in Theorem 1. Code and documentation can be found at https://github.com/max-hill/MSCR-simulator.git. In all simulations, sequence data is generated according to the MSCR process on an ultrametric species phylogeny S with three species A, B, C, and rooted topology $AB|C$. In all cases, $k = 500$, $\tau_A = 1$ and θ does not vary among populations. We use the notation $\hat{p}_{XY|Z}$ to denote the proportion of the m samples from which the rooted triple $XY|Z$ was inferred, and $\hat{t}$ to denote the R^* uniquely favored rooted triple of the m samples. By the strong law of large numbers, $\hat{p}_{XY|Z}$ serves as an estimate of $\mathbb{P}[E_{XY|Z}]$ for large m, where $E_{XY|Z}$ is defined as in Lemma 2.

The range of recombination rates considered in these simulations are comparable to those in [14], who suggest they encompass biologically plausible values. As for mutation rates, typical rates in eukaryotes are on the order of $\mu = 10^{-9}$ to 10^{-8} per site per generation [11,17] and effective eukaryotic population sizes N_e range from 10^4 to 10^8 [18], making the values considered here of $\theta = 2N_e\mu \in \{0.01, 0.1\}$ plausible as well. Computational constraints limited the ability to consider mutation rates lower than these, as doing so would have necessitated an increase in k or m to compensate; however the analytic results here predict that the inconsistency zone will persist, and may grow, for smaller values of θ: the computed difference $\bar{\alpha} = \mathbb{P}[F|R_1C_0] - \mathbb{P}[E|R_1C_0]$ actually increases as $\theta \to 0$, suggesting that phylogenetic conflict may be greater under regimes with smaller mutation rates than those simulated here.

In the first experiment, we simulated the MSCR-JC(k) process under a variety of parameter regimes in order to characterize the anomaly zone and evaluate the robustness of triplet-based inference in the presence of intralocus recombination. In particular $m = 10^5$ replicates were generated independently under each parameter regime, with the aim of estimating how frequently the correct topology was inferred. The parameters used were $\theta = 0.1$, $\tau_{AB} \in \{0.01, 0.02, \ldots, 0.15\}$, $\rho_A \in \{0, \ldots, 20\}$, and $\rho_X =$ for all $X \neq A$, so that recombination occurred only in population A. Figure 4 shows the value of $\hat{t}$ for each simulated parameter regime, and Fig. 5 plots the surface $z = \hat{p}_{AB|C} - \hat{p}_{BC|A}$ as a function of ρ_A and τ_{AB}, so that parameter regimes with negative z values indicates inconsistent inference.

We also evaluated R^* inference with rooted triples inferred not by equation (1), but rather by maximum-likelihood under the (false) assumption of no intralocus recombination; in this mode, which we call $\boldsymbol{R^*}$ **with maximum likelihood**, binary sequences were simulated and the maximum likelihood rooted triple was computed analytically using the method in [31]. A plot almost identical to Fig. 4 was obtained. For the very short internal branch length $\tau_{AB} = 0.01$, simulations were run with similar parameters and higher number of replicates ($m = 15,000$), with inference performed using both R^* with sequence distances and R^* with maximum likelihood. Figure 6 plots the difference $y = \hat{p}_{BC|A} - \hat{p}_{AB|C}$ as a function of ρ_A obtained from these simulations.

These results show that the combination of intralocus recombination in population A along with a very short internal branch length τ_{AB} resulted in the rooted triple $BC|A$ being more slightly likely to be inferred than the correct topology $AB|C$. Figure 6 shows clearly that this effect increases for larger values of ρ_A. Nonetheless, as both Figs. 5 and 6 show, the magnitude of this effect is relatively small: even when $\hat{p}_{BC|A} - \hat{p}_{AB|C}$ is positive, it is never greater than 0.1. Moreover, as Figs. 4 and 5 show, this effect disappears when τ_{AB} is increased

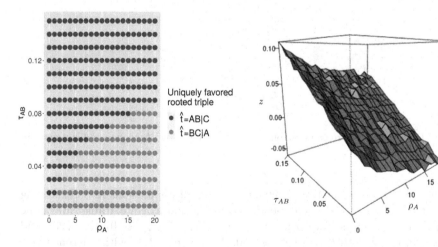

Fig. 4. R^* inconsistency zone. The color of each dot represents a simulation of $m = 10^5$ replicates.

Fig. 5. The surface $z = \hat{p}_{AB|C} - \hat{p}_{BC|A}$ as a function of τ_{AB} and ρ_A.

(ILS being less likely to occur on longer edges of S). Notably, even for high rates of recombination, R^* under both sequence distance mode and maximum likelihood mode always correctly inferred the topology of S when $\tau_{AB} > 0.1$ coalescent units.

In our second experiment, we relaxed the assumption that recombination occurs only in population A by allowing for recombination in population B as well. For this simulation, $\tau_{AB} = 0.01$ and $\theta = 0.01$, with inference performed using R^* with sequence distances. Figure 7 shows the uniquely favored rooted triple for each choice of ρ_A and ρ_B, with each estimate obtained from $m = 10^5$ samples. When this experiment was repeated with $\tau_{AB} = 0.1$, all but one parameter regimes resulted in correct inference; the exception was when $\rho_A = 0$ and $\rho_B = 20$, in which case $\hat{t} = AC|B$. These results support the hypothesis that taxa exhibiting higher rates of recombination relative to other taxa are more likely to be inferred as more distantly related, but that the effect is small and manifests only in species triplets with very short internal branches.

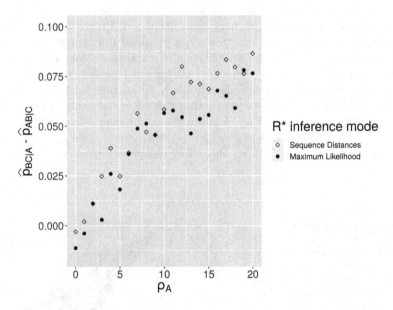

Fig. 6. The effect of increasing ρ_A on inference using R^* with sequence distances and maximum likelihood.

The third experiment tested the effect when all populations in S (excluding the root population ABC) experience recombination at comparable rates. The simulation parameters were $\rho := \rho_A = \rho_B = \rho_C = \rho_{AB} \in \{0, 1, \ldots, 20\}$ and $\rho_{ABC} = 0$, along with $\theta = 0.1$, $\tau_{AB} = 0.01$, and $m = 10^6$, with inference performed using R^* with sequence distances. The results, shown in Fig. 8, suggest that when recombination rates are similar on the edges of S, greater recombination rates does *not* lead to incorrect inference of rooted triples: in all cases,

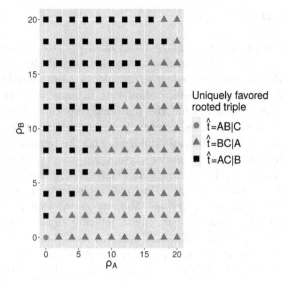

Fig. 7. R^* inference with recombination in both populations A and B.

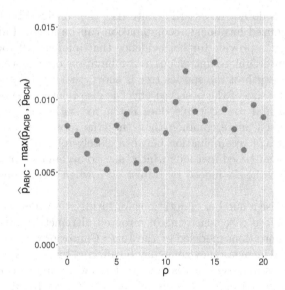

Fig. 8. Equal recombination rates in A, B, C and AB.

$\hat{p}_{AB|C} > \hat{p}_{AC|B} \vee \hat{p}_{BC|A}$, suggesting consistent inference despite the very short internal branch length, a result which agrees with the conclusions of [14] that even high recombination rates are not a significant source of error, at least when rates are comparable across species. Thus, the existence of differential rates of recombination between closely related taxa appears to be a necessary condition for a species tree S to lie in the inconsistency zone.

4   Discussion

The primary focus of this study is the effect of intralocus recombination on the inference of rooted triples. In contrast to previous simulation studies [3,14,29], the current work considers the effect of intralocus recombination on inference of species phylogenies *with recombination rate heterogeneity across taxa*. Our main result is a proof that within the parameter space of species phylogenies there exists a subset—the inconsistency zone—in which phylogenetic conflict between the topology of the species phylogeny and the topology of inferred gene trees is of a sufficient level to render certain majority vote methods statistically inconsistent. We further quantify and characterize this inconsistency zone through simulations, showing that it includes biologically plausible recombination and mutation rates for eukaryotes, and suggesting that it arises on species phylogenies exhibiting both (1) very short internal branch lengths (less than 0.1 coalescent units) and (2) differential rates of recombination between closely related taxa. These results highlight a way in which intralocus recombination can exacerbate ILS and lead to overestimation of the divergence times of those taxa exhibiting disproportionately high intralocus recombination rates relative to other taxa.

These findings do not necessarily contradict the conclusions of [14] that the effect of unrecognized intralocus recombination can be minor. Indeed, our simulation experiments provide further evidence that inference of rooted triples is hampered by unrecognized intralocus recombination only in cases where the internal branch length of the species tree is short, that is in cases where ILS is already high. The size of the observed effect is also relatively small; even when the uniquely favored rooted triple does not agree with the species tree, it is usually only slightly more common than the true rooted triple. Furthermore, if differential rates of recombination between closely-related taxa are rare, then summary coalescent-based methods which take no account of intralocus recombination may nonetheless indeed be robust even when recombination rates are high.

Our results raise a number of questions for future study. Our analysis focused on a simple idealized case consisting of a rooted ultrametric three-taxa species phylogeny with mutations modeled by the Jukes-Cantor process. The nature and significance of the inconsistency zone may be affected by factors such as variable population sizes as well as elements of mutation and recombination rate heterogeneity not considered here. In addition, our theoretical results only consider distance-based gene tree estimation. Extending these results to likelihood-based inference would be of interest.

Acknowledgements. MH was supported by supported by NSF grants DMS-1902892 (to SR) and DMS-2023239 (TRIPODS Phase II). SR was supported by NSF grants DMS-1902892 and DMS-2023239 (TRIPODS Phase II). MH and SR are grateful for the feedback from Cecile Ane and her lab members as well as Claudia Solis-Lemus.

References

1. Arenas, M.: The importance and application of the ancestral recombination graph. Front. Genet. **4**, 206 (2013). https://doi.org/10.3389/fgene.2013.00206
2. Bryant, D., Hahn, M.W.: The concatenation question. In: Scornavacca, C., Delsuc, F., Galtier, N. (eds.) Phylogenetics in the Genomic Era, Chap. 3.4, pp. 3.4:1–3.4:23 (2020). No commercial publisher—Authors open access book. https://hal.inria.fr/PGE
3. Conry, M.: Determining the impact of recombination on phylogenetic inference. Ph.D. thesis, The Florida State University (2020)
4. Dasarathy, G., Mossel, E., Nowak, R., Roch, S.: Coalescent-based species tree estimation: a stochastic Farris transform. arXiv preprint arXiv:1707.04300 (2017)
5. Degnan, J.H.: Anomalous unrooted gene trees. Syst. Biol. **62**(4), 574–590 (2013). https://doi.org/10.1093/sysbio/syt023
6. Degnan, J.H., DeGiorgio, M., Bryant, D., Rosenberg, N.A.: Properties of consensus methods for inferring species trees from gene trees. Syst. Biol. **58**(1), 35–54 (2009). https://doi.org/10.1093/sysbio/syp008
7. Degnan, J.H., Rosenberg, N.A.: Discordance of species trees with their most likely gene trees. PLoS Genet. **2**(5), e68 (2006). https://doi.org/10.1371/journal.pgen.0020068
8. Degnan, J.H., Rosenberg, N.A.: Gene tree discordance, phylogenetic inference and the multispecies coalescent. Trends Ecol. Evol. **24**(6), 332–340 (2009). https://doi.org/10.1016/j.tree.2009.01.009
9. Edwards, S.V., et al.: Implementing and testing the multispecies coalescent model: a valuable paradigm for phylogenomics. Mol. Phylogene. Evol. **94**, 447–462 (2016). https://doi.org/10.1016/j.ympev.2015.10.027
10. Griffiths, R.C., Marjoram, P.: An ancestral recombination graph. In: Donnelly, P., Tavaré, S. (eds.) Progress in Population Genetics and Human Evolution. vol. 87, p. 257. Springer New York (1997)
11. Hahn, M.W.: Molecular Population Genetics. Oxford University Press, Oxford (2018)
12. Jukes, T.H., Cantor, C.R.: Evolution of protein molecules. Mammalian Protein Metab. **3**, 21–132 (1969). https://doi.org/10.1016/B978-1-4832-3211-9.50009-7
13. Kapli, P., Yang, Z., Telford, M.J.: Phylogenetic tree building in the genomic age. Nat. Rev. Gene. **21**(7), 428–444 (2020). https://doi.org/10.1038/s41576-020-0233-0
14. Lanier, H.C., Knowles, L.L.: Is recombination a problem for species-tree analyses? Syst. Biol. **61**(4), 691–701 (2012). https://doi.org/10.1093/sysbio/syr128
15. Larget, B.R., Kotha, S.K., Dewey, C.N., Ané, C.: BUCKy: Gene tree/species tree reconciliation with Bayesian concordance analysis. Bioinformatics **26**(22), 2910–2911 (2010). https://doi.org/10.1093/bioinformatics/btq539
16. Liu, L., Yu, L., Edwards, S.V.: A maximum pseudo-likelihood approach for estimating species trees under the coalescent model. BMC Evol. Biol. **10**(1), 1–18 (2010). https://doi.org/10.1186/1471-2148-10-302
17. Lynch, M.: Evolution of the mutation rate. TRENDS Genet. **26**(8), 345–352 (2010). https://doi.org/10.1016/j.tig.2010.05.003
18. Lynch, M., Marinov, G.K.: The bioenergetic costs of a gene. Proc. Nat. Acad. Sci. **112**(51), 15690–15695 (2015). https://doi.org/10.1073/pnas.1514974112
19. Mendes, F.K., Livera, A.P., Hahn, M.W.: The perils of intralocus recombination for inferences of molecular convergence. Philos. Trans. R. Soc. B **374**(1777), 20180244 (2019). https://doi.org/10.1098/rstb.2018.0244

20. Mirarab, S., Reaz, R., Bayzid, M.S., Zimmermann, T., Swenson, M.S., Warnow, T.: Astral: genome-scale coalescent-based species tree estimation. Bioinformatics **30**(17), i541–i548 (2014). https://doi.org/10.1093/bioinformatics/btu462
21. Rannala, B., Edwards, S.V., Leaché, A., Yang, Z.: The multi-species coalescent model and species tree inference. In: Scornavacca, C., Delsuc, F., Galtier, N. (eds.) Phylogenetics in the Genomic Era, Chap. 3.3, pp. 3.3:1–3.3:21 (2020). No commercial publisher—Authors open access book. https://hal.inria.fr/PGE
22. Rannala, B., Yang, Z.: Bayes estimation of species divergence times and ancestral population sizes using DNA sequences from multiple loci. Genetics **164**(4), 1645–1656 (2003). https://doi.org/10.1093/genetics/164.4.1645
23. Schrempf, D., Szöllősi, G.: The sources of phylogenetic conflicts. In: Scornavacca, C., Delsuc, F., Galtier, N. (eds.) Phylogenetics in the Genomic Era, Chap. 3.1, pp. 3.1:1–3.1:23 (2020). No commercial publisher—Authors open access book. https://hal.inria.fr/PGE
24. Semple, C., Steel, M., et al.: Phylogenetics, vol. 24. Oxford University Press on Demand, London(2003)
25. Springer, M.S., Gatesy, J.: The gene tree delusion. Mol. Phylogene. Evol. **94**, 1–33 (2016). https://doi.org/10.1016/j.ympev.2015.07.018
26. Springer, M.S., Gatesy, J.: Delimiting coalescence genes (c-genes) in phylogenomic data sets. Genes **9**(3), 123 (2018). https://doi.org/10.3390/genes9030123
27. Steel, M.: Phylogeny: Discrete and Random Processes in Evolution. SIAM (2016)
28. Wang, K.C.: Phylogenetic reconstruction accuracy in the face of heterogeneity, recombination, and reticulate evolution. The University of Wisconsin-Madison (2017)
29. Wang, Z., Liu, K.J.: A performance study of the impact of recombination on species tree analysis. BMC genomics **17**(10), 165–174 (2016). https://doi.org/10.1186/s12864-016-3104-5
30. Warnow, T.: Computational Phylogenetics: An Introduction to Designing Methods for Phylogeny Estimation. Cambridge University Press, New York (2017)
31. Yang, Z.: Complexity of the simplest phylogenetic estimation problem. Proc. R. Soc. Lond. Ser. B Biol. Sci. **267**(1439), 109–116 (2000). https://doi.org/10.1098/rspb.2000.0974
32. Zhu, T., Yang, Z.: Complexity of the simplest species tree problem. Mol. Biol. Evol. **38**(9), 3993–4009 (2021). https://doi.org/10.1093/molbev/msab009

QT-GILD: Quartet Based Gene Tree Imputation Using Deep Learning Improves Phylogenomic Analyses Despite Missing Data

Sazan Mahbub[1,2], Shashata Sawmya[1], Arpita Saha[1], Rezwana Reaz[1], M. Sohel Rahman[1], and Md. Shamsuzzoha Bayzid[1(✉)]

[1] Computer Science and Engineering, Bangladesh University of Engineering and Technology, Dhaka 1205, Bangladesh
shams_bayzid@cse.buet.ac.bd
[2] Computer Science, University of Maryland, College Park, MD 20742, USA

Abstract. Species tree estimation is frequently based on phylogenomic approaches that use multiple genes from throughout the genome. However, for a combination of reasons (ranging from sampling biases to more biological causes, as in gene birth and loss), gene trees are often incomplete, meaning that not all species of interest have a common set of genes. Incomplete gene trees can potentially impact the accuracy of phylogenomic inference. We, for the first time, introduce the problem of imputing the quartet distribution induced by a set of incomplete gene trees, which involves adding the missing quartets back to the quartet distribution. We present QT-GILD, an automated and specially tailored unsupervised deep learning technique, accompanied by cues from natural language processing (NLP), which learns the quartet distribution in a given set of incomplete gene trees and generates a complete set of quartets accordingly. QT-GILD is a general-purpose technique needing no explicit modeling of the subject system or reasons for missing data or gene tree heterogeneity. Experimental studies on a collection of simulated and empirical data sets suggest that QT-GILD can effectively impute the quartet distribution, which results in a dramatic improvement in the species tree accuracy. Remarkably, QT-GILD not only imputes the missing quartets but it can also account for gene tree estimation error. Therefore, QT-GILD advances the state-of-the-art in species tree estimation from gene trees in the face of missing data. QT-GILD is freely available in open source form at https://github.com/pythonLoader/QT-GILD.

Keywords: Phylogenomic analysis · Species tree · Gene tree · Gene tree discordance · Incomplete lineage sorting · Missing data · Deep learning

S. Mahbub and S. Sawmya—These authors contributed equally to this work.
The preprint version of the paper (available at https://doi.org/10.1101/2021.11.03.467204) contains the supplementary material.

I. Pe'er (Ed.): RECOMB 2022, LNBI 13278, pp. 159–176, 2022.
https://doi.org/10.1007/978-3-031-04749-7_10

1   Introduction

High-throughput DNA sequencing is generating new genome-wide data sets for phylogenetic analyses, potentially including hundreds or even thousands of loci. The estimation of species trees from multiple genes is necessary since true gene trees can differ from each other and from the true species tree due to various processes, including gene duplication and loss, horizontal gene transfer, and incomplete lineage sorting (ILS) [32]. ILS (also known as deep coalescence) is considered to be a dominant cause for gene tree heterogeneity, which is best understood under the coalescent model [10,12,21,38,39,45,53,54]. In the presence of gene tree heterogeneity, standard methods for estimating species trees, such as concatenation (which combines sequence alignments from different loci into a single "supermatrix", and then computes a tree on the supermatrix) can be statistically inconsistent [9,44], and produce incorrect trees with high support [24]. Therefore, *summary methods* that operate by combining estimated gene trees and can explicitly take gene tree discordance into account are gaining increasing attention from the systematists, and some of the coalescent based summary methods are statistically consistent [7,22,23,25,29–31,35,37,56]. Other coalescent species tree estimation methods include BEST [28] and *BEAST [19], which co-estimate gene trees and species tree from input sequence alignments. These co-estimation methods can produce substantially more accurate trees than other methods, but are computationally intensive and do not scale up for genome-level analyses [4,5,26,49].

There has been notable progress in constructing large scale phylogenetic trees by analyzing genome-wide data, but substantial challenges remain. Assembling a complete dataset with hundreds of orthologous genes from a large set of species remains a difficult task which has downstream impact on species tree inference [20,51,59]. Gene trees can be incomplete due to various reasons ranging from biological processes as in gene birth and loss [2] to biases in taxon and gene sampling, stochasticity inherent in collecting data across thousands of loci, and difficulty in sequencing and assembling the complete set of taxa of interest (see [6,20,27,51] for more elaborate discussion). Incomplete gene trees may decrease the species tree accuracy [4,40,59] and introduce ambiguity in the tree search [15,46]. Indeed, as we will show in our experimental results, incomplete gene trees substantially reduce the accuracy of the best existing coalescent based summary methods.

We address the problem of missing data in phylogenomic data sets by formulating the *Quartet Distribution Imputation* (QDI) problem, where we seek to add the missing quartets to the quartet distribution induced by a given set of incomplete gene trees. Quartet-based methods have gained substantial interest as quartets (4-leaf unrooted gene trees) do not contain the "anomaly zone" [10,11,13], a condition where the most probable gene tree topology may not be identical to the species tree topology. ASTRAL [35], which is one of the most accurate and widely used coalescent-based methods, seeks to infer a species tree so that the number of induced quartets in the gene trees that are consistent with the species tree is maximized. Another approach is to infer individual quartets (with or without weights), and then amalgamate them into a single coherent species tree [1, 7, 33, 41, 42, 48, 50, 52]. wQFM [33] and wQMC [1] represent the latter category of quartet-based methods.

Existing methods for adding missing taxa to the gene trees use a "reference tree" and attempt to add the missing species in such a way that various distance metrics (e.g., "extra-lineage score", Robinson-Foulds distance) are optimized with respect to the given reference tree [4, 8]. However, obtaining a reasonably accurate reference tree in the presence of missing data is difficult [8], which subsequently affects the tree completion steps as the reference tree can be far from the true tree. In this study, we present QT-GILD (**Q**uartet based **G**ene tree **I**mputation using **D**eep **L**earning) – a novel deep learning based technique which learns the quartet distribution induced by the given set of incomplete gene trees using an especially tailored autoencoder model, and subsequently generates a complete set of quartets. In doing so, it not only imputes the missing quartets, but also attempts to correct the quartets present in the given set of estimated and incomplete gene trees by leveraging the underlying quartet distribution. Therefore, the complete set of quartets generated by QT-GILD is often closer to the set of quartets present in the true gene trees than to those present in the estimated gene trees. Thus, QT-GILD obviates the need for a reference tree as well as accounts for gene tree estimation error. Our experimental results, on a collection of simulated and real biological data sets covering a wide range of model conditions, indeed show that amalgamating the imputed set of quartets generated by QT-GILD can remarkably improve the accuracy of species tree inference.

2   Quartet Imputation Problem

2.1   Problem Definition

Let $\mathcal{G} = \{g_1, g_2, \ldots, g_k\}$ be a set of k gene trees, where each g_i is a tree on taxon set $S_i \subseteq S$ (i.e., any gene tree g_i can be on the full set S of n taxa or can be on a subset S_i of taxa, making the gene tree incomplete). For a set of four taxa $a, b, c, d \in S$, the quartet tree $ab|cd$ denotes the unrooted quartet tree in which the pair a, b is separated from the pair c, d by an edge.

Let $\mathcal{Q}_i$ be the set of quartets in g_i. Therefore, $\mathcal{Q} = \mathcal{Q}_1 \cup \mathcal{Q}_2 \cup \ldots \cup \mathcal{Q}_k$ is the multi-set of quartets present in $\mathcal{G}$. Note that there are $\binom{n_i}{4}$ quartets in $\mathcal{Q}_i$, where $n_i = |S_i|$. When all the trees in $\mathcal{G}$ are complete, $|\mathcal{Q}| = k\binom{n}{4}$. Let $\mathcal{MT}_i = S - S_i$ be the set of missing taxa in g_i. Therefore, any quartet q involving any subset of taxa in $\mathcal{MT}_i$ is a missing quartet in g_i. In the presence of missing taxa in the gene trees, $|\mathcal{Q}| < k\binom{n}{4}$. We now define the quartet imputation problem as follows.

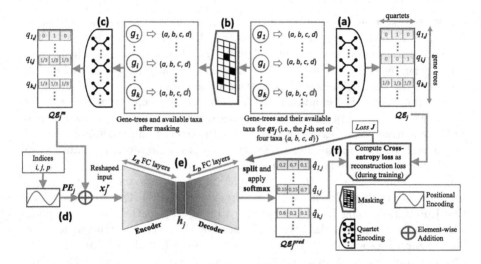

Fig. 1. Overall pipeline of QT-GILD. (a) Generation of *Quartet Encoding* matrix $\mathcal{QE}_j$ for a set of four taxa, qs_j. (b) Masking (in this example, taxon c was masked out from gene tree g_i). (c) Generation of *masked quartet encoding* matrix $\mathcal{QE}_j^m$ from the output of masking. (d) Positional encoding PE_j generation from indices i, j, p, which are the index of a gene tree, index of a set of four taxa, and index of a possible quartet topology, respectively. (e) The reshaped version of the element-wise summation of $\mathcal{QE}_j^m$ and PE_j (i.e., x_j^r) is passed through the autoencoder and the output is split and normalized using softmax, generating $\mathcal{QE}_j^{pred}$. (f) Cross-entropy loss J is computed between $\mathcal{QE}_j^{pred}$ and $\mathcal{QE}_j$ and optimized to update the parameters of the autoencoder (the backward arrow represents backpropagation through the network).

Problem	**Quartet Distribution Imputation (QDI)**
INPUT	A set $\mathcal{G}$ of k gene trees and the multi-set $\mathcal{Q} = \mathcal{Q}_1 \cup \mathcal{Q}_2 \cup \ldots \cup \mathcal{Q}_k$ of quartets induced by $\mathcal{G}$ such that $|\mathcal{Q}| < k\binom{n}{4}$.
OUTPUT	A multi-set $\mathcal{Q}' = \mathcal{Q}_1' \cup \mathcal{Q}_2' \cup \ldots \cup \mathcal{Q}_k'$ of quartets on S, where $\mathcal{Q}_i'$ is a complete set of quartets on S generated for g_i (i.e., for every set of four taxa in S, there is a quartet topology $q_i \in \mathcal{Q}_i'$, and therefore there is no missing quartet in $\mathcal{Q}_i'$).

A natural optimization criteria to solve QDI is to impute the missing quartets in $\mathcal{G}$ in such a way that the total number of consistent quartets in $\mathcal{Q}'$ is maximized (a set of quartets is consistent if they can co-exist in a single tree). In this study, we do not approach QDI as a discrete optimization problem; instead, we propose a machine learning approach to learn the underlying quartet distribution in $\mathcal{G}$ and infer the missing quartets accordingly.

Masking for Self-supervision. We particularly aimed for developing an unsupervised autoencoder based model, but in order to increase the learning capability of our model in a supervised manner, we make it a *self-supervised* model where a portion of the input is used as a supervisory signal to the autoencoder fed with the remaining portion of the input. In this context, we leverage a technique called "masking", which has been used in several state-of-the-art language modeling approaches, where we mask out a certain percentage of non-missing quartets in a quartet encoding matrix $\mathcal{QE}_j$, and ask the autoencoder to predict the masked quartets.

We mask out randomly selected $\lambda|S|$ ($0 \leq \lambda < 1$) non-missing quartets by replacing the one-hot encoded quartet vectors corresponding to the randomly selected non-missing quartets by $(\frac{1}{3}, \frac{1}{3}, \frac{1}{3})$. In this study, we set $\lambda = 0.1$. Thus, we create a masked quartet encoding matrix $\mathcal{QE}_j^m$ from the original matrix $\mathcal{QE}_j$. During the training, the autoencoder takes $\mathcal{QE}_j^m$ as input and tries to predict the masked quartet vectors. The corresponding cross entropy loss J as the reconstruction loss (error) of predicting the masked entries with respect to the original quartet encoding matrix $\mathcal{QE}_j$ is computed and used as a feedback. Thus, the autoencoder is trained iteratively by backpropagating the loss J within a feedback-loop, transforming a fully unsupervised approach into a self-supervised autoencoder and thereby helping the model to effectively learn the quartet distribution.

Positional Encoding. Masked quartet encoding matrices $\mathcal{QE}_j^m$, $1 \leq j \leq \binom{n}{4}$ defined on $\binom{n}{4}$ subsets of four taxa $qs_1, qs_2, \ldots, qs_{\binom{n}{4}}$ encode the information regarding three possible quartets on different four taxa set across all the gene trees in $\mathcal{G}$. However, $\mathcal{QE}_j^m$ does not encode any information about qs_j, making it impossible for the autoencoder to identify which set of four taxa is represented by a particular $\mathcal{QE}_j^m$. In order to inject some information about different sets of four taxa, we incorporate an extra non-learnable signal, *positional-encoding* (PE) [57]. PE was originally proposed for NLP by Vaswani *et al.* [57] and is widely used to encode positional information of the tokens in sequence data. Recently, it has also been successfully applied to computational biology [55]. For the j-th four-taxa-set qs_j, we generate a $k \times 3$ positional-encoding matrix $PE_j \in \mathbb{R}^{k \times 3}$ according to Eq. 1.

$$PE_j(i,p) = \begin{cases} sin(j/10000^{i/k}), & \text{if i is even} \\ cos(j/10000^{(i-1)/k}) & \text{if i is odd} \end{cases} \tag{1}$$

Here, $PE_j(i,p)$ represents the positional-encoding of the p-th possible quartet ($1 \leq p \leq 3$) on qs_j for the i-th gene tree $g_i \in \mathcal{G}$ and $|\mathcal{G}| = k$. Note that this function is constant with respect to p so that we get the same positional-encoding for three possible quartets on qs_j for a particular gene tree g_i. The output of positional encoding PE_j is element-wise added with $\mathcal{QE}_j^m$, resulting in new representations $X_j \in \mathbb{R}^{k \times 3}$ which contain not only the information about three possible quartets across all the gene trees, but also the information about different sets of four-taxa qs_j, $1 \leq j \leq \binom{n}{4}$ (see Eq. 2 and Fig. 1(d)).

$$X_j = PE_j + \mathcal{QE}_j^m \tag{2}$$

Deep Autoencoder. Autoencoder (AE) [17] is a type of artificial neural network that learns to copy its input to its output. The deep autoencoder architecture consists of two modules – encoder and decoder – stacked one after the other and built using fully-connected layers (Fig. 1(e)). The output X_j of the element-wise summation of $\mathcal{QE}_j^m$ and PE_j (as discussed in Sect. 2.1) is reshaped into a vector $x_j^r \in \mathbb{R}^{k*3}$, which is used as the input of the encoder module. By using subsequent L_E fully-connected layers, the dimension of x_j^r is reduced and a condensed latent representation h_j is generated, which is subsequently used as the input of the decoder module. In the decoder, the dimension of h_j is expanded by L_D fully-connected layers (here, L_E and L_D are two hyperparameters). The output of the L_D-th layer is divided into k equal segments of length 3 and softmax normalization is applied on each of them. Here, softmax on the i-th segment generates the predicted quartet-vector $\hat{q}_{i,j}$, which is the estimated probability distribution over the possible quartets on qs_j for the gene-tree g_i. Thus, $\forall g_i \in \mathcal{G}$, we get $\mathcal{QE}_j^{pred} = [\hat{q}_{1,j}, \hat{q}_{2,j}, \dots, \hat{q}_{k,j}]$, which is the predicted quartet-encoding matrix corresponding to qs_j.

Overall Pipeline of QT-GILD. Figure 1 shows the overall end-to-end pipeline of QT-GILD comprising the individual components described in Sections 2.2–2.5. For each set of four taxa qs_j, we first create a quartet encoding matrix $\mathcal{QE}_j$, which is subsequently masked to produce $\mathcal{QE}_j^m$. Next, these masked encoding matrices are added to positional encoding matrices and reshaped to generate the input x_i^r of the autoencoder. The autoencoder network produces the predicted quartet-encoding matrix $\mathcal{QE}_j^{pred}$. During the training phase, considering $\mathcal{QE}_j^{pred}$ as the prediction and $\mathcal{QE}_j$ as the ground truth, we compute the cross-entropy loss J as the reconstruction loss for the non-missing quartets, i.e., for the quartet-vectors $\{q_{i,j} \mid q_{i,j} \in \mathcal{QE}_j \text{ and } q_{i,j} \neq (\frac{1}{3}, \frac{1}{3}, \frac{1}{3})\}$. The autoencoder is trained by backpropagating this loss in a self-supervised fashion (see Fig. 1(f) and Sect. 2.1). This training with backpropagation is run for a predefined number of epochs. Finally, it is run once without the loss function to produce the final encoding matrix $\mathcal{QE}_j^{pred}$. Next, we generate the set of imputed quartets $\mathcal{Q}_i'$ – representing the quartets in the i-th gene-tree g_i – where the j-th quartet $\mathcal{Q}_i'(j)$ is computed according to Eq. 3. This imputed set of quartets is subsequently used by quartet amalgamation techniques, such as wQFM and wQMC.

$$\mathcal{Q}_i'(j) = \begin{cases} \text{quartet corresponding to } Argmax(\mathcal{QE}_i(j)), & \text{if } \mathcal{QE}_i(j) \neq (\frac{1}{3}, \frac{1}{3}, \frac{1}{3}) \\ \text{quartet corresponding to } Argmax(\mathcal{QE}_i^{pred}(j)), & \text{otherwise} \end{cases}$$
$$(3)$$

3   Experimental Study

3.1   Datasets

We used previously studied simulated and biological datasets to evaluate the performance of QT-GILD. We studied three collections of simulated datasets:

one based on a biological dataset (37-taxon mammalian dataset) and a 15-taxon dataset that were generated in some prior studies [3,34]. These datasets consist of gene sequence alignments generated under a multi-stage simulation process that begins with a species tree, simulates gene trees down the species tree under the multi-species coalescent model (and so can differ topologically from the species tree), and then simulates gene sequence alignments down the gene trees. These datasets vary from moderately low to extremely high levels of ILS, and also vary in terms of number of genes and gene-tree estimation errors (controlled by sequence lengths). Thus, the simulated datasets provide a range of conditions in which we explored the performance of QT-GILD and investigate the impact of quartet distribution imputation in species tree inference. Table S1 in the Supplementary Material (available at https://doi.org/10.1101/2021.11.03. 467204) presents a summary of these datasets. We also evaluated QT-GILD on a challenging biological dataset comprising 42 angiosperms from Xi *et al.* [60].

3.2   Generating Incomplete Gene Trees

We deleted taxa randomly, varying the number of taxa removed from each gene tree, thus producing incomplete gene trees. Instead of a fixed number of taxa, we removed different ranges of taxa. For a particular gene tree g_i in the input set $\mathcal{G}$ of gene trees and for a particular range x–y of missing taxa, we randomly select an integer $mt \in [x, y]$, and randomly select and delete mt taxa from g_i. For example, for a range of 3–4 missing taxa, we remove 3 or 4 (selected randomly) from the gene trees in $\mathcal{G}$. We varied the number of missing taxa (2–40%) for different datasets – creating model conditions with 13%-80% missing quartets. In all the datasets analyzed in this study, this random taxa deletion protocol did not remove any particular taxa from all the gene trees, i.e., each taxa remained present in at least one incomplete gene tree.

3.3   Species Tree Estimation Methods

We used wQFM [33], a highly accurate weighted quartet amalgamation technique, to estimate species trees from weighted quartets. We also used wQMC, which is another well known weighted quartet amalgamation technique, in order to show the usability of the imputed quartets generated by QT-GILD across various quartet amalgamation techniques as well as to show that the improvements in species tree inference resulting from amalgamating imputed quartets is not due to wQFM, rather mostly due to the effective imputation of the quartet distribution by QT-GILD. We ran wQFM and wQMC using the embedded quartets in the gene trees with weights reflecting the frequencies of the quartets (i.e., number of gene trees that induce a particular quartet). We compared wQFM with ASTRAL-III [35,61] (version 5.7.3), which is one of the most accurate and widely used quartet amalgamation techniques. These methods were evaluated on both complete and incomplete gene trees, showing the impact of missing data. wQFM and and wQMC were also run on imputed set of quartets, generated by QT-GILD, to demonstrate the impact of quartet imputation on species tree estimation. Therefore, we evaluated the following variants of different methods.

- *ASTRAL-complete*: ASTRAL, when run on a given set of complete gene trees.
- *ASTRAL-incomplete*: ASTRAL, run on a given set of incomplete gene trees.
- *wQFM-complete*: wQFM, run on the set of weighted quartets induced by a given set of complete gene trees.
- *wQFM-incomplete*: wQFM, run on the set of weighted quartets induced by a given set of incomplete gene trees.
- *wQFM-imputed*: wQFM, run on the imputed set of weighted quartets generated by QT-GILD from a given set of incomplete gene trees.
- wQMC-complete, wQMC-incomplete and wQMC-imputed: defined similarly as wQFM-complete, wQFM-incomplete and wQFM-imputed.

Note that ASTRAL cannot take a set of quartets as input and as such we cannot evaluate ASTRAL on imputed quartet distributions. For brevity, we denote by *complete quartet distributions* and *incomplete quartet distributions* the weighted quartet distributions induced by complete and incomplete estimated gene trees, respectively. Because wQFM generally produces better trees than wQMC [33] and to keep the figures and relevant discussion readable and easy to follow, the results for ASTRAL and wQFM are presented here, while the results for wQMC are presented in the Supplementary Material of the preprint version of the paper available at https://doi.org/10.1101/2021.11.03.467204.

3.4   Measurements

We compared the estimated trees (on simulated datasets) with the model species tree using normalized Robinson-Foulds (RF) distance [43] to measure the tree error. The RF distance between two trees is the sum of the bipartitions (splits) induced by one tree but not by the other, and vice versa. We also compared the quartet scores (the number of quartets in the gene trees that agree with a species tree) of the trees estimated by different methods. All the trees estimated in this study are binary and so False Positive (FP), and False Negative (FN) and RF rates are identical. For the biological dataset, we compared the estimated species trees to the scientific literature. We analyzed multiple replicates of data for various model conditions and performed two-sided Wilcoxon signed-rank test (with $\alpha = 0.05$) to measure the statistical significance of the differences between two methods. We assessed the quality of the quartet distributions, induced by complete, incomplete, and imputed gene tree distributions, by comparing them with the true quartet distribution (i.e., the quartets induced by true gene trees).

4   Results and Discussion

4.1   Results on 15-Taxon Dataset

The average RF rates of different variants of ASTRAL and wQFM on various model conditions in 15-taxon dataset are shown in Fig. 2. We have investigated the performance on varying gene tree estimation errors using 100bp and 1000bp sequence lengths and on varying numbers of genes (100 and 1000). In general, wQFM is more accurate than ASTRAL both on complete and incomplete gene trees. For all levels of missing data, certain trends were clearly seen. As expected, tree accuracy of ASTRAL and wQFM deteriorated in the presence of missing taxa, and the RF rates increased with increasing levels of missing data. However, tree accuracy of ASTRAL is more impacted by increasing levels of missing data than that of wQFM as wQFM-incomplete was notably better than ASTRAL-incomplete on most of the model conditions (especially, see the results on 100gene-1000bp and 1000gene-100bp model conditions) and in many cases, the differences are statistically significant ($p \ll 0.05$). ASTRAL and wQFM improves in accuracy as the number of genes increased (from 100 to 1000) and often achieved competitive tree accuracy (compared to the accuracy of the trees estimated on complete gene trees) even when the amount of missing data was large. This is mostly due to the fact that a large number of gene trees provide more unique bipartitions (and hence more quartets) than a relatively smaller number of genes. Similar trends were observed in case of wQMC (see Sect. 5 in Supplementary Material). wQMC is, in general, less accurate than wQFM both on complete and incomplete gene trees. Notably, WQMC-incomplete is slightly better than ASTRAL-incomplete in some model conditions (e.g., 100gene–100bp and 1000gene–100bp model conditions with 4–5 and 5–6 missing taxa in Supplementary Fig. S3).

The most important and interesting results were observed on imputed sets of quartets. wQFM-imputed substantially outperformed not only wQFM-incomplete and ASTRAL-incomplete but also wQFM-complete and ASTRAL-complete, showing the efficacy of QT-GILD in imputing quartet distributions. The improvements are remarkable as in most of the model conditions, wQFM-imputed returned the true species tree (RF-rate = 0), whereas ASTRAL-incomplete and wQFM-incomplete incurred as high as ~45% errors (note the higher taxa removal rates on 100gene-100bp model condition). This dramatic improvement was also observed in case of wQMC-imputed. Even though wQMC-imputed is not as good as wQFM-imputed trees in some cases (see 100gene and 1000gene-1000bp model conditions with smaller numbers of missing taxa in Fig. S3 in the Supplementary Material), wQMC-imputed trees returned the true species trees in most of the cases and consistently outperformed ASTRAL-incomplete as well as ASTRAL-complete. Note that, even with a reasonably well imputed set of quartets, the expected trend is for the accuracy of the tree estimated on an imputed set of quartets to be higher than that of the tree estimated on the corresponding incomplete quartet distribution but lower than or as good as the tree accuracy obtained on the complete quartet distribution. But, surprisingly, wQFM-imputed and wQMC-imputed improves upon wQFM-complete and wQMC-complete (as well as ASTRAL-complete) on this particular dataset. These results suggest, and as we will show in the following, that the imputed quartet distributions are closer to true quartet distributions (i.e., the quartet distribution induced by true gene trees) than the complete and incomplete quartet distributions are to true distributions. Note that, on the 100gene–100bp model condition, the RF rates of the trees estimated on imputed quartets with 1-2 missing taxa were higher than those with larger amounts of missing data. We believe this is because there are only 18272.8 missing quartets with 1-2 missing taxa, and thus QT-GILD could not account for gene tree estimation error (while imputing the missing quartets) as much as it did in other model conditions with larger numbers of missing quartets.

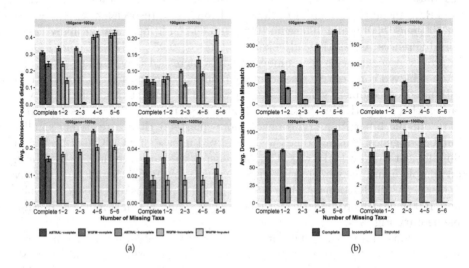

(a)   (b)

Fig. 2. Results on 15-taxon dataset. (a) Comparison of different variants of ASTRAL and wQFM on 15-taxon dataset. We show the average RF rates with standard errors over 10 replicates. For each model condition, we varied the taxa removal rate from ~6% to ~40%, resulting in 13–80% missing quartets. (b) Average numbers of dominant quartets (out of $\binom{15}{4}$) in different quartets distributions that differ from the true dominant quartets.

We performed a series of experiments to show the impact and quality of the quartet distributions produced by QT-GILD. First, we measure the divergence between true quartet distributions and different sets of quartet distributions in estimated gene trees (e.g., complete, incomplete and imputed) in terms of the number of "dominant" quartets that differ between two quartet distributions. For a set of four taxa a, b, c, d, the dominant quartet (out of three possible quartets $ab|cd$, $ac|bd$, and $bc|ad$) is defined to be the quartet with the highest weight (i.e., the most frequent quartet topology). A dominant quartet topology is the statistically consistent estimate for the true evolutionary history among a group of four taxa since there are no *anomalous* 4-leaf unrooted gene trees [11,13]. For a set of n taxa, there are $\binom{n}{4}$ different four-taxa sets, and thus $\binom{n}{4}$ dominant quartets. In Fig. 2(b), we show the numbers of dominant quartets (out of $\binom{15}{4}$) in complete, incomplete and imputed set of quartets that differ from the dominant quartets in true gene trees. Incomplete quartet distributions had the highest numbers of mismatches, followed by complete quartet distribution and the imputed sets of quartets incurred the lowest amount of mismatch (in most cases, the numbers of mismatches are ∼0), explaining why the RF rates of the trees estimated from imputed quartets are ∼0. The numbers of mismatches in incomplete quartet distributions increased as we increase the amount of missing data. In contrast, the numbers of mismatch in imputed quartet distributions decreased with increasing amounts of missing data (especially on model conditions with higher amounts of gene tree estimation error, i.e., the 100-bp model condition). This is because, in the presence of higher numbers of missing quartets, QT-GILD had the opportunity to impute more quartets and in doing so it accounted for larger amounts of gene tree estimation error. This explains the seemingly counter-intuitive trend that RF rates of wQFM-imputed decreased with increasing amounts of missing data. Next, we measured the divergence between the quartet distribution of estimated gene trees and the quartet distribution of true gene trees using Jensen-Shannon divergence [16]. We represent the gene tree distribution by the frequency of each of the three possible alternative topologies for all the $\binom{n}{4}$ quartets of taxa. Jensen-Shannon divergence of complete, incomplete and imputed quartet distributions from true quartet distributions are shown in Fig. S2 in Supplementary Material. The fact that the difference (both in terms of numbers of mismatch in dominant quartet topologies and Jensen-Shannon distance) between true and complete quartet distributions is higher than the difference between true and imputed quartet distributions suggests that QT-GILD not only imputes the missing quartets, but also accounts for gene tree estimation error. Next, in order to show the efficacy of QT-GILD in imputing missing quartets, we report the proportion of missing quartets that were correctly imputed, with respect to both estimated and true gene trees, by QT-GILD (see Table S2 in the Supplementary Material). QT-GILD was able to correctly impute around 51–58% and 51–63% of missing quartets with respect to estimated and true gene trees, respectively.

Finally, we computed the quartet scores of different species trees estimated by different methods with respect to both estimated and true gene trees (see Supplementary Tables S3 and S4). Interestingly, on estimated gene trees, although ASTRAL obtained higher quartet scores than wQFM-incomplete, and wQFM-imputed in most of the model conditions, the scores of wQFM-incomplete and wQFM-imputed are closer to the true quartet score than the scores of ASTRAL-estimated trees are to the true score. This is mostly a result of the presence of gene tree estimation error. The statistical consistency of estimating species trees by maximizing the quartet score criterion holds when we have a sufficiently large number of true gene trees (with no estimation error). Unfortunately, in practice, the number of genes is limited and the estimated gene trees are not error-free. Therefore, the optimal tree with respect to the quartet score may not be the true species tree. As such, quartet-based methods may "overshoot" the quartet score as they return trees with higher quartet scores than the true quartet score, especially when we have a limited number of estimated gene trees (with estimation error) [15]. The improved performance of wQFM and the efficacy

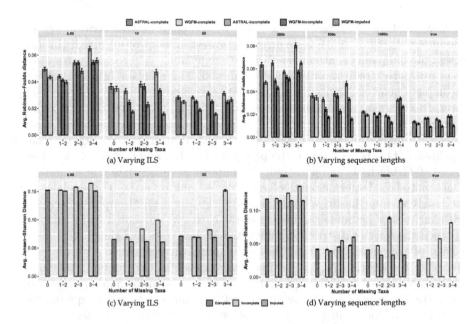

Fig. 3. Results on 37-taxon dataset. (a)–(b) We show the average RF rates with standard errors over 20 replicates. (a) The level of ILS was varied from 0.5X (highest) to 2X (lowest) amount, keeping the sequence length fixed at 500 bp and the number of genes at 200. (b) The sequence length was varied from 250 bp to 1500 bp, keeping the number of genes fixed at 200, and ILS at 1X (moderate ILS). We also analyzed the true gene trees. (c)–(d) Jansen-Shannon divergence between true quartet distributions and complete, incomplete and imputed quartet distributions.

of QT-GILD as an imputation method are even more evident from the quartet scores when they are computed with respect to the true gene trees (i.e., no estimation error). Across all the model conditions, wQFM-imputed obtained the highest quartet scores (w.r.t true gene trees), followed by wQFM-complete and ASTRAL-complete (Table S4 in Supplementary Material). The lowest scores were obtained when the gene trees are incomplete. Notably, wQFM-incomplete consistently achieved higher quartet scores than ASTRAL-incomplete. Note that the quartet consistency score is a statistical consistent measure meaning that the higher the quartet scores, the more accurate the species trees will be (given a sufficiently large number of true gene trees). Therefore, these quartet scores with respect to true gene trees support the trends observed in RF rates (see Fig. 2).

4.2   Results on 37-Taxon Mammalian Simulated Dataset

In the simulated mammalian data, we explored the impact of varying numbers of genes (25–800), varying amounts of gene tree estimation error (i.e., the amount of phylogenetic signal by varying the sequence length for the markers: 250 bp–1500 bp). We also investigated three levels of ILS (shorter branches increases ILS) by multiplying or dividing all internal branch lengths in the model species tree by two – producing three model conditions that are referred to as 1X (moderate ILS), 0.5X (high ILS) and 2X (low ILS). For each model condition, we varied the taxa removal rate from ~2% to ~10%, resulting in 6–25% missing quartets. Because the number of quartets grows exponentially with the number of taxa, even with 25% missing quartets, the 37-taxon dataset contained a substantial number (~7,00,000–~33,00,000) of missing quartets.

Figure 3(a) shows the RF rates of various methods on varying ILS levels (0.5X, 1X, 2X) with 200 genes and a fixed sequence length (500 bp). As expected, species tree error rates of various methods increased as ILS level increased. Similar to 15-taxon dataset, wQFM is better than ASTRAL (both on complete and incomplete datasets), and wQFM-imputed is consistently better than wQFM-incomplete. Remarkably, for moderate (1X) to low ILS (2X) datasets, wQFM-imputed is substantially better than ASTRAL-complete and wQFM-complete (except for 3-4 taxa removal rate on 2X model condition). On the high-ILS dataset (0.5X), wQFM-imputed is better than wQFM-incomplete and ASTRAL-incomplete. Moreover, for small amount of missing taxa (1-2), wQFM-imputed is even better than ASTRAL-complete and wQFM-complete. In some cases with small numbers of missing taxa, wQFM-incomplete is better than wQFM-complete (albeit the differences are small and not statistically significant). The improved performance resulting from the imputed set of quartets was also observed for wQMC. wQMC-imputed was as good as or better than wQMC-incomplete (see Fig. S4 in the Supplementary Material).

RF rates on varying gene tree estimation error (controlled by sequence length; 500–1500 bp) are shown in Figs. 3(b). All methods showed improved accuracy with increasing sequence lengths, and best results were obtained on true gene trees. These results clearly show that wQFM-imputed is better than wQFM-incomplete and ASTRAL-complete and, in many cases, it is even better than wQFM-complete and ASTRAL-complete. Similar to 15-taxon dataset, we investigated the divergence of various quartet distributions from true quartet distributions (see Fig. 3(c)–(d)) and the quartet scores of different estimated species trees with respect to both estimated and true gene trees (see Sect. 4.2 in Supplementary Material) and observed similar trends, supporting our claim that QT-GILD effectively imputes missing data and accounts for gene tree estimation error while imputing the missing quartets.

4.3   Results on Biological Dataset

Angiosperm Dataset. We re-analyzed the angriosperm (the most diverse plant clade) dataset from [60] containing 310 genes sampled from 42 angiosperms and 4 outgroups (three gymnosperms and one lycophyte). The goal is to resolve the position of *Amborella trichopoda Baill*. This dataset has a high level of missing data containing gene trees with as low as 16 taxa. On average, there are 12.9 missing taxa in the gene trees, resulting in a large number (30, 129, 163) of missing quartets. These missing quartets were imputed by QT-GILD. We estimated trees using ASTRAL and wQFM (with and without imputation). These three

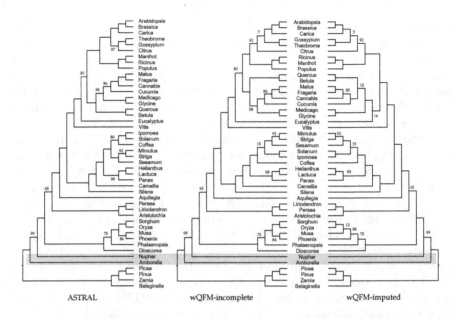

Fig. 4. Analyses of the angiosperm dataset using ASTRAL and wQFM (with and without imputation). Branch supports (BS) represent quartet based local posterior probability [47] (multiplied by 100). All BS values are 100% except where noted.

trees (Fig. 4) are highly congruent and placed *Amborella* as sister to water lilies (i.e., Nymphaeales) and rest of the angiosperms with high support. This relationship is consistent to the tree estimated by concatenation using maximum likelihood (reported in [60] as well as other molecular studies [33,36,58,62]). However, alternative relationship (e.g., the placement of *Amborella* plus water lilies as sister to all other angiosperms) have also been reported [14,18,60]. wQFM-incomplete and wQFM-imputed recovered highly similar trees, differing only on a few branches with low support (e.g., the relative position of Fabales (*Glycine, Medicago*) and Fagales (*Betula, Quercus*) and the relationships within the clade containing *Musa, Phoenix, Oryza,* and *Sorghum*). The fact that there were a large proportion (71.33%) of missing quartets and QT-GILD imputes them by leveraging remaining 28.66% quartets, and yet the imputed distribution resulted in a meaningful angiosperm tree shows the efficacy of QT-GILD.

4.4   Running Time

The running time of QT-GILD mostly depends on the number of taxa, number of genes and the predefined number of epochs. We ran QT-GILD for 2000 epochs in our study. All analyses were run on the same machine with Intel Core i7-10700K CPU (16 cores), 64GB RAM, NVIDIA GeForce RTX 3070 GPU (8GB memory). For a single epoch, it takes $25 \times 10^{-3} - 800 \times 10^{-3}$ s on the 15-taxon dataset with varying numbers of genes. For 37-taxon datasets, it takes 1–6 s and, on the 46-taxon angiosperm dataset with 310 genes, QT-GILD takes 10–11 s to run (see Table S9 in Supplementary Material). The running time of QT-GILD is more sensitive to the number of taxa, as the size of quartet distributions grows exponentially with the number of taxa.

5 Conclusions

This study introduces the quartet distribution imputation problem and shows the power and feasibility of applying deep learning techniques for imputing quartet distributions. Our proposed QT-GILD not only imputes the missing quartets but it can also "correct" the estimated quartets – resulting in a set of quartets that is often closer to the quartet distribution of true gene trees than that of the estimated gene trees, and thereby accounting for gene tree estimation error. Experimental studies using both simulated and real biological dataset suggest that QT-GILD may result in dramatic improvements in species tree accuracy. Therefore, the idea of estimating species trees by imputing quartet distributions has merit and should be pursued and used in future phylogenomic studies. As an immediate extension of the current study we plan to evaluate QT-GILD on a diverse set of real biological datasets as the pattern of missing data is sufficiently complex and heterogeneous across various datasets. Automatic selection of various hyper parameters of this architecture is another important research avenue. Because QT-GILD tries to learn the overall quartet distribution guided by a self-supervised feedback loop and correct for gene tree estimation error, investigating its application beyond imputing incomplete gene trees where the goal would be to improve estimated gene tree distributions is another interesting direction to take. In this regard, we can mask a reasonably large number of quartets in the estimated complete quartet distribution and then impute them with QT-GILD, utilizing its error correcting utility. QT-GILD is currently not scalable to large datasets (both in terms of the number of taxa and genes), therefore this study was limited to small to moderate size datasets. Future study needs to develop a scalable variant of QT-GILD and assess its performance on large datasets.

References

1. Avni, E., Cohen, R., Snir, S.: Weighted quartets phylogenetics. Syst. Biol. **64**(2), 233–242 (2015)
2. Bayzid, M.S., Warnow, T.: Gene tree parsimony for incomplete gene trees: addressing true biological loss. Algor. Molecul. Biol. **13**, 1 (2018)
3. Bayzid, M.S., Mirarab, S., Boussau, B., Warnow, T.: Weighted statistical binning: enabling statistically consistent genome-scale phylogenetic analyses. PLoS ONE **10**(6) (2015)
4. Bayzid, M.S., Warnow, T.: Estimating optimal species trees from incomplete gene trees under deep coalescence. J. Comput. Biol. **19**(6), 591–605 (2012)
5. Bayzid, M.S., Warnow, T.: Naive binning improves phylogenomic analyses. Bioinformatics **29**(18), 2277–2284 (2013)

6. Burleigh, J.G., Hilu, K.W., Soltis, D.E.: Inferring phylogenies with incomplete data sets: a 5-gene, 567-taxon analysis of angiosperms. BMC Evol. Biol. **9**(1), 1–11 (2009)
7. Chifman, J., Kubatko, L.: Quartet from SNP data under the coalescent model. Bioinformatics **30**(23), 3317–3324 (2014)
8. Christensen, S., Molloy, E.K., Vachaspati, P., Warnow, T.: Octal: Optimal completion of gene trees in polynomial time. Algor. Molecul. Biol. **13**(1), 1–18 (2018)
9. Degnan, J.H., DeGiorgio, M., Bryant, D., Rosenberg, N.A.: Properties of consensus methods for inferring species trees from gene trees. Syst. Biol. **58**, 35–54 (2009)
10. Degnan, J.H., Rosenberg, N.A.: Discordance of species trees with their most likely gene trees. PLoS Genet. **2**, 762–768 (2006)
11. Degnan, J.H., Rosenberg, N.A.: Gene tree discordance, phylogenetic inference and the multispecies coalescent. Trends Ecol. Evol. **26**(6) (2009)
12. Degnan, J.H., Salter, L.A.: Gene tree distributions under the coalescent process. Evolution **59**(1), 24–37 (2005). http://view.ncbi.nlm.nih.gov/pubmed/15792224
13. Degnan, J.H.: Anomalous unrooted gene trees. Syst. Biol. **62**(4), 574–590 (2013)
14. Drew, B.T., et al.: Another look at the root of the angiosperms reveals a familiar tale. Syst. Biol. **63**(3), 368–382 (2014)
15. Farah, I.T., Islam, M., Zinat, K.T., Rahman, A.H., Bayzid, S.: Species tree estimation from gene trees by minimizing deep coalescence and maximizing quartet consistency: a comparative study and the presence of pseudo species tree terraces. System. Biol. **70**(6), 1213–1231 (04 2021). https://doi.org/10.1093/sysbio/syab026, https://doi.org/10.1093/sysbio/syab026
16. Fuglede, B., Topsoe, F.: Jensen-shannon divergence and hilbert space embedding. In: International Symposium on Information Theory, 2004. ISIT 2004. Proceedings, p. 31. IEEE (2004)
17. Goodfellow, I., Bengio, Y., Courville, A.: Deep Learning. MIT Press (2016)
18. Goremykin, V.V., et al.: The evolutionary root of flowering plants. Syst. Biol. **62**(1), 50–61 (2013)
19. Heled, J., Drummond, A.J.: Bayesian inference of species trees from multilocus data. Mol. Biol. Evol. **27**, 570–580 (2010)
20. Hosner, P.A., Faircloth, B.C., Glenn, T.C., Braun, E.L., Kimball, R.T.: Avoiding missing data biases in phylogenomic inference: an empirical study in the landfowl (aves: Galliformes). Mol. Biol. Evol. **33**(4), 1110–1125 (2016)
21. Hudson, R.R.: Testing the constant-rate neutral allele model with protein sequence data. Evolution **37**, 203–217 (1983)
22. Islam, M., Sarker, K., Das, T., Reaz, R., Bayzid, M.S.: Stelar: a statistically consistent coalescent-based species tree estimation method by maximizing triplet consistency. BMC Genom. **21**(1), 1–13 (2020)
23. Kubatko, L.S., Carstens, B.C., Knowles, L.L.: Stem: species tree estimation using maximum likelihood for gene trees under coalescence. Bioinformatics **25**, 971–973 (2009)
24. Kubatko, L.S., Degnan, J.H.: Inconsistency of phylogenetic estimates from concatenated data under coalescence. Syst. Biol. **56**, 17 (2007)
25. Larget, B., Kotha, S.K., Dewey, C.N., Ané, C.: BUCKy: gene tree/species tree reconciliation with the Bayesian concordance analysis. Bioinformatics **26**(22), 2910–2911 (2010)
26. Leaché, A.D., Rannala, B.: The accuracy of species tree estimation under simulation: a comparison of methods. Syst. Biol. **60**(2), 126–137 (2011)

27. Lemmon, A.R., Brown, J.M., Stanger-Hall, K., Lemmon, E.M.: The effect of ambiguous data on phylogenetic estimates obtained by maximum likelihood and bayesian inference. Syst. Biol. **58**(1), 130–145 (2009)
28. Liu, L.: BEST: Bayesian estimation of species trees under the coalescent model. Bioinformatics **24**, 2542–2543 (2008)
29. Liu, L., Yu, L.: Estimating species trees from unrooted gene trees. Syst. Biol. **60**(5), 661–667 (2011). https://doi.org/10.1093/sysbio/syr027
30. Liu, L., Yu, L., Edwards, S.V.: A maximum pseudo-likelihood approach for estimating species trees under the coalescent model. BMC Evol. Biol. **10**, 302 (2010)
31. Liu, L., Yu, L., Pearl, D.K., Edwards, S.V.: Estimating species phylogenies using coalescence times among sequences. Syst. Biol. **58**(5), 468–477 (2009)
32. Maddison, W.P.: Gene trees in species trees. Syst. Biol. **46**, 523–536 (1997)
33. Mahbub, M., Wahab, Z., Reaz, R., Rahman, M.S., Bayzid, M.S.: wQFM: highly accurate genome-scale species tree estimation from weighted quartets. Bioinformatics **37**(21), 3734–3743 (2021)
34. Mirarab, S., Bayzid, M.S., Boussau, B., Warnow, T.: Statistical binning enables an accurate coalescent-based estimation of the avian tree. Science **346**(6215), 1250463 (2014)
35. Mirarab, S., Reaz, R., Bayzid, M.S., Zimmermann, T., Swenson, M.S., Warnow, T.: ASTRAL: genome-scale coalescent-based species tree estimation. Bioinformatics **30**(17), i541–i548 (2014)
36. Mirarab, S., Warnow, T.: Astral-ii: coalescent-based species tree estimation with many hundreds of taxa and thousands of genes. Bioinformatics **31**(12), i44–i52 (2015)
37. Mossel, E., Roch, S.: Incomplete lineage sorting: consistent phylogeny estimation from multiple loci. IEEE/ACM Trans. Comput. Biol. Bioinf. **7**(1), 166–171 (2011)
38. Nei, M.: Stochastic errors in DNA evolution and molecular phylogeny. In: Gershowitz, H., Rucknagel, D.L., Tashian, R.E. (eds.) Evolutionary Perspectives and the New Genetics, pp. 133–147 (1986)
39. Nei, M.: Molecular evolutionary genetics. Columbia University Press, New York (1987)
40. Nute, M., Chou, J., Molloy, E.K., Warnow, T.: The performance of coalescent-based species tree estimation methods under models of missing data. BMC Genom. **19**(5), 1–22 (2018)
41. Ranwez, V., Gascuel, O.: Quartet-based phylogenetic inference: improvements and limits. Mol. Biol. Evol. **18**(6), 1103–1116 (2001)
42. Reaz, R., Bayzid, M.S., Rahman, M.S.: Accurate phylogenetic tree reconstruction from quartets: a heuristic approach. PLoS ONE **9**(8), e104008 (2014)
43. Robinson, D., Foulds, L.: Comparison of phylogenetic trees. Math. Biosci. **53**, 131–147 (1981)
44. Roch, S., Steel, M.: Likelihood-based tree reconstruction on a concatenation of aligned sequence data sets can be statistically inconsistent. Theor. Popul. Biol. **100**, 56–62 (2015)
45. Rosenberg, N.: The probability of topological concordance of gene trees and species trees. Theor. Popul. Biol. **61**(2), 225–247 (2002)
46. Sanderson, M.J., McMahon, M.M., Steel, M.: Terraces in phylogenetic tree space. Science **333**(6041), 448–450 (2011)
47. Sayyari, E., Mirarab, S.: Fast coalescent-based computation of local branch support from quartet frequencies. Mol. Biol. Evol. **33**(7), 1654–1668 (2016)

48. Schmidt, H.A., Strimmer, K., Vingron, M., von Haeseler, A.: Tree-puzzle: maximum likelihood phylogenetic analysis using quartets and parallel computing. Bioinformatics **18**(3), 502–504 (2002)
49. Smith, B.T., Harvey, M.G., Faircloth, B.C., Glenn, T.C., Brumfield, R.T.: Target capture and massively parallel sequencing of ultraconserved elements for comparative studies at shallow evolutionary time scales. Syst. Biol. **63**(1), 83–95 (2013)
50. Snir, S., Rao, S.: Quartets MaxCut: a divide and conquer quartets algorithm. IEEE/ACM Trans. Comput. Biol. Bioinf. **7**(4), 704–718 (2010)
51. Streicher, J.W., Schulte, J.A., Wiens, J.J.: How should genes and taxa be sampled for phylogenomic analyses with missing data? an empirical study in iguanian lizards. Syst. Biol. **65**(1), 128–145 (2016)
52. Strimmer, K., von Haeseler, A.: Quartet puzzling: a quartet maximim-likelihood method for reconstructing tree topologies. Mol. Biol. Evol. **13**(7), 964–969, e104008 (1996)
53. Tajima, F.: Evolutionary relationship of DNA sequences in finite populations. Genetics **105**(2), 437–460 (1983). http://www.genetics.org/cgi/content/abstract/105/2/437
54. Takahata, N.: Gene geneaology in three related populations: consistency probability between gene and population trees. Genetics **122**, 957–966 (1989)
55. Uddin, M.R., Mahbub, S., Rahman, M.S., Bayzid, M.S.: SAINT: self-attention augmented inception-inside-inception network improves protein secondary structure prediction. Bioinformatics **36**(17), 4599–4608 (2020)
56. Vachaspati, P., Warnow, T.: Astrid: accurate species trees from internode distances. BMC Genom. **16**(10), S3, e104008 (2015)
57. Vaswani, A., et al.: Attention is all you need. In: NIPS (2017)
58. Wickett, N.J., et al.: Phylotranscriptomic analysis of the origin and early diversification of land plants. Proc. Natl. Acad. Sci. **111**(45), E4859–E4868 (2014)
59. Xi, Z., Liu, L., Davis, C.C.: The impact of missing data on species tree estimation. Mol. Biol. Evol. **33**(3), 838–860 (2016)
60. Xi, Z., Liu, L., Rest, J.S., Davis, C.C.: Coalescent versus concatenation methods and the placement of amborella as sister to water lilies. Syst. Biol. **63**(6), 919–932 (2014)
61. Zhang, C., Rabiee, M., Sayyari, E., Mirarab, S.: Astral-iii: polynomial time species tree reconstruction from partially resolved gene trees. BMC Bioinf. **19**(6), 153, e104008 (2018)
62. Zhang, N., Zeng, L., Shan, H., Ma, H.: Highly conserved low-copy nuclear genes as effective markers for phylogenetic analyses in angiosperms. New Phytol. **195**(4), 923–937 (2012)

Safety and Completeness in Flow Decompositions for RNA Assembly

Shahbaz Khan[1,2]([✉]) [iD], Milla Kortelainen[2] [iD], Manuel Cáceres[2] [iD],
Lucia Williams[3] [iD], and Alexandru I. Tomescu[2] [iD]

[1] Department of Computer Science and Engineering, IIT Roorkee, Roorkee, India
shahbaz.khan@cs.iitr.ac.in
[2] Department of Computer Science, University of Helsinki, Helsinki, Finland
{shahbaz.khan,milla.kortelainen,manuel.caceresreyes,
alexandru.tomescu}@helsinki.fi
[3] School of Computing, Montana State University, Bozeman, USA
luciawilliams@montana.edu

Abstract. Flow decomposition has numerous applications, ranging from networking to bioinformatics. Some applications require any valid decomposition that optimizes some property as number of paths, robustness, or path lengths. Many bioinformatic applications require the specific decomposition which relates to the underlying data that generated the flow. Thus, no optimization criteria guarantees to identify the correct decomposition for real inputs. We propose to instead report the *safe* paths, which are subpaths of at least one path in every flow decomposition.

Ma et al. [WABI 2020] addressed the existence of multiple optimal solutions in a probabilistic framework, which is referred to as *non-identifiability*. Later, they gave a quadratic-time algorithm [RECOMB 2021] based on a *global* criterion for solving a problem called AND-Quant, which generalizes the problem of reporting whether a given path is safe.

We present the first *local* characterization of safe paths for flow decompositions in directed acyclic graphs, giving a practical algorithm for finding the *complete* set of safe paths. We also evaluated our algorithm against the trivial safe algorithms (unitigs, extended unitigs) and a popular heuristic (greedy-width) for flow decomposition on RNA transcripts datasets. Despite maintaining perfect precision our algorithm reports ≈50% higher coverage over trivial safe algorithms. Though greedy-width reports better coverage, it has significantly lower precision on complex graphs. On a unified metric (F-Score) of coverage and precision, our algorithm outperforms greedy-width by ≈20%, when the evaluated dataset has significant number of complex graphs. Also, it has superior time (3–5×) and space efficiency (1.2–2.2×), resulting in a better and more practical approach for bioinformatics applications of flow decomposition.

We thank Romeo Rizzi and Edin Husić for helpful discussions. This work was partially funded by the European Research Council (ERC) under the European Union's Horizon 2020 research and innovation programme (grant agreement No. 851093, SAFEBIO) and partially by the Academy of Finland (grants No. 322595, 328877) and the US NSF (award 1759522). The full version of the paper is available at [15].

I. Pe'er (Ed.): RECOMB 2022, LNBI 13278, pp. 177–192, 2022.
https://doi.org/10.1007/978-3-031-04749-7_11

Keywords: Safety · Flow decomposition · DAGs · RNA assembly

1 Introduction

Network flows are a central topic in computer science, that define problems with countless practical applications. Assuming that the flow network has a unique source s and a unique sink t, every flow can be decomposed into a collection of weighted s-t paths and cycles [11]; for directed acyclic graphs (DAGs), such a decomposition contains only paths. Such a path (and cycle) view of a flow is used to optimally route information or goods from s to t, where flow decomposition is a key step in problems such as network routing [13] and transportation [26]. Finding the decomposition with the minimum number of paths and *possibly* cycles (or *minimum flow decomposition*) is NP-hard, even for a DAG [37]. On the theoretical side, this hardness result led to research on approximation algorithms [13,30], and FPT algorithms [17,34]. On the practical side, many approaches employ a standard *greedy-width* heuristic [37], of repeatedly removing an s-t path carrying the most flow. Another pseudo-polynomial-time heuristic called *Catfish* [32] tries to iteratively simplify the graph so that smaller decompositions can be found.

However, for a flow network built by superimposing a set of weighted paths, and one may seek the decomposition corresponding to that set of weighted paths. Such a decomposition is used by the more recent and prominent application of reconstructing biological sequences (*RNA transcripts* [34,35,40] or *viral quasi-species genomes* [4,5]). Each flow path represents a reconstructed sequence, and so a different set of flow paths encodes a different set of biological sequences, which may differ from the real ones. If there are multiple optimal solutions, then the reconstructed sequences may not match the original ones, and thus be incorrect. While many popular multiassembly tools use minimum flow decompositions, Williams et al. [41] reported that in an error-free transcript dataset 20% of human genes admit multiple minimum flow decomposition solutions.

1.1 Safety Framework for Addressing Multiple Solutions

Motivated by such an RNA assembly application, Ma et al. [20] were the first to address the issue of multiple solutions to the flow decomposition problem under a probabilistic framework. Later, they [21] solve a problem (*AND-Quant*), which, in particular, leads to a quadratic-time algorithm for the following problem: given a flow in a DAG, and edges $e_1, e_2, \ldots, e_k$, decide if in *every* flow decomposition there is always a decomposed flow path passing through all of $e_1, e_2, \ldots, e_k$. Thus, by taking the edges $e_1, e_2, \ldots, e_k$ to be a path P, the AND-Quant problem can decide if P (i.e., a given biological sequence) appears in all flow decompositions. This indicates that P is likely part of some original RNA transcript.

We build upon the AND-Quant problem, by addressing the flow decomposition problem under the *safety* framework [36], first introduced for genome assembly. For a problem admitting multiple solutions, a partial solution is said to be *safe* if it appears in all solutions to the problem. For example, a path P

is safe for the flow decomposition problem, if for *every* flow decomposition into paths $\mathcal{P}$, it holds that P is a subpath of some path in $\mathcal{P}$. Further, P is called *w-safe* if in *every* flow decomposition, P is a subpath of some weighted path(s) in $\mathcal{P}$ whose total weight is at least w. Bioinformatics applications [17,32,35] commonly use a minimum cardinality path decomposition (or path cover [19]). We consider *any* flow decomposition as a valid solution, not only the ones of minimum cardinality, which is motivated by both theory and practice. On the one hand, since minimum-cardinality flow decomposition is NP-hard [37], we believe that finding its safe paths is also intractable. On the other hand, given the issues with sequencing data, practical methods usually incorporate different variations of the minimality criterion [4,5]. Thus, safe paths for *all* flow decompositions are likely correct for many practical variations of the flow decomposition problem.

Safety has precursors in combinatorial optimization, as *persistency*. Costa [10] studied the persistent edges in all maximum bipartite matchings. Incidentally, for the maximum flow problem persistent edges always having a non-zero flow value in any maximum flow solution were studied [9]. In bioinformatics, safety has been previously studied for the genome assembly problem which at its core solves the problem of computing arc-covering walks on the assembly graph. Again since the problem admits multiple solutions where only one is correct, practical genome assemblers output only those solutions likely to be correct. The prominent approach dating back to 1995 [14] is to compute trivially correct *unitigs* (having internal nodes with *unit* indegree and unit outdegree), which can be computed in linear time. Unitigs were generalised first in [29], and later [16,23] to be *extended* by adding their unique incoming and outgoing paths. These *extended unitigs*, though safe, are not guaranteed to report *everything* that can be correctly assembled, presenting an important open question [25] about the *assembly limit* (if any). This was finally resolved by Tomescu and Medvedev [36] for a specific genome assembly formulation (single circular walk) by introducing *safe and complete* algorithms, which report everything that is theoretically reported as safe. Its running time was later optimized in [7] and [8]. Safe and complete algorithms were also studied by Acosta et al. [1] under a different genome assembly formulation of multiple circular walks. Recently, Cáceres et al. [6] studied safe and complete algorithms for path covers in an application on RNA Assembly.

1.2 Safety in Flow Decomposition for RNA Assembly

In bioinformatics, flow decomposition is prominently used in RNA transcript assembly, which is described as follows. In complex organisms, a gene may produce multiple RNA molecules (*RNA transcripts*, i.e., strings over an alphabet of four characters), each having a different abundance. Given a sample, one can partially read the RNA transcripts and find their abundances using *high-throughput sequencing* [38]. This technology produces short overlapping substrings of the RNA transcripts. The main approach for recovering the RNA transcripts from such data is to build an edge-weighted DAG from these fragments, then to transform the weights into flow values by various optimization criteria, and finally to decompose the resulting flow into an "optimal" set of weighted paths (i.e., the

RNA transcripts and their abundances in the sample) [22]. A common strategy for choosing the optimal set of weighted paths is to look for the parsimonious solution, i.e., the solution with the fewest paths. Since this problem is NP-hard, in practice many tools use the popular *greedy-width* heuristic [28,35]. Greedy-width is also used in the assemblers for the related problem of viral quasispecies assembly [4]. Further, some tools attempt to incorporate additional information into the flow decomposition process, such as by using longer reads or super reads [28,41]. Despite the large number of tools and methods that have been developed for RNA transcript assembly, there is no method that consistently reports the correct set of transcripts [28,42]. This suggests that the addressing the problem under the safety framework may be a promising approach. However, while a safe and complete solution clearly gives the maximally reportable correct solution, it is significant to evaluate whether such a solution covers a large part of the true transcript, to be useful in practice. A possible application of such partial and reliable solution is to consider them as constrains (see e.g. [41]) of real RNA transcript assemblers, to guide the assembly process of such heuristics. Another possible application could be to evaluate the accuracy of assemblers: does the output of the assembler include the safe and complete solution?

1.3   Our Results

Our contributions can be succinctly described as follows.

1. **Simple local characterization and optimal verification algorithm**: We characterize a safe path P using its local property called *excess flow*.

Theorem 1. *For $w > 0$, a path P is w-safe iff its excess flow $f_P \geq w$.*

The previous work [21] on AND-Quant describes a *global* characterization using the maximum flow of the entire graph transformed according to P, requiring $O(mn)$ time. Instead, the excess flow is a *local* property of P which is computable in time linear in the length of P. This also directly gives a simple verification algorithm which is optimal.

Theorem 2. *Given a flow graph (DAG) having n vertices and m edges, it can be preprocessed in $O(m)$ time to verify the safety of a path P in $O(|P|) = O(n)$ time.*

2. **Simple enumeration algorithm:** The above characterization also results in a simple algorithm for reporting all maximal safe paths by using an arbitrary flow decomposition of the graph.

Theorem 3. *Given a flow graph (DAG) having n vertices and m edges, all its maximal safe paths can be reported in $O(|\mathcal{P}_f|) = O(mn)$ time, where $\mathcal{P}_f$ is some flow decomposition.*

This approach starts with a candidate solution and uses the characterization on its subpaths in an efficient manner (a similar approach was previously used by [1,6,10]).

3. **Empirically improved approach for RNA assembly:** On simulated RNA splice graphs, safe and complete paths for flow decomposition provide precise RNA assemblies while covering most of RNA transcripts. They have ≈50% better coverage over previous notions of safe paths, while maintaining the perfect precision ensured by safety. Further, for the combined metric of coverage and precision (F-Score), they outperform the popular greedy-width heuristic significantly (≈20%) and previous safety algorithms appreciably (≈13%). Though our approach takes 1.2–2.5× time than the previous safety algorithms requiring equivalent memory, the greedy-width approach takes roughly 3–5× time and 1.2–2.2× memory than our approach. The significance of our approach in quality parameters increases with the increase in complex graph instances in the dataset, with significantly better performance over greedy-width, without losing a lot over previous safe algorithms.

2   Preliminaries and Notations

We consider a DAG $G = (V, E)$ with n vertices and m edges, where each edge e has a positive flow (or *weight*) $f(e)$ passing through it. We assume the graph is connected and hence $m \geq n$. For each vertex u, $f_{in}(u)$ and $f_{out}(u)$ denote the total flow on its incoming edges and outgoing edges, respectively. A vertex v is called a *source* if $f_{in}(v) = 0$ and a *sink* if $f_{out}(v) = 0$. Every other vertex v satisfies the *conservation of flow* $f_{in}(v) = f_{out}(v)$, making the graph a *flow graph*. For a path P, $|P|$ denotes the number of its edges. For a set of paths $\mathcal{P} = \{P_1, \cdots, P_k\}$ we denote its total size (number of edges) by $|\mathcal{P}| = |P_1| + \cdots + |P_k|$.

For any flow graph (DAG), its *flow decomposition* is a set of weighted *paths* $\mathcal{P}_f$ such that the flow on each edge of the flow graph equals the sum of the weights of the paths containing the edge. A flow decomposition of a graph can be computed in $O(|\mathcal{P}_f|) = O(mn)$ time using the simple path decomposition algorithm [3]. A path P is called *w-safe* if, in every possible flow decomposition, P is a subpath of some paths in $\mathcal{P}_f$ whose total weight is at least w. If P is w-safe with $w > 0$, we call P a *safe flow path*, or simply *safe path*. Intuitively, for any edge e with non-zero flow, we consider *where did the flow on e come from?* We would like to report all the maximal paths ending with e along which some $w > 0$ weight always "flows" to e (see Fig. 1). A safe path is *left maximal* (or *right maximal*) if extending it to the left (or right) with any edge makes it unsafe (i.e. not safe). A safe path is *maximal* if it is both left and right maximal. A set of safe paths is called *complete* if it consists of *all* the maximal safe paths.

Some previous notions of safety for other problems also naturally extend to the flow decomposition problem as follows. The paths having internal nodes with unit indegree and unit outdegree are called *unitigs* [14], which are trivially safe because every source-to-sink path which passes through an edge of unitig, also passes through the entire unitig contiguously. Further, a unitig can naturally be *extended* to include its unique incoming path (having nodes with unit indegree), and its unique outgoing path (having nodes with unit outdegree). This maximal extension of a unitig is called the *extended unitig* [16,23], which is similarly safe.

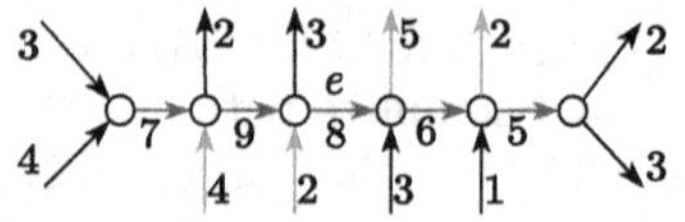

Fig. 1. The prefix of the path (blue) up to e contributes at least 2 units of flow to e, as the rest may enter the path by the edges (red) with flow 4 and 2. Similarly, the suffix of the path (blue) from e maintains at least 1 unit of flow from e, as the rest may exit the path from the edges (red) with flow 5 and 2. Both these safe paths are *maximal* as they cannot be extended left or right. (Color figure online)

For some graphs the above notions already define the safety of flow decomposition *completely*. Millani et al. [24] defined a class of DAGs called *funnels*, where every source-to-sink path is uniquely identifiable by at least one edge, which is not used by any other source-to-sink path. Considering such an edge as a trivial unitig (having a single edge), its extended unitig is exactly the corresponding source-to-sink path, making it safe. Thus, in a funnel all the source-to-sink paths are naturally safe and hence trivially complete. Moreover, it implies that a funnel has a unique flow decomposition, making the problem trivial for funnel instances.

However, for non-funnel graphs unitigs and extended unitigs are safe but potentially not complete. Note that both unitigs and extended unitigs are also safe for problems dealing with unweighted graphs (e.g. path cover). Hence, they do not use the flows on the edges of the graph, potentially missing some paths that are safe for flow decomposition but not for problems like path cover.

3   Characterization of Safe and Complete Paths

Safety of a path can be characterized by its *excess flow* defined as follows.

Definition 1 (Excess flow). Excess flow f_P of a path $P = \{u_1, u_2, ..., u_k\}$ is

$$f_P = f(u_1, u_2) - \sum_{\substack{u_i \in \{u_2, ..., u_{k-1}\} \\ v \neq u_{i+1}}} f(u_i, v) = f(u_{k-1}, u_k) - \sum_{\substack{u_i \in \{u_2, ..., u_{k-1}\} \\ v \neq u_{i-1}}} f(v, u_i)$$

the former and later formulations are diverging *and* converging, *respectively.*

Remark 1. Alternatively, the converging and diverging formulations are

$$f_P = \sum_{i=1}^{k-1} f(u_i, u_{i+1}) - \sum_{i=2}^{k-1} f_{out}(u_i) = \sum_{i=1}^{k-1} f(u_i, u_{i+1}) - \sum_{i=2}^{k-1} f_{in}(u_i).$$

The converging and diverging formulations are equivalent by the conservation of flow on internal vertices. The idea behind excess flow f_P (see Fig. 2) is that even in the worst case, the maximum *leakage* , or the flow leaving (or entering) P

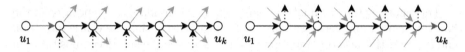

Fig. 2. The excess flow of a path P (left) is the incoming flow (blue) that necessarily pass through the whole P despite the flow (red) leaving P at its internal vertices. Analogously (right), it is the outgoing flow (blue) that necessarily came through the whole P despite the flow (red) entering P at its internal vertices. (Color figure online)

at the internal nodes, is the sum of the flow on the outgoing (or incoming) edges of the internal nodes of P, that are not in P. Hence, if the value of incoming flow (or outgoing flow) is higher than this maximum leakage, then this excess value f_P necessarily passes through the entire P. The following results give the simple characterization and an additional property (see [15] for proof) of safe paths.

Theorem 1. *For $w > 0$, a path P is w-safe iff its excess flow $f_P \geq w$.*

Proof. The excess flow f_P of a path P trivially makes it $w \leq f_P$-safe by definition. If $f_P < w$, we can prove that P is not w-safe by modifying any flow decomposition having w flow on P to leave only f_P flow (or 0, if $f_P < 0$) on P as follows. In Fig. 2 (diverging), consider a flow path P' entering P through edge e_1 (except first edge (blue)) and leaving P at an edge e_2 (red) except last edge of P. Since $f_P < w$, it is not possible that every path leaving P using a red edge starts at the first blue edge (by definition of f_P), hence P' always exists. We modify P' by using flow on P to form two paths, which enter from e_1 and leave at the last edge, and which enter from the first edge and leave at e_2. We can repeat such modifications until flow on P is f_P (or 0, if $f_P < 0$) due to conservation of flow. Additionally, for a path to be safe, it must hold that $w > 0$. $\square$

Lemma 1. *Adding an edge (u,v) to the start or the end of a path in the flow graph, reduces its excess flow by $f_{in}(v) - f(u,v)$, or $f_{out}(u) - f(u,v)$, respectively.*

4   Simple Verification and Enumeration Algorithms

The characterization of a safe path in a flow graph (Theorem 1) can be directly adapted to simple algorithms for verification and enumeration of all maximal safe paths. We preprocess the graph to compute the incoming flow $f_{in}(u)$ and outgoing flow $f_{out}(u)$ for each vertex u in $O(m)$ time. Using Remark 1 we can verify if a path P is safe in $O(|P|) = O(n)$ time, proving the following theorem.

Theorem 2. *Given a flow graph (DAG) having n vertices and m edges, it can be preprocessed in $O(m)$ time to verify the safety of a path P in $O(|P|) = O(n)$ time.*

For reporting the maximal safe paths we use a candidate decomposition of the flow into paths, and verify the safety of its subpaths using the characterization and a scan with the two-pointer approach. The candidate flow decomposition can be computed in $O(mn)$ time using the classical flow decomposition algorithm [11] resulting in $O(m)$ paths $\mathcal{P}_f$ each of $O(n)$ length. Now, we use the two-pointer scan along each path $P \in \mathcal{P}_f$ as follows. We start with the subpath containing the first two edges of the path P. We compute its excess flow f, and if $f > 0$ we append the next edge to the path on the right and incrementally compute its excess flow by Lemma 1. Otherwise, if $f \leq 0$ we remove the first edge of the path from the left and incrementally compute the excess flow similarly by Lemma 1 (removing an edge (u, v) would conversely modify the flow by $f_{in}(v) - f(u, v)$). We stop when the end of P is reached with a positive excess flow.

The excess flow can be updated in $O(1)$ time when adding an edge to the subpath on the right or removing an edge from the left. If the excess flow of a subpath P' is positive and on appending it with the next edge it ceases to be positive, we report P' as a maximal safe path by reporting only its two indices on the path P. Thus, given a path of length $O(n)$, all its maximal safe paths can be reported in $O(n)$ time, and hence require total $O(mn)$ time for the $O(m)$ paths in the flow decomposition $\mathcal{P}_f$, resulting in the following theorem.

Theorem 3. *Given a flow graph (DAG) having n vertices and m edges, all its maximal safe paths can be reported in $O(|\mathcal{P}_f|) = O(mn)$ time, where $\mathcal{P}_f$ is some flow decomposition.*

5   Experimental Evaluation

We now evaluate the performance of our safe and complete algorithm by comparing it with the most promising algorithms for flow decomposition. Since the performance of various algorithms closely depend on the input graphs, we consider some practically relevant datasets to evaluate their true impact. As the most significant application of flow decomposition derives from RNA assembly, we consider the flow networks extracted as splice graphs of simulated RNA-Seq experiments. That is, starting from a set of RNA transcripts, we simulate their expression levels and superimpose the transcripts to create a flow graph. Evaluating our approach in such *perfect* scenario allows us to remove the biases introduced by real RNA-Seq experiments [33] and focus the features offered by the each technique instead. Further, the performance of algorithms also closely depend on the complexity k of a graph, that we measure as the number of paths in the ground truth decomposition of the graph. Thus, we present our results with regards to the complexity k of the input graph instances.

We first investigate the practical significance of *safety* by comparing our safe solution to a popularly used flow decomposition heuristic that is also scalable. The greedy-width [37] heuristic decomposes the flow by sequentially selecting the heaviest possible path, resulting in a simple algorithm that is both scalable and performs well in practice. However, any flow decomposition algorithm may not always report the ground truth paths that originally built the instance of the flow graph. Thus, it is important to measure the reported solution using a *precision* metric which evaluates the correctness of the solution. We thus investigate how the precision of greedy-width varies particularly as the value of k increases.

We then investigate the practical significance of *completeness* as reported by our solution, over the previously known safe solutions as reported by unitigs and extended unitigs (recall Sect. 2). Note that every safe solution would always score perfectly in a precision metric by definition. Hence, all safe solutions would always outperform greedy-width (or any flow decomposition algorithm) in precision metrics. However, this perfect precision comes at the cost of the amount of the solution that is reported. Intuitively, this can be measured using some *coverage* metrics which describe how much of the ground truth sequence is included in the reported paths. Note that any flow decomposition algorithm will perform better than any safe algorithm by definition, as the safe paths are always subpaths of the paths reported by any flow decomposition algorithm. Further, the extended unitigs would clearly outperform unitigs, and our safe paths would clearly outperform both unitigs and extended unitigs. We thus investigate how the coverage of various algorithms varies with respect to the greedy-width particularly as the value of k increases.

Finally, to understand the overall impact of different algorithms, where safe algorithms as compared to greedy-width clearly outperform in precision metrics and underperform in coverage metrics, we address both coverage and precision measures using a single metric, i.e., F-score. We thus investigate the variation in F-score over different values of k. In addition, to understand the practical utility of the algorithms we also investigate their performance measures in terms of running time and space requirements.

5.1   Datasets

We consider two RNA transcripts datasets, generated based on approach of Shao et al. [32]. They create "perfect" flow graphs where the true set of transcripts and abundances is always a flow decomposition of the graph (hence satisfy conservation of flow). They start with this flow decomposition and create the input instances by superimposing them into a single graph, adding a single source s (and sink t) with an edge to the beginning (and end) of each transcript.

Funnel Instances: In funnels [24] all paths are safe and the problem is trivial (recall Sect. 2). Our evaluation thus ignores these trivial funnel instances. For the sake of completeness we address the funnels in our full paper [15].

Catfish Dataset: We consider the dataset first used by Shao and Kingsford [32], which includes 100 simulated human transcriptomes for human, mouse, and zebrafish using Flux-Simulator [12]. Additionally, it includes 1,000 experiments from the Sequence Read Archive, with simulated abundances for transcripts using Salmon [27]. In both cases, the weighted transcripts are superimposed to build splice graphs as described above. This dataset has also been used in other flow decomposition benchmarking studies [17,41]. There are 17,335,407 graphs in total in this dataset, of which 8,301,682 are non-trivial (47.89%). However, in this dataset the details about the number of bases on each node (exons or pseudo-exons) are omitted, which results in an incomplete measure of coverage and precision. Moreover, this dataset has negligible complex graph instances (having large k). Hence, we do not include its evaluation in the main paper, rather defer it to the full paper [15] for the sake of completeness.

Reference-Sim Dataset: We consider a dataset [39] containing a single simulated transcriptome as follows. For each transcript on the positive strand in the GRCh.104 *homo sapiens* reference genome, it samples a value from the lognormal distribution with mean and variance both equal to -4, as done in the default settings of RNASeqReadSimulator [18]. It then multiplies the resulting values by 1000 and round to the nearest integer. Then it excludes any transcript that is rounded to 0. There are 17,941 total graphs in this dataset, of which 10,323 are non-trivial (57.54%). In this dataset, we also have access to the genomic coordinates (and hence number of bases) represented by nodes, allowing us to compute more practically relevant coverage and precision metrics.

5.2   Evaluation Metrics

All metrics are computed in terms of bases for the Reference-Sim dataset. However, since the Catfish dataset omits the base information its metrics are computed in terms of exons or pseudo-exons (vertices in the flow graph). For every algorithm, R denotes a reported path (for Catfish) or a reported safe subpath (for unitigs, extended unitigs, and safe complete) of the solution. In addition, T denotes a path in the set of ground truth transcripts provided in the dataset. For each graph, we compute the following metrics which were also used earlier by [6] for safety in constrained path covers:

Weighted precision: We classify a reported path R as *correct* if R is a subpath of some ground truth transcript T of the flow graph. Weighted precision is the total length of correctly reported paths divided by the total length of reported paths. The commonly used precision metric [28,31] for measuring the accuracy of RNA assembly methods considers only those paths as correct which are (almost) exactly contained in the ground truth decomposition. Further, the precision is computed as the number of correctly reported paths divided by the total reported paths. However, since all the safe algorithms reports (possibly) partial transcripts, we use subpaths instead of (almost) exactly same paths. To highlight *how much* is reported correctly instead of *how many*, we use weighted precision to give a better score for longer correctly reported paths.

Maximum relative coverage: Given a ground truth transcript T and a reported path R, we define a *segment* of R inside T as a maximal subpath of R that is also subpath of T. We define the maximum relative coverage of a ground truth transcript as the length of the longest segment of a reported path inside T, divided by the length of T. The corresponding value for the entire graph is the average of the values over all T. While it is common in the literature [28, 31] to report *sensitivity* (the proportion of ground truth transcripts that are correctly predicted), we measure correctness based on coverage since all the safe algorithms report paths that (possibly) do not cover an entire transcript.

F-score: The standard measure to combine precision and sensitivity is using an *F-score*, which is the harmonic mean of the two. In our evaluation we correspondingly use the weighted precision and the maximum relative coverage for computing the F-score.

5.3   Implementation and Environment Details

We evaluate the following algorithms in our experiments.

Unitigs: It computes the unitigs, by considering each unvisited edge in the topological order and extending it towards the right as long as the internal nodes have unit indegree and unit outdegree. The result ignores single edges.

ExtUnitigs: It computes the extended unitigs, by considering each unitig and single edges, and extending it towards the left and right as long as the internal nodes have unit indegree and unit outdegree, respectively.

Safe&Comp: It computes the safe and complete paths for flow decomposition using our enumeration algorithm described in Sect. 4. Since the metrics evaluation scripts uses each safe path individually (similar to other algorithms), we output all safe paths completely which requires output size (and hence time) of $O(mn^2)$ instead of $O(mn)$ as stated in Theorem 1.

Greedy: It computes the greedy-width heuristic using Catfish [32] with the `-a greedy` parameter.

All algorithms are implemented in C++, whereas the scripts for evaluating metrics are implemented in Python. The Unitigs, ExtUnitigs, and Safe&Comp implementations use optimization level 3 of GNU C++ (compiled with $-O3$ flag), whereas the Greedy uses the optimizations of the Catfish pipeline. The Unitigs, ExtUnitigs, and Safe&Comp additionally require a post processing step using Aho Corasick Trie [2] for removing duplicates, and prefix/suffixes to make the set of safe paths minimal. However, the time and memory requirements are evaluated considering only the algorithm, and not post processing and metric evaluations which are not optimized. All performances were evaluated on a laptop using a single core (i5-8265U CPU 1.60 GHZ) having 15.3 GB memory. The source code of our project is available on Github[1] under GNU Genral Public License v3 license.

5.4   Results

We first evaluate the significance of *safety* among the reported solution. Figure 3a compares the weighted precision of all the algorithms on the Reference-Sim

---

[1] https://github.com/algbio/flow-decomposition-safety.

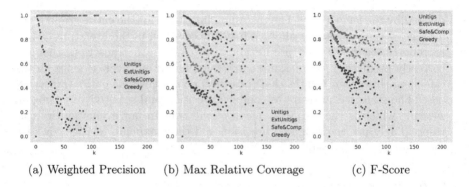

(a) Weighted Precision   (b) Max Relative Coverage   (c) F-Score

Fig. 3. Evaluation metrics on graphs w.r.t. k for the Reference-Sim dataset.

dataset distributed over k. All the safe algorithms clearly report perfect precision as expected. However, the precision of the Greedy algorithm sharply declines with the increase in k, almost linearly to 30% for $k = 35$. This may be explained by the sharp increase in the number of possible paths in graphs with increase in k, which can be used by any flow decomposition algorithm. Hence, the significance of safety becomes very prominent as k increases.

We then evaluate the significance of *completeness* of the safe algorithms. Figure 3b compares the maximum relative coverage of all the algorithms on the Reference-Sim dataset distributed over k. As expected, Greedy outperforms all the other, followed by Safe&Comp, ExtUnitigs and Unitigs. However, note that as k reaches 20 Safe&Comp, ExtUnitigs and Unitigs sharply fall to 75%, 60% and 40%, while Greedy maintains around 95% coverage. Overall, Safe&Comp is almost always ≈85–90% of that of Greedy, whereas ExtUnitigs and Unitigs falls to 60% and 40% respectively. Hence, the Safe&Comp manages to maintain perfect precision without losing a lot on coverage, demonstrating the importance of *completeness* among the safe algorithms.

Figure 3c supports the above inference by evaluating the combined metric F-Score, where Safe&Comp dominates Unitigs and ExtUnitigs by definition. Safe&Comp also dominates Greedy as k approaches 10. It is also important to note that both ExtUnitigs and Unitigs eventually dominate Greedy for a slightly larger value of $k > 20$ and $k > 30$, respectively. This shows the significance of considering Safe algorithms for complex graphs. However, the significance of the Safe&Comp as the number of graphs with such higher complexities also reduces sharply (see full paper [15]).

Hence, we evaluate a summary of the above results averaged over all graphs irrespective of k. Table 1 summarizes the evaluation metrics for all the algorithms for simple graphs ($k < 10$) and complex graphs ($k > 10$), and both. While on the simpler graphs Greedy dominates Safe&Comp mildly (≈3%), for complex graphs it is dominated significantly (≈20%) by Safe&Comp and appreciably (≈8%) by ExtUnitigs. However, despite the larger ratio of simpler graphs, the collective

Table 1. Summary of evaluation metrics for the Reference-Sim dataset.

Graphs	Algorithm	Max. coverage	Wt. precision	F-score
$k \geq 2$ (100%)	Unitigs	0.51	1.00	0.66
	ExtUnitigs	0.69	1.00	0.81
	Safe&Comp	0.82	1.00	0.90
	Greedy	0.98	0.81	0.86
$2 \leq k \leq 10$ (68%)	Unitigs	0.55	1.00	0.70
	ExtUnitigs	0.73	1.00	0.84
	Safe&Comp	0.84	1.00	0.91
	Greedy	0.99	0.91	0.94
$k > 10$ (32%)	Unitigs	0.41	1.00	0.58
	ExtUnitigs	0.61	1.00	0.75
	Safe&Comp	0.76	1.00	0.86
	Greedy	0.95	0.60	0.69

Table 2. Time (s) and Memory (MB) taken by different algorithms on datasets.

Algorithm	Reference-Sim		Catfish							
	Human		Zebrafish		Mouse		Human		Human (salmon)	
	25.6 MB		122 MB		137 MB		157 MB		2.5 GB	
	Time	Mem	Time	Mem	Time	Mem	Time	Mem	Time	Mem
Unitigs	0.68	3.58	13.82	3.51	15.62	3.53	18.22	3.54	303.72	3.66
ExtUnitigs	0.99	3.63	18.31	3.52	20.87	3.57	23.64	3.56	404.50	3.68
Safe&Comp	2.56	4.47	20.17	3.56	25.76	3.66	28.59	3.54	667.27	3.84
Greedy	7.71	4.88	108.30	6.00	127.38	6.29	148.46	6.34	2684.30	8.47

F-score over all graphs is still ($\approx$4%) better for Safe&Comp over Greedy which signifies the applicability of Safe&Comp over Greedy.

Finally, we evaluate the applicability of the above algorithms in practice, by comparing their running time and peak memory requirements. Since all the algorithms are implemented in the same language (C++) and evaluated on the same machine, it is reasonable to directly compare these measures. In Table 2, we see that Unitigs clearly are the fastest, where ExtUnitigs takes roughly 1.3–1.5× time. Safe&Comp takes upto roughly 1.2–2.5× time than ExtUnitigs, and Greedy requires roughly 3–5× time than Safe&Comp. The peak memory requirements of the safe algorithms are very close (within 5%–25%), whereas Greedy requires roughly 1.1–2.2× more memory than Safe&Comp. Overall, for the performance measures Safe&Comp shows a significant improvement over Greedy, without losing a lot over the trivial algorithms.

6   Conclusion

We study the flow decomposition in DAGs under the Safe and Complete paradigm, which has numerous applications including the more prominent multi-assembly of biological sequences. Previous work characterized such paths (and their generalizations) using a global criterion. Instead, we present a simpler characterization based on a more efficiently computable local criterion, which is directly adapted into an optimal verification algorithm, and a simple enumeration algorithm. Intuitively, it is a *weighted* adaptation of *extended unitigs* which is a prominent approach for computing safe paths.

Our experiments show that our algorithm outperform the popularly used greedy-width heuristic for RNA assembly instances having significant complex graph instances, both on quality (F-score) and performance (running time and memory) parameters. On simple graphs, Greedy outperforms Safe&Comp and Safe&Comp outperforms ExtUnitigs mildly ($\approx$3–5%). However, on complex graphs, Safe&Comp outperforms Greedy significantly ($\approx$20%) and ExtUnitigs appreciably ($\approx$13%). While the Reference-Sim dataset shows the overall dominance of Safe&Comp since complex graphs are appreciable (32%), Greedy dominates Safe&Comp in Catfish dataset since complex graphs are negligible ($\approx$2%). Another significant reason for the dominance of Greedy over Safe&Comp on Catfish datasets is the absence of base information on nodes (see full paper [15]). Hence, the importance of Safe&Comp algorithms increases with the increase in complex graph instances in the dataset, and prominently when we consider information about the genetic information represented by each node. In terms of performance, ExtUnitigs are 1.3–1.5$\times$ slower than the fastest approach (Unitigs), while Safe&Comp further takes roughly 1.2–2.5$\times$ time than ExtUnitigs, both requiring equivalent memory. However, Greedy requires roughly 3–5$\times$ time and 1.1–2.2$\times$ memory than Safe&Comp. Overall, Safe&Comp performs significantly better than Greedy, without losing a lot over the trivial algorithms.

References

1. Acosta, N.O., Mäkinen, V., Tomescu, A.I.: A safe and complete algorithm for metagenomic assembly. Algorithms Mol. Biol. **13**(1), 3:1–3:12 (2018). https://doi.org/10.1186/s13015-018-0122-7
2. Aho, A.V., Corasick, M.J.: Efficient string matching: an aid to bibliographic search. Commun. ACM **18**(6), 333–340 (1975). https://doi.org/10.1145/360825.360855
3. Ahuja, R.K., Magnanti, T.L., Orlin, J.B.: Network Flows - Theory, Algorithms and Applications. Prentice Hall, Upper Saddle River (1993)
4. Baaijens, J.A., der Roest, B.V., Köster, J., Stougie, L., Schönhuth, A.: Full-length de novo viral quasispecies assembly through variation graph construction. Bioinformatics **35**(24), 5086–5094 (2019). https://doi.org/10.1093/bioinformatics/btz443
5. Baaijens, Jasmijn A.., Stougie, Leen, Schönhuth, Alexander: Strain-aware assembly of genomes from mixed samples using flow variation graphs. In: Schwartz, Russell (ed.) RECOMB 2020. LNCS, vol. 12074, pp. 221–222. Springer, Cham (2020). https://doi.org/10.1007/978-3-030-45257-5_14

6. Caceres, M., et al.: Safety in multi-assembly via paths appearing in all path covers of a DAG. IEEE/ACM Trans. Comput. Biol. Bioinform. (2021)

7. Cairo, M., Medvedev, P., Acosta, N.O., Rizzi, R., Tomescu, A.I.: An optimal O(nm) algorithm for enumerating all walks common to all closed edge-covering walks of a graph. ACM Trans. Algorithms 15(4), 48:1–48:17 (2019). https://doi.org/10.1145/3341731

8. Cairo, M., Rizzi, R., Tomescu, A.I., Zirondelli, E.C.: Genome assembly, from practice to theory: safe, complete and linear-time. In: Bansal, N., Merelli, E., Worrell, J. (eds.) 48th International Colloquium on Automata, Languages, and Programming, ICALP 2021, 12–16 July 2021, Glasgow, Scotland (Virtual Conference). LIPIcs, vol. 198, pp. 43:1–43:18. Schloss Dagstuhl - Leibniz-Zentrum für Informatik (2021)

9. Cechlárová, K., Lacko, V.: Persistency in combinatorial optimization problems on matroids. Discret. Appl. Math. 110(2–3), 121–132 (2001). https://doi.org/10.1016/S0166-218X(00)00279-1

10. Costa, M.C.: Persistency in maximum cardinality bipartite matchings. Oper. Res. Lett. 15(3), 143–149 (1994). https://doi.org/10.1016/0167-6377(94)90049-3

11. Ford, D.R., Fulkerson, D.R.: Flows Netw. Princeton University Press, Princeton (2010)

12. Griebel, T., et al.: Modelling and simulating generic RNA-seq experiments with the flux simulator. Nucleic Acids Res. 40(20), 10073–10083 (2012)

13. Hartman, T., Hassidim, A., Kaplan, H., Raz, D., Segalov, M.: How to split a flow? In: 2012 Proceedings IEEE INFOCOM, pp. 828–836. IEEE (2012)

14. Kececioglu, J.D., Myers, E.W.: Combinatorial algorithms for DNA sequence assembly. Algorithmica 13(1/2), 7–51 (1995)

15. Khan, S., Kortelainen, M., Cáceres, M., Williams, L., Tomescu, A.I.: Safety and completeness in flow decompositions for RNA assembly. CoRR abs/2201.10372 (2022)

16. Kingsford, C., Schatz, M.C., Pop, M.: Assembly complexity of prokaryotic genomes using short reads. BMC Bioinform. 11(1), 21 (2010)

17. Kloster, K., et al.: A practical fpt algorithm for flow decomposition and transcript assembly. In: 2018 Proceedings of the Twentieth Workshop on Algorithm Engineering and Experiments (ALENEX), pp. 75–86. SIAM (2018)

18. Li, W.: RNASeqReadSimulator: a simple RNA-seq read simulator (2014)

19. Liu, R., Dickerson, J.: Strawberry: fast and accurate genome-guided transcript reconstruction and quantification from RNA-seq. PLoS Comput. Biol. 13(11), e1005851 (2017)

20. Ma, C., Zheng, H., Kingsford, C.: Exact transcript quantification over splice graphs. In: Kingsford, C., Pisanti, N. (eds.) 20th International Workshop on Algorithms in Bioinformatics, WABI 2020, 7–9 September 2020, Pisa, Italy (Virtual Conference). LIPIcs, vol. 172, pp. 12:1–12:18. Schloss Dagstuhl - Leibniz-Zentrum für Informatik (2020). https://doi.org/10.4230/LIPIcs.WABI.2020.12

21. Ma, C., Zheng, H., Kingsford, C.: Finding ranges of optimal transcript expression quantification in cases of non-identifiability. bioRxiv (2020). https://doi.org/10.1101/2019.12.13.875625 to appear at RECOMB 2021

22. Mäkinen, V., Belazzougui, D., Cunial, F., Tomescu, A.I.: Genome-Scale Algorithm Design: Biological Sequence Analysis in the Era of High-Throughput Sequencing. Cambridge University Press, London (2015). https://doi.org/10.1017/CBO9781139940023

23. Medvedev, P., Georgiou, K., Myers, G., Brudno, M.: Computability of models for sequence assembly. In: WABI, pp. 289–301 (2007)

24. Millani, M.G., Molter, H., Niedermeier, R., Sorge, M.: Efficient algorithms for measuring the funnel-likeness of DAGs. J. Comb. Optim. **39**(1), 216–245 (2020)
25. Nagarajan, N., Pop, M.: Parametric complexity of sequence assembly: theory and applications to next generation sequencing. J. Comput. Biol. **16**(7), 897–908 (2009)
26. Olsen, N., Kliewer, N., Wolbeck, L.: A study on flow decomposition methods for scheduling of electric buses in public transport based on aggregated time–space network models. Central Eur. J. Oper. Res. 1–37 (2020). https://doi.org/10.1007/s10100-020-00705-6
27. Patro, R., Duggal, G., Kingsford, C.: Salmon: accurate, versatile and ultrafast quantification from RNA-seq data using lightweight-alignment. BioRxiv p. 021592 (2015)
28. Pertea, M., Pertea, G.M., Antonescu, C.M., Chang, T.C., Mendell, J.T., Salzberg, S.L.: Stringtie enables improved reconstruction of a transcriptome from RNA-seq reads. Nat. Biotechnol. **33**(3), 290–295 (2015)
29. Pevzner, P.A., Tang, H., Waterman, M.S.: An Eulerian path approach to DNA fragment assembly. Proc. Natl. Acad. Sci. **98**(17), 9748–9753 (2001)
30. Pieńkosz, K., Kołtyś, K.: Integral flow decomposition with minimum longest path length. Eur. J. Oper. Res. **247**(2), 414–420 (2015)
31. Shao, M., Kingsford, C.: Accurate assembly of transcripts through phase-preserving graph decomposition. Nat. Biotechnol. **35**(12), 1167–1169 (2017)
32. Shao, M., Kingsford, C.: Theory and a heuristic for the minimum path flow decomposition problem. IEEE/ACM Trans. Comput. Biol. Bioinform. **16**(2), 658–670 (2017)
33. Srivastava, A., et al.: Alignment and mapping methodology influence transcript abundance estimation. Genome Biol. **21**(1), 1–29 (2020)
34. Tomescu, A.I., Gagie, T., Popa, A., Rizzi, R., Kuosmanen, A., Mäkinen, V.: Explaining a weighted DAG with few paths for solving genome-guided multi-assembly. IEEE ACM Trans. Comput. Biol. Bioinform. **12**(6), 1345–1354 (2015). https://doi.org/10.1109/TCBB.2015.2418753
35. Tomescu, A.I., Kuosmanen, A., Rizzi, R., Mäkinen, V.: A novel min-cost flow method for estimating transcript expression with RNA-seq. BMC bioinform. **14**(S5), S15 (2013)
36. Tomescu, A.I., Medvedev, P.: Safe and complete contig assembly through omnitigs. J. Comput. Biol. **24**(6), 590–602 (2017), preliminary version appeared in RECOMB 2016
37. Vatinlen, B., Chauvet, F., Chrétienne, P., Mahey, P.: Simple bounds and greedy algorithms for decomposing a flow into a minimal set of paths. European Journal of Operational Research **185**(3), 1390–1401 (2008)
38. Wang, Z., Gerstein, M., Snyder, M.: RNA-Seq: a revolutionary tool for transcriptomics. Nat. Rev. Genet **10**(1), 57–63 (2009)
39. Williams, L.: Reference-sim. e1005851 (2021). https://doi.org/10.5281/zenodo.5646910
40. Williams, L., Reynolds, G., Mumey, B.: Rna transcript assembly using inexact flows. In: 2019 IEEE International Conference on Bioinformatics and Biomedicine (BIBM), pp. 1907–1914. IEEE (2019)
41. Williams, L., Tomescu, A., Mumey, B.M., et al.: Flow decomposition with subpath constraints. In: 21st International Workshop on Algorithms in Bioinformatics (WABI 2021). Schloss Dagstuhl-Leibniz-Zentrum für Informatik (2021)
42. Yu, T., Mu, Z., Fang, Z., Liu, X., Gao, X., Liu, J.: TransBorrow: genome-guided transcriptome assembly by borrowing assemblies from different assemblers. Genome Res. **30**(8), 1181–1190 (2020)

NetMix2: Unifying Network Propagation and Altered Subnetworks

Uthsav Chitra[1], Tae Yoon Park[1,2], and Benjamin J. Raphael[1,2(✉)]

[1] Department of Computer Science, Princeton University, Princeton, NJ 08544, USA
[2] Lewis-Sigler Institute for Integrative Genomics, Princeton University, Princeton, NJ 08544, USA
braphael@princeton.edu

Abstract. A standard paradigm in computational biology is to use interaction networks to analyze high-throughput biological data. Two common approaches for leveraging interaction networks are: (1) *network ranking*, where one ranks vertices in the network according to both vertex scores and network topology; (2) *altered subnetwork* identification, where one identifies one or more subnetworks in an interaction network using both vertex scores and network topology. The dominant approach in network ranking is network propagation which smooths vertex scores over the network using a random walk or diffusion process, thus utilizing the global structure of the network. For altered subnetwork identification, existing algorithms either restrict solutions to subnetworks in *subnetwork families* with simple topological constraints, such as connected subnetworks, or utilize ad hoc heuristics that lack a rigorous statistical foundation. In this work, we unify the network propagation and altered subnetwork approaches. We derive a subnetwork family which we call the *propagation family* that approximates the subnetworks ranked highly by network propagation. We introduce NetMix2, a principled algorithm for identifying altered subnetworks from a wide range of subnetwork families, including the propagation family, thus combining the advantages of the network propagation and altered subnetwork approaches. We show that NetMix2 outperforms network propagation on data simulated using the propagation family. Furthermore, NetMix2 outperforms other methods at recovering known disease genes in pan-cancer somatic mutation data and in genome-wide association data from multiple human diseases. NetMix2 is publicly available at https://github.com/raphael-group/netmix2.

Keywords: Interaction networks · Network anomaly · Network propagation · Cancer · GWAS

1 Introduction

Biological systems consist of interactions between many components. These interactions are often represented with networks, e.g., protein-protein interaction

---

U. Chitra and T. Y. Park—Contributed equally to the manuscript.

© The Author(s), under exclusive license to Springer Nature Switzerland AG 2022
I. Pe'er (Ed.): RECOMB 2022, LNBI 13278, pp. 193–208, 2022.
https://doi.org/10.1007/978-3-031-04749-7_12

networks or gene regulatory networks. A standard paradigm in computational biology is to use an interaction network as prior knowledge for interpreting high-throughput, genome-scale data. Interaction networks have informed the analysis of biological data in many different applications including protein function prediction [19,26,60,68,72], differential expression analysis [17,25,28,43,79,85,88], prioritization of germline variants [12,40,42,51,53,71], identification of driver mutations in cancer [24,39,52,61,76,81], and more [9,15,20,21,33,35,38,51,56, 58,58,66,86].

Numerous methods that use interaction networks in interpreting high-throughput omics data have been developed (reviewed in [9,23,24,27,45,59]). While the algorithmic details of these methods are diverse, nearly all of them employ one of two different strategies. The first strategy is *network ranking*, where one is given either a subset of vertices (genes/proteins) or a score for each vertex (gene/protein), and the goal is to rank all vertices according to both the subset/scores and the positions of vertices in the network. Early network ranking algorithms relied on the "guilt-by-association" principle, or the idea that genes/proteins with similar functions are directly connected in the interaction network. These "direct connection" algorithms were typically applied in *semi-supervised* settings, e.g., protein function prediction or disease-gene prioritization [48,87], where only a subset of vertices are known to have a specific biological function. Later, inspired by the success of random walk, diffusion, and graph kernel methods in statistics and machine learning (e.g., the PageRank algorithm [63]), *network propagation*—also known as label propagation [89]— became the dominant approach for network ranking [23]. Briefly, network propagation involves using a random walk or diffusion process to "smooth" vertex scores across a network. Following [23], we use the term network propagation to refer to the broad class of methods that smooth scores over a network using a random walk or diffusion process. This includes popular processes like the random walk with restart (i.e., PageRank) [63], but also other processes including diffusion state distance [13,22] or the heat kernel [81,82]. By using these random walk/diffusion processes, network propagation methods simultaneously account for all possible paths between vertices. Thus, in contrast to the early methods which only use "direct connections" (edges) between vertices, network propagation methods fully utilize the *global* structure of the network. Indeed, network propagation has even been shown to be asymptotically optimal for network ranking for some random graph models [46].

The second strategy is the identification of *altered subnetworks*, also called *network modules* or *active subnetworks*. Here, the input is a measurement or a score for each vertex of the interaction network (e.g., p-values from differential gene expression), and the goal is to identify subnetworks (modules) that contain high scoring vertices that are "close" in the network[1]. Altered subnetwork approaches rely on the specification of a *subnetwork family*, or a family of possible

---

[1] A related problem is the identification of altered subnetworks according to network topology alone. Many of the leading methods for this problem were benchmarked in a recent DREAM competition [18].

subnetworks; sometimes the family is stated explicitly – e.g., the early approaches such as jActiveModules [44] or heinz [28] identify connected subnetworks—but in other methods, the subnetwork family is implicitly specified—e.g., the optimization problems of [5] and [55] penalize subnetworks with large cut-size and small edge-density, respectively. The altered subnetwork approach is closely related to the identification of *network anomalies* in the data mining and machine learning literature [1–4,73–75]. In contrast to ranking algorithms that yield a ranking of all vertices, altered subnetwork approaches output one or more *subsets* of vertices, and thus explicitly estimate the number of vertices in the altered subnetworks. A major challenge with altered subnetwork approaches is to choose an appropriate subnetwork family. For example, connectivity is often too weak of an assumption for biological networks, e.g., some methods that use connectivity identify large subnetworks [62] because of a statistical bias in a commonly used test statistic [16,69].

There have been a few attempts to bridge the gap between network propagation approaches and altered subnetwork approaches, combining the modeling of global network topology from network propagation with the optimization over subnetwork families from altered subnetwork approaches. For example, PRINCE [83] first propagates the vertex scores and identifies altered subnetworks as edge-dense subnetworks whose vertices have large network propagated scores. The HotNet algorithms [52,70,80,81] identify altered subnetworks by finding clusters in a weighted and directed graph derived from network propagation. TieDIE [65] propagates two sets of vertex scores and aims to find high-scoring subnetworks for both sets of propagated scores. More recently, the NetCore algorithm [7] finds subnetworks whose vertices have large node "coreness" and large propagated scores. However, none of these approaches give an explicit definition of the subnetwork family, instead relying on heuristics to identify the altered subnetwork after performing network propagation. Because of these heuristics, network propagation approaches typically do not have provable guarantees for altered subnetwork identification. In contrast, methods that explicitly rely on a well-defined subnetwork family often have statistical or theoretical guarantees, e.g., jActiveModules [44] computes a maximum likelihood estimator while our recent estimator NetMix is asymptotically unbiased [16,69].

Another practical issue is the evaluation of altered subnetwork methods. Most network algorithms demonstrate their performance by benchmarking their algorithm against existing network algorithms. While these comparisons are useful, they may also hide biases shared between algorithms. For example, Lazareva et al. [50] observed that some well-known network algorithms have a bias towards high-degree vertices in the interaction network, while Levi et al. [54] similarly observed a bias in GO term enrichment. In order to quantify the potential biases of altered subnetwork algorithms, these algorithms need to be compared against carefully selected baselines, including baselines that do not use the interaction network and baselines that do not use the vertex scores.

In this paper, we introduce NetMix2, an algorithm which unifies the network propagation and altered subnetwork approaches. NetMix2 generalizes NetMix [69] to a wide range of subnetwork families and vertex score distributions.

Specifically, NetMix2 takes as input a wide variety of subnetwork families, including not only the "connected family" used in existing altered subnetwork methods [28,43,69] but also any subnetwork family defined by linear or quadratic constraints, such as subnetworks with high edge density or subnetworks with small cut-size. We use this flexibility to investigate the topology of subnetworks identified by network propagation methods. We show empirically that network propagation does not correspond to standard topological constraints on altered subnetworks such as connectivity [28,43,69], cut-size [5], or edge-density [55]. Instead, we derive the *propagation family*, a subnetwork family that we show "approximates" the subnetworks identified by network propagation approaches and thereby unifies the two major network approaches in the literature: network propagation and altered subnetwork identification. NetMix2 also uses local false discovery rate (local FDR) methods [30–32] to flexibly model vertex score distributions, in contrast to the strict parametric assumptions made by existing methods [28,69].

On simulated data we show that NetMix2 outperforms network propagation for subnetworks from the propagation family and other common subnetwork families. Interestingly, NetMix2 outperforms network propagation by the largest margin for the propagation family. We then apply NetMix2 with the propagation family to cancer mutation data and genome-wide association studies (GWAS) data from several complex diseases. On cancer data, we show that NetMix2 outperforms existing network propagation and altered subnetwork methods in identifying cancer driver genes. On GWAS data, we demonstrate that network propagation often has similar performance to simple baselines that only use the vertex scores or only use the network. However, in cases where network propagation outperforms these baselines, we show that NetMix2 outperforms network propagation. The simulated data and GWAS experiments will be available in the full version of the paper.

2   Methods

2.1   Altered Subnetwork Problem

We start by formalizing the problem of altered subnetwork identification. Let $G = (V, E)$ be an interaction network with a score X_v for each vertex v. We assume there is an *altered subnetwork* $A \subseteq V$ whose scores $\{X_v\}_{v \in A}$ are drawn independently from a different distribution than the scores $\{X_v\}_{v \notin A}$ of vertices not in the altered subnetwork A. The topology of the altered subnetwork is described by membership in a *subnetwork family* $S \subseteq \mathcal{P}(V)$, where $\mathcal{P}(V)$ denotes the power set of all subsets of vertices V.

Following the exposition in [16,69], we model the distribution of the scores $\mathbf{X} = \{X_v\}_{v \in V}$ as the Altered Subnetwork Distribution (ASD).

Altered Subnetwork Distribution (ASD). *Let $G = (V, E)$ be a graph, let $S \subseteq \mathcal{P}(V)$ be a subnetwork family, and let $A \in S$. We say $\mathbf{X} = (X_v)_{v \in V}$ is* *distributed according to the* Altered Subnetwork Distribution $ASD_{\mathcal{S}}(A, \mathcal{D}_a, \mathcal{D}_b)$ *provided the X_v are independently distributed as*

$$X_v \sim \begin{cases} \mathcal{D}_{\mathrm{a}}, & \textit{if } i \in A, \\ \mathcal{D}_{\mathrm{b}}, & \textit{otherwise,} \end{cases} \tag{1}$$

where $\mathcal{D}_{\mathrm{a}}$ is the altered *distribution and $\mathcal{D}_{\mathrm{b}}$ is the* background *distribution.*

The distribution $ASD_{\mathcal{S}}(A, \mathcal{D}_a, \mathcal{D}_b)$ is parameterized by four quantities: the altered subnetwork A, the subnetwork family $\mathcal{S}$, the altered distribution $\mathcal{D}_a$, and the background distribution $\mathcal{D}_{\mathrm{b}}$.

Given the measurements $\mathbf{X} \sim ASD_{\mathcal{S}}(A, \mathcal{D}_a, \mathcal{D}_b)$ and the subnetwork family $\mathcal{S} \subseteq \mathcal{P}(V)$, the goal of the *Altered Subnetwork Problem* is to identify the altered subnetwork A. We formalize this problem below.

Altered Subnetwork Problem (ASP). Given $\mathbf{X} \sim ASD_{\mathcal{S}}(A, \mathcal{D}_a, \mathcal{D}_b)$ and subnetwork family $\mathcal{S}$, find A.

The ASP describes a broad class of problems that are studied in many fields including computational biology [28,43,69], statistics [1,2,34,47] and machine learning [10,16,73], with different problems making different choices for (1) the distributions $\mathcal{D}_a$, $\mathcal{D}_b$ and (2) the subnetwork family $\mathcal{S}$. Two prominent examples of distributions $\mathcal{D}_a$, $\mathcal{D}_b$ that have been previously studied in the biological literature are the following.

- **Normal distributions:** $\mathcal{D}_{\mathrm{a}} = N(\mu, 1)$ and $\mathcal{D}_{\mathrm{b}} = N(0, 1)$. Normal distributions are often used to model z-scores [11,29,57,64,69]. We call the ASP and ASD with these distributions the *normally distributed ASP* and *normally distributed ASD*, respectively; for notational convenience, we use $\mathrm{NASD}_{\mathcal{S}}(A, \mu)$ to refer to the normally distributed ASD.
- **Beta-uniform distributions:** $\mathcal{D}_{\mathrm{a}} = \mathrm{Beta}(a, 1)$ and $\mathcal{D}_{\mathrm{b}} = \mathrm{Uni}(0, 1)$. Beta-uniform mixture distributions are another common model for p-value distributions [28,67]. We call the ASP, ASD with these distributions the *Beta-Uniform ASP* and *Beta-Uniform ASD*, respectively.

We also list several examples of subnetwork families $\mathcal{S}$, with each subnetwork family corresponding to a different topological assumption on the altered subnetwork A. Some of these families have been explicitly applied in biological settings, while other families formalize topological constraints that are implicitly made in the biological literature.

- $\mathcal{S} = \mathcal{C}_G$, the *connected family*, or the set of all connected subgraphs S of an interaction network G. [28,43,69] identify altered subnetworks by solving the ASP for the connected family $\mathcal{C}_G$.
- $\mathcal{S} = \mathcal{E}_{G,p}$, the *edge-dense family*, or the set of all subgraphs S of G with edge-density $\frac{E(S)}{\binom{|S|}{2}} \geq p$, where $E(S) = |\{(u, v) \in E : u \in S, v \in S\}|$ is the number of edges between vertices in S. The edge-dense family $\mathcal{E}_{G,p}$ formalizes the topological constraints made by [36,55,83], which identify subnetworks with large edge-density.

- $\mathcal{S} = \mathcal{T}_{G,\rho}$, the *cut family*, or the set of all subgraphs S of G with $\frac{\text{cut}(S)}{|S|} \leq \rho$, where $\text{cut}(S) = |\{(u,v) \in E : u \in S, v \notin S\}|$ is the number of edges with exactly one endpoint in S. The cut family $\mathcal{T}_{G,\rho}$ formalizes the topological constraints made by [5], which identifies subnetworks with small cut.

We note that the ASP—with the subnetwork families $\mathcal{S}$ described above— describes the problem of identifying a single altered subnetwork in a network G. By creating a new subnetwork family consisting of the union of k disjoint subnetworks in family $\mathcal{S}$, the ASP also describes the problem of identifying *multiple* altered subnetworks.

Early methods for identifying altered subnetwork solved the ASP for the connected family $\mathcal{S} = \mathcal{C}_G$ and different choices of vertex score distributions $\mathcal{D}_a$, $\mathcal{D}_b$. For example, two seminal methods, jActiveModules [43] and heinz [28], solve the normally distributed and Beta-Uniform ASP, respectively, with the connected family $\mathcal{S} = \mathcal{C}_G$. Recently we showed that many existing methods, including jActiveModules and heinz, are *biased*, in the sense that they typically estimate subnetworks $\widehat{A}$ that are much larger than the altered subnetwork A [16,69]. To this end, we derived the NetMix algorithm, which finds an asymptotically unbiased $\widehat{A}_{\text{NetMix}}$ of the altered subnetwork A for the connected family $\mathcal{S} = \mathcal{C}_G$. However, as we demonstrate in [69] and Sect. 3 below, many of these methods—including NetMix—have comparable performance to a naive "scores-only" baseline that does not use the network G.

2.2   Network Propagation and the Propagation Family

Another strategy often used to incorporate interaction networks G with high throughput biological data is *network propagation*. Network propagation involves the use of random walk or diffusion processes to smooth or "propagate" vertex scores X_v across a network [23]. Formally, given vertex scores X_v, the *network propagated scores* Y_v are computed as $Y_v = \sum_{w \in V} M_{v,w} X_w$ where $M \in \mathbb{R}^{|V| \times |V|}$ is a similarity matrix on the vertices V typically derived from a random walk on the network G. One popular choice for the similarity matrix M is the random walk with restart (personalized PageRank) similarity matrix $M_{\text{PPR}} = r(I - (1 - r)P)^{-1}$, where $r \in (0,1)$ is the restart probability, I is the identity matrix, and P is the transition matrix for a random walk with restart on G.

A few methods have attempted to use network propagation to identify the altered subnetwork A from propagated scores Y_v, e.g., PRINCE [83] finds edge-dense subnetwork with large propagated scores Y_v. These methods implicitly assume that the propagated scores Y_v are larger for vertices $v \in A$ in the altered subnetwork A compared to vertices $v \notin A$ not in the altered subnetwork A. However, we empirically find that this assumption generally does not hold for altered subnetworks $A \in \mathcal{S}$ from the connected family $\mathcal{S} = \mathcal{C}_G$, the edge-dense family $\mathcal{S} = \mathcal{E}_{G,p}$ and the cut family $\mathcal{S} = \mathcal{T}_{G,\rho}$, which suggests that network propagation methods do not solve the ASP with these subnetwork families $\mathcal{S}$.

Thus, we derive a subnetwork family $\mathcal{S}$ that approximates the subnetworks identified by network propagation methods. Informally, we first note that

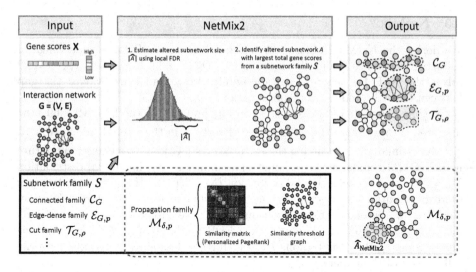

Fig. 1. Overview of the NetMix2 algorithm. The inputs to NetMix2 are a graph G, gene scores $\{X_v\}_{v \in V}$, and a subnetwork family $\mathcal{S}$. First, NetMix2 computes an estimate $|\widehat{A}|$ of the size $|A|$ of the altered subnetwork A using local false discovery rate (local FDR). Next, NetMix2 solves an optimization problem to identify the subnetwork $S \in \mathcal{S}$ size $|S| = |\widehat{A}|$ from the input subnetwork family $\mathcal{S}$ and with the largest total vertex score $\sum_{v \in S} X_v$. By default, NetMix2 uses the *propagation family* $\mathcal{S} = \mathcal{M}_{\delta,p}$. In this case, NetMix2 constructs an additional graph (the similarity threshold graph) based on vertex similarities quantified by Personalized PageRank from the input graph. The choice of subnetwork family $\mathcal{S}$ for NetMix2 is flexible and can be generalized to other families defined by linear or quadratic constraints including the *connected family* $\mathcal{C}_G$, *edge-dense family* $\mathcal{E}_{G,p}$, and *cut family* $\mathcal{T}_{G,\rho}$.

network propagation methods identify altered subnetworks A whose vertices $v \in A$ have large propagated scores Y_v. We observe that—by making the simplifying assumption that the vertex scores $X_v = 1_{\{v \in A\}}$ are binary—the propagated score $Y_v = \sum_{w \in A} M_{v,w}$ of a vertex v is large if the similarities $M_{v,w}$ is large for many $w \in A$. Intuitively, one natural way to enforce that the similarities $M_{v,w}$ are large is to lower-bound them, i.e., require that $M_{v,w} \geq \delta$ for many $w \in A$ and for some (large) constant $\delta > 0$.

This intuition motivates the formal definition of the *propagation family* $\mathcal{M}_{\delta,p}$, or the set of all subgraphs S with $M_{u,v} \geq \delta$ and $M_{v,u} \geq \delta$ for p fraction of tuples $(u,v) \in S$. (Because the similarity matrix M may not be symmetric, we constrain both $M_{u,v}$ and $M_{v,u}$.) We note that the propagation family $\mathcal{M}_{\delta,p}$ is equal to the edge-dense family $\mathcal{E}_{G_\delta,\rho}$ for the similarity threshold graph $G_\delta = (V, E_\delta)$, which has edge $(u,v) \in E_\delta$ if and only if $M_{u,v} \geq \delta$ and $M_{v,u} \geq \delta$.

In the full version of the paper, we partially formalize our derivation of the propagation family $\mathcal{M}_{\delta,p}$ with a bound on the false discovery rate (FDR) of the subnetwork consisting of the largest propagated scores.

2.3   NetMix2

We derive the NetMix2 algorithm, which solves the ASP for a wide range of subnetwork families $\mathcal{S}$ and distributions $\mathcal{D}_a$, $\mathcal{D}_b$ (Fig. 1). In particular, NetMix2 solves the ASP for the propagation family $\mathcal{M}_{\delta,p}$, and thus bridges the gap between the ASP and network propagation.

NetMix2 consists of two steps. As in our previous method NetMix [69], the first step is to estimate the number $\widehat{|A|}$ of vertices in the altered subnetwork A. NetMix estimated $\widehat{|A|}$ by fitting the vertex scores $\{X_v\}_{v \in V}$ to a Gaussian mixture model (GMM), under strict parametric assumptions on the altered distribution $\mathcal{D}_a = N(\mu, 1)$ and background distribution $\mathcal{D}_b = N(0, 1)$. However, not all vertex score distributions are well-fit by normal distributions of this form. Thus, in NetMix2, we extend NetMix by using local false discovery rate (local FDR) methods [30–32] to estimate α, as local FDR methods make mild assumptions on the forms of the distributions $\mathcal{D}_a$, $\mathcal{D}_b$ (further details will be available in the full version of the paper).

The second step of NetMix2 is to compute the subnetwork $S \in \mathcal{S}$ with size $|S| = \widehat{|A|}$ and largest total vertex score X_v:

$$\widehat{A}_{\text{NetMix2}} = \operatorname*{argmax}_{\substack{S \in \mathcal{S} \\ |S| \leq \widehat{|A|}}} \sum_{v \in S} X_v. \tag{2}$$

(2) can be computed using an integer linear program or integer quadratic program solver (e.g., Gurobi [37]) for a number of subnetwork families, including the edge-dense family $\mathcal{E}_{G,p}$, the cut family $\mathcal{T}_{G,\rho}$, the connected family $\mathcal{C}_G$, and the propagation family $\mathcal{M}_{\delta,p}$. We note that for the propagation family $\mathcal{M}_{\delta,p}$, the run-time for solving (2) depends on both the number $|E_\delta|$ of edges in the similarity threshold graph G_δ and the density p.

Note that (2) involves maximizing the sum $\sum_{v \in S} X_v$ of the vertex scores X_v, while the objective in the NetMix optimization problem [69] is the sum $\sum_{v \in S} r_v$ of the vertex responsibilities $r_v = P(v \in A \mid X_v)$. In practice, we observe that maximizing the sum of the vertex scores X_v yields slightly better performance than that of the responsibilities r_v.

2.4   Scores-Only and Network-Only Baselines

When evaluating any algorithm for the identification of altered subnetworks, we argue that it is essential to compare against two baselines: a *"scores-only"* baseline that only uses the vertex scores X_v, and a *"network-only"* baseline that only uses the interaction network G. These two baselines quantify whether the altered subnetwork algorithm is outperforming simpler approaches that do not integrate vertex scores with a network; moreover, these baselines should be evaluated on each dataset and match as closely as possible the inputs to the altered subnetwork problem. A scores-only baseline is straightforward: we rank the vertices v by their vertex scores X_v. Because this baseline outputs a ranked list of all vertices in the graph, we threshold the ranking when evaluating against other altered subnetwork algorithms by taking the k most highly ranked vertices for some integer k.

Defining a network-only baseline is a more subtle issue, and was discussed in two recent papers [50,54]. Levi et al. [54] benchmarks altered subnetwork algorithms on randomly permuted vertex scores $\widetilde{X}_v$ while keeping the network G fixed. The authors find that many existing methods output similar altered subnetworks (in terms of GO enrichment) on their permuted data, which suggests that these methods are utilizing the network G more than the vertex scores X_v. Lazareva et al. [50] benchmarks altered subnetwork algorithms on randomly permuted networks with the same degree distribution as G while keeping the vertex scores X_v fixed. The authors find that many existing algorithms output similar altered subnetworks on permuted networks, indicating a degree bias in these methods. We propose a more direct network-only baseline: we rank vertices v by their network centrality score $N(v)$ for a network centrality measure N that is derived from the topological constraints used by the altered subnetwork algorithm. For example, for an algorithm that relies on the connected subfamily, we propose that degree centrality $N(v) = d_v$ is an appropriate measure, as in [50]. However, for network propagation algorithms that use random walk with restart, we claim that the PageRank centrality $N(v) = (M_{\text{PPR}} \cdot \mathbf{1})_v$, where $\mathbf{1} \in \mathbb{R}^n$ is an all-ones vector, is the more appropriate network-only baseline. This is because compared to degree centrality, PageRank centrality better captures how network propagation methods use the interaction network G.

3   Results

We evaluated NetMix2 on simulated data, somatic mutations in cancer, and genome-wide association studies (GWAS) from several diseases. Details of our analyses on simulated data and GWAS data will be available in the full version of the paper. Unless indicated otherwise, we ran NetMix2 with the propagation family $\mathcal{M}_{\delta,p}$ using the personalized PageRank matrix M_{PPR} with restart probability $r = 0.4$. We solved the integer program in (2) using the Gurobi optimizer [37]. We ran Gurobi for up to 24 h, which typically results in a near-optimal solution. For all ranking methods (e.g., network propagation, scores-only, and network-only baselines), we estimated the altered subnetwork $\widehat{A}$ as the $|\widehat{A}_{\text{NetMix2}}|$ highest ranked vertices, where $\widehat{A}_{\text{NetMix2}}$ is the output of NetMix2.

3.1   Somatic Mutations in Cancer

Next, we compared the performance of NetMix2 against several other methods for identification of altered subnetworks [28,41,69,70] on the task of identifying cancer driver genes. For each vertex (gene) v, the vertex score X_v is a z-score computed from p-values from MutSig2CV [49], a statistical method that predicts cancer driver genes based on the frequency that the gene is mutated in a cohort of cancer patients. We obtained these scores for 10,437 samples across

Table 1. Results of altered subnetwork identification methods using MutSig2CV cancer driver gene *p*-values from TCGA tumor samples. Subnetworks are evaluated using reference sets of cancer genes from CGC, OncoKB, and TCGA. The best scores are colored in bold red. ** GMM from NetMix overestimated the size of the subnetwork, thus we excluded genes with outlier scores as described in [69].*

Method	STRING network						
	Subnetwork size	CGC		OncoKB		TCGA	
		Number	*F-measure*	*Number*	*F-measure*	*Number*	*F-measure*
NetMix2	280	132	**0.3**	133	**0.313**	151	**0.546**
NetMix	313*	129	0.282	130	0.295	147	0.502
Heinz (FDR = 0.01)	335	139	0.297	138	0.306	156	0.513
NetSig	773	145	0.211	172	0.257	84	0.161
Hierarchical HotNet	246	73	0.172	70	0.172	74	0.285
Network propagation	280	86	0.195	89	0.210	98	0.354
Scores-only	280	126	0.286	127	0.3	145	0.524
Network-only	280	77	0.175	83	0.196	55	0.199

33 cancer types from the TCGA PanCanAtlas project [6]. We also compared to three ranking methods: network propagation, scores-only, and network-only (PageRank centrality). We ran each method using the STRING protein-protein interaction network [77] and evaluated the performance by computing the overlap between genes in their reported subnetworks and reference lists of cancer driver genes from the COSMIC Cancer Gene Census (CGC) [78], OncoKB [14], and TCGA [6]. Further details on datasets and procedures for running each method will be available in the full version of the paper.

We found that NetMix2 using the propagation family outperformed other methods in F-measure for all three reference gene sets (Table 1). In addition, comparing NetMix2 using the propagation family and NetMix2 using the connected family (the second best method) shows that the altered subnetwork found using the propagation family contains several genes that are not found by using the connected family. For example, NetMix2 using the propagation family identifies 9 CGC driver genes that are not found by using the connected family including *PDGFRA*, an oncogene whose gain-of-function mutations promote cancer growth [84] and *NCOR2*, a well-known tumor suppressor implicated in breast and prostate cancers [8]; none of these genes are found by the baseline methods.

In the full version of the paper, we include a comparison against altered subnetwork methods with different parameter settings as well as a network-only baseline with degree centrality. We also observe that many network approaches, including NetSig and Hierarchical HotNet, have lower F-measure than the scores-only baseline. While it is possible we are not using the optimal parameters for these methods, our results suggest that these methods are over-utilizing the network compared to the vertex scores.

4   Discussion

We introduced NetMix2, an algorithm that unifies the network propagation and altered subnetwork approaches to analyze biological data using interaction networks. NetMix2 is inspired by network propagation, a standard approach for solving the *network ranking* problem, and attempts to bridge the gap between two paradigms for using networks in the analysis of high-throughput genomic data—*network ranking* and the identification of *altered subnetworks*—in a principled way by explicitly deriving a new family of subnetworks called the *propagation family* that approximates the altered subnetworks found by network propagation methods. We showed that NetMix2 is effective in finding disease-associated genes using somatic mutation data in cancer. At the same time, our evaluation revealed that simple baseline methods that use either only the vertex scores or only the interaction network sometimes perform surprisingly well, often outperforming more sophisticated network methods. While publications describing new network methods typically benchmark against other network methods, they are wildly inconsistent in benchmarking against scores-only and network-only baselines. It is rare to see a paper that benchmarks against both baselines. Moreover, the network-only baseline should be calibrated to use the same network information as the method under evaluation; e.g., PageRank centrality is a more appropriate benchmark for network propagation methods than vertex degree.

There are several directions for future work. The first direction is to extend NetMix2 to identify *multiple* altered subnetworks simultaneously. This can be done by running NetMix2 iteratively, or by modifying the integer program to output multiple solutions. However, solving the corresponding model selection procedure to choose the number and sizes of altered subnetworks without overfitting is a difficult problem. A second direction is to extend NetMix2 with an appropriate permutation test to evaluate the statistical significance of the altered subnetwork(s). Finally, while we evaluated several network methods and simple baselines, there are numerous other network methods that could be included in these benchmarks. However, there are few gold standards to perform such a comprehensive evaluation as the reference disease gene sets remain relatively limited and potentially biased by their source. Thus, a useful extension would be deriving a reliable evaluation scheme for network methods that accounts for various sources of bias including the ascertainment bias in current interaction networks and disease gene sets.

Acknowledgement. The authors would like to thank Jasper C. H. Lee and Christopher Musco for helpful discussions, as well as Matthew A. Myers and Palash Sashittal for reviewing early versions of the manuscript. U.C. is supported by NSF GRFP DGE 2039656. B.J.R. is supported by grant U24CA264027 from the National Cancer Institute (NCI).

References

1. Addario-Berry, L., Broutin, N., Devroye, L., Lugosi, G.: On combinatorial testing problems. Ann. Stat. **38**(5), 3063–3092 (2010)
2. Arias-Castro, E., Candès, E.J., Durand, A.: Detection of an anomalous cluster in a network. Ann. Stat. **39**(1), 278–304 (2011)
3. Arias-Castro, E., Candès, E.J., Helgason, H., Zeitouni, O.: Searching for a trail of evidence in a maze. Ann. Stat. **36**(4), 1726–1757 (2008)
4. Arias-Castro, E., Donoho, D.L., Huo, X.: Adaptive multiscale detection of filamentary structures in a background of uniform random points. Ann. Stat. **34**(1), 326–349 (2006)
5. Azencott, C.A., Grimm, D., Sugiyama, M., Kawahara, Y., Borgwardt, K.M.: Efficient network-guided multi-locus association mapping with graph cuts. Bioinformatics **29**(13), i171–i179 (2013)
6. Bailey, M.H., et al.: Comprehensive characterization of cancer driver genes and mutations. Cell **173**(2), 371–385 (2018)
7. Barel, G., Herwig, R.: NetCore: a network propagation approach using node coreness. Nucleic Acids Res. **48**(17), e98–e98 (2020)
8. Battaglia, S., Maguire, O., Campbell, M.J.: Transcription factor co-repressors in cancer biology: roles and targeting. Int. J. Cancer **126**(11), 2511–2519 (2010)
9. Berger, B., Peng, J., Singh, M.: Computational solutions for omics data. Nature Rev. Genet. **14**(5), 333–346 (2013)
10. Cadena, J., Chen, F., Vullikanti, A.: Near-optimal and practical algorithms for graph scan statistics with connectivity constraints. ACM Trans. Knowl. Discov. Data **13**(2), 20:1-20:33 (2019)
11. Cai, T.T., Jin, J., Low, M.G.: Estimation and confidence sets for sparse normal mixtures. Ann. Stat. **35**(6), 2421–2449 (2007)
12. Califano, A., Butte, A.J., Friend, S., Ideker, T., Schadt, E.: Leveraging models of cell regulation and GWAS data in integrative network-based association studies. Nat. Genet. **44**(8), 841–847 (2012)
13. Cao, M., et al.: Going the distance for protein function prediction: a new distance metric for protein interaction networks. PLoS One **8**(10), 1–12 (2013)
14. Chakravarty, D., et al.: OncoKB: a precision oncology knowledge base. JCO Precis. Oncol. **1**, 1–16 (2017)
15. Chasman, D., Siahpirani, A.F., Roy, S.: Network-based approaches for analysis of complex biological systems. Curr. Opin. Biotech. **39**, 157–166 (2016)
16. Chitra, U., Ding, K., Lee, J.C., Raphael, B.J.: Quantifying and reducing bias in maximum likelihood estimation of structured anomalies. In: Proceedings of the 38th International Conference on Machine Learning, pp. 1908–1919. PMLR, 18–24 July 2021
17. Cho, D.Y., Kim, Y.A., Przytycka, T.M.: Chapter 5: network biology approach to complex diseases. PLoS Comput. Biol. **8**(12), 1–11 (2012)
18. Choobdar, S., et al.: Assessment of network module identification across complex diseases. Nat. Methods **16**(9), 843–852 (2019)
19. Chua, H.N., Sung, W.K., Wong, L.: Exploiting indirect neighbours and topological weight to predict protein function from protein-protein interactions. Bioinformatics **22**(13), 1623–1630 (2006)
20. modENCODE Consortium, Roy, S., Ernst, J., Kharchenko, P.V., Kheradpour, P., et al.: Identification of functional elements and regulatory circuits by drosophila modencode. Science **330**(6012), 1787–1797 (2010)

21. Cornish, A.J., Markowetz, F.: SANTA: Quantifying the functional content of molecular networks. PLoS Comput. Biol. **10**(9), e1003808 (2014)
22. Cowen, L., Devkota, K., Hu, X., Murphy, J.M., Wu, K.: Diffusion state distances: Multitemporal analysis, fast algorithms, and applications to biological networks. SIAM J. Math. Data Sci. **3**(1), 142–170 (2021)
23. Cowen, L., Ideker, T., Raphael, B.J., Sharan, R.: Network propagation: a universal amplifier of genetic associations. Nat. Rev. Genet. **18**(9), 551–562 (2017)
24. Creixell, P., et al.: Pathway and network analysis of cancer genomes. Nat. Methods **12**(7), 615–621 (2015)
25. de la Fuente, A.: From 'differential expression' to 'differential networking' - identification of dysfunctional regulatory networks in diseases. Trends Genet. **26**(7), 326–333 (2010)
26. Deng, M., Zhang, K., Mehta, S., Chen, T., Sun, F.: Prediction of protein function using protein-protein interaction data. J. Comput. Biol. **10**(6), 947–960 (2003)
27. Dimitrakopoulos, C.M., Beerenwinkel, N.: Computational approaches for the identification of cancer genes and pathways. WIREs Syst. Biol. Med. **9**(1), e1364 (2017)
28. Dittrich, M.T., Klau, G., Rosenwald, A., Dandekar, T., Muller, T.: Identifying functional modules in protein-protein interaction networks: an integrated exact approach. Bioinformatics **24**(13), i223–i231 (2008)
29. Donoho, D., Jin, J.: Higher criticism for detecting sparse heterogeneous mixtures. Ann. Stat. **32**(3), 962–994 (2004)
30. Efron, B.: Large-scale simultaneous hypothesis testing: the choice of a null hypothesis. J. Am. Stat. Assoc. **99**(465), 96–104 (2004)
31. Efron, B.: Correlation and large-scale simultaneous significance testing. J. Am. Stat. Assoc. **102**(477), 93–103 (2007)
32. Efron, B.: Size, power and false discovery rates. Ann. Stat. **35**(4), 1351–1377 (2007)
33. Ghiassian, S.D., Menche, J., Barabási, A.L.: A DIseAse MOdule Detection (DIA-MOnD) algorithm derived from a systematic analysis of connectivity patterns of disease proteins in the human interactome. PLoS Comput. Biol. **11**(4), e1004120 (2015)
34. Glaz, J., Naus, J., Wallenstein, S.: Scan Statistics. Springer-Verlag, New York (2001). https://doi.org/10.1007/978-1-4757-3460-7
35. Gligorijević, V., Pržulj, N.: Methods for biological data integration: perspectives and challenges. J. Roy. Soc. Interface **12**(112), 20150571 (2015)
36. Guo, Z., et al.: Edge-based scoring and searching method for identifying condition-responsive protein-protein interaction sub-network. Bioinformatics **23**(16), 2121–2128 (2007)
37. Gurobi Optimization, LLC: Gurobi Optimizer Reference Manual (2021)
38. Halldórsson, B.V., Sharan, R.: Network-based interpretation of genomic variation data. J. Mol. Biol. **425**(21), 3964–3969 (2013)
39. Hofree, M., Shen, J.P., Carter, H., Gross, A., Ideker, T.: Network-based stratification of tumor mutations. Nat. Methods **10**(11), 1108–1115 (2013)
40. Hormozdiari, F., Penn, O., Borenstein, E., Eichler, E.E.: The discovery of integrated gene networks for autism and related disorders. Genome Res. **25**(1), 142–154 (2015)
41. Horn, H., Lawrence, M.S., Chouinard, C.R., Shrestha, Y., Hu, J.X., et al.: NetSig: network-based discovery from cancer genomes. Nat. Methods **15**(1), 61–66 (2018)
42. Huang, J.K., et al.: Systematic evaluation of molecular networks for discovery of disease genes. Cell Syst. **6**(4), 484–495 (2018)

43. Ideker, T., Ozier, O., Schwikowski, B., Siegel, A.F.: Discovering regulatory and signalling circuits in molecular interaction networks. Bioinformatics **18**(suppl 1), S233–S240 (2002)
44. Ideker, T., et al.: Integrated genomic and proteomic analyses of a systematically perturbed metabolic network. Science **292**(5518), 929–934 (2001)
45. Jia, P., Zhao, Z.: Network assisted analysis to prioritize GWAS results: principles, methods and perspectives. Hum. Genet. **133**(2), 125–138 (2014). https://doi.org/10.1007/s00439-013-1377-1
46. Kloumann, I.M., Ugander, J., Kleinberg, J.: Block models and personalized PageRank. Proc. Natl. Acad. Sci. **114**(1), 33–38 (2017)
47. Kulldorff, M.: A spatial scan statistic. Commun. Stat. Theory Methods **26**(6), 1481–1496 (1997)
48. Köhler, S., Bauer, S., Horn, D., Robinson, P.N.: Walking the interactome for prioritization of candidate disease genes. Am. J. Hum. Genet. **82**(4), 949–958 (2008)
49. Lawrence, M.S., et al.: Discovery and saturation analysis of cancer genes across 21 tumour types. Nature **505**(7484), 495–501 (2014)
50. Lazareva, O., Baumbach, J., List, M., Blumenthal, D.B.: On the limits of active module identification. Briefings Bioinf. **22**(5), bbab066 (2021)
51. Lee, I., Blom, U.M., Wang, P.I., Shim, J.E., Marcotte, E.M.: Prioritizing candidate disease genes by network-based boosting of genome-wide association data. Genome Res. **21**(7), 1109–1121 (2011)
52. Leiserson, M.D.M., Vandin, F., Wu, H.T., Dobson, J.R., et al.: Pan-cancer network analysis identifies combinations of rare somatic mutations across pathways and protein complexes. Nat. Genetics **47**(2), 106–114 (2015)
53. Leiserson, M.D., Eldridge, J.V., Ramachandran, S., Raphael, B.J.: Network analysis of GWAS data. Curr. Opin. Genet. Dev. **23**(6), 602–610 (2013)
54. Levi, H., Elkon, R., Shamir, R.: DOMINO: a network-based active module identification algorithm with reduced rate of false calls. Mol. Syst. Biol. **17**(1), e9593 (2021)
55. Liu, Y., et al.: SigMod: an exact and efficient method to identify a strongly interconnected disease-associated module in a gene network. Bioinformatics **33**(10), 1536–1544 (2017)
56. Luo, Y., et al.: A network integration approach for drug-target interaction prediction and computational drug repositioning from heterogeneous information. Nat. Commun. **8**(1), 573 (2017)
57. McLachlan, G., Bean, R.W., Jones, L.B.T.: A simple implementation of a normal mixture approach to differential gene expression in multiclass microarrays. Bioinformatics **22**(13), 1608–1615 (2006)
58. Menche, J., et al.: Uncovering disease-disease relationships through the incomplete human interactome. Science **347**(6224), 1257601 (2015)
59. Mitra, K., Carvunis, A.R., Ramesh, S.K., Ideker, T.: Integrative approaches for finding modular structure in biological networks. Nat. Rev. Genet. **14**(10), 719–732 (2013)
60. Nabieva, E., Jim, K., Agarwal, A., Chazelle, B., Singh, M.: Whole-proteome prediction of protein function via graph-theoretic analysis of interaction maps. Bioinformatics **21**, i302–i310 (2005)
61. Nibbe, R.K., Koyutürk, M., Chance, M.R.: An integrative-omics approach to identify functional sub-networks in human colorectal cancer. PLoS Comput. Biol. **6**(1), e1000639 (2010)
62. Nikolayeva, I., Pla, O.G., Schwikowski, B.: Network module identification-a widespread theoretical bias and best practices. Methods **132**, 19–25 (2018)

63. Page, L., Brin, S., Motwani, R., Winograd, T.: The PageRank citation ranking: Bringing order to the web. Technical report 1999-66, Stanford InfoLab, November 1999

64. Pan, W., Lin, J., Le, C.T.: A mixture model approach to detecting differentially expressed genes with microarray data. Funct. Integr. Genomics **3**(3), 117–124 (2003). https://doi.org/10.1007/s10142-003-0085-7

65. Paull, E.O., Carlin, D.E., Niepel, M., Sorger, P.K., Haussler, D., Stuart, J.M.: Discovering causal pathways linking genomic events to transcriptional states using Tied Diffusion Through Interacting Events (TieDIE). Bioinformatics **29**(21), 2757–2764 (2013)

66. Picart-Armada, S., Barrett, S.J., Willé, D.R., Perera-Lluna, A., Gutteridge, A., Dessailly, B.H.: Benchmarking network propagation methods for disease gene identification. PLoS Comput. Biol. **15**(9), 1–24 (2019)

67. Pounds, S., Morris, S.W.: Estimating the occurrence of false positives and false negatives in microarray studies by approximating and partitioning the empirical distribution of p-values. Bioinformatics **19**(10), 1236–1242 (2003)

68. Radivojac, P., et al.: A large-scale evaluation of computational protein function prediction. Nat. Methods **10**(3), 221–227 (2013)

69. Reyna, M.A., Chitra, U., Elyanow, R., Raphael, B.J.: NetMix: a network-structured mixture model for reduced-bias estimation of altered subnetworks. J. Computat. Biol. **28**(5), 469–484 (2021)

70. Reyna, M.A., Leiserson, M.D., Raphael, B.J.: Hierarchical HotNet: identifying hierarchies of altered subnetworks. Bioinformatics **34**(17), i972–i980 (2018)

71. Robinson, S., Nevalainen, J., Pinna, G., Campalans, A., Radicella, J.P., Guyon, L.: Incorporating interaction networks into the determination of functionally related hit genes in genomic experiments with Markov random fields. Bioinformatics **33**(14), i170–i179 (2017)

72. Sharan, R., Ulitsky, I., Shamir, R.: Network-based prediction of protein function. Mol. Syst. Biol. **3**, 88 (2007)

73. Sharpnack, J., Krishnamurthy, A., Singh, A.: Near-optimal anomaly detection in graphs using Lovász extended scan statistic. In: Proceedings of the 26th International Conference on Neural Information Processing Systems, NIPS 2013, vol. 2. pp. 1959–1967 (2013)

74. Sharpnack, J., Rinaldo, A., Singh, A.: Detecting anomalous activity on networks with the graph Fourier scan statistic. IEEE Trans. Signal Process. **64**(2), 364–379 (2016)

75. Sharpnack, J., Singh, A., Rinaldo, A.: Changepoint detection over graphs with the spectral scan statistic. In: Artificial Intelligence and Statistics, pp. 545–553 (2013)

76. Shrestha, R., et al.: HIT'nDRIVE: patient-specific multidriver gene prioritization for precision oncology. Genome Res. **27**(9), 1573–1588 (2017)

77. Szklarczyk, D., et al.: STRING v10: protein-protein interaction networks, integrated over the tree of life. Nucleic Acids Res. **43**(D1), D447–D452 (2015)

78. Tate, J.G., et al.: COSMIC: the catalogue of somatic mutations in cancer. Nucleic Acids Res. **47**(D1), D941–D947 (2019)

79. Ulitsky, I., Shamir, R.: Identification of functional modules using network topology and high-throughput data. BMC Syst. Biol. **1**(1), 8 (2007). https://doi.org/10.1186/1752-0509-1-8

80. Vandin, F., Clay, P., Upfal, E., Raphael, B.J.: Discovery of mutated subnetworks associated with clinical data in cancer. In: Pacific Symposium on Biocomputing, vol. 17, pp. 55–66 (2012)

81. Vandin, F., Upfal, E., Raphael, B.J.: Algorithms for detecting significantly mutated pathways in cancer. J. Comput. Biol. **18**(3), 507–522 (2011)
82. Vandin, F., Upfal, E., Raphael, B.J.: De novo discovery of mutated driver pathways in cancer. Genome Res. **22**(2), 375–385 (2012)
83. Vanunu, O., Magger, O., Ruppin, E., Shlomi, T., Sharan, R.: Associating genes and protein complexes with disease via network propagation. PLoS Comput. Biol. **6**(1), e1000641 (2010)
84. Velghe, A., et al.: PDGFRA alterations in cancer: characterization of a gain-of-function V536E transmembrane mutant as well as loss-of-function and passenger mutations. Oncogene **33**(20), 2568–2576 (2014)
85. Vlaic, S., et al.: ModuleDiscoverer: identification of regulatory modules in protein-protein interaction networks. Sci. Rep. **8**(1), 433 (2018)
86. Wang, X., Terfve, C., Rose, J.C., Markowetz, F.: HTSanalyzeR: an R/Bioconductor package for integrated network analysis of high-throughput screens. Bioinformatics **27**(6), 879–880 (2011)
87. Weston, J., Elisseeff, A., Zhou, D., Leslie, C.S., Noble, W.S.: Protein ranking: from local to global structure in the protein similarity network. Proc. Nat. Acad. Sci. **101**(17), 6559–6563 (2004)
88. Xia, J., Gill, E.E., Hancock, R.E.W.: NetworkAnalyst for statistical, visual and network-based meta-analysis of gene expression data. Nat. Protoc. **10**(6), 823–844 (2015)
89. Zhou, D., Bousquet, O., Lal, T., Weston, J., Schölkopf, B.: Learning with local and global consistency. In: Advances in Neural Information Processing Systems, vol. 16. MIT Press (2004)

Multi-modal Genotype and Phenotype Mutual Learning to Enhance Single-Modal Input Based Longitudinal Outcome Prediction

Alireza Ganjdanesh[1], Jipeng Zhang[2], Wei Chen[2,3,4(✉)], and Heng Huang[1(✉)]

[1] Department of Electrical and Computer Engineering, University of Pittsburgh, Pittsburgh, PA, USA
{alireza.ganjdanesh,heng.huang}@pitt.edu

[2] Department of Biostatistics, University of Pittsburgh, Pittsburgh, PA, USA
jiz214@pitt.edu

[3] Department of Pediatrics, UPMC Children's Hospital of Pittsburgh, Pittsburgh, PA, USA
wei.chen@chp.edu

[4] Department of Human Genetics, University of Pittsburgh, Pittsburgh, PA, USA

Abstract. In recent years, due to the advance of modern sensory devices, the collection of multiple biomedical data modalities such as imaging genetics has gotten feasible, and multimodal data analysis has attracted significant attention in bioinformatics. Although existing multimodal learning methods have shown superior ability in combining data from multiple sources, they are not directly applicable for many real-world biological and biomedical studies that suffer from missing data modalities due to the high expenses of collecting all modalities. Thus, in practice, usually, only a *main* modality containing a major 'diagnostic signal' is used for decision making as *auxiliary* modalities are not available. In addition, during the examination of a subject regarding a chronic disease (with longitudinal progression) in a visit, typically, two diagnosis-related questions are of main interest that are what their status currently is (diagnosis) and how it will change before their next visit (longitudinal outcome) if they maintain their disease trajectory and lifestyle. Accurate answers to these questions can distinguish vulnerable subjects and enable clinicians to start early treatments for them. In this paper, we propose a new adversarial mutual learning framework for longitudinal prediction of disease progression such that we properly leverage several modalities of data available in training set to develop a more accurate model using single-modal for prediction. Specifically, in our framework, a single-modal

This work was partially supported by NSF IIS 1845666, 1852606, 1838627, 1837956, 1956002, IIA 2040588.

Supplementary Information The online version contains supplementary material available at https://doi.org/10.1007/978-3-031-04749-7_13.

model (that utilizes the *main* modality) learns from a pretrained multimodal model (which takes both *main* and *auxiliary* modalities as input) in a mutual learning manner to 1) infer outcome-related representations of the *auxiliary* modalities based on its own representations for the *main* modality during adversarial training and 2) effectively combine them to predict the longitudinal outcome. We apply our new method to analyze the retinal imaging genetics for the early diagnosis of Age-related Macular Degeneration (AMD) disease in which we formulate prediction of longitudinal AMD progression outcome of subjects as a classification problem of simultaneously grading their current AMD severity as well as predicting their condition in their next visit with a preselected time duration between visits. Our experiments on the Age-Related Eye Disease Study (AREDS) dataset demonstrate the superiority of our model compared to baselines for simultaneously grading and predicting future AMD severity of subjects.

1   Introduction

Recent advances in multimodal biomedical imaging and high throughput genotyping and sequencing techniques allow us to study integrative imaging genetics and provide exciting new opportunities to ultimately improve our understanding of different disease mechanisms. Although many multimodal learning methods have been developed and shown superior ability in integrative analysis of imaging genetics data, the following two challenging problems are still desired to address for practical applications:

Input Data with Missing Modalities: An ideal case is that the researchers or clinicians have access to all of the informative data modalities for decision making, *i.e.*, be able to perform multimodal data based diagnosis. However, due to the high cost of collecting all data modalities, typically, only a single *main* modality that provides the majority of 'signal' about a subject's status is examined in practice. For instance, it has been established that genetic factors play an essential role in the progression of Age-related Macular Degeneration (AMD) pathogenesis [20,21,71,77]. Thanks to advances in sequencing technologies [1,48,49], the determination of whole-genome sequence is feasible nowadays and can provide valuable information for AMD diagnosis, but AMD severity score [19] is usually only determined by exploring characteristics of subjects' Color Fundus Photographs (CFP) - that is the most accessible retinal image modality globally - in practice due to lack of expensive facilities required for sequencing, especially in low-resourced areas.

Diagnosis and Prediction of Longitudinal Outcome: Many diseases have several stages in terms of severity, and a subject may progress to advanced ones through time. Predicting the disease progression can help understand the disease's dynamics and thus, advise physicians on medication intake. Two questions of main interest when studying a subject's condition in clinical practice are that given their examination records, "how is current severity status of them?" (*diagnosis*), and "how will their disease severity change until their next visit?" (*i.e.*, *longitudinal outcome prediction*) Accurate answers to these questions can comprehensively predict a subject's current status as well as their future disease trajectory and enable clinicians to start early treatment for highly vulnerable ones to decelerate their disease progression. However, it is often prohibited to collect the time series biomedical data (from multiple years visits) to predict the disease progression in practical applications, especially for low-resourced areas. The researchers and clinicians often want to make the diagnosis and longitudinal outcome prediction only using the data at the current visit, which makes the disease progression prediction more challenging.

We aim to solve both challenging tasks in the second aspect while considering the constraints mentioned in the first one. To do so, firstly, our intuition is that single-modal input based models that benefit from the *main* and *auxiliary* data modalities collected in multi-modal datasets during training and rely on the *main* modality in their inference phase better mimic clinical practice. Therefore, we train such a model in our framework. Secondly, we can overcome the longitudinal prediction challenge by leveraging records collected at the current visit to make predictions for the current and next visits if the time gap between them is not too large compared to the typical pace of the disease progression.

Multimodal learning (MML) [22,23,44,69,79] and Deep Mutual Learning (DML) [31,81] methods have shown significant results recently. On the one hand, MML methods can effectively utilize the supervision from several modalities to improve the classification performance in tasks such as visual question answering and video categorization. However, they require that all input modalities be available for their inference, which limits their practicality for biomedical applications that usually suffer from missing modalities. On the other hand, DML methods have demonstrated that two models that are trained together and get feedback from their peers have better generalization performance compared to their baseline models that are trained separately. Thus, our intuition is to overcome the missing modality problem of multimodal learning methods for our task by developing a single-modal model while leveraging the benefits of mutual learning by training the model mutually with a multimodal one.

In this paper, we introduce a novel framework based on deep mutual learning [31,81] in which a single-modal model – our model only need the *main* diagnostic modality (*e.g.* CFP) of a target disease (*e.g.* AMD) to conduct the predictions – and a pretrained multimodal model that takes the *main* and *auxiliary* (genetics and age) data modalities as input evolve together during training. Both models learn to solve our formulated classification problem to simultaneously 1) grade the current disease status of a subject (Advanced or not) and 2) predict their future condition in their next visit (Advanced or not, with a predefined time-gap between visits, *e.g.* 3 years). Further, we hypothesize that genetics and demographics (age) information can provide 'complementary knowledge' for a model for longitudinal outcome prediction, especially in the subjects with similar fundus images that may have different future trajectories due to their genetic differences. Therefore, we design our framework such that the single-modal model learns to infer outcome-related representations of *auxiliary* modalities using its representations for the *main* modality from its multimodal colleague using a

Riemannian adversarial training scheme. After that, it combines them to make the predictions. In addition, we use entropy regularization during the pretraining stage of the multimodal model to prevent it from neglecting noisy auxiliary modalities and focusing only on the main one. Our contributions can be summarized as follows:

- We introduce a new framework to simultaneously diagnose current status and predict the longitudinal outcome of subjects for disease progression by developing a model that only requires the *main* diagnostic modality – collected at current visit – for its predictions while properly leveraging *auxiliary* modalities available in the training set to enhance final model's performance.
- We propose to model the complex relationship of representations of the main modality and auxiliary ones by Riemannian Generative Adversarial Networks.
- We design a functional entropy regularized pretraining scheme for the multimodal model to prevent it from shortcut learning to discard the auxiliary modality and only use the more informative main modality.

2   Related Work

Multi-Modal Learning (MML): MML combines knowledge from several modalities to enhance predictions for a target task. It has achieved significant results in domains such as video understanding and visual question answering that leverage several types of visual, audial, or textual data [2,17,22–24,28,35,39,44,50,56,67,70,79]. However, these works assume that all modalities are present during training and inference which limits their direct application in medical problems that missing modalities are a common challenge in them. A popular workaround is to reconstruct and impute missing modalities using available ones [14,47,57,61,64,66,76]. However, reconstruction of extremely high-dimensional modalities such as genetics ($\sim 1.6 \times 10^5$ dimensional in our problem) is not practical in healthcare problems with limited training data. Further, predicting some modalities from others may not always be feasible. For instance, prediction of one of RGB and thermal images [76] from the other is sensible, but reconstruction of whole-genome sequence from fundus images of eyes is not. Another group of methods proposes variational approaches to deal with missing modalities and model the joint posterior of representations of modalities as a product-of-experts [74]. Lee and Van der Schaar [42] use this method to integrate multi-omics data and train modality-specific predictors to ensure representations of individual modalities are learned faithfully. Nevertheless, a modality-specific predictor is not reasonable in the longitudinal prediction of disease outcome for modalities such as genetics that are *static* while the disease status of a subject may change in time. This is the case for the method of Wang et al. [69] as well that trains modality-specific classifiers with incomplete data pairs and train a final multi-modal model using limited complete pairs while distilling [27,34,45] the knowledge of pretrained models in it.

Deep Mutual Learning (DML): In a nutshell, two or several models are trained simultaneously in DML such that each model gets supervision from training labels and predictions/representations of other models. Zhang et al. [81] introduced DML and showed it has better image classification performance compared to knowledge distillation [27,34,45] methods. Since then, different types of DML for various applications such as image classification [31,41,59,73], semi-supervised learning [75], self-supervised learning [8,68], and object detection [54] have been proposed. These models are not suitable for our problem as they train two models with the same input modality. Recently, Zhang et al. [80] proposed a multimodal image segmentation model to train two single-modal models in a DML manner. However, their multimodal DML idea is designed for problems that their modalities are two 'views' of the same phenomenon, not 'complementary' modalities such as CFP and genetics for AMD that CFP contains the majority of the diagnostic signal while noisy genetics input only complements the knowledge from CFP.

Age-Related Macular Degeneration (AMD): In this paper, we analyze the retinal imaging genetics data which were collected to study the AMD disease and are a good testing platform to evaluate our new method. AMD is a chronic disease [46] that causes the progressive decline of vision due to the dysfunction of the central retina in older adults and is the major root of blindness in elder Caucasians [9,16,65]. Based on a scale called AMD severity score, three stages are defined for AMD: early, intermediate, and late (advanced) [19]. The severity score is determined by exploring characteristics of the Color Fundus Photographs (CFP) of subjects. The main symptom of the early and intermediate stages is the presence of yellowish deposits called 'drusen' in the retina, and most patients are asymptomatic in them [5,29]. The irreversible stage that is accompanied by severe vision loss is late AMD that appears in two forms: 'Dry' and 'Wet'. In Dry AMD (Geographic Atrophy), accumulation of drusen in the retina decreases its sensitivity to light stimuli and causes gradual loss of central vision. In Wet AMD (Choroidal Neovascularization), the growth of leaky blood vessels under the retina damages photoreceptor cells and affects visual acuity. GWAS studies have shown that genetic and environmental factors are critical elements associated with AMD [20,21,71] and its progression time [77]. In recent years, multiple deep learning based predictive models are proposed for AMD. They have two categories: 1) diagnostic models that predict AMD severity of a subject based on their CFP taken at their current visit [11–13,29,38,52]. Although these models have shown convincing performance for the *diagnosis* task, they cannot predict subjects' *longitudinal outcome* that is crucial information for clinicians to start preventive treatments for vulnerable subjects. 2) Models predicting whether a subject progresses into late AMD in less than 'n' years [10,53,78], where 'n' is a predefined value. Nonetheless, if their answer is yes, they do not provide any information about whether the subject is already in advanced AMD or they will progress to it in the future. Furthermore, the majority of previous works are single-modal based on CFPs that waste genetic modality in training datasets or they are multi-modal [53,78] taking CFPs and 52 AMD-associated variants [77] which limits their practicality because they need genetic modality in their inference phase.

3   Proposed Method

We develop an adversarial mutual framework capable of utilizing *auxiliary* modalities (genetics and age) available in training set to improve the training of a single-modal model (using only *main* modality (CFP)) that simultaneously addresses main queries regarding a subject's status when a chronic disease is concerned that are: 1) the current status of a subject (*e.g.*, current AMD severity) and 2) how their status will change until their next visit (*e.g.*, how their AMD severity score will change in the near future, i.e., longitudinal outcome) if they maintain their current lifestyle and disease progression trajectory. This knowledge empowers practitioners to start early treatment to decelerate the disease progression for susceptible subjects. We explain the intuitions behind our model step by step in the following subsections using AMD terminologies, but as we noted, it is applicable for similar diseases as well. Our procedure can be seen in Fig. 1.

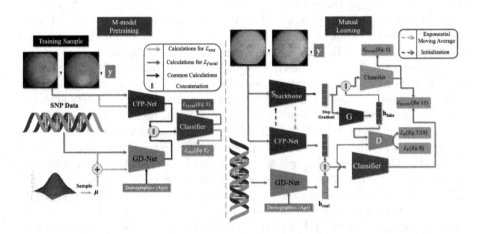

Fig. 1. Overview of our framework. **Left:** pretraining of our multimodal M-model. Color Fundus Photographs (CFP) and genetics information of subjects and are used to train the model. CFP contains the majority of the 'diagnostic signal' related to AMD. Thus, to prevent the model to get biased toward CFP and discard the genetic modality, we impose entropy regularization using a Gaussian measure on the model during training. (Sect. 3.3) **Right:** mutual learning of our single-modal S-model (top) with the pretrained M-model (bottom). S-model learns from the M-model to infer joint AMD-related representations of the genetics and demographics modalities - using its representations for an input CFP - using a Riemannian GAN model. The backbone of the S-model gets initialized by the weights of the CFP-Net of the M-model, and the M-model evolves during training by updating its CFP-Net using the exponential moving average of the weights of the S-model's backbone. (Sect. 3.3)

3.1   Problem Formulation

We formulate our prediction task as a classification problem. Considering AMD severity condition of a subject in their current and next visits (with a pre-defined time gap T_{gap} between them e.g., $T_{gap} = 3$ years), we define three classes: 1) $y = 0$ if a subject is not in the advanced AMD condition and will not progress to it until their next visit. 2) $y = 1$ if they are not currently in the advanced stage but will progress to advanced AMD until their next visit. 3) $y = 2$ if they have already progressed to the advanced phase. As there is no treatment for late AMD yet, [65] the fourth case for (current, next) $\sim$ (advanced, not advanced) is not possible. Our goal is to develop a model that accurately classifies subjects into one of the mentioned classes based on their current visit's CFP images. This formulation enables us to overcome the challenge of heterogeneity of time gaps between consecutive visits for subjects in longitudinal datasets. For instance, we can use records of a subject at visit numbers $\{1, 3, 7, 9\}$ to train a model with $T_{gap} = 2$ with pairs $\{(1, 3), (7, 9)\}$, but a sequence model should handle uneven time gaps $(2, 4, 2)$ between successive visits.

3.2   Notation

Let us assume that we have a longitudinal dataset such as AREDS [60] in which each subject has a random number of records corresponding to the visit time points that their data is collected during the study. We denote the training dataset as $D = \{(x_{i_g}, \{(x_{i_f, t_j}, y_{i, t_j})\}) | i \in [N], t_j \in T_i, T_i \subseteq T\}$ where N is the number of subjects, T is the set of all possible visit indices during the study, T_i is the set of available visit indices for the i-th subject, x_{i_f, t_j} is the fundus image of the subject taken during the visit with index t_j, and x_{i_g} is the genetic modality of the subject, which is static. For example, in the AREDS dataset [60], examinations are performed every six months, and the maximum follow-up study length for a subject is 13 years (26 visits). Thus, $T = \{1, 2, \cdots, 26\}$ is the set of all possible visit numbers. In addition, we denote our single-modal model as S-model and multimodal on as M-model in the rest of the paper.

3.3   Longitudinal Predictive Model

We introduce an adversarial mutual learning framework in which the single-modal S-model learns from a pretrained multi-modal M-model model to 1) infer outcome-related joint representation of genetics and demographics (age) from its representations for input CFPs using a Riemannian GAN model - inspired by studies [20,21,71,77] that have established high association between these modalities and AMD severity outcome that make it reasonable to incorporate such prior in our model - and 2) combining the predicted representation and the one for the visual modality to solve longitudinal outcome classification task in the course of a mutual training scheme [31,81] that benefits both models. In summary, our algorithm consists of pretraining the multimodal M-model and Mutual training the S-model along with the M-model. We describe details of each one in the following.

M-model Pretraining: We use a multimodal M-model to guide the training process of the S-model in a mutual learning fashion. The architecture of the M-model is shown in Fig.1. It consists of two sub-networks: 1) **CFP-net:** ResNet [33] backbone for CFP modality and 2) **GD-net:** a feed-forward model that combines genetics as well as demographics (age) modalities to obtain a joint outcome-related representation for them. Finally, obtained representations are combined in an early fusion [7] scheme and passed to a classifier to perform prediction.

As the number of samples in the case group (advanced AMD condition) is far less than the control group in our problem, our classification problem is imbalanced. We use Focal loss [43] to train the M-model because it down-weights the contribution of 'simple' examples from majority classes (e.g., control cases without any symptoms that the model can easily classify) in the loss function that the model is already confident about them. Formally, given y_i is the correct class corresponding to a sample x and $p_i = \mathcal{P}_{model}(\mathbf{y} = y_i | \mathbf{x} = x)$ be the predicted conditional probability of our teacher model for class y_i given x, Focal loss for the training sample (x, y) is calculated as

$$L_{focal}(x, y) = -(1 - p_i)^\gamma \log(p_i) \tag{1}$$

where γ is a hyperparameter controlling the down-weighting factor. As can be seen, Focal loss is a scaled version of Cross-Entropy loss that has a lower value for confident predictions of the model.

As we mentioned, the CFP of subjects contains the majority of the 'diagnostic singal' regarding their AMD status, and the genetics modality provides complementary knowledge with a much lower signal-to-noise ratio compared to the CFP modality. Therefore, directly training the model with Focal loss and standard regularization schemes for deep learning training such as ℓ_2-norm of weights that prefers networks with simpler structures may bias the model to discard the genetic modality and only focus on the CFP one. This phenomenon has been observed in the literature for domains such as visual question answering [2,17,28]. To overcome this problem, we use functional entropy regularization that balances the contribution of modalities. The intuition is that if our model's predictions show high entropy when we perturb a modality, then it is not bypassing the modality. Formally, given a probability measure μ over the space of input x of a non-negative function $g(x)$, functional entropy of g is defined as [6]:

$$Ent(g) = \int g(x) \log(g(x)) d\mu(x) - \int g(x) d\mu(x) \log(\int g(x) d\mu(x)) \tag{2}$$

However, the calculation of the RHS of this equation is intractable. As a workaround, Logarithmic Sobolev Inequality [6,24] is calculated as an upper bound of the functional entropy for Gaussian measures μ:

$$Ent(g) \leq \frac{1}{2} \int \frac{||\nabla g(x)||^2}{g(x)} d\mu(x) \tag{3}$$

In our problem, we define g as a measure of a discrepancy between the softmax output distribution of the M-model when the original genetics modality and its

Gaussian perturbed version of it are inputted to the model while keeping the input CFP fixed. In other words, given an input sample $x = (x_f, x_g)$:

$$\mathcal{P}_{model}(\mathbf{y}|\mathbf{x} = (x_f, x_g)) = (p_1, \cdots, p_K)$$
$$\mathcal{P}_{model}(\mathbf{y}|\mathbf{x} = (x_f, x_g + \epsilon)) = (p'_1, \cdots, p'_K), \ \epsilon \sim \mathcal{N}(0, \Sigma_{x_g}),$$
$$g(x, \epsilon) \triangleq \frac{1}{K} \sum_{j=1}^{K} BCE(p_j, p'_j) \tag{4}$$

The function g defined in Eq. (4) can represent the sensitivity of the model's predictions to Gaussian perturbations of the genetic modality. Now, we plug g into Eq. (3) and define a loss function $\mathcal{L}_{ent}$ which encourages the model to have high functional entropy $w.r.t$ its genetics input:

$$\mathcal{L}_{ent} = -\frac{1}{2} \int \frac{||\nabla g(x, \epsilon)||^2}{g(x, \epsilon)} d\mu(\epsilon). \tag{5}$$

In practice, we estimate the integral using Monte Carlo sampling, i.e., we approximate it with one σ for each sample. In addition, we set Σ_{x_g} as a diagonal covariance matrix with diagonal elements being the empirical variance of samples in the batch in each iteration.

Mutual Learning of S-model and M-model: After pretraining the M-model, we develop a training scheme based on mutual learning to train the S-model. As shown in Fig. 1, S-model has a backbone identical to CFP-net in M-model and a 'predictor' module. We aim to embed two prior medical knowledge into the inductive bias of our model that are: 1) high association between AMD severity and genetic variants [20,21,71,77]. 2) the ability of fundus images to predict the age of subjects [72]. To do so, we use the predictor module inside the S-model to predict representations of GD-net of the M-model. This prediction will be in a much lower dimensional space than reconstructing/imputing the whole genetic and age modalities together [14,47,57,61,64,76], and thus, is more sample efficient. The distribution of joint representation of genetics and age given the representation of CFP images may be multimodal, i.e., the mapping between them not necessarily be bijective. Thus, we train the predictor sub-network of the S-model using Generative Adversarial Networks (GAN) that are capable of modeling complex high dimensional distributions [3,26,30].

Modeling Interactions Between Representation of a CFP and Corresponding Joint Representation of Genetics and Age: We formulate learning such complex interaction with Riemannian GAN [51,58] training. In summary, GAN [3,26,30,51,58] models are trained using a two-player game in which a generator model G aims to learn the underlying distribution of a set of samples in the training set to trick a discriminator model D that distinguishes whether its input is real or a fake one generated by G. As the training process advances, the generator learns the distribution of training samples, and the discriminator will not be able to differentiate between real and fake samples generated by G. Conventional GAN models' discriminators [26] measure the distance between real and fake samples using Euclidean distance between their low dimensional embeddings. However, it is shown that [4,18] such distance may not faithfully reflect distances of data points as it is well-known that high dimensional real-world data is not randomly distributed in the ambient space and are often restricted to a nonlinear low-dimensional manifold [63] with unknown intrinsic dimension. Therefore, Riemannian GAN models' discriminators, project low dimensional representations of samples on a Riemannian manifold such as hypersphere [51,58] and calculate distances between them with the length of geodesics connecting them on the manifold. Distances on hypersphere are limited which makes the training stable, and it is shown that [51] training GAN with geodesic distances on hypersphere is equivalent to minimizing high order Wasserstein distances between real and fake distributions and generalizes methods that minimize the 1-Wasserstein distance [3,30].

Formally, we define a unit hypersphere with a center c and the main axis direction u ($c, u \in \mathbb{R}^d$) that are learnable. Given a joint representation on genetics and age (can be real predicted by GD-net of M-model or fake one by predictor of S-model) input $h \in \mathbb{R}^D$ ($D > d$) to the discriminator, it projects h into a d-dimensional space using nonlinear layers to obtain an embedding g. Then, it projects g on the unit sphere with center c such that $g_{proj} = \frac{g-c}{||g-c||}$. Now, let's consider circular cross-sections of the hypersphere that the main axis u of the hypersphere is the normal vector of the surface that they lie in. The idea is that if the discriminator gets designed to distinguish between real and fake samples based on the closeness of the cross-section that they lie on to the greatest circle of the hypersphere - i.e., the larger the radius of the cross-section that a sample lies on, more realness score is assigned to it - then the generator will attempt to generate samples that are on the largest circle of the hypersphere. Therefore, it will be able to generate more diverse samples, which prevents mode collapse. Given a batch of samples $H = \{h^i\}_{i=1}^B$, we calculate g_{proj}^j for each sample h^j and decompose it as $g_{proj}^j = g_{proj,u}^j + g_{proj,u\perp}^j$. The output score of the discriminator for a sample h^j is calculated as:

$$D(h_j) = -\frac{||g_{proj,u}^j||}{\sigma_{proj,u}} + \frac{||g_{proj,u\perp}^j||}{\sigma_{proj,u\perp}} \tag{6}$$

where $\sigma_{proj,u}$ and $\sigma_{proj,u\perp}$ are empirical variances of $||g_{proj,u}^j||$ and $||g_{proj,u\perp}^j||$ respectively. We use the relativistic objective [37] to train the GAN model. In a nutshell, it is designed such that the generator not only attempts to increase the score of the discriminator for fake samples, but also aims to decrease its score for real samples. If we denote joint representations of GD-net in M-model by $h \sim \mathcal{P}_{GD}$ and the ones predicted by the predictor model of S-model with $h' \sim \mathcal{P}_{pred}$, objectives of G (predictor in S-model) and discriminator D are as follows:

$$\mathcal{L}_D = \max_D \mathbb{E}_{h \sim \mathcal{P}_{GD}}[\log(f(D(h) - \mathbb{E}_{h' \sim \mathcal{P}_{pred}}[D(h')]))]$$

$$+ \mathbb{E}_{h' \sim \mathcal{P}_{pred}}[\log(f(\mathbb{E}_{h \sim \mathcal{P}_{GD}}[D(h)] - D(h')))] \quad (7)$$

$$\mathcal{L}_G = \max_G \mathbb{E}_{h' \sim \mathcal{P}_{pred}}[\log(f(D(h') - \mathbb{E}_{h \sim \mathcal{P}_{GD}}[D(h)]))]$$

$$+ \mathbb{E}_{h \sim \mathcal{P}_{GD}}[\log(f(\mathbb{E}_{h' \sim \mathcal{P}_{pred}}[D(h')] - D(h)))] \quad (8)$$

where $f(z) = sigmoid(\lambda z)$ calculates the discriminator's estimated probability that one/batch of real sample[s] is/are more realistic than a batch/one fake one[s], and λ is a hyperparameter [37]. We train the parameters for the main axis u and center c as follows. In each iteration, given a batch of real and fake samples $H = \{h^i\}_{i=1}^B$, at first, we update the center parameter with:

$$\mathcal{L}_c = \frac{1}{|B|} \sum_{j=1}^{|B|} \mathcal{H}(||g_{proj}^j - c||_2) \quad (9)$$

$\mathcal{H}$ is the Huber function [36], and the objective estimates the center of the hypersphere given a batch of samples. Then, we fix the center parameter, and to make the training of the center parameter stable, we encourage the discriminator to map samples to embeddings with similar distances relative to the center, i.e.,

$$\mathcal{L}_{dist} = \frac{1}{|B|} \sum_{j=1}^{|B|} \mathcal{H}(||g_{proj}^j - c||_2 - \sigma_h) \quad (10)$$

where σ_h is the empirical standard deviation of $||g_{proj}^j - c||_2$ distances from projected embeddings to the center. Parameters of the main axis u and discriminator are updated with backpropagated gradients from loss functions in Eqs. (7, 10).

We train the S-model's classifier to combine its representation for CFP and the predicted joint one for genetics and demographics modalities to accurately classify subjects' status. Firstly, we use Focal loss [43] defined in Eq. (1) to leverage training labels. Secondly, we use a distillation loss [34] to guide the S-model using predictions of the M-model:

$$\mathcal{L}_{distill} = KL(\mathcal{P}_S(\mathbf{y}|\mathbf{x}; T), \mathcal{P}_M(\mathbf{y}|\mathbf{x}; T)) \quad (11)$$

where the parameter T is a temperature parameter that controls the sharpness of output softmax distributions of models. In summary, the training objective for S-model's training is:

$$\mathcal{L}_S = \mathcal{L}_{focal} + \lambda_1 \mathcal{L}_{distill} + \lambda_2 \mathcal{L}_G \quad (12)$$

Before starting training the S-model, we initialize its backbone with the weights of the pretrained M-model's CFP-net to make the convergence faster.

As adversarial training may cause instability and degradation of the backbone's representations [15,25,62], we do not backpropagate gradients from adversarial training for the backbone's weights. Instead, we train them using supervision from Focal loss and distillation loss. Finally, as shown that mutual learning benefits from both models getting feedback from their peers, we update M-model's CFP-net's weights with exponential moving average (EMA) of the backbone of the S-model, i.e., after each iteration, we update CFP-net's weights as:

$$\theta_{CFP} \leftarrow \alpha\theta_{CFP} + (1 - \alpha)\theta_{Backbone} \tag{13}$$

Doing so prevents corruption of the weights of pre-trained M-model happening when using well-known distillation loss from S-model to M-model [31,81] in the starting phase of training as S-model's predictions are not reliable yet. We summarize our training algorithm in supplementary materials.

4 Experiments

In this section, we evaluate the effectiveness of our proposed adversarial mutual learning method on the task of simultaneously grading the current AMD severity of a subject as well as predicting their longitudinal outcome in their next visit when the predefined time gap between visits are 2, 3, and 4 years respectively. We compare our model with baseline methods, provide its interpretations, and perform an ablation study to analyze the effect of its different components.

4.1 Experimental Setup

Data Description: We use Age-related Eye Disease Study (AREDS) dataset [60] for our experiments, which is the largest longitudinal dataset available for AMD collected and maintained by National Eye Institute (NEI). It is available at the dbGaP[1] AREDS contains longitudinal CFPs of 4628 participants, and a subject may have up to 13-year follow-up visits since the baseline. For preprocessing step, we cropped each CFP to a square that encompasses the Macula [13,52] and resized it to 224×224 pixels resolution. As mentioned in Sect. 1, the yellowish color of drusen in the Macula and the red color of leaky blood vessels are important characteristics of dry and wet AMD respectively. Thus, we use a nonlinear Bézier augmentation [82] - previously proposed for CT scans and X-ray data - followed by random vertical and horizontal flip to augment CFPs. In addition to CFPs, genome sequence of 2780 (~60%) subjects is available in AREDS. We use all the genetic variants that are in the 34 loci regions [21] associated with advanced AMD with minor allele frequency (MAF) > 0.01 [21], and 156,864 SNPs remain after filtering. We then partition the AREDS dataset on the subject level and take all subjects that their genetics information is available

---

[1] https://www.ncbi.nlm.nih.gov/projects/gap/cgi-bin/study.cgi?study_id=phs000001. v3.p1.

as our train set. We randomly partition the rest into two halves for our validation and test sets. We refer to supplementary materials for more details about our data preparation.

Baselines: We compare our method against previous mutual learning and knowledge distillation methods in the literature. **DML** [81] trains two models from scratch with different initialization such that each model is trained with a loss function that is the sum of two terms, namely Cross-Entropy loss and KL-divergence between the distributions predicted by the model and its peer. **KDCL** [31] improves DML by using 'ensemble' of models' predictions instead of prediction of the peer model in the KL-divergence term. We use two ensemble schemes for KDCL, namely 'min-logit' and 'mean'. **KD** [34] distills the knowledge in the powerful large pretrained model, called teacher model, into a model, student, by training the student model using KL-divergence loss between its predictions and the ones for the teacher model. In addition, to show the effectiveness of leveraging 'complementary' knowledge in the genetics modality, we compare our model with single-modal baselines such that we train a ResNet architecture with Focal loss and Cross-Entropy loss. We denote these two cases in our experiments as **Base-Focal** and **Base-CE**.

Training and Evaluation: We use multi-class Area Under Curve (AUC) introduced by Hand and Till [32] as our evaluation metric because it is suitable for imbalance classification problems and has been used in AMD literature [13,52,53,78]. We pretrain our M-model for 10 epochs with batch size 128. Then, we train S-model mutually with M-model for 10 epochs with batch size 32. We use the same architectures for two sub-networks of all other mutual learning and knowledge distillation methods, and we use the architecture of our S-model for Base-CE/Focal. By doing so, we reduce the effect of architectural design and can more readily compare the methods. For a fair comparison, we train all baseline models for 20 epochs with batch size 128. We use Adam optimizer [40] with learning rate 0.0003, exponential decay rates $(\beta_1, \beta_2) = (0.9, 0.99)$, and weight decay 0.0001 for all models except for the parameters of the S-model's predictor and discriminator that we set $(\beta_1, \beta_2) = (0.5, 0.999)$, and also, initialize their parameters with normal distribution with zero mean and std of 0.02. We refer to supplementary materials for more details of experiments.

4.2 Experimental Results

Comparison with Baselines Models. Table 1 summarizes the performance of baseline methods and our adversarial mutual learning scheme for simultaneously grading and longitudinal prediction of AMD status of subjects. We explore baseline methods in two settings: 1) genetics modality is incorporated in their training where a multimodal network is trained along with a single-modal one, and we denote them with (M ↔ S). 2) only CFP is used in their training, and two single-modal models are trained together that are shown by (S ↔ S). It can be seen that mutual learning models consistently outperform knowledge distillation and standard single-network training baselines Base-CE/Focal, which is consistent with observations for natural image classification tasks. [31,81] Interestingly, Base-Focal has a competitive or even better performance compared to KD (S ↔ S) and shows better results compared to Base-CE, which shows the superior ability of the Focal loss [43] to handle long-tailed distributions compared to Cross-Entropy loss. In all cases except KDCL-MinLogit with 2 years gap, incorporating the genetics modality in the training procedure of the methods enhances the performance of the final single-modal model in inference, which supports our hypothesis that the genetics modality can provide supervision that is beneficial to the model's training. Furthermore, our model outperforms mutual learning models in all three cases of 2, 3, and 4 years gap between visits that demonstrates our model can more effectively 'denoise' the highly noisy genetics modality during training compared to other baselines and properly learn to predict AMD related joint representation of genetics and demographics modalities from its own one for an input CFP and combine them to perform longitudinal prediction.

Table 1. Comparison of our proposed method with baseline methods. Mean and standard deviation of 5 runs with different initialization are reported.

Time gap		2 years	3 years	4 years
Method	Using auxiliary modality	AUC		
KDCL - MinLogit (M ↔ S) [31]	✓	0.882 ± 0.003	0.881 ± 0.004	0.889 ± 0.003
KDCL - MinLogit (S ↔ S) [31]	✗	0.883 ± 0.004	0.880 ± 0.003	0.886 ± 0.004
KDCL - Mean (M ↔ S) [31]	✓	0.876 ± 0.005	0.881 ± 0.003	0.889 ± 0.002
KDCL - Mean (S ↔ S) [31]	✗	0.869 ± 0.004	0.874 ± 0.003	0.886 ± 0.005
DML (M ↔ S) [81]	✓	0.879 ± 0.002	0.877 ± 0.004	0.898 ± 0.003
DML (S ↔ S) [81]	✗	0.872 ± 0.004	0.874 ± 0.004	0.896 ± 0.004
KD (M ↔ S) [34]	✓	0.872 ± 0.002	0.877 ± 0.003	0.888 ± 0.003
KD (S ↔ S) [34]	✗	0.867 ± 0.003	0.873 ± 0.001	0.884 ± 0.001
Base-CE	✗	0.862 ± 0.005	0.867 ± 0.005	0.877 ± 0.005
Base-focal	✗	0.866 ± 0.003	0.877 ± 0.005	0.881 ± 0.008
AdvML (ours)	✓	$\mathbf{0.896 \pm 0.001}$	$\mathbf{0.899 \pm 0.001}$	$\mathbf{0.914 \pm 0.001}$

Interpretation of S-model's Predictions Figure 2 demonstrates Grad-CAM [55] saliency maps of our S-model. As mentioned in Sect. 1, the main characteristics of AMD in CFPs are the accumulation of yellow deposits called drusen in the Macula of an eye as well as the growth of leaky blood vessels under the retina that cause leakage of blood on photoreceptor cells. Saliency maps in Fig. 2 indicate that our S-model looks for these characteristics in the Macula for decision making, which is aligned with the clinical practice.

Ablation Study: In this section, we perform an ablation study to explore the effect of each component of our model. We remove entropy regularization in M-model's pretraining and the GAN training component in the mutual learning both separately and simultaneously. Table 2 summarizes the results. We can observe that removing entropy regularization for the genetics modality causes more severe performance degradation for our model, which highlights its importance to properly 'debias' the multimodal model to not neglect the genetics modality and only rely on the CFPs and effectively denoise it to extract its discriminative features for classification.

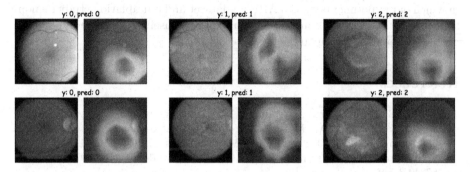

Fig. 2. Grad-CAM [55] saliency maps of our S-model's decisions. It focuses on the Macula region of the eyes and AMD symptoms, namely leaky blood vessels in the retina and yellow deposits in the Macula called drusen, which is aligned with clinical practice. **Left:** neither drusen nor leaky vessels are present in the Macula. **Middle:** Small areas of accumulation of drusen are observable. **Right:** leaked blood in the retina (top) and large areas of drusen (bottom) in the Macula exist.

Table 2. Ablation experiments' results for different components of our method.

Time gap	2 years	3 years	4 years
Ablation experiment	AUC		
W/O Ent Reg	0.880 ± 0.000	0.885 ± 0.001	0.887 ± 0.002
W/O GAN	0.881 ± 0.001	0.889 ± 0.002	0.903 ± 0.002
W/O Ent Reg & GAN	0.871 ± 0.002	0.879 ± 0.003	0.882 ± 0.001

5   Conclusion

In this paper, we introduced a new adversarial mutual learning framework that is capable of leveraging several *auxiliary* diagnostic modalities (containing complementary diagnostic signals that are collected in the training set and missing in inference) to train a more accurate single-modal model which uses the *main* modality (that provides the majority of diagnostic signal and is available in both training and inference) for inference. To do so, the single-modal model is trained with a pretrained multimodal model in a mutual learning manner. We imposed entropy regularization on the multimodal model during its pretraining to encourage it not to neglect the auxiliary modality in its decisions and learn to 'denoise' them to keep their discriminative information. Our single-modal model learns from the multimodal one to infer joint representation of the auxiliary modalities from its representation for the main modality and effectively combine them for its predictions. We modeled the complex interaction between modalities using a Riemannian GAN model and defined our classification task as simultaneously diagnosis of the current status of a subject as well as predicting their longitudinal outcome. We applied our method to the problem of early detection of AMD in which our experiments on the AREDS dataset and our ablation study demonstrated the superiority of our model compared to baselines and the importance of each component for our model.

References

1. Aakur, S.N., Narayanan, S., Indla, V., Bagavathi, A., Laguduva Ramnath, V., Ramachandran, A.: MG-NET: leveraging pseudo-imaging for multi-modal metagenome analysis. In: de Bruijne, M., et al. (eds.) MICCAI 2021. LNCS, vol. 12905, pp. 592–602. Springer, Cham (2021). https://doi.org/10.1007/978-3-030-87240-3_57
2. Agrawal, A., Batra, D., Parikh, D., Kembhavi, A.: Don't just assume; look and answer: Overcoming priors for visual question answering. In: Proceedings of the IEEE Conference on Computer Vision and Pattern Recognition, pp. 4971–4980 (2018)
3. Arjovsky, M., Chintala, S., Bottou, L.: Wasserstein generative adversarial networks. In: International Conference on Machine Learning, pp. 214–223. PMLR (2017)
4. Arvanitidis, G., Hauberg, S., Schölkopf, B.: Geometrically enriched latent spaces. In: Banerjee, A., Fukumizu, K. (eds.) The 24th International Conference on Artificial Intelligence and Statistics, AISTATS 2021, 13–15 April 2021, Virtual Event. Proceedings of Machine Learning Research, vol. 130, pp. 631–639. PMLR (2021). http://proceedings.mlr.press/v130/arvanitidis21a.html
5. Ayoub, T., Patel, N.: Age-related macular degeneration. J. R. Soc. Med. **102**(2), 56–61 (2009)
6. Bakry, D., Gentil, I., Ledoux, M., et al.: Analysis and Geometry of Markov Diffusion Operators, vol. 103. Springer, Cham (2014). https://doi.org/10.1007/978-3-319-00227-9
7. Baltrušaitis, T., Ahuja, C., Morency, L.P.: Multimodal machine learning: a survey and taxonomy. IEEE Trans. Pattern Anal. Mach. Intell. **41**(2), 423–443 (2018)
8. Bhat, P., Arani, E., Zonooz, B.: Distill on the go: online knowledge distillation in self-supervised learning. In: IEEE Conference on Computer Vision and Pattern Recognition Workshops, CVPR Workshops 2021, virtual, June 19–25, 2021. pp. 2678–2687. Computer Vision Foundation/IEEE (2021). https://doi.org/10.1109/CVPRW53098.2021.00301
9. Bird, A.C., et al.: An international classification and grading system for age-related maculopathy and age-related macular degeneration. Surv. Ophthalmol. **39**(5), 367–374 (1995)
10. Bridge, J., Harding, S., Zheng, Y.: Development and validation of a novel prognostic model for predicting AMD progression using longitudinal fundus images. BMJ Open Ophthal. **5**(1), e000569 (2020)

11. Burlina, P., Freund, D.E., Joshi, N., Wolfson, Y., Bressler, N.M.: Detection of age-related macular degeneration via deep learning. In: 2016 IEEE 13th International Symposium on Biomedical Imaging (ISBI), pp. 184–188. IEEE (2016)

12. Burlina, P.M., Joshi, N., Pacheco, K.D., Freund, D.E., Kong, J., Bressler, N.M.: Use of deep learning for detailed severity characterization and estimation of 5-year risk among patients with age-related macular degeneration. JAMA Ophthalmol. **136**(12), 1359–1366 (2018)

13. Burlina, P.M., Joshi, N., Pekala, M., Pacheco, K.D., Freund, D.E., Bressler, N.M.: Automated grading of age-related macular degeneration from color fundus images using deep convolutional neural networks. JAMA Ophthalmol. **135**(11), 1170–1176 (2017)

14. Cai, L., Wang, Z., Gao, H., Shen, D., Ji, S.: Deep adversarial learning for multi-modality missing data completion. In: Proceedings of the 24th ACM SIGKDD International Conference on Knowledge Discovery & Data Mining, pp. 1158–1166 (2018)

15. Chavdarova, T., Fleuret, F.: SGAN: an alternative training of generative adversarial networks. In: Proceedings of the IEEE Conference on Computer Vision and Pattern Recognition, pp. 9407–9415 (2018)

16. Congdon, N., et al.: Causes and prevalence of visual impairment among adults in the united states. Arch. Ophthalmol. (Chicago, Ill.: 1960) **122**(4), 477–485 (2004)

17. Dancette, C., Cadene, R., Teney, D., Cord, M.: Beyond question-based biases: assessing multimodal shortcut learning in visual question answering. In: Proceedings of the IEEE/CVF International Conference on Computer Vision (ICCV), pp. 1574–1583, October 2021

18. Edraki, M., Qi, G.J.: Generalized loss-sensitive adversarial learning with manifold margins. In: Proceedings of the European Conference on Computer Vision (ECCV), pp. 87–102 (2018)

19. Ferris III, F.L., et al.: Clinical classification of age-related macular degeneration. Ophthalmology **120**(4), 844–851 (2013)

20. Fritsche, L.G., et al.: Seven new loci associated with age-related macular degeneration. Nat. Geneti. **45**(4), 433–439 (2013)

21. Fritsche, L.G., et al.: A large genome-wide association study of age-related macular degeneration highlights contributions of rare and common variants. Nat. Genet. **48**(2), 134–143 (2016)

22. Gao, R., Oh, T.H., Grauman, K., Torresani, L.: Listen to look: action recognition by previewing audio. In: Proceedings of the IEEE/CVF Conference on Computer Vision and Pattern Recognition, pp. 10457–10467 (2020)

23. Garcia, N., Nakashima, Y.: Knowledge-based video question answering with unsupervised scene descriptions. In: Vedaldi, A., Bischof, H., Brox, T., Frahm, J.-M. (eds.) ECCV 2020. LNCS, vol. 12363, pp. 581–598. Springer, Cham (2020). https://doi.org/10.1007/978-3-030-58523-5_34

24. Gat, I., Schwartz, I., Schwing, A.G., Hazan, T.: Removing bias in multimodal classifiers: regularization by maximizing functional entropies. In: Larochelle, H., Ranzato, M., Hadsell, R., Balcan, M., Lin, H. (eds.) Advances in Neural Information Processing Systems 33: Annual Conference on Neural Information Processing Systems 2020, NeurIPS 2020, 6–12 December 2020, virtual (2020). https://proceedings.neurips.cc/paper/2020/hash/20d749bc05f47d2bd3026ce457dcfd8e-Abstract.html

25. Goodfellow, I.: NIPS 2016 tutorial: generative adversarial networks. arXiv preprint arXiv:1701.00160 (2016)

26. Goodfellow, I., et al.: Generative adversarial nets. Adv. Neural Inf. Process. Syst. **27** (2014)
27. Gou, J., Yu, B., Maybank, S.J., Tao, D.: Knowledge distillation: a survey. Int. J. Comput. Vis. **129**(6), 1789–1819 (2021)
28. Goyal, Y., Khot, T., Summers-Stay, D., Batra, D., Parikh, D.: Making the V in VQA matter: elevating the role of image understanding in visual question answering. In: Proceedings of the IEEE Conference on Computer Vision and Pattern Recognition, pp. 6904–6913 (2017)
29. Grassmann, F., et al.: A deep learning algorithm for prediction of age-related eye disease study severity scale for age-related macular degeneration from color fundus photography. Ophthalmology **125**(9), 1410–1420 (2018)
30. bibitemch13DBLP:confspsnipsspsGulrajaniAADC17 Gulrajani, I., Ahmed, F., Arjovsky, M., Dumoulin, V., Courville, A.C.: Improved training of Wasserstein GANs. In: Guyon, I., et al. (eds.) Annual Conference on Neural Information Processing Systems 2017, vol. 30, 4–9 December 2017, Long Beach, CA, USA. pp. 5767–5777 (2017). https://proceedings.neurips.cc/paper/2017/hash/892c3b1c6dccd52936e27cbd0ff683d6-Abstract.html
31. Guo, Q., et al.: Online knowledge distillation via collaborative learning. In: Proceedings of the IEEE/CVF Conference on Computer Vision and Pattern Recognition, pp. 11020–11029 (2020)
32. Hand, D.J., Till, R.J.: A simple generalisation of the area under the roc curve for multiple class classification problems. Mach. Learn. **45**(2), 171–186 (2001)
33. He, K., Zhang, X., Ren, S., Sun, J.: Deep residual learning for image recognition. In: Proceedings of the IEEE Conference on Computer Vision and Pattern Recognition, pp. 770–778 (2016)
34. Hinton, G., Vinyals, O., Dean, J.: Distilling the knowledge in a neural network. arXiv preprint arXiv:1503.02531 (2015)
35. Hou, J.C., Wang, S.S., Lai, Y.H., Tsao, Y., Chang, H.W., Wang, H.M.: Audio-visual speech enhancement using multimodal deep convolutional neural networks. IEEE Trans. Emerg. Topics Comput. Intell. **2**(2), 117–128 (2018)
36. Huber, P.J.: Robust estimation of a location parameter. In: Kotz, S., Johnson, N.L. (eds.) Breakthroughs in Statistics. Springer Series in Statistics, pp. 492–518. Springer, New York (1992). https://doi.org/10.1007/978-1-4612-4380-9_35
37. Jolicoeur-Martineau, A.: The relativistic discriminator: a key element missing from standard GAN. In: 7th International Conference on Learning Representations, ICLR 2019, New Orleans, LA, USA, 6–9 May 2019. OpenReview.net (2019). https://openreview.net/forum?id=S1erHoR5t7
38. Keenan, T.D., et al.: A deep learning approach for automated detection of geographic atrophy from color fundus photographs. Ophthalmology **126**(11), 1533–1540 (2019)
39. Kim, J., Jun, J., Zhang, B.: Bilinear attention networks. In: Bengio, S., Wallach, H.M., Larochelle, H., Grauman, K., Cesa-Bianchi, N., Garnett, R. (eds.) Annual Conference on Neural Information Processing Systems 2018, NeurIPS 2018, vol. 31, 3–8 December 2018, Montréal, Canada, pp. 1571–1581 (2018), https://proceedings.neurips.cc/paper/2018/hash/96ea64f3a1aa2fd00c72faacf0cb8ac9-Abstract.html
40. Kingma, D.P., Ba, J.: Adam: a method for stochastic optimization. In: Bengio, Y., LeCun, Y. (eds.) Proceedings of the 3rd International Conference on Learning Representations, ICLR 2015, San Diego, CA, USA, 7–9 May 2015 (2015). http://arxiv.org/abs/1412.6980

41. Lan, X., Zhu, X., Gong, S.: Knowledge distillation by on-the-fly native ensemble. In: Bengio, S., Wallach, H.M., Larochelle, H., Grauman, K., Cesa-Bianchi, N., Garnett, R. (eds.) Annual Conference on Neural Information Processing Systems, 2018, vol. 31 NeurIPS 2018, 3–8 December 2018, Montréal, Canada. pp. 7528–7538 (2018). https://proceedings.neurips.cc/paper/2018/hash/94ef7214c4a90790186e255304f8fd1f-Abstract.html

42. Lee, C., Schaar, M.: A variational information bottleneck approach to multi-omics data integration. In: International Conference on Artificial Intelligence and Statistics, pp. 1513–1521. PMLR (2021)

43. Lin, T.Y., Goyal, P., Girshick, R., He, K., Dollár, P.: Focal loss for dense object detection. In: Proceedings of the IEEE International Conference on Computer Vision, pp. 2980–2988 (2017)

44. Lin, X., Bertasius, G., Wang, J., Chang, S.F., Parikh, D., Torresani, L.: Vx2text: end-to-end learning of video-based text generation from multimodal inputs. In: Proceedings of the IEEE/CVF Conference on Computer Vision and Pattern Recognition (CVPR), pp. 7005–7015, June 2021

45. Liu, Y., et al.: Unbiased teacher for semi-supervised object detection. In: 9th International Conference on Learning Representations, ICLR 2021, Virtual Event, Austria, 3–7 May 2021. OpenReview.net (2021). https://openreview.net/forum?id=MJIve1zgR_

46. Luu, J., Palczewski, K.: Human aging and disease: lessons from age-related macular degeneration. Proc. Natil. Acad. Sci. **115**(12), 2866–2872 (2018)

47. Ma, M., Ren, J., Zhao, L., Tulyakov, S., Wu, C., Peng, X.: SMIL: multimodal learning with severely missing modality. In: Proceedings of the AAAI Conference on Artificial Intelligence, vol. 35, pp. 2302–2310 (2021)

48. Metzker, M.L.: Sequencing technologies-the next generation. Nat. Rev. Genet. **11**(1), 31–46 (2010)

49. Mikheyev, A.S., Tin, M.M.: A first look at the oxford nanopore minion sequencer. Mol. Ecol. Resour. **14**(6), 1097–1102 (2014)

50. Panda, R., et al.: AdaMML adaptive multi-modal learning for efficient video recognition. In: Proceedings of the IEEE/CVF International Conference on Computer Vision (ICCV), pp. 7576–7585, October 2021

51. Park, S.W., Kwon, J.: Sphere generative adversarial network based on geometric moment matching. In: Proceedings of the IEEE/CVF Conference on Computer Vision and Pattern Recognition, pp. 4292–4301 (2019)

52. Peng, Y., et al.: DeepSeeNet: a deep learning model for automated classification of patient-based age-related macular degeneration severity from color fundus photographs. Ophthalmology **126**(4), 565–575 (2019)

53. Peng, Y., et al.: Predicting risk of late age-related macular degeneration using deep learning. NPJ Digit. Med. **3**(1), 1–10 (2020)

54. Qi, L., et al,: Multi-scale aligned distillation for low-resolution detection. In: Proceedings of the IEEE/CVF Conference on Computer Vision and Pattern Recognition, pp. 14443–14453 (2021)

55. Selvaraju, R.R., Cogswell, M., Das, A., Vedantam, R., Parikh, D., Batra, D.: Grad-CAM: visual explanations from deep networks via gradient-based localization. In: Proceedings of the IEEE International Conference on Computer Vision, pp. 618–626 (2017)

56. Seo, A., Kang, G., Park, J., Zhang, B.: Attend what you need: motion-appearance synergistic networks for video question answering. In: Zong, C., Xia, F., Li, W., Navigli, R. (eds.) Proceedings of the 59th Annual Meeting of the Association for Computational Linguistics and the 11th International Joint Conference on Natural Language Processing, ACL/IJCNLP 2021, (Volume 1: Long Papers), Virtual Event, 1–6 August 2021. pp. 6167–6177. Association for Computational Linguistics (2021). https://doi.org/10.18653/v1/2021.acl-long.481

57. Shi, Y., Narayanaswamy, S., Paige, B., Torr, P.H.S.: Variational mixture-of-experts autoencoders for multi-modal deep generative models. In: Wallach, H.M., Larochelle, H., Beygelzimer, A., d'Alché-Buc, F., Fox, E.B., Garnett, R. (eds.) Annual Conference on Neural Information Processing Systems 2019, NeurIPS 2019, vol. 32, 8–14 December 2019, Vancouver, BC, Canada. pp. 15692–15703 (2019). https://proceedings.neurips.cc/paper/2019/hash/0ae775a8cb3b499ad1fca944e6f5c836-Abstract.html

58. Shim, W., Cho, M.: CircleGAN: generative adversarial learning across spherical circles. In: Larochelle, H., Ranzato, M., Hadsell, R., Balcan, M.F., Lin, H. (eds.) Advances in Neural Information Processing Systems, vol. 33, pp. 21081–21091. Curran Associates, Inc. (2020). https://proceedings.neurips.cc/paper/2020/file/f14bc21be7eaeed046fed206a492e652-Paper.pdf

59. Son, W., Na, J., Choi, J., Hwang, W.: Densely guided knowledge distillation using multiple teacher assistants. In: Proceedings of the IEEE/CVF International Conference on Computer Vision, pp. 9395–9404 (2021)

60. Study, T.A.R.E.D., et al.: The age-related eye disease study (AREDS): design implications AREDS report no. 1. Control. Clin. Trials 20(6), 573–600 (1999)

61. Suo, Q., Zhong, W., Ma, F., Yuan, Y., Gao, J., Zhang, A.: Metric learning on healthcare data with incomplete modalities. In: IJCAI, pp. 3534–3540 (2019)

62. Tao, S., Wang, J.: Alleviation of gradient exploding in GANs: fake can be real. In: Proceedings of the IEEE/CVF Conference on Computer Vision and Pattern Recognition, pp. 1191–1200 (2020)

63. Tenenbaum, J.B., De Silva, V., Langford, J.C.: A global geometric framework for nonlinear dimensionality reduction. Science 290(5500), 2319–2323 (2000)

64. Tran, L., Liu, X., Zhou, J., Jin, R.: Missing modalities imputation via cascaded residual autoencoder. In: Proceedings of the IEEE Conference on Computer Vision and Pattern Recognition, pp. 1405–1414 (2017)

65. Trucco, E., MacGillivray, T., Xu, Y.: Computational retinal image analysis: tools. In: Trucco, E., MacGillivray, T., Xu, Y. (eds.) Applications and Perspectives, Academic Press, New York (2019)

66. Tsai, Y.H., Liang, P.P., Zadeh, A., Morency, L., Salakhutdinov, R.: Learning factorized multimodal representations. In: 7th International Conference on Learning Representations, ICLR 2019, New Orleans, LA, USA, 6–9 May 2019. OpenReview.net (2019). https://openreview.net/forum?id=rygqqsA9KX

67. Uppal, S., Bhagat, S., Hazarika, D., Majumder, N., Poria, S., Zimmermann, R., Zadeh, A.: Multimodal research in vision and language: a review of current and emerging trends. Inf. Fusion 77, 149–171 (2021)

68. Wang, J., Li, Y., Hu, J., Yang, X., Ding, Y.: Self-supervised mutual learning for video representation learning. In: 2021 IEEE International Conference on Multimedia and Expo (ICME), pp. 1–6. IEEE (2021)

69. Wang, Q., Zhan, L., Thompson, P., Zhou, J.: Multimodal learning with incomplete modalities by knowledge distillation. In: Proceedings of the 26th ACM SIGKDD International Conference on Knowledge Discovery & Data Mining, pp. 1828–1838 (2020)
70. Wang, W., Tran, D., Feiszli, M.: What makes training multi-modal classification networks hard? In: Proceedings of the IEEE/CVF Conference on Computer Vision and Pattern Recognition, pp. 12695–12705 (2020)
71. Wei, Y., Liu, Y., Sun, T., Chen, W., Ding, Y.: Gene-based association analysis for bivariate time-to-event data through functional regression with copula models. Biometrics 76(2), 619–629 (2020)
72. Wen, Y., Chen, L., Qiao, L., Deng, Y., Zhou, C.: On the deep learning-based age prediction of color fundus images and correlation with ophthalmic diseases. In: 2020 IEEE International Conference on Bioinformatics and Biomedicine (BIBM), pp. 1171–1175. IEEE (2020)
73. Wu, G., Gong, S.: Peer collaborative learning for online knowledge distillation. In: Thirty-Fifth AAAI Conference on Artificial Intelligence, AAAI 2021, Thirty-Third Conference on Innovative Applications of Artificial Intelligence, IAAI 2021, The Eleventh Symposium on Educational Advances in Artificial Intelligence, EAAI 2021, Virtual Event, 2–9 February 2021, pp. 10302–10310. AAAI Press (2021). https://ojs.aaai.org/index.php/AAAI/article/view/17234
74. Wu, M., Goodman, N.D.: Multimodal generative models for scalable weakly-supervised learning. In: Bengio, S., Wallach, H.M., Larochelle, H., Grauman, K., Cesa-Bianchi, N., Garnett, R. (eds.) Advances in Neural Information Processing Systems 31: Annual Conference on Neural Information Processing Systems 2018, NeurIPS 2018, 3–8 December 2018, Montréal, Canada. pp. 5580–5590 (2018). https://proceedings.neurips.cc/paper/2018/hash/1102a326d5f7c9e04fc3c89d0ede88c9-Abstract.html
75. Wu, S., Li, J., Liu, C., Yu, Z., Wong, H.S.: Mutual learning of complementary networks via residual correction for improving semi-supervised classification. In: Proceedings of the IEEE/CVF Conference on Computer Vision and Pattern Recognition, pp. 6500–6509 (2019)
76. Xu, D., Ouyang, W., Ricci, E., Wang, X., Sebe, N.: Learning cross-modal deep representations for robust pedestrian detection. In: Proceedings of the IEEE Conference on Computer Vision and Pattern Recognition, pp. 5363–5371 (2017)
77. Yan, Q., et al.: Genome-wide analysis of disease progression in age-related macular degeneration. Hum. Mol. Genet. 27(5), 929–940 (2018)
78. Yan, Q., et al.: Deep-learning-based prediction of late age-related macular degeneration progression. Nat. Mach. Intell. 2(2), 141–150 (2020)
79. Zellers, R., Bisk, Y., Farhadi, A., Choi, Y.: From recognition to cognition: Visual commonsense reasoning. In: Proceedings of the IEEE/CVF Conference on Computer Vision and Pattern Recognition, pp. 6720–6731 (2019)
80. Zhang, Y., et al.: Modality-aware mutual learning for multi-modal medical image segmentation. In: de Bruijne, M., et al. (eds.) MICCAI 2021. LNCS, vol. 12901, pp. 589–599. Springer, Cham (2021). https://doi.org/10.1007/978-3-030-87193-2_56
81. Zhang, Y., Xiang, T., Hospedales, T.M., Lu, H.: Deep mutual learning. In: Proceedings of the IEEE Conference on Computer Vision and Pattern Recognition, pp. 4320–4328 (2018)
82. Zhou, Z., et al.: Models genesis: generic autodidactic models for 3D medical image analysis. In: Shen, D., et al. (eds.) MICCAI 2019. LNCS, vol. 11767, pp. 384–393. Springer, Cham (2019). https://doi.org/10.1007/978-3-030-32251-9_42

Fast, Flexible, and Exact Minimum Flow Decompositions via ILP

Fernando H. C. Dias[1], Lucia Williams[2], Brendan Mumey[2], and Alexandru I. Tomescu[1]([✉])

[1] Department of Computer Science, University of Helsinki, Helsinki, Finland
{fernando.cunhadias,alexandru.tomescu}@helsinki.fi
[2] School of Computing, Montana State University, Bozeman, MT, USA
{luciawilliams,brendan.mumey}@montana.edu

Abstract. Minimum flow decomposition (MFD)—the problem of finding a minimum set of paths that perfectly decomposes a flow—is a classical problem in Computer Science, and variants of it are powerful models in multiassembly problems in Bioinformatics (e.g. RNA assembly). However, because this problem and its variants are NP-hard, practical multiassembly tools either use heuristics or solve simpler, polynomial-time solvable versions of the problem, which may yield solutions that are not minimal or do not perfectly decompose the flow. Many RNA assemblers also use integer linear programming (ILP) formulations of such practical variants, having the major limitation they need to encode all the potentially exponentially many solution paths. Moreover, the only exact solver for MFD does not scale to large instances, and cannot be efficiently generalized to practical MFD variants.

In this work, we provide the first practical ILP formulation for MFD (and thus the first fast and exact solver for MFD), based on encoding *all* of the exponentially many solution paths using only a *quadratic* number of variables. On both simulated and real flow graphs, our approach runs in under 13 s on average. We also show that our ILP formulation can be easily and efficiently adapted for many practical variants, such as incorporating longer or paired-end reads, or minimizing flow errors.

We hope that our results can remove the current tradeoff between the complexity of a multiassembly model and its tractability, and can lie at the core of future practical RNA assembly tools. Our implementations are freely available at github.com/algbio/MFD-ILP.

Keywords: Network flow · Flow decomposition · Integer linear programming · Multiassembly · RNA assembly

This work was partially funded by the European Research Council (ERC) under the European Union's Horizon 2020 research and innovation programme (grant agreement No. 851093, SAFEBIO), partially by the Academy of Finland (grants No. 322595, 328877, 308030), and partially by the US NSF (award 1759522). The full version of this work is available at [8].

I. Pe'er (Ed.): RECOMB 2022, LNBI 13278, pp. 230–245, 2022.
https://doi.org/10.1007/978-3-031-04749-7_14

1   Introduction

Flow decomposition (FD), the problem of decomposing a network flow into a set of source-to-sink paths and associated weights that perfectly explain the flow values on the edges, is a classical and well-studied concept in Computer Science. For example, it is a standard result that any flow in a directed acyclic graph (DAG) with m edges can be decomposed into at most m weighted paths (see, e.g., [1]). However, finding an FD with a *minimum* number of paths (MFD) is NP-hard [47], even on DAGs. This result was later strengthened by [14] who proved that MFD is hard to approximate (i.e., there is some $\epsilon > 0$ such that MFD cannot be approximated to within a $(1+\epsilon)$ factor, unless P = NP). More recent work has shown that the problem is FPT in the size of the minimum decomposition [21], and that it can be approximated with an exponential factor [31]. It is also possible to decompose all but a ε-fraction of the flow within a $O(1/\varepsilon)$ factor of the optimal number of paths [14]. Heuristic approaches to the problem have also been developed, particularly greedy methods based on choosing the widest or longest paths [47], which can be improved by making iterative modifications to the flow graph before finding a greedy decomposition [40]. But despite this history of work on algorithms for MFD, an exact solver that is fast for instances with large optimal solutions or large flow values remains elusive.

FD is also a key step in numerous applications. For example, some network routing problems (e.g. [7,14,15,31]) and transportation problems (e.g. [33,34]) require FDs that are optimal with respect to various measures. MFDs in particular are used to reconstruct biological sequences such as RNA transcripts [4,10,36,43,44,53], and viral quasispecies [3]. However, because MFD is NP-hard, all of these tools in fact use heuristics or solve some simpler version of the problem ignoring some information that is available from the sequencing process, resulting in tools that may not reconstruct the correct sequence, even if no other errors are present. More broadly, it has been noted [32] that the lack of exact solvers for many of the sub-problems involved in DNA sequencing has led to heuristic and ad hoc tools with no provable guarantees on the quality of solutions. Additionally, some authors [4,6] have noted that there is a tradeoff between the complexity of the model for RNA assembly (i.e., how much of the true possible solution space that it supports) and its tractability. But if a fast exact solver for MFD exists, this tradeoff may not be necessary.

1.1   Minimum Flow Decomposition in Multiassembly

The main bioinformatics motivation for this paper is *multiassembly* [55]. In this problem, we seek to reconstruct multiple genomic sequences from mixed samples using short substrings (called *reads*) generated cheaply and accurately from next-generation sequencing technology. The two major multiassembly problems are RNA assembly and viral quasispecies assembly, which we describe below.

One mechanism by which complex organisms create a vast array of proteins is alternative splicing of gene sequences, where multiple different RNA transcripts (that are then used to produce different proteins) can be created from the same gene [41]. In humans, over 90% of genes are believed to produce multiple transcripts [51]. Reconstructing the specific RNA transcripts has proved essential in characterizing gene regulation and function, and in studying development and diseases, including cancer; see, e.g., [20,38]. A second multiassembly problem is the reconstruction of viral quasispecies, for example, the different HIV or hepatitis strains present in a single patient sequencing sample, or the different SARS-CoV-2 strains present in a sewage water sample. Because viruses evolve quickly, there can be many distinct strains present at one time, and this diversity can be an important factor in the success or effect of the virus [48].

While the biological realities underlying the different multiassembly problems may yield some differences in how the problems can be solved, at their heart many approaches contain the algorithmic step of decomposing a network flow into weighted paths. The basic setup and approach for multiassembly is as follows. Given a sample of unknown sequences, each with some unknown abundance (for example, a set of RNA transcripts or virus strains), all sequences are multiplied and then broken into fragments that can be read by next-generation sequencers to produce millions of sequence reads ranging from hundreds to tens of thousands DNA characters in length. Many approaches are reference-based (e.g., [4,22,24,30,36,43,46] for RNA assembly and [45,56] for viral quasispecies assembly), meaning that they use a previously-constructed reference genome to guide the assembly process. These approaches construct a graph using the sequences contained in the reads where nodes are strings, edges represent overlaps, and weights on edges give the counts of reads that support each overlap. Because a reference is used, these graphs are always DAGs. In the non-reference case (called *de novo*), graphs may have cycles; we address this further at the end of the paper. If errors are minimal, the weights on the edges should form a flow on the network, and the underlying sequences and their abundances must be some decomposition of the flow into weighted paths. For RNA assembly, recent work [21,54] has confirmed the common assertion (e.g., by [22,26,28,29,39,43,58]) that the true transcripts and abundances should be *minimum* flow decomposition. No such study has been done for viral quasispecies assembly, but existing tools seek minimum-sized decompositions [3,52]. However, while the abovementioned tools seek minimum-sized flow decompositions, since MFD is NP-hard, they in fact compute decompositions which are not guaranteed to be minimum (and thus may not give the correct assembly, even if no other errors are present).

1.2   Limitations of Current ILP Solutions

One promising direction for fast exact solvers for MFD is integer linear programming (ILP). Existing ILP solvers like Gurobi [11] and CPLEX [17] incorporate optimizations that allow for fast runtimes in practice for problems that should be hard in general (see also [12] for various applications of ILP in Bioinformatics). Indeed, many existing multiassembly tools do use ILP to solve MFD as one step in their process. The basic idea behind these existing formulations is to consider some set of source-to-sink paths through the graph and assign each a binary variable indicating whether or not it is selected in the optimal solution, along with constraints to fully encode the FD problem (i.e. that the selected set of paths—with the weights derived for them by the ILP—form an FD) and to model further practical aspects of the specific multiassembly problem. However, the number of paths in a DAG is exponential, meaning that if the tools enumerate all paths (and thus can be guaranteed to find the true optimal solution) they are impractical for larger instances (e.g., Toboggan [21]). The most common strategy is to pre-select some set of paths, either for all instances (e.g., vg-flow [3] and CLIIQ [26]), or only when the input is large (e.g., MultiTrans [58] and SSP [37]). But by pre-selecting paths, these formulations may not find the optimal MFD solution for the instance.

1.3 Our Contributions

We give a new ILP approach to the MFD problem on DAGs, and we show that it can be used on both simulated and real RNA assembly graphs under conditions used in many reference-based multiassembly tools. Our approach is:

Fast and Exact. We show in Sect. 3.1 that it is not necessary to enumerate all paths in order to encode them in an ILP. The key idea is that any path must have a conserved (unit) flow from its start to its end, and that this concept can be encoded using only a number of variables that is linear in the size of the graph (rather than exponential, as is the case when the model enumerates all possible paths). This is a standard integer programming method for expressing paths in DAGs, used for example in [42]. An implementation of our ILP formulation using CPLEX finds optimal flow decomposition solutions on RNA assembly graphs (simulated and assembled from real reads) in under 13 s on average, over all the datasets tested. This is several times faster than the state-of-the-art MFD solver Toboggan [21], depending on the dataset. While heuristic solvers such as Catfish [40] or CoasterHeuristic [54] finish withing a few seconds, we show that they do not provide optimum solutions. Another benefit of our ILP solutions is that *all* optimum solutions can be reported by the ILP solver, thus potentially helping in "identifying" the correct RNA multiassembly solution (practical issue acknowledged by e.g. [18,27]).

Flexible. In practice, many multiassembly tools in fact solve variants of MFD. For example, many tools account for paired-end reads by requiring that they be included in the same path. Another common strategy is to incorporate longer reads as subpath constraints or phasing paths [36,39,54] which again must be covered by some predicted transcript (i.e. flow path). In Sect. 3.2, we give additional constraints that are expressive enough to not only encode paired-end reads and subpath constraints, but also any generic set of edges that must be covered by a single path (e.g., as when modelling the recent Smart-seq3 protocol producing RNA multi-end reads [13]). Additionally, due to sequencing or read mapping errors, the weights on edges may not be a flow (i.e. flow conservation might not hold). One approach in this case is to consider intervals of edge weights instead, as in [37,53]. We give a formulation to handle this approach in Sect. 3.3. Our implementation solves subpath constraint instances in similar time to standard instances, while the existing exact solver could not complete on many instances in under 60 s. Moreover, while the existing interval heuristic is fast, it finds decompositions that are far from optimum. While all these additional constraints are naturely expressed in ILP (further underlining the flexibility of our approach), the novelty here is their integration with the ILP encoding of all possible paths in the DAG from Sect. 3.1.

2 Preliminaries

Given a graph $G = (V, E)$, with vertex set V and edge set $E \subseteq V \times V$, we say that $s \in V$ is a *source* if s has no in-coming edges. Analogously, we say that $t \in V$ is a *sink* if t has no out-going edges. Moreover, we say that G is a *directed acyclic graph (DAG)* if G contains no directed cycles.

Definition 1 (Flow network). *A tuple $G = (V, E, f)$ is said to be a* flow network *if (V, E) is a DAG with unique source s and unique sink t, where for every edge $(u, v) \in E$ we have an associated positive integer flow value f_{uv}, satisfying* conservation of flow *for every $v \in V \setminus \{s, t\}$, namely:*

$$\sum_{(u,v) \in E} f_{uv} = \sum_{(v,w) \in E} f_{vw}. \tag{1}$$

Given a flow network, a *flow decomposition* for it consists of a set of source-to-sink *flow paths*, and associated weights strictly greater than 0, such that the flow value of each edge equals the sum of the weights of the paths passing through that edge. In other words, the superposition of the weighted paths of the flow decomposition equals the flow of the network (see also Fig. 1). Formally:

Definition 2 (k-Flow Decomposition). *A k-flow decomposition $(\mathcal{P}, w)$ for a flow network $G = (V, E, f)$ is a set of k s-t flow paths $\mathcal{P} = (P_1, \ldots, P_k)$ and associated weights $w = (w_1, \ldots, w_k)$, with each $w_i \in \mathbb{Z}^+$, such that for each edge $(u, v) \in E$ it holds that:*

$$\sum_{\substack{i \in \{1,\ldots,k\} \ s.t. \\ (u,v) \in P_i}} w_i = f_{uv}. \tag{2}$$

Our above definitions assume integer flow values in the network and integer weights of the flow paths, as is natural since these values count the number of sequenced reads traversing the edges, and are also consistent with previous works such as [21]. However, in practical applications, one could have both fractional flow values and flow path weights, as in e.g., [36]. Note also that the integer and fractional decompositions to the problem may differ, for example [47] observes that are integer flow networks which admit a k-flow decomposition with fractional weights, but no k'-flow decomposition with integer weights, for any $k' \leq k$.

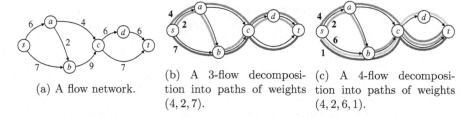

(a) A flow network.

(b) A 3-flow decomposition into paths of weights $(4, 2, 7)$.

(c) A 4-flow decomposition into paths of weights $(4, 2, 6, 1)$.

Fig. 1. Example of a flow network and of two flow decompositions of it.

3 ILP Formulations

3.1 Minimum Flow Decomposition

In this section we consider the following problem of finding a minimum-size flow decomposition.

Problem 1 (Minimum flow decomposition (MFD)). Given a flow network $G = (V, E, f)$, the *minimum flow decomposition (MFD) problem* is to find a flow decomposition $(\mathcal{P}, w)$ such that $|\mathcal{P}|$ is minimized.

Our solution for Problem MFD is based on an ILP formulation of a flow decomposition with a given number k of paths (a k-flow decomposition). Using this, one can easily solve the MFD problem by finding smallest k such that the flow network admits a k-flow decomposition. Notice that any DAG admits a flow decomposition of size at most $|E|$, see e.g., [1] (since one can iteratively take the edge with smallest flow value and create an s-t path of weight equaling this flow value). Moreover, if assuming integer weights, another trivial upper bound on the size of any flow decomposition is $|f|$, namely the flow exiting s, and there is always a flow decomposition with $|f|$ paths of weight one. Thus, if there is a k-flow decomposition, there is also a k'-flow decomposition, for all $k < k' \leq \min\{|E|, |f|\}$ (just duplicate a path of weight greater than one, and move weight one from the old copy to the new one). This shows that when searching for the smaller k such that the graph admits a k-flow decomposition we can either do a linear scan in increasing order, or binary search. Since k is usually small in our applications, we just do a linear scan. As mentioned at the end of Sect. 2, the problem can also be defined as allowing real flow values and/or weights. Our ILP formulation can also handle this variant by just changing the domain of the corresponding variables (in which case we will obtain a Mixed Integer Linear Program (MILP))[1].

---

[1] We note that this version has one subtlety to address: as discussed below, it is necessary to linearize products in the formulation to make it a true ILP (or MILP, in this case). To linearize products of the *real* variables, it is required that the real variables have closed bounds. However, if we solve k-FD for increasing k (and not binary search), we can use $w_i \geq 0$, since no weight 0 path will be included. This introduces the limitation that this formulation could not be used to solve flow decomposition for a fixed k, but only if k is an upper bound on the solution size.

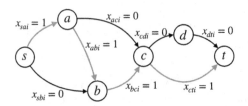

Fig. 2. Example of the edge variables of the ith Eqs. (3a) to (3c).

We start by recalling the standard formulation of a path used for example by [42] for the shortest path problem. If an s-t path repeats no edge (which is always the case if the graph is a DAG) then we can interpret it simply as the set of edges belonging to the path. If we assign value 1 for each edge on the path, and value 0 for each edge not on the path, then these binary values correspond to a conceptual flow in the graph (V, E) (different from the input flow). Moreover, this conceptual flow induced by the (single) path is such that the flow out-going from s is 1 and the flow in-coming to t is 1. It can be easily checked (cf. e.g., [42]) that if the graph is a DAG, then this is a precise characterization of an s-t path.

Thus, for every path $i \in \{1, \dots, k\}$, and every edge $(u, v) \in E$, we can introduce a binary variable x_{uvi} indicating whether the edge (u, v) belongs to the i-th path. The above characterization of a path can be expressed by the following equations (see also Fig. 2):

$$\sum_{(s,v)\in E} x_{svi} = 1, \qquad \forall i \in \{1, \dots, k\}, \tag{3a}$$

$$\sum_{(u,t)\in E} x_{uti} = 1, \qquad \forall i \in \{1, \dots, k\}, \tag{3b}$$

$$\sum_{(u,v)\in E} x_{uvi} - \sum_{(v,w)\in E} x_{vwi} = 0, \quad \forall i \in \{1, \dots, k\}, \forall v \in V \setminus \{s,t\}. \tag{3c}$$

Having expressed a set of k s-t paths with already known ILP constraints, we need to introduce the new constraints tailored for the k-flow decomposition problem. That is, we need to state that the superposition of their weights equals the given flow in the network (2). Thus, for each path i we introduce a positive integer variable w_i corresponding to its weight, and add the constraint:

$$\sum_{i\in\{1,\dots,k\}} x_{uvi} w_i = f_{uv}, \qquad \forall (u, v) \in E. \tag{4}$$

To get the ILP formulation, it remains to linearize equation (4), which is nonlinear because it involves a product of two decision variables. Let us remark that even though non-linear programming solvers exist (such as IPOPT [50]), they are inefficient, do not scale to a large number of variables and are non-professional grade. Instead, having an ILP formulation means that we can make use of popular solvers such as CPLEX [17] and Gurobi [5].

Since the decision variables involved in the product in Eq. (4) are bounded (x_{uvi} is binary and w_i is at most the largest flow value of any edge), this equation can be linearized by standard techniques as in e.g., [9] and [25]. For that, we introduce the integer decision variable π_{uvi} which represents the product between w_i and x_{uvi}, and a constant $\overline{w}$ that is a large enough upper bound for any variable w_i (e.g., the largest flow value of any edge). As such, Eq. (4) can be replaced by the following equations:

$$f_{uv} = \sum_{i \in \{1,\ldots,k\}} \pi_{uvi}, \qquad \forall (u,v) \in E, \tag{5a}$$

$$\pi_{uvi} \leq \overline{w} x_{uvi}, \qquad \forall (u,v) \in E, \forall i \in \{1,\ldots,k\}, \tag{5b}$$

$$\pi_{uvi} \leq w_i, \qquad \forall (u,v) \in E, \forall i \in \{1,\ldots,k\}, \tag{5c}$$

$$\pi_{uvi} \geq w_i - (1 - x_{uvi})\overline{w}, \qquad \forall (u,v) \in E, \forall i \in \{1,\ldots,k\}. \tag{5d}$$

In these constraints, Eq. (5b) ensures that π_{uvi} is 0 if x_{uvi} is 0, and Eqs. (5c) and (5d) ensure that π_{uvi} is w_i if x_{uvi} is 1.

3.2 Subpath Constraints

In this section we consider the flow decomposition variant where we are also given a set of *subpath constraints* that must appear (as a subpath of some path) in any flow decomposition. Among all such decompositions we must find of one with the minimum number of paths. In multiassembly, subpath constraints represent longer reads that span three or more vertices; they are used in popular RNA assembly tools such as StringTie [22] and Scallop [39] and their usefulness for that problem was confirmed empirically in [54]. Such subpath constraints can also naturally model long RNA-seq reads, and we note that, as several authors also acknowledge [2,49,57], long reads do not render the RNA assembly problem obsolete, because they do not always capture full-length transcripts (due to the conversion from RNA to cDNA), and do not fully capture low-expressed transcripts. Formally, the problem can be defined as follows (see also Fig. 3a).

Definition 3 (Flow decomposition with subpath constraints). *Let $G = (V, E, f)$ be a flow network. Subpath constraints are defined to be a set of simple paths $\mathcal{R} = \{R_1, \ldots, R_\ell\}$ in G (not necessarily s-t paths). A flow decomposition $(\mathcal{P}, w)$ satisfies the subpath constraints if and only if*

$$\forall R_j \in \mathcal{R}, \exists P_i \in \mathcal{P} \text{ such that } R_j \text{ is a subpath of } P_i. \tag{6}$$

Problem 2 (Min. flow decomposition with subpath constraints (MFDSC) [54]). Given a flow network $G = (V, E, f)$ and subpath constraints $\mathcal{R}$, the *minimum flow decomposition with subpath constraints* problem is to determine if there exists, and if so, find a flow decomposition $(\mathcal{P}, w)$ satisfying (6) such that $|\mathcal{P}|$ is minimized.

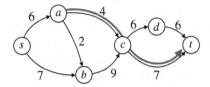 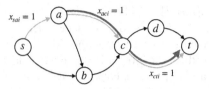

(a) A flow network with a single sub-path constraint $R_1 = (a, c, t)$.

(b) Constraint R_1 is satisfied because for the ith path we can set $r_{i1} = 1$ (and satisfy Eq. (7b)) so that $x_{aci} + x_{cti} \geq 2r_{i1}$ holds (and satisfy Eq. (7a)).

Fig. 3. The flow network from Fig. 1 with a subpath constraint (which is satisfied by the 4-flow decomposition from Fig. 1c, but not by the one in Fig. 1b), and example of a path satisfying the constraint.

For this, we can expand the previous ILP formulation for k-Flow Decomposition to incorporate the conditions necessary to represent the subpath constraints. Let $\mathcal{R}$ be the set of simple paths that are required to be part of at least one path of the flow decomposition. For each $R_j \in \mathcal{R}$, we introduce an additional binary variable r_{ij} denoting the presence of the subpath R_j in the ith path. It clearly holds that $r_{ij} = 1$ if and only if for each edge (u, v) in R_j we have that $x_{uvi} = 1$. Let $|R_j|$ denote the length (i.e., number of edges) of subpath constraint R_j, which is a parameter (i.e. constant). The following inequalities guarantee that each subpath constraint is satisfied by the flow decomposition (see also Fig. 3b):

$$\sum_{(u,v) \in R_j} x_{uvi} \geq |R_j| r_{ij}, \qquad \forall i \in \{1, \ldots, k\}, \forall R_j \in \mathcal{R}, \qquad (7a)$$

$$\sum_{i \in \{1, \ldots, k\}} r_{ij} \geq 1, \qquad \forall R_j \in \mathcal{R}. \qquad (7b)$$

Remark 1. In the above ILP formulation we do not use the fact that the edges of subpath constraint R_j are consecutive (i.e., form a path). Thus, the same formulation applies also if the constraint consists of a *pair* of edge-disjoint paths that must all occur in the same transcript, modelling paired-end Illumina reads, or if it consists of a *set* of edge-disjoint paths (or simply of a set of edges), modelling multi-end Smart-seq3 RNA reads [13]. More specifically, Eq. (7a) simply characterizes when all edges of constraint R_j are covered by some flow path i, and Eq. (7b) requires that at least one flow path satisfies the constraint R_j.

3.3 Inexact Flow

Another variant of the flow decomposition problem is when the given values on the edges of the flow network do not satisfy the conservation of flow property.

Instead, they are required to belong to a given interval, for each edge. Thus, we are looking for an *inexact flow decomposition*, namely one such that the superposition of its weights belongs to the given interval of each edge. This model was studied in [53] and is used in the practical RNA assembler SSP [37], which seeks a set of transcripts explaining the read coverage within some user-defined error tolerance (i.e., interval around the observed weights) on all edges.

The problem is formally stated as follows.

Definition 4 (Inexact flow network). *A tuple $G = (V, E, \underline{f}, \overline{f})$ is said to be an* inexact flow network *if (V, E) is a DAG with unique source s and unique sink t, where for every edge $(u, v) \in E$ we have associated two positive integer values $\underline{f}_{uv}$ and $\overline{f}_{uv}$, satisfying $\underline{f}_{uv} \leq \overline{f}_{uv}$.*

Problem 3 (Minimum inexact flow decomposition (MIFD) [53]). Given an inexact flow network $G = (V, E, \underline{f}, \overline{f})$ the *minimum inexact flow decomposition* problem is to determine if there exists, and if so, find a minimum-size set of s-t paths $\mathcal{P} = (P_1, \ldots, P_k)$ and associated weights $w = (w_1, \ldots, w_k)$ with $w_i \in \mathbb{Z}^+$ such that for each edge $(u, v) \in E$ it holds that:

$$\underline{f}_{uv} \leq \sum_{\substack{i \in \{1, \ldots, k\} \text{ s.t.} \\ (u,v) \in P_i}} w_i \leq \overline{f}_{uv}. \tag{8}$$

In this variant, the same formulation as presented k-Flow Decomposition can be expanded to accommodate the inexact flow component. By simply replacing the flow conservation expressed in Eq. (4) (in the linearized form in Eq. (5a)), with the following two constraints:

$$\underline{f}_{uv} \leq \sum_{i \in \{1, \ldots, k\}} \pi_{uvi} \leq \overline{f}_{uv}, \qquad \forall (u, v) \in E. \tag{9}$$

Remark 2. Notice that Eq. (9) can be combined with Eqs. (7a) and (7b) to obtain a solution if one needs to solve an inexact flow decomposition with subpath constraints problem, further underscoring the versatility of the ILP solution in handling various practical variants of the flow decomposition problem.

4 Experiments

Solvers. We denote by StandardILP and SubpathConstraintsILP our ILP formulations for Problems 1 (MFD) and 2 (MFDSC), respectively.[2] We implemented these using the CPLEX Python API under default settings. We compare StandardILP with Toboggan, the implementation by [21] for their exact FPT

---

[2] We refer the reader to the full version of this paper [8] for experiments on Problem 3 (MIFD) in comparison with IFDSolver, which is an implementation of a heuristic algorithm for MIFD by [53].

algorithm for MFD, and with Catfish, the implementation by [40] of their heuristic algorithm for MFD. We compare SubpathConstraintsILP with Coaster, the implementation by [54] for MFDSC, which is an exact FPT algorithm extending Toboggan, and also with CoasterHeuristic, which is a heuristic for MFDSC also by [54]. Given the size of the datasets, we set a time limit for each graph, as also done by [21,54] (we use 1 min in all cases, except that we also include a run of Toboggan with a 5 min time limit). The runtimes of our ILP implementations include the linear scan in increasing order to find the smallest k for which there is a k-flow decomposition.

Datasets. To test the performance of the solvers under a range of biologically-occurring graph topologies and flows weights, we used three human transcriptomic datasets containing a perfect (i.e., the edge weights satisfy conservation of flow) splice graph for each gene of the human genome. The first dataset, produced by the authors of [39] and also used in a number of flow decomposition benchmarking studies [21,54], was built using publicly available RNA transcripts from the Sequence Read Archive with quantification using the tool Salmon [35]. We use one of the larger transcriptomes[3] and call this dataset **SRR020730-Salmon**. We also produce perfect splice graphs by running HiSat2 [19] with the provided GRCh38 reference index and then popular RNA assembly tool StringTie [22] on real RNA reads from SRR307903, and superimposing the resulting transcripts and abundances (after rounding abundances to the nearest integer). We call this dataset **SRR307903-StringTie**. Finally, we create another dataset by directly simulating expression values for all reference transcripts of all genes in the reference genome GRCh.104 *homo sapiens* by sampling weights from the lognormal distribution with mean and variance both equal to -4, as in the default setting of the RNASeqReadSimulator tool [23]. We multiply the simulated values by 1000 and round to the nearest integer. We call this dataset **Reference-Sim**. For both the **Reference-Sim** and **SRR307903-StringTie** datasets, we use only genes on the positive strand.

For the subpath constraint experiments, we simulate four subpath constraints in each graph as in [54]. For four of the groundtruth paths, we take the prefix of the path that includes three nontrivial junctions (equivalent to three edges in the contracted graph described in [21, Lemma 13]) as a subpath constraint. If a splice graph has fewer than four groundtruth paths, it is excluded from this experiment.

From all datasets, the trivial graphs made up of a single path (i.e. admitting a trivial flow decomposition) are excluded.

Metrics. For each dataset and each FD variant, we report **min** k, the number of paths in a minimum flow decomposition for each problem variant; **Amount**, namely the number of graphs having that specifc value of **min** k; **Avg.**, the average time (in seconds) for each instance solved within the time limit; Σ, the total time (in seconds) required to solve all instances (this included also the

---

[3] The full dataset from [40] is available at https://zenodo.org/record/1460998. We use the file `rnaseq/sparse_quant_SRR020730.graph`.

Table 1. Results for problem MFD.

	min k	Amount	StandardILP			Toboggan (1 min)			Toboggan (5 min)			Catfish			
			Avg.	Σ	Solved	Avg.	Σ	Solved	Avg.	Σ	Solved	Avg.	Σ	Solved	Diff.
SRR020730	2–5	34371	0.091	4093	100	0.002	68	100	0.002	68	100	0.001	34	100	0.00
Salmon	6–10	2291	0.204	458	100	0.023	52	100	0.024	54	100	0.031	71	100	0.00
	11–15	95	4.692	448	100	2.361	225	100	2.612	248	100	3.582	342	100	2.85
	16–20	16	5.891	97	100	10.453	287	86	22.531	671	93	8.451	139	100	3.75
	21–max	7	10.222	75	100	16.564	281	50	33.221	643	78	11.621	85	100	4.56
Reference	2–5	14513	0.089	1300	100	0.002	28	100	0.003	43	100	0.058	14	100	0.00
Sim	6–10	1506	0.352	301	100	0.124	34	100	0.123	186	100	0.124	46	100	0.00
	11–15	261	4.564	1225	100	24.132	4365	75	29.312	6575	92	1.299	935	100	2.79
	16–20	63	10.332	375	100	36.344	1753	65	46.444	3759	83	10.45	538	100	3.75
	21–max	41	12.833	419	100	54.732	1553	51	57.672	4268	73	31.65	478	100	4.56
SRR30790	2–5	7335	0.122	660	100	0.022	14	100	0.022	212	100	0.029	7	100	0.00
StringTie	6–10	768	1.051	153	100	1.191	17	100	1.191	911	100	0.172	23	100	0.00
	11–15	133	4.855	625	100	5.063	2535	71	10.343	5998	88	3.871	447	100	2.53
	16–20	55	6.895	328	100	12.451	1764	57	21.561	5167	74	5.452	471	100	3.75
	21–max	37	10.512	384	100	20.562	1433	51	32.211	4362	68	9.651	437	100	4.56

running time of the instances that did not finish within the time limit); **Solved**, the percentage all instances solved within the time limit; **Diff.**, the average difference between the number of paths obtained with a heuristic algorithm and the optimum one.

Results. The results for Problem MFD are shown in Table 1. For all three datasets, the average time and the total time of Toboggan and Catfish outperform StandardILP for less complex genes, where the number of flow-paths is at most 10 or 15. However, as genes becomes more complex (larger optimum flow decompositions), StandardILP is capable of solving all instances within an average of 10 s, while Toboggan and Catfish require on average 16 and 11 s for the solved instances, respectively. In addition, Toboggan does not solve all instances even within the 5 min time limit. Recall also that Catfish is a heuristic, and thus it does not always return optimum solutions (see **Diff.**).

Among the different datasets, **SRR020730-Salmon** has fewer complex genes and most instances are solved more easily. However for **SRR307903-StringTie** (constructed from real RNA reads) and **Reference-Sim** datasets, there is a larger amount of complex genes and consequently fewer instances can be solved by Toboggan and Catfish, while StandardILP remains efficient and scalable. In these results, although StandardILP does not perform as fast as on **SRR020730-Salmon**, its runtime is still competitive, it can be scaled to graphs with larger k without compromising its efficiency. On the other hand, Toboggan's runtime is exponential in the size of the optimum decomposition, which hinders its usage on larger instances. Moreover, notice that in some applications (e.g. cancer transcriptomics [16]) the graphs of interest do have a large number of RNA transcripts because of the genetic mechanism driving the disease. Hence, in such applications the need to find a flow decomposition is even greater for large k.

Table 2. Results for problem MFDSC.

	min k	Amount	SubpathConstraintsILP			Coaster			CoasterHeuristic			
			Avg.	Σ	Solved	Avg.	Σ	Solved	Avg.	Σ	Solved	Diff.
SRR020730	4–10	5691	0.192	1082	100	30.123	196823	85	0.005	0.51	100	2.14
Salmon	11–15	95	1.475	139	100	45.121	3367	44	0.014	1.42	100	3.04
	16–20	16	3.461	55	100	60.000	960	0	0.025	2.51	100	3.91
	21–max	8	10.452	83	100	60.000	480	0	0.067	6.73	100	4.51
Reference	4–10	6512	0.18	1367	100	37.132	263963	84	0.006	0.61	100	3.13
Sim	11–15	260	1.10	379	100	46.211	15097	14	0.031	3.12	100	4.12
	16–20	78	2.58	303	100	60.000	4680	0	0.041	4.32	100	5.12
	21–max	40	11.51	672	100	60.000	3000	0	0.064	6.54	100	8.13
SRR30790	4–10	864	0.181	156	100	28.241	32013	86	0.006	0.61	100	2.98
StringTie	11–15	104	1.124	115	100	45.142	7484	25	0.032	3.21	100	3.07
	16–20	70	2.578	181	100	60.000	3240	0	0.083	8.31	100	4.14
	21–max	27	11.51	321	100	60.000	2160	0	0.091	9.13	100	5.78

Lastly, one of the key steps in the Toboggan implementation is a reduction of the graph (to simplify nodes with in-degree *or* out-degree equal to one, see [21]), which is a key insight behind its efficiency. However, this observation is highly tailored to the MFD problem, and cannot be easily extended to other FD variants (in fact, it is not used by real RNA assemblers).

The results for Problem MFDSC are shown in Table 2. For all three datasets, SubpathConstraintsILP is capable of solving instances of any size within a few seconds. As an ILP formulation, the addition of the constraints corresponding to the subpath constraints do not hinder its scalability or efficiency. On the other hand, Coaster is both slow on small instances, and does not solve large instances. This shows that the Toboggan implementation is optimized to use many properties of the standard MFD problem, that are not generalizable to variants of it of practical applicability, such as Problem MFDSC. Moreover, similarly to the Catfish heuristic, CoasterHeuristic does not return optimum solutions.

5   Conclusions

In this paper we showed an efficient quadratic-size ILP for MFD, avoiding for the first time the current limitation of (exhaustively) enumerating candidate s-t paths. Many constraints inside state-of-the-art RNA assemblers can be easily modeled on top of our basic ILP (i.e. subpath constraints, inexact flows). Further flexibility also comes from the fact that all our ILPs are based on modeling a specific type of flow decomposition with a *given*, or *upper bounded* number k of paths (thus, they do not need to solve the minimum version of the problem). On both simulated and real datasets, our ILP formulations finish within 13 s on average on any dataset, and within a few seconds on most instances.

On the practical side, we hope that our flexible ILP formulations can lie at the core of future reference-based RNA assemblers employing *exact* solutions. Thus, the current tradeoff between the complexity of the model and its tractability might not be necessary anymore. On the theoretical side, our ILP formulation represents the first *exact* solver for MFD scaling to large values of k, and it could be a reference when benchmarking other heuristic or approximation algorithms.

References

1. Ahuja, R.K., et al.: Network Flows. Alfred P. Sloan School of Management, Cambridge (1988)
2. Amarasinghe, S.L., et al.: Opportunities and challenges in long-read sequencing data analysis. Genome Biol. **21**(1), 1–16 (2020)
3. Baaijens, J.A., Stougie, L., Schönhuth, A.: Strain-aware assembly of genomes from mixed samples using flow variation graphs. In: Schwartz, R. (ed.) RECOMB 2020. LNCS, vol. 12074, pp. 221–222. Springer, Cham (2020). https://doi.org/10.1007/978-3-030-45257-5_14
4. Bernard, E., et al.: Efficient RNA isoform identification and quantification from RNA-Seq data with network flows. Bioinformatics **30**(17), 2447–2455 (2014)
5. Bixby, B.: The Gurobi optimizer. Transp. Res. Part B **41**(2), 159–178 (2007)
6. Canzar, S., et al.: CIDANE: comprehensive isoform discovery and abundance estimation. Genome Biol. **17**(1), 1–18 (2016)
7. Cohen, R., et al.: On the effect of forwarding table size on SDN network utilization. In: IEEE INFOCOM 2014-IEEE conference on computer communications, pp. 1734–1742. IEEE (2014)
8. Dias, F.H.C.: Fast, Flexible, and Exact Minimum Flow Decompositions via ILP. arXiv arXiv:2201.10923 (2022)
9. Furini, F., Traversi, E.: Theoretical and computational study of several linearisation techniques for binary quadratic problems. Ann. Oper. Res. **279**(1), 387–411 (2019). https://doi.org/10.1007/s10479-018-3118-2
10. Gatter, T., Stadler, P.F.: Ryūtō: network-flow based transcriptome reconstruction. BMC Bioinf. **20**(1), 1–14 (2019). https://doi.org/10.1186/s12859-019-2786-5
11. Gurobi Optimization, LLC: Gurobi Optimizer Reference Manual (2021). https://www.gurobi.com
12. Gusfield, D.: Integer Linear Programming in Computational and Systems Biology: An Entry-Level Text and Course. Cambridge University Press, New York (2019)
13. Hagemann-Jensen, M., et al.: Single-cell RNA counting at allele and isoform resolution using Smart-seq3. Nat. Biotechnol. **38**(6), 708–714 (2020)
14. Hartman, T., et al.: How to split a flow? In: 2012 Proceedings IEEE INFOCOM, pp. 828–836. IEEE (2012)
15. Hong, C.Y., et al.: Achieving high utilization with software-driven wan. In: Proceedings of the ACM SIGCOMM 2013 conference on SIGCOMM, pp. 15–26 (2013)
16. Huang, K.K., et al.: Long-read transcriptome sequencing reveals abundant promoter diversity in distinct molecular subtypes of gastric cancer. Genome Biol. **22**(1), 1–24 (2021). https://doi.org/10.1186/s13059-021-02261-x
17. IBM ILOG CPLEX Optimization Studio: CPLEX Users Manual, ver. 12.7 (2017)
18. Khan, S., et al.: Safety and Completeness in Flow Decompositions for RNA Assembly. arXiv arXiv:2201.10372 (2022)
19. Kim, D., et al.: Graph-based genome alignment and genotyping with HISAT2 and HISAT-genotype. Nat. Biotechnol. **37**(8), 907–915 (2019)
20. Kim, P.M., et al.: Analysis of copy number variants and segmental duplications in the human genome: Evidence for a change in the process of formation in recent evolutionary history. Genome Res. **18**(12), 1865–1874 (2008)

21. Kloster, K., et al.: A practical FPT algorithm for flow decomposition and transcript assembly. In: 2018 Proceedings of the Twentieth Workshop on Algorithm Engineering and Experiments (ALENEX), pp. 75–86. SIAM (2018)

22. Kovaka, S., et al.: Transcriptome assembly from long-read RNA-seq alignments with StringTie2. Genome Biol. **20**(1), 1–13 (2019). https://doi.org/10.1186/s13059-019-1910-1

23. Li, W.: RNASeqReadSimulator: a simple RNA-seq read simulator (2014)

24. Li, W., et al.: IsoLasso: a LASSO regression approach to RNA-Seq based transcriptome assembly. J. Comput. Biol. **18**(11), 1693–1707 (2011)

25. Liberti, L.: Compact linearization for binary quadratic problems. **4OR**(3), 31–245 (2007)

26. Lin, Y.-Y., et al.: CLIIQ: accurate comparative detection and quantification of expressed isoforms in a population. In: Raphael, B., Tang, J. (eds.) WABI 2012. LNCS, vol. 7534, pp. 178–189. Springer, Heidelberg (2012). https://doi.org/10.1007/978-3-642-33122-0_14

27. Ma, C., et al.: Finding ranges of optimal transcript expression quantification in cases of non-identifiability. bioRxiv (2020). https://doi.org/10.1101/2019.12.13.875625, to appear at RECOMB 2021

28. Mangul, S., et al.: An integer programming approach to novel transcript reconstruction from paired-end RNA-Seq reads. In: Proceedings of the ACM Conference on Bioinformatics, Computational Biology and Biomedicine, pp. 369–376 (2012)

29. Mao, S., et al.: Refshannon: a genome-guided transcriptome assembler using sparse flow decomposition. PLoS One **15**(6), e0232946 (2020)

30. Maretty, L., et al.: Bayesian transcriptome assembly. Genome Biol. **15**(10), 1–11 (2014)

31. Mumey, B., Shahmohammadi, S., McManus, K., Yaw, S.: Parity balancing path flow decomposition and routing. In: 2015 IEEE Globecom Workshops (GC Wkshps), pp. 1–6. IEEE (2015)

32. Nagarajan, N., Pop, M.: Sequence assembly demystified. Nat. Rev. Genet. **14**(3), 157–167 (2013)

33. Ohst, J.P.: On the Construction of Optimal Paths from Flows and the Analysis of Evacuation Scenarios. Ph.D. thesis, University of Koblenz and Landau, Germany (2015)

34. Olsen, N., et al.: A study on flow decomposition methods for scheduling of electric buses in public transport based on aggregated time–space network models. Cent. Eur. J. Oper. Res. (2020). https://doi.org/10.1007/s10100-020-00705-6

35. Patro, R., et al.: Salmon: accurate, versatile and ultrafast quantification from RNA-seq data using lightweight-alignment. BioRxiv, p. 021592 (2015)

36. Pertea, M., et al.: StringTie enables improved reconstruction of a transcriptome from RNA-seq reads. Nat. Biotechnol. **33**(3), 290–295 (2015)

37. Safikhani, Z., et al.: SSP: an interval integer linear programming for de novo transcriptome assembly and isoform discovery of RNA-seq reads. Genomics **102**(5–6), 507–514 (2013)

38. Shah, S.P., et al.: The clonal and mutational evolution spectrum of primary triple-negative breast cancers. Nature **486**(7403), 395–399 (2012)

39. Shao, M., Kingsford, C.: Accurate assembly of transcripts through phase-preserving graph decomposition. Nat. Biotechnol. **35**(12), 1167–1169 (2017)

40. Shao, M., Kingsford, C.: Theory and a heuristic for the minimum path flow decomposition problem. IEEE/ACM Trans. Comput. Biol. Bioinf. **16**(2), 658–670 (2017)

41. Stamm, S., et al.: Function of alternative splicing. Gene **344**, 1–20 (2005)

42. Taccari, L.: Integer programming formulations for the elementary shortest path problem. Eur. J. Oper. Res. **252**(1), 122–130 (2016)
43. Tomescu, A.I., et al.: A novel min-cost flow method for estimating transcript expression with RNA-Seq. BMC Bioinf. **14**, S1:51-S1:51 (2013). https://doi.org/10.1186/1471-2105-14-S5-S15
44. Tomescu, A.I., et al.: Explaining a weighted DAG with few paths for solving genome-guided multi-assembly. IEEE/ACM Trans. Comput. Biol. Bioinf. **12**(6), 1345–1354 (2015)
45. Töpfer, A., et al.: Probabilistic inference of viral quasispecies subject to recombination. J. Comput. Biol. **20**(2), 113–123 (2013)
46. Trapnell, C., et al.: Transcript assembly and quantification by RNA-Seq reveals unannotated transcripts and isoform switching during cell differentiation. Nature Biotechnol. **28**(5), 511–515 (2010)
47. Vatinlen, B., et al.: Simple bounds and greedy algorithms for decomposing a flow into a minimal set of paths. Eur. J. Oper. Res. **185**(3), 1390–1401 (2008)
48. Vignuzzi, M., et al.: Quasispecies diversity determines pathogenesis through cooperative interactions in a viral population. Nature **439**(7074), 344–348 (2006)
49. Voshall, A., Moriyama, E.N.: Next-generation transcriptome assembly: strategies and performance analysis. In: Bioinformatics in the Era of Post Genomics and Big Data, pp. 15–36 (2018)
50. Wächter, A., Biegler, L.T.: On the implementation of an interior-point filter linesearch algorithm for large-scale nonlinear programming. Math. Program. **106**(1), 25–57 (2006). https://doi.org/10.1007/s10107-004-0559-y
51. Wang, E.T., et al.: Alternative isoform regulation in human tissue transcriptomes. Nature **456**(7221), 470–476 (2008)
52. Westbrooks, K., Astrovskaya, I., Campo, D., Khudyakov, Y., Berman, P., Zelikovsky, A.: HCV Quasispecies assembly using network flows. In: Măndoiu, I., Sunderraman, R., Zelikovsky, A. (eds.) ISBRA 2008. LNCS, vol. 4983, pp. 159–170. Springer, Heidelberg (2008). https://doi.org/10.1007/978-3-540-79450-9_15
53. Williams, L., et al.: RNA transcript assembly using inexact flows. In: 2019 IEEE International Conference on Bioinformatics and Biomedicine (BIBM), pp. 1907–1914. IEEE (2019)
54. Williams, L., et al.: Flow decomposition with subpath constraints. In: 21st International Workshop on Algorithms in Bioinformatics (WABI 2021). Schloss Dagstuhl-Leibniz-Zentrum für Informatik (2021)
55. Xing, Y., et al.: The multiassembly problem: reconstructing multiple transcript isoforms from EST fragment mixtures. Genome Res. **14**(3), 426–441 (2004)
56. Zagordi, O., et al.: ShoRAH: estimating the genetic diversity of a mixed sample from next-generation sequencing data. BMC Bioinf. **12**(1), 1–5 (2011). https://doi.org/10.1186/1471-2105-12-119
57. Zhang, Q., et al.: Scallop2 enables accurate assembly of multiple-end RNA-seq data. bioRxiv (2021). https://doi.org/10.1101/2021.09.03.458862
58. Zhao, J., et al.: Multitrans: an algorithm for path extraction through mixed integer linear programming for transcriptome assembly. IEEE/ACM Trans. Comput. Biol. Bioinf. (2021). https://doi.org/10.1109/TCBB.2021.3083277

Co-linear Chaining with Overlaps and Gap Costs

Chirag Jain[1], Daniel Gibney[2(✉)], and Sharma V. Thankachan[3]

[1] Department of Computational and Data Sciences, Indian Institute of Science, Bangalore, India
`chirag@iisc.ac.in`
[2] School of Computer Science and Engineering, Georgia Institute of Technology, Atlanta, USA
`daniel.j.gibney@gmail.com`
[3] Department of Computer Science, University of Central Florida, Orlando, USA
`sharma.thankachan@gmail.com`

Abstract. Co-linear chaining has proven to be a powerful heuristic for finding near-optimal alignments of long DNA sequences (e.g., long reads or a genome assembly) to a reference. It is used as an intermediate step in several alignment tools that employ a seed-chain-extend strategy. Despite this popularity, efficient subquadratic-time algorithms for the general case where chains support anchor overlaps and gap costs are not currently known. We present algorithms to solve the co-linear chaining problem with anchor overlaps and gap costs in $\tilde{O}(n)$ time, where n denotes the count of anchors. We also establish the first theoretical connection between co-linear chaining cost and edit distance. Specifically, we prove that for a fixed set of anchors under a carefully designed chaining cost function, the optimal 'anchored' edit distance equals the optimal co-linear chaining cost. Finally, we demonstrate experimentally that optimal co-linear chaining cost under the proposed cost function can be computed orders of magnitude faster than edit distance, and achieves correlation coefficient above 0.9 with edit distance for closely as well as distantly related sequences.

Keywords: Edit distance · Alignment · Co-linear chaining

1 Introduction

Computing an optimal alignment between two sequences is one of the most fundamental problems in computational biology. Unfortunately, conditional lower-bounds suggest that an algorithm for computing an optimal alignment, or edit distance, in strongly subquadratic time is unlikely [3,10]. This lower-bound indicates a challenge for scaling the computation of edit distance to high-throughput sequencing data. Instead, heuristics are often used to obtain an approximate solution in less time and space. One such popular heuristic is co-linear chaining. This technique involves precomputing fragments between the two sequences

I. Pe'er (Ed.): RECOMB 2022, LNBI 13278, pp. 246–262, 2022.
https://doi.org/10.1007/978-3-031-04749-7_15

that closely agree (in this work, exact matches called *anchors*), then determining which of these anchors should be kept within the alignment (see Fig. 1). Techniques along these lines are used in long-read mappers [6,12,15,16,24,25,27] and generic sequence aligners [2,5,14,19,23]. We will focus on the following problem (described formally in Sect. 2): Given a set of n anchors, determine an optimal ordered subset (or chain) of these anchors.

Several algorithms have been developed for the co-linear chaining [1,17,28,31] and even more in the context of sparse dynamic programming [8,9,18,20,22,33]. Solutions with different time complexities exist for different variations of this problem. These depend on the cost-function assigned to a chain and the types of chains permitted. Solutions include an algorithm running in $O(n \log n \log \log n)$ time for a simpler variant of the problem where anchors used in a solution must be non-overlapping [1]. More recently, Mäkinen and Sahlin gave an algorithm running in $O(n \log n)$ time where anchor overlaps are allowed, but gaps between anchors are not considered in the cost-function [17]. None of the solutions introduced thus far provide a subquadratic time algorithm for variations that use both overlap and gap costs. However, including overlaps and gaps into a cost-function is a more realistic model for anchor chaining. For example, consider a simple scenario where minimizers [26] are used to identify anchors. Suppose query and reference sequences are identical, then adjacent minimizer-anchors will likely overlap. Not allowing anchor overlaps during chaining will lead to a penalty cost associated with gaps between chained anchors despite the two strings being identical. Therefore, depending on the type of anchor, there may be no reason to assume that in an optimal alignment the anchors would be non-overlapping. At the same time, not penalizing long gaps between the anchors is unlikely to produce correct alignments. This is why both anchor overlaps and gap costs are supported during chaining in widely-used aligners, e.g., Minimap2 [13,15], Nucmer4 [19]. This work's contribution is the following:

- We provide the first algorithm running in subquadratic, $\widetilde{O}(n)$ time for chaining with overlap and gap costs[1]. Refinements based on the specific type of anchor and chain under consideration are also given. These refinements include an $O(n \log^2 n)$ time algorithm for the case where all anchors are of the same length, as is the case with k-mers.
- When n is not too large (less than the sequence lengths), we present an algorithm with $O(n \cdot OPT + n \log n)$ average-case time where OPT is the optimal solution value. This provides a simple algorithm that is efficient in practice.
- Using a carefully designed cost-function, we mathematically relate the optimal chaining cost with a generalized version of edit distance, which we call *anchored edit distance*. This is equivalent to the usual edit distance with the modification that matches performed without the support of an anchor have unit cost. A more formal definition appears in Sect. 2. With our cost function, we prove that the optimal chaining cost is equal to the anchored edit distance.

---

[1] $\widetilde{O}(\cdot)$ hides poly-logarithmic factors.

– We empirically demonstrate that computing optimal chaining cost is orders of magnitude faster than computing edit distance, especially in semi-global comparison mode. We also demonstrate a strong correlation between optimal chaining cost and edit distance. The correlation coefficients are favorable when compared to suboptimal chaining methods implemented in Minimap2 and Nucmer4.

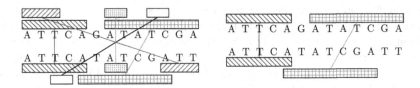

Fig. 1. (Left) Anchors representing a set of exact matches are shown as rectangles. The co-linear chaining problem is to find an optimal ordered subset of anchors subject to some cost function. (Right) A chain of overlapping anchors.

2   Concepts and Definitions

For a given pair of strings S_1 and S_2, an anchor interval pair $([a..b], [c..d])$ signifies an exact match between $S_1[a..b]$ and $S_2[c..d]$. For an anchor I, we denote these values as $I.a$, $I.b$, $I.c$, and $I.d$. Here $b - a = d - c$ and $S_1[a + j] = S_2[c + j]$ for all $0 \leq j \leq b - a$. We say that the character match $S_1[a + j] = S_2[c + j]$, $0 \leq j \leq b - a$, is *supported* by the anchor $([a..b], [c..d])$. Maximal exact matches (MEMs), maximal unique matches (MUMs), or k-mer matches are some of the common ways to define anchors. Maximal unique matches [7] are a subset of maximal exact matches, having the added constraint that the pattern involved occurs only once in both strings. If all intervals across all anchors have the same length (e.g., using k-mers), we say that the *fixed-length* property holds.

Our algorithms will make use of dynamic range minimum queries (RmQs). For a set of n d-dimensional points, each with an associated weight, a 'query' consists of an orthogonal d-dimensional range. The query response is the point in that range with the smallest weight. Using known techniques in computational geometry, a data structure can be built in $O(n \log^{d-1} n)$ time and space, that can both answer queries and modify a point's weight in $O(\log^d n)$ time [4].

2.1   Co-linear Chaining Problem with Overlap and Gap Costs

Given a set of n anchors $\mathcal{A}$ for strings S_1 and S_2, we assume that $\mathcal{A}$ already contains two *end-point* anchors $\mathcal{A}_{left} = ([0,0], [0,0])$ and $\mathcal{A}_{right} = ([|S_1| + 1, |S_1| + 1], [|S_2| + 1, |S_2| + 1])$. We define the strict precedence relationship $\prec$ between two anchors $I' := \mathcal{A}[j]$ and $I := \mathcal{A}[i]$ as $I' \prec I$ if and only if $I'.a \leq I.a$, $I'.b \leq I.b$, $I'.c \leq I.c$, $I'.d \leq I.d$, and strict inequality holds for at least one of the four inequalities. In other words, the interval belonging to I' for S_1 (resp. S_2) should start before or at the starting position of the interval belonging to I for S_1 (resp.

S_2) and should not extend past it. We also define the weak precedence relation $\prec_w$ as $I' \prec_w I$ if and only if $I'.a \leq I.a$, $I'.c \leq I.c$ and strict inequality holds for at least one of the two inequalities, i.e., intervals belonging to I' should start before or at the starting position of intervals belonging to I, but now intervals belonging to I' can be extended past the intervals belonging to I. The aim of the problem is to find a totally ordered subset (a chain) of $\mathcal{A}$ that achieves the minimum cost under the cost function presented next. We specify whether we mean a chain under strict precedence or under weak precedence when necessary.

Cost Function. For $I' \prec I$, the function $connect(I', I)$ is designed to indicate the cost of connecting anchor I' to anchor I in a chain. The chaining problem asks for a chain of $m \leq n$ anchors, $\mathcal{A}'[1], \mathcal{A}'[2], \ldots, \mathcal{A}'[m]$, such that the following properties hold: (i) $\mathcal{A}'[1] = \mathcal{A}_{left}$, (ii) $\mathcal{A}'[m] = \mathcal{A}_{right}$, (iii) $\mathcal{A}'[1] \prec \mathcal{A}'[2] \prec \ldots \prec \mathcal{A}'[m]$, and (iv) the cost $\sum_{i=1}^{m-1} connect(\mathcal{A}'[i], \mathcal{A}'[i+1])$ is minimized.

We next define the function $connect$. In Sect. 4, we will see that this definition is well motivated by the relationship with anchored edit distance. For a pair of anchors I', I such that $I' \prec I$:

- The gap in string S_1 between anchors I' and I is $g_1 = \max(0, I.a - I'.b - 1)$. Similarly, the gap between the anchors in string S_2 is $g_2 = \max(0, I.c - I'.d - 1)$. Define the gap cost $g(I', I) = \max(g_1, g_2)$.
- The overlap o_1 is defined such that $I'.b - o_1$ reflects the non-overlapping prefix of anchor I' in string S_1. Specifically, $o_1 = \max(0, I'.b - I.a + 1)$. Similarly, define $o_2 = \max(0, I'.d - I.c + 1)$. We define the overlap cost as $o(I', I) = |o_1 - o_2|$.
- Lastly, define $connect(I', I) = g(I', I) + o(I', I)$.

The same definitions are used for weak precedence, only using $\prec_w$ in the place of $\prec$. Regardless of the definition of $connect$, the above problem can be trivially solved in $O(n^2)$ time and $O(n)$ space. First sort the anchors by the component $\mathcal{A}[\cdot].a$ and let $\mathcal{A}'$ be the sorted array. The chaining problem then has a direct dynamic programming solution by filling an n-sized array C from left-to-right, such that $C[i]$ reflects the cost of an optimal chain that ends at anchor $\mathcal{A}'[i]$. The value $C[i]$ is computed using the recursion: $C[i] = \min_{\mathcal{A}'[k] \prec \mathcal{A}'[i]} (C[k] + connect(\mathcal{A}'[k], \mathcal{A}'[i]))$ where the base case associated with anchor $\mathcal{A}_{left}$ is $C[1] = 0$. The optimal chaining cost will be stored in $C[n]$ after spending $O(n^2)$ time. We will provide an $O(n \log^4 n)$ time algorithm for this problem.

2.2 Anchored Edit Distance

The edit distance problem is to identify the minimum number of operations (substitutions, insertions, or deletions) that must be applied to string S_2 to transform it to S_1. Edit operations can be equivalently represented as an alignment (a.k.a. edit transcript) that specifies the associated matches, mismatches and gaps while placing one string on top of another. The *anchored edit distance problem* is as follows: given strings S_1 and S_2 and a set of n anchors $\mathcal{A}$, compute the optimal edit distance subject to the condition that a match supported by an anchor has edit cost 0, and a match that is not supported by an anchor has edit cost 1.

The above problem is solvable in $O(|S_1||S_2|)$ time and space. We can assume that input does not contain redundant anchors, therefore, the count of anchors is $\leq |S_1||S_2|$. Next, the standard dynamic programming recursion for solving the edit distance problem can be revised. Let $D[i,j]$ denote anchored edit distance between $S_1[1,i]$ and $S_2[1,j]$, then $D[i,j] = \min(D[i-1,j-1] + x, D[i-1,j] + 1, D[i,j-1] + 1)$, where $x = 0$ if $S_1[i] = S_2[j]$ and the match is supported by some anchor, and $x = 1$ otherwise.

2.3   Graph Representation of Alignment

It is useful to consider the following representation of an alignment of two strings S_1 and S_2. As illustrated in Fig. 2, we have a set of $|S_1|$ top vertices and $|S_2|$ bottom vertices. There are two types of edges between the top and bottom vertices: (i) A solid edge from ith top vertex to the jth bottom vertex. This represents an anchor supported character match between the ith character in S_1 and the jth character in S_2; (ii) A dashed edge from the ith top vertex to the jth bottom vertex. This represents a character being substituted to form a match between $S_1[i]$ and $S_2[j]$ or a character match not supported by an anchor. All unmatched vertices are labeled with an 'x' to indicate that the corresponding character is deleted. An important observation is that no two edges cross.

In a solution to the anchored edit distance problem every solid edge must be 'supported' by an anchor. By 'supported' here we mean that the match between the corresponding characters in S_1 and S_2 is supported by an anchor. In Fig. 2, these anchors are represented with rectangles above and below the vertices. We use $\mathcal{M}$ to denote a particular alignment. We also associate an edit cost with the alignment, denoted as $EDIT(\mathcal{M})$. This is equal to the number of vertices marked with x in $\mathcal{M}$ plus the number of dashed edges in $\mathcal{M}$.

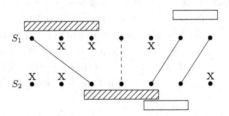

Fig. 2. The graph representation of an alignment. Solid edges represent anchor-supported character matches, dashed edges represent substitutions and unsupported matches, and x's represent deletions. We use $\mathcal{M}$ to denote an alignment. Here $EDIT(\mathcal{M}) = 7$, the total number of x's and dashed edges.

3 Our Algorithms

Theorem 1. *The co-linear chaining problem with overlap and gap costs can be solved in time $\widetilde{O}(n)$. In particular, in time $O(n \log^2 n)$ for chains with fixed-length anchors; in time $O(n \log^3 n)$ for chains under weak precedence; and in time $O(n \log^4 n)$ for chains under strict precedence.*

The proposed algorithm still uses the recursive formula given in Sect. 2.1. However, it uses range minimum query (RmQ) data structures to avoid having to check every anchor $\mathcal{A}[k]$ where $\mathcal{A}[k].a < \mathcal{A}[i].a$. We achieve this by considering six cases concerning the optimal choice of the prior anchor. We use the best of the six distinct possibilities to determine the optimal $C[i]$ value. This $C[i]$ value is then used to update the RmQ data structures. For the strict precedence case, some of the six cases require up to four dimensions for the range minimum queries. When only weak precedence is required, we reduce this to at most three dimensions. When the fixed-length property holds (e.g., k-mers), we reduce this to two dimensions.

Algorithm for Chains Under Strict Precedence. The first step is to sort the set of anchors $\mathcal{A}$ using the key $\mathcal{A}[\cdot].a$. Let $\mathcal{A}'$ be the sorted array. We will next use six RmQ data structures labeled $T_{1a}, T_{1b}, T_{2a}, T_{2b}, T_{3a}, T_{3b}$. These RmQ data structures are initialized with the following points for every anchor: For anchor $I \in \mathcal{A}'$: T_{1a} is initialized with the point $(I.b, I.d - I.b)$, T_{1b} with $(I.d, I.d - I.b)$, T_{2a} with $(I.b, I.c, I.d)$, T_{2b} with $(I.b, I.d)$, T_{3a} with $(I.b, I.c, I.d, I.d - I.b)$, and T_{3b} with $(I.b, I.d, I.d - I.b)$. All weights are initially set to ∞ except for $I = \mathcal{A}_{left}$, where the corresponding points are given weight 0. We then process the anchors in sorted order and update the RmQ data structures after each iteration. On the ith iteration, for $j < i$, we let $C[j]$ be the optimal co-linear chaining cost of any ordered subset of $\mathcal{A}'[1], \mathcal{A}'[2], ..., \mathcal{A}'[j]$ that ends with $\mathcal{A}'[j]$. For $i > 1$, RmQ queries are used to find the optimal $j < i$ by considering six different cases. We let $I = \mathcal{A}'[i]$, $I' = \mathcal{A}'[j]$, and $C[I'] = C[j]$.

The query for each RmQ structure is determined by the different inequalities relating $I.a$, $I.b$, $I.c$, and $I.d$ to previous anchors in the case considered. For example, in Case 1.a (Fig. 3), it can be seen that $I'.b < I.a$ and $I.a - I'.b < I.c - I'.d$, making $I'.b \in [0, I.a - 1]$ and $I'.d - I'.b \in [-\infty, I.c - I.a]$, motivating the query input $[0, I.a - 1] \times [-\infty, I.c - I.a]$. At the same time, the values stored in these RmQ structures are determined by the expression for the co-linear chaining cost in that case, $C[I'] + I.c - I'.d - 1$. Note that the values stored in each RmQ structure depend only on previously processed anchors and are combined with the values $I.a$, $I.b$, $I.c$, and $I.d$ for the current anchor I being processed to obtain the appropriate cost. Hence, for T_{1a} we store values of the form $C[I'] - I'.d$ and combine this with $I.c$ to obtain the cost. The other cases can be similarly analyzed.

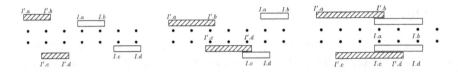

Fig. 3. (Left) Case 1.a. Colinear chaining cost is $C[I'] + g_2 = C[I'] + I.c - I'.d - 1$. (Middle) Case 2.a. Chaining cost is $C[I'] + g_1 + o_2 = C[I'] + I.a - I'.b + I'.d - I.c$. (Right) Case 3.a. Chaining cost is $C[I'] + o_2 - o_1 = C[j] + I'.d - I.c - (I'.b - I.a)$.

1. Case: I' disjoint from I.
   (a) Case: The gap in S_1 is less or equal to gap in S_2 (Fig. 3 (Left)). The range minimum query (query input) is of the form: $[0, I.a - 1] \times [-\infty, I.c - I.a]$. Let the query response (weight) from T_{1a} be $v_{1a} = \min\{C[I'] - I'.d : (I'.b, I'.d - I'.b) \in [0, I.a - 1] \times [-\infty, I.c - I.a]\}$ and let $C_{1a} = v_{1a} + I.c - 1$.
   (b) Case: The gap in S_2 is less than gap in S_1. The range minimum query is of the form $[0, I.c - 1] \times [I.c - I.a + 1, \infty]$. Let the query response from T_{1b} be $v_{1b} = \min\{C[I'] - I'.b : (I'.d, I'.d - I'.b) \in [0, I.c - 1] \times [I.c - I.a + 1, \infty]\}$ and let $C_{1b} = v_{1b} + I.a - 1$.
2. Case: I' and I overlap in only one dimension.
   (a) Case: I' and I overlap only in S_2 (Fig. 3 (Middle)). The range minimum query is of the form $[0, I.a - 1] \times [0, I.c] \times [I.c, I.d]$. Let the query response from T_{2a} be $v_{2a} = \min\{C[I'] - I'.b + I'.d : (I'.b, I'.c, I'.d) \in [0, I.a - 1] \times [0, I.c] \times [I.c, I.d]\}$ and let $C_{2a} = v_{2a} + I.a - I.c$.
   (b) Case: I' and I overlap only in S_1. Since the anchors are sorted on $\mathcal{A}[\cdot].a$, this can be done with a two dimensional RmQ structure. The range minimum query is of the form $[I.a, I.b] \times [0, I.c - 1]$. Let the query response from T_{2b} be $v_{2b} = \min\{C[I'] + I'.b - I'.d : (I'.b, I'.d) \in [I.a, I.b] \times [0, I.c - 1]\}$ and let $C_{2b} = v_{2b} + I.c - I.a$.
3. Case: I' and I overlap in both dimensions.
   (a) Case: Greater overlap in S_2 (Fig. 3 (Right)). Here, $|o_1 - o_2| = o_2 - o_1 = I'.d - I.c - (I'.b - I.a)$. The range minimum query is of the form $[I.a, I.b] \times [0, I.c] \times [I.c, I.d] \times [I.c - I.a + 1, \infty]$. Let the query response from T_{3a} be $v_{3a} = \min\{C[I'] - I'.b + I'.d : (I'.b, I'c, I'.d, I'.d - I'.b) \in [I.a, I.b] \times [0, I.c] \times [I.c, I.d] \times [I.c - I.a + 1, \infty]\}$ and let $C_{3a} = v_{3a} + I.a - I.c$.
   (b) Case: Greater or equal overlap in S_1. Here, $|o_1 - o_2| = o_1 - o_2 = I'.b - I.a - (I'.d - I.c)$. If $o_1 \geq o_2 > 0$, $I'.b \in [I.a, I.b]$, and $I'.a \in [0, I.a]$, then $I'.c \in [0, I.c]$. Hence, the range minimum query is of the form $[I.a, I.b] \times [I.c, I.d] \times [-\infty, I.c - I.a]$. Let the query response from T_{3b} be $v_{3b} = \min\{C[I'] + I'.b - I'.d : (I'.b, I'.d, I'.d - I'.b) \in [I.a, I.b] \times [I.c, I.d] \times [-\infty, I.c - I.a]\}$ and let $C_{3b} = v_{3b} - I.a + I.c$.

Finally, let $C[i] = \min(C_{1a}, C_{1b}, C_{2a}, C_{2b}, C_{3a}, C_{3b})$ and update the RmQ structures accordingly (see full version [11] for details and pseudocode). Every RmQ structure T has the query method $T.RmQ()$ which takes as arguments an interval for each dimension. It also has the method $T.update()$, which takes a point and a weight and updates the point to have the new weight. The four-dimensional RmQ structures for Case 3.a require $O(\log^4 n)$ time per query and update, causing an overall time complexity that is $O(n \log^4 n)$. We defer the modifications for weak precendence and fixed-length anchors to the full version [11].

4   Proof of Equivalence

Theorem 2. *For a fixed set of anchors $\mathcal{A}$, the following quantities are equal: the anchored edit distance, the optimal co-linear chaining cost under strict precedence, and the optimal co-linear chaining cost under weak precedence.*

The optimal co-linear chaining cost is defined using the cost function described in Sect. 2.1. An implication of Theorems 1 and 2 is that if only the anchored edit distance is desired (and not an optimal strictly ordered anchor chain), there exists a $O(n \log^3 n)$ for computing this value.

Theorem 2 will follow from Lemmas 1 and 2.

Lemma 1. *Anchored edit distance $\leq$ optimal co-linear chaining cost under weak precedence $\leq$ optimal co-linear chaining cost under strict precedence.*

Proof. The second inequality follows from the observation that every set of anchors ordered under strict precedence is also ordered under weak precedence. We now focus on the inequality between anchored edit distance and co-linear chaining cost under weak precedence. Starting with an anchor chain under weak precedence, $\mathcal{A}[1]$, $\mathcal{A}[2]$, ... with associated co-linear chaining cost x, we provide an alignment with an anchored edit distance that is at most x. This alignment is obtained using a greedy algorithm that works from left-to-right, always taking the closest exact match when possible, and when not possible, a character substitution or unsupported exact match, or if none of these are possible, a deletion. We now present the details.

Greedy Algorithm. Assume inductively that all symbols in $S_1[1, \mathcal{A}[i].b]$ and $S_2[1, \mathcal{A}[i].d]$ have been processed, that is, either matched, substituted, or deleted (represented by check-marks in Figs. 4, 5 and 6). The base case of this induction holds trivially for $\mathcal{A}_{left}$. We consider the anchor $\mathcal{A}[i + 1]$ and the possible cases regarding its position relative to $\mathcal{A}[i]$. Symmetric cases that only swap the roles of S_1 and S_2 are ignored. To ease notation, let $I' = \mathcal{A}[i]$ and $I = \mathcal{A}[i + 1]$.

1. **Case $I'.b \geq I.b$ and $I'.d \geq I.c$ (Fig 4):** To continue the alignment, delete the substring $S_2[I'.d+1, I.d]$ from S_2. This has edit cost $I.d - I'.d$. We can assume both intervals of I' are not nested in intervals of I, hence $connect(I', I) = o_1 - o_2 = I'.b - I.a - I'.d + I.c \geq I.c + I.b - I.a - I'.d = I.d - I'.d$.

2. **Case $I'.b \geq I.b$ and $I'.d < I.c$ (Fig 4):** Delete the substring $S_2[I'.d + 1, I.d]$ from S_2, with edit cost $I.d - I'.d$. Also $connect(I', I) = o_1 + g_2 = I'.b - I.a + I.c - I'.d \geq I.c + I.b - I.a - I'.d = I.d - I'.d$.

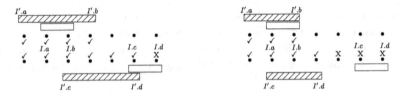

Fig. 4. Cases in Proof of Lemma 1. The ✓ symbol indicates symbols processed prior to considering I. (Left) Case $I', b \geq I.b$ and $I'.d \geq I.c$ (Right) $I'.b \geq I.b$ and $I'.d < I.c$.

3. **Case** $I.b > I'.b$, $I.a \leq I'.b$, $I.c \leq I'.d$ (Fig. 5): Supposing wlog that $o_1 > o_2$, delete $S_2[I'.d+1, I'.d+o_1-o_2]$, and match $S_1[I'.b+1, I.b]$ and $S_2[I'.d+o_1 - o_2 + 1, I.d]$. This has edit cost $o_1 - o_2$ and $connect(I', I) = o_1 - o_2$.
4. **Case** $I.b > I'.b$, $I.a \leq I'.b$, $I.c > I'.d$ (Fig. 5): We delete $S_2[I'.d + 1, I'.d + o_1 + g_2]$ and match $S_1[I'.b + 1, I.b]$ with $S_2[I'.d + o_1 + g_2 + 1, I.d]$. This has edit cost $o_1 + g_2$ and $connect(I', I) = o_1 + g_2$.
5. **Case** $I.a > I'.b$, $I.c > I'.d$ (Fig. 6): Supposing wlog $g_2 \geq g_1$, match with substitutions or unsupported exact matches $S_1[I'.b + 1, I'.b + g_1]$ and $S_2[I'.d + 1, I'.d + g_1]$. Delete the substring $S_2[I'.d + g_1 + 1, I.c - 1]$. Finally, match $S_1[I.a, I.b]$ and $S_2[I.c, I.d]$. The edits consist of g_1 of substitutions or unsupported exact matches and $g_2 - g_1$ deletions, which is g_2 edits in total. Also, $connect(I', I) = \max\{g_1, g_2\} = g_2$.

Continuing this process until $\mathcal{A}_{right}$, all symbols in S_1 and S_2 become included in the alignment. □

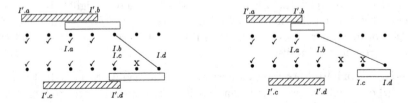

Fig. 5. Cases in Proof of Lemma 1. (Left) Case $I.b > I'.b$, $I.a \leq I'.b$, $I.c \leq I'.d$. (Right) Case $I.b > I'.b$, $I.a \leq I'.b$, $I.c > I'.d$.

We delay the details of Lemma 2's proof to Sect. 4.1.

Lemma 2. *For a set of anchors $\mathcal{A}$, optimal chaining cost under strict precedence $\leq$ anchored edit distance.*

Proof. We start with an arbitrary alignment $\mathcal{M}$ supported by $\mathcal{A}$. We will show in Lemma 3 how to obtain a subset $\mathcal{B} \subseteq \mathcal{A}$ totally ordered under strict precedence and supporting an alignment $\mathcal{M}'$ where $EDIT(\mathcal{M}') \leq EDIT(\mathcal{M})$. We will then show in Lemma 4 that the edit cost of $\mathcal{M}'$ is greater or equal to the edit cost of the alignment $\mathcal{M}_G$ given by the greedy algorithm on $\mathcal{B}$. Finally, in Lemma 5 we show that the co-linear chaining cost of $\mathcal{B}$ is equal to the edit cost of $\mathcal{M}_G$. Combining, we have $EDIT(\mathcal{M}) \geq EDIT(\mathcal{M}') \geq EDIT(\mathcal{M}_G) =$ the co-linear chaining cost on $\mathcal{B} \geq$ optimal co-linear chaining cost under strict precedence for $\mathcal{A}$. The result follows from the fact that $EDIT(\mathcal{M})$ equals the anchored edit distance when $\mathcal{M}$ is an optimal alignment for $\mathcal{A}$. □

4.1   Details of Lemma 2 Proof

We apply Algorithm (i) followed by Algorithm (ii) to convert a supporting set of anchors $\mathcal{A}$ for $\mathcal{M}$ into the totally ordered subset of anchors $\mathcal{B}$ supporting $\mathcal{M}'$. Note that these algorithms are only for the purpose of the proof. Moving forward, we call an edge $e = (S_1[h], S_2[k])$ contained but not supported by I if $h \in [I.a, I.b]$ or $k \in [I.c, I.d]$ and $h - I.a \neq k - I.c$. We define for e the two edges $e' = (S_1[h], S_2[I.c + h - I.a])$ and $e'' = (S_1[I.a + k - I.c], S_2[k])$, which are supported by I.

Algorithm (i). Algorithm for Removing Incomparable Anchors. Let I and I' be two incomparable anchors under weak precedence (Fig. 6). The anchor that has the rightmost supported solid edge will be the anchor we keep. Suppose wlog it is I. Working from right-to-left, starting with that rightmost edge, for any edge e that is contained but not supported by I, we replace e with the rightmost of e' and e''. Note that at least one side of every edge supported by I' is within an interval of I. Hence, all edges supported by I' are eventually replaced. We then remove I'. This algorithm is repeated until a total ordering under weak precedence is possible.

Algorithm (ii). Algorithm for Removing Anchors with Nested Intervals. Consider two anchors I and I' where wlog I' has an interval nested in one of the intervals belonging to I. Let e_R be the rightmost edge supported by I. Working from right-to-left, we replace any edge e to the left of e_R that is contained but not supported by I with the rightmost of e' and e''. Next, working from left-to-right, we replace any edge e to the right of e_R that is contained but not supported by I with the leftmost of e' and e''. These procedures combined will replace all edges supported by I' with those supported by I. We repeat this until there are no two nested intervals amongst all remaining anchors. Finally, remove all anchors that do not support any edge. We call such an anchor chain where every anchor supports at least one edge *minimal*.

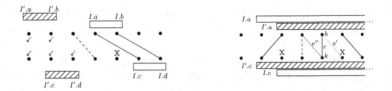

Fig. 6. (Left) Case $I.a > I'.b$, $I.c > I'.d$. (Right) Anchors I and I' are incomparable. The current alignment is shown with black solid and dashed edges. To remove I' we sweep from right-to-left, replacing edges not supported by I with edges supported by I. Here, $e = (S_1[h], S_2[k])$ is not supported by I and will be replaced with $e' = (S_1[h], S_2[I.c + h − I.a])$ (in red), which is supported by I.

Lemma 3. $EDIT(\mathcal{M}') \leq EDIT(\mathcal{M})$.

Proof. For Algorithm (i), suppose we are replacing an edge e not supported by the anchor I, the anchor we wish to keep. Suppose wlog that e' is the rightmost of e' and e'', so we replace e with e'. Because the edge immediately to the right of e is also aligned with I, deleting $S_2[k]$ and matching $S_2[I.c + h − I.a]$, does not require modifying any additional edges. If e was a solid edge the edit cost is unaltered, since the total number of deletions and matches is unaltered. If e was a dashed edge, replacing e with e' converts a substitution or unsupported exact match at $S_2[k]$ to a deletion, and removes a deletion at $S_2[I.c + h − I.a]$, decreasing the edit cost by 1. The same arguments hold for Algorithm (ii) when we replace edges from right-to-left. In Algorithm (ii) when we process edges from left-to-right, since any edges to left of the edge e being replaced are supported by I, replacing e with the leftmost of e' and e'' does not require modifying any additional edges. Again, if e is solid, the edit cost is unaltered, and if e is dashed, the edit cost is decreased by 1.   □

Lemma 4. *The greedy algorithm described in the proof of Lemma 1 produces an optimal alignment for a 'minimal' anchor chain under strict precedence.*

Proof. Similar to proof of Lemma 3 (see full version [11]).   □

Lemma 5. *For an anchor chain under strict precedence, the edit cost of the alignment produced by the greedy algorithm described in the proof of Lemma 1 is equal to the chaining cost.*

Proof. This follows from induction on the number of anchors processed, using the same arguments used in the proof of Lemma 1. However, only $I'.b = I.b$ needs to be considered in Cases 1 and 2 leading to equality in these cases.   □

5   Implementation

In multi-dimensional RmQs, $O(n\log^{d-1} n)$ storage requirement and irregular memory access during a query can limit their efficacy in practice [4]. We can take advantage of two observations to design a more practical algorithm. First, if sequences are highly similar, their edit distance will be relatively small. Hence the anchored edit distance, denoted in this section as OPT, will be relatively small for MUM or MEM anchors. Second, if the sequences are dissimilar, then the number of MUM or MEM anchors, n, will likely be small. These observations allow us to design an alternative algorithm (Algorithm 1) that requires $O(n)$ worst-case space and $O(n \cdot OPT + n \log n)$ average-case time over all possible inputs where $n \leq \max(|S_1|, |S_2|)$, i.e., the number of anchors is less than the longer sequence length (proof is deferred to full version [11]). This property always holds when the anchors are MUMs and is typically true for MEMs as well. This makes the algorithm presented here a practical alternative.

As before, let $\mathcal{A}$ be the initial (possibly unsorted) set of anchors, but with $\mathcal{A}_{left} = \mathcal{A}[1]$ and $\mathcal{A}_{right} = \mathcal{A}[n]$. We assume wlog $|S_1| \geq |S_2|$. We begin by sorting anchor set $\mathcal{A}$ by the component $\mathcal{A}[\cdot].a$ and making a guess for the optimal solution, B (Algorithm 1). The value B is used at every step to bound the range of $\mathcal{A}[\cdot].a$ values that need to be examined. This bounds the number of anchors that need to be considered (on average). If $C[n]$ is greater than our current guess B after processing all n anchors, we update our guess to $B_2 \cdot B$.

Input: n anchors $\mathcal{A}$ and parameters B_1 and B_2.
Output: $C[1, n]$ s.t. $C[i]$ is optimal co-linear chaining cost for any
ordered subset of $\mathcal{A}[1, i]$ ending at $\mathcal{A}[i]$.
Let $\mathcal{A}'[1], \dots \mathcal{A}'[n]$ be the set of anchors $\mathcal{A}$ sorted on $\mathcal{A}[\cdot].a$;
Initialize array C of size n to 0 and $B \leftarrow B_1$;
do
    $j \leftarrow 1$;
    for $i \leftarrow 1$ **to** n **do**
        while $\mathcal{A}'[i].a - \mathcal{A}'[j].a > B$ **do**
            $j \leftarrow j + 1$;
        end
        $C[i] \leftarrow \min\{C[k] + connect(\mathcal{A}'[k], \mathcal{A}'[i]) \mid j \leq k < i$ and $\mathcal{A}'[k] \prec \mathcal{A}'[i]\}$;
    end
    $B_{last} \leftarrow B$;
    $B \leftarrow B_2 \cdot B$;
while $C[n] > B_{last}$;
return $C[1, n]$
      Algorithm 1: $O(OPT \cdot n + n \log n)$ average-case algorithm.

Extending the above pseudo-code to enable semi-global chaining, i.e., free anchor gap on both ends of reference sequences, is also simple. In each i-loop, the connection to anchor $\mathcal{A}_{left}$ must be always considered, and for last iteration when $i = n$, j must be set to 1. Second, a revised cost function must be used when connecting to either $\mathcal{A}_{left}$ or $\mathcal{A}_{right}$ where a gap penalty is used only for anchor gap over the query sequence. The experiments in the next section use an implementation of this algorithm.

6   Evaluation

There are multiple open-source libraries/tools that implement edit distance computation. Edlib (v1.2.7) [29] uses Myers's bit-vector algorithm [21] and Ukkonen's banded algorithm [30], and is known to be the fastest implementation currently. In this section, we aim to show that: (i) the proposed algorithm as well as existing chaining methods achieve significant speedup compared to computing exact edit distance using Edlib, and (ii) in contrast to existing chaining methods, our implementation consistently achieves high Pearson correlation (> 0.90) with edit distance while requiring modest time and memory resources.

We implemented Algorithm 1 in C++, and refer to it as ChainX. The code is available at https://github.com/at-cg/ChainX. Inputs are a target string, query strings, comparison mode (global or semi-global), anchor type preferred, i.e., maximal unique matches (MUMs) or maximal exact matches (MEMs), and a minimum match length. We include a pre-processing step to index target string using the same suffix array-based algorithm [32] used in Nucmer4 [19]. Chaining costs computed using ChainX for each query-target pair are provably-optimal.

Existing Co-linear Chaining Implementations. Co-linear chaining has been implemented previously as a stand-alone utility [2,23] and also used as a heuristic inside widely used sequence aligners [5,15,19]. Out of these, Clasp (v1.1), Nucmer4 (v4.0.0rc1) and Minimap2 (v2.22-r1101) tools are available as open-source, and used here for comparison purpose. Unlike our algorithm where the optimization problem involves minimizing a cost function, these tools execute their respective chaining algorithms using a score maximization objective function. Clasp, being a stand-alone chaining method returns chaining scores in its output, whereas we modified Minimap2 and Nucmer4 to print the maximum chaining score for each query-target string pair, and skip subsequent steps. To enable a fair comparison, all methods were run with single thread and same minimum anchor size 20. Accordingly, ChainX, Clasp and Nucmer4 were run with MUMs of length ≥ 20, and Minimap2 was configured to use minimizer k-mers of length 20. For these tests, we made use of an Intel Xeon Processor E5-2698 v3 processor with 32 cores and 128 GB RAM. All tools were required to match only the forward strand of each query string. ChainX and Clasp are both exact solvers of co-linear chaining problem, but use different gap-cost functions. Clasp only permits non-overlapping anchors in a chain, and supports two cost functions which were referred to as *sum-of-pair* and *linear* gap cost functions in their paper [23]. We tested Clasp with both of its gap-cost functions, and refer to these two versions as Clasp-sop and Clasp-linear respectively. Both the versions solve co-linear chaining using RmQ data structures, requiring $O(n \log^2 n)$ and $O(n \log n)$ time respectively. Both require a set of anchors as input, therefore, we supplied the same set of anchors, i.e., MUMs of length ≥ 20 as used by ChainX. Minimap2 and Nucmer4 use co-linear chaining as part of their seed-chain-extend pipelines. Both Minimap2 and Nucmer2 support anchor overlaps in a chain, as well as penalize gaps using custom functions. However, both these tools employ heuristics (e.g., enforce a maximum gap between adjacent chained anchors) for faster processing which can result in suboptimal chaining scores.

Runtime and Memory Comparison. We downloaded the same set of query and target strings that were used for benchmarking in Edlib paper [29][2]. These test strings, ranging from 10 kbp to 5000 kbp in length, allowed us to compare tools for end-to-end global sequence comparisons as well as semi-global comparisons at various degrees of similarity levels. To test end-to-end comparisons, the target string had been artificially mutated at various rates using mutatrix (https://github.com/ekg/mutatrix), whereas for the semi-global comparisons, a substring of the target string had been sampled and mutated. Table 1 presents runtime and memory comparison of all tools. Columns of the table are organized to show tools in three categories: edit distance solver (Edlib); optimal co-linear chaining solvers (ChainX, Clasp-sop, Clasp-linear); and heuristic implementations (Nucmer4, Minimap2). We make the following observations here. First, chaining methods (both optimal and heuristic-based) are significantly faster than Edlib in most cases, and we see up to three orders of magnitude speedup. Second, within optimal chaining methods, Clasp-sop's time and memory consumption increases quickly with increase in count of anchors, which is likely due to irregular memory access and storage overhead of its algorithm that uses a 2d-RmQ data structure. Finally, we note that Minimap2 and Nucmer4 are often faster than exact algorithms during global string comparisons due to their fast heuristics.

All tools (except Edlib) use an indexing step such as building a k-mer hash table (Minimap2) or computing suffix array (ChainX, Clasp-sop, Clasp-linear, Nucmer4). Indexing time was excluded from reported results, and was found to be comparable. For instance, in the case of semi-global comparisons, ChainX, Nucmer4, Minimap2 required 590 ms, 736 ms, 236 ms for index computation respectively.

Correlation with Edit Distance. We checked how well the chaining cost (or score) correlates with edit distance. We use absolute value of Pearson correlation coefficients for a comparison. In this experiment, we simulated 100 query strings within three similarity ranges: 90–100%, 80–90% and 75–80%. Table 2 shows the correlation achieved by all the tools. Here we observe that ChainX and Clasp-sop are more consistent in terms of maintaining high correlation across all similarity ranges. Between the two, ChainX was shown to offer superior scalability in terms of runtime and memory usage (Table 1). Hence, ChainX can be useful in practice when good performance and accuracy is desired across a wide similarity range.

---

[2] https://github.com/Martinsos/edlib/tree/master/test_data.

Table 1. Runtime and memory usage comparison of edit distance solver Edlib and co-linear chaining methods ChainX, Clasp, Nucmer4 and Minimap2. Runtime is measured in milliseconds across the columns, and memory usage (Mem) is noted in MBs. In this experiment, ChainX, Clasp-sop, Clasp-linear and Nucmer4 used maximal unique matches (MUMs) of length ≥ 20 as input anchors, while Minimap2 used fixed-length minimizer k-mers of size 20.

Similarity	No. of MUMs	Edlib Time (Mem)	ChainX Time (Mem)	Clasp-sop Time (Mem)	Clasp-linear Time (Mem)	Nucmer4 Time (Mem)	Minimap2 Time (Mem)
Semi-global pairwise sequence comparisons, sequence sizes $10^4 \times 5 * 10^6$							
99%	67	190 (17)	2.0 (57)	1.8 (57)	**0.9** (57)	1.8 (60)	1.9 (75)
97%	160	642 (17)	2.9 (57)	4.8 (57)	**1.8** (57)	4.1 (60)	2.3 (75)
94%	176	1165 (17)	3.0 (57)	5.9 (57)	2.1 (57)	3.2 (60)	**1.6** (75)
90%	135	2168 (17)	5.6 (57)	4.7 (57)	2.0 (57)	5.5 (60)	**1.9** (75)
80%	28	2360 (17)	4.2 (57)	2.5 (57)	**2.2** (57)	3.4 (60)	4.3 (75)
70%	3	4297 (17)	3.7 (57)	2.2 (57)	2.3 (57)	5.5 (60)	**1.1** (75)
Global pairwise sequence comparisons, sequence sizes $10^6 \times 10^6$							
99%	7012	949 (8)	**47.2** (24)	1236.8 (1800)	182.8 (257)	68.7 (26)	193.5 (35)
97%	15862	1308 (8)	490.4 (24)	5363.7 (8742)	765.4 (1278)	**87.8** (26)	179.0 (36)
94%	18389	2613 (8)	677.9 (24)	11737.1 (20501)	1021.0 (1694)	**113.5** (27)	116.9 (34)
90%	14472	6233 (8)	851.5 (24)	5110.3 (8277)	115.3 (27)	121.8 (26)	**94.8** (33)
80%	2964	12506 (8)	158.8 (24)	504.8 (572)	133.7 (24)	148.9 (26)	**69.5** (32)
70%	195	29602 (8)	136.5 (23)	140.6 (23)	139.6 (23)	167.3 (26)	**55.6** (32)

Table 2. Absolute Pearson correlation coefficients of chaining costs (or scores) computed by various methods with the corresponding edit distances. 100 query strings were simulated and matched to the target string within each similarity range.

Seq. sizes	Similarity	Correlation coefficient				
		ChainX	Clasp-sop	Clasp-linear	Nucmer4	Minimap2
Semi-global sequence comparisons						
$10^4 \times 5 * 10^6$	90%–100%	**0.996**	0.994	0.986	0.968	0.995
$10^4 \times 5 * 10^6$	80%–90%	0.975	**0.976**	0.786	0.864	0.958
$10^4 \times 5 * 10^6$	75%–80%	**0.927**	0.915	0.732	0.733	0.808
Global sequence comparisons						
$10^6 \times 10^6$	90%–100%	**0.999**	0.997	0.994	0.991	**0.999**
$10^6 \times 10^6$	80%–90%	**0.998**	**0.998**	0.922	0.955	0.996
$10^6 \times 10^6$	75%–80%	0.992	**0.993**	0.871	0.907	0.952

Table 3. Effect of anchor pre-computation method on the performance of ChainX. Total runtime to do 100 pairwise semi-global sequence comparisons (sequence size: $10^4 \times 5 * 10^6$) is measured in seconds, and correlation (corr.) with the corresponding edit distances is computed using Pearson correlation coefficient.

Similarity	Using MUMs			Using MEMs		
	$len \geq 20$ Time (corr.)	$len \geq 10$ Time (corr.)	$len \geq 7$ Time (corr.)	$len \geq 20$ Time (corr.)	$len \geq 10$ Time (corr.)	$len \geq 7$ Time (corr.)
90%–100%	7.2 (0.996)	**2.9** (0.997)	3.5 (0.997)	5.1 (0.996)	8.1 (0.997)	2652 (**0.998**)
80%–90%	4.5 (0.975)	5.6 (0.992)	**3.2** (0.992)	4.5 (0.975)	7.4 (0.993)	5413 (**0.995**)
75%–80%	5.3 (0.927)	5.9 (0.977)	**1.9** (0.977)	5.0 (0.927)	10.9 (0.987)	9221 (**0.992**)

Effect of Anchor Type and Minimum Match Length. How many anchors are given as input will naturally affect the performance and output quality of a chaining algorithm. We tested runtime and correlation with edit distance achieved by ChainX while varying the anchor type (MUMs/MEMs) and minimum match-length l_{min} parameter (Table 3). When MUMs are used as anchors, we observe good scalability, and lowering l_{min} from 20 to 10 improves the correlation, but the correlation saturates afterwards. This is because very short exact matches will unlikely be unique and won't be selected as MUMs. However, when MEMs are used as anchors, correlation continues to improve with decreasing minimum length parameter, however, runtime grows exponentially. Excessive count of anchors improves the correlation but then anchor chaining becomes computationally demanding.

Acknowledgements. This research is supported in part by the U.S. National Science Foundation (NSF) grants CCF-1704552, CCF-1816027, CCF-2112643, CCF-2146003, and funding from the Indian Institute of Science.

References

1. Abouelhoda, M., Ohlebusch, E.: Chaining algorithms for multiple genome comparison. J. Discrete Algorithms **3**(2–4), 321–341 (2005)
2. Abouelhoda, M.I., Kurtz, S., Ohlebusch, E.: CoCoNUT: an efficient system for the comparison and analysis of genomes. BMC Bioinf. **9**(1), 476 (2008). https://doi.org/10.1186/1471-2105-9-476
3. Backurs, A., Indyk, P.: Edit distance cannot be computed in strongly subquadratic time (unless SETH is false). In: Proceedings of the Forty-Seventh Annual ACM on Symposium on Theory of Computing, STOC 2015, pp. 51–58 (2015)
4. de Berg, M., Cheong, O., van Kreveld, M.J., Overmars, M.H.: Computational Geometry: Algorithms and applications, 3rd edn. Springer, Heidelberg (2008). https://doi.org/10.1007/978-3-540-77974-2
5. Bray, N., Dubchak, I., Pachter, L.: AVID: a global alignment program. Genome Res. **13**(1), 97–102 (2003)
6. Chaisson, M.J., Tesler, G.: Mapping single molecule sequencing reads using basic local alignment with successive refinement ((BLASR): application and theory. BMC Bioinf. **13**(1), 238 (2012). https://doi.org/10.1186/1471-2105-13-238
7. Delcher, A.L., Kasif, S., et al.: Alignment of whole genomes. Nucleic Acids Res. **27**(11), 2369–2376 (1999)
8. Eppstein, D., Galil, Z., Giancarlo, R., Italiano, G.F.: Sparse dynamic programming i: linear cost functions. J. ACM (JACM) **39**(3), 519–545 (1992)
9. Eppstein, D., Galil, Z., et al.: Sparse dynamic programming ii: convex and concave cost functions. J.. ACM (JACM) **39**(3), 546–567 (1992)
10. Hoppenworth, G., Bentley, J.W., Gibney, D., Thankachan, S.V.: The fine-grained complexity of median and center string problems under edit distance. In: 28th Annual European Symposium on Algorithms, ESA 2020, Pisa, Italy, vol. 173, pp. 61:1–61:19. Schloss Dagstuhl - Leibniz-Zentrum für Informatik (2020)
11. Jain, C., Gibney, D., Thankachan, S.V.: Co-linear chaining with overlaps and gap costs. bioRxiv (2021). https://doi.org/10.1101/2021.02.03.429492

12. Jain, C., Rhie, A., Hansen, N., Koren, S., Phillippy, A.M.: A long read mapping method for highly repetitive reference sequences. bioRxiv (2020)
13. Kalikar, S., Jain, C., Md, V., Misra, S.: Accelerating long-read analysis on modern CPUs. bioRxiv (2021)
14. Kurtz, S., et al.: Versatile and open software for comparing large genomes. Genome Biol. **5**(2), R12 (2004)
15. Li, H.: Minimap2: pairwise alignment for nucleotide sequences. Bioinformatics **34**(18), 3094–3100 (2018)
16. Li, H., Feng, X., Chu, C.: The design and construction of reference pangenome graphs with minigraph. Genome Biol. **21**(1), 265 (2020). https://doi.org/10.1186/s13059-020-02168-z
17. Mäkinen, V., Sahlin, K.: Chaining with overlaps revisited. In: 31st Annual Symposium on Combinatorial Pattern Matching, CPM 2020, 17–19 June 2020, Copenhagen, Denmark, vol. 161, pp. 25:1–25:12. Schloss Dagstuhl - Leibniz-Zentrum für Informatik (2020)
18. Mäkinen, V., Tomescu, A.I., Kuosmanen, A., Paavilainen, T., Gagie, T., Chikhi, R.: Sparse dynamic programming on DAGs with small width. ACM Trans. Algorithms **15**(2), 29:1-29:21 (2019). https://doi.org/10.1145/3301312
19. Marçais, G., Delcher, A.L., et al.: MUMmer4: a fast and versatile genome alignment system. PLoS Comput. Biol. **14**(1), e1005944 (2018)
20. Morgenstern, B.: A simple and space-efficient fragment-chaining algorithm for alignment of DNA and protein sequences. Appl. Math. Lett. **15**(1), 11–16 (2002)
21. Myers, G.: A fast bit-vector algorithm for approximate string matching based on dynamic programming. J. ACM (JACM) **46**(3), 395–415 (1999)
22. Myers, G., Miller, W.: Chaining multiple-alignment fragments in sub-quadratic time. In: SODA. vol. 95, pp. 38–47 (1995)
23. Otto, C., Hoffmann, S., Gorodkin, J., Stadler, P.F.: Fast local fragment chaining using sum-of-pair gap costs. Algorithms Mol. Biol. **6**(1), 4 (2011). https://doi.org/10.1186/1748-7188-6-4
24. Ren, J., Chaisson, M.J.: lra: a long read aligner for sequences and contigs. PLOS Comput. Biol. **17**(6), e1009078 (2021)
25. Sahlin, K., Mäkinen, V.: Accurate spliced alignment of long RNA sequencing reads. Bioinformatics **37**(24), 4643–4651 (2021)
26. Schleimer, S., Wilkerson, D.S., Aiken, A.: Winnowing: local algorithms for document fingerprinting. In: Proceedings of the 2003 ACM SIGMOD International Conference on Management of Data, pp. 76–85 (2003)
27. Sedlazeck, F.J., et al.: Accurate detection of complex structural variations using single-molecule sequencing. Nat. Methods **15**(6), 461–468 (2018)
28. Shibuya, T., Kurochkin, I.: Match chaining algorithms for cDNA mapping. In: Benson, G., Page, R.D.M. (eds.) WABI 2003. LNCS, vol. 2812, pp. 462–475. Springer, Heidelberg (2003). https://doi.org/10.1007/978-3-540-39763-2_33
29. Šošić, M., Šikić, M.: Edlib: a C/C++ library for fast, exact sequence alignment using edit distance. Bioinformatics **33**(9), 1394–1395 (2017)
30. Ukkonen, E.: Algorithms for approximate string matching. Inf. Control **64**(1–3), 100–118 (1985)
31. Uricaru, R., et al.: Novel definition and algorithm for chaining fragments with proportional overlaps. J. Comput. Biol. **18**(9), 1141–1154 (2011)
32. Vyverman, M., De Baets, B., et al.: essaMEM: finding maximal exact matches using enhanced sparse suffix arrays. Bioinformatics **29**(6), 802–804 (2013)
33. Wilbur, W.J., Lipman, D.J.: Rapid similarity searches of nucleic acid and protein data banks. Proc. Natl. Acad. Sci. **80**(3), 726–730 (1983)

The Complexity of Approximate Pattern Matching on de Bruijn Graphs

Daniel Gibney[1]([⊠]), Sharma V. Thankachan[2], and Srinivas Aluru[1]

[1] School of Computational Science and Engineering, Georgia Institute of Technology, Atlanta, USA
daniel.j.gibney@gmail.com, aluru@cc.gatech.edu
[2] Department of Computer Science, University of Central Florida, Orlando, USA
sharma.thankachan@gmail.com

Abstract. Aligning a sequence to a *walk* in a labeled graph is a problem of fundamental importance to Computational Biology. For finding a walk in an arbitrary graph with $|E|$ edges that exactly matches a pattern of length m, a lower bound based on the Strong Exponential Time Hypothesis (SETH) implies an algorithm significantly faster than $\mathcal{O}(|E|m)$ time is unlikely [Equi et al., ICALP 2019]. However, for many special graphs, such as de Bruijn graphs, the problem can be solved in linear time [Bowe et al., WABI 2012]. For approximate matching, the picture is more complex. When edits (substitutions, insertions, and deletions) are only allowed to the pattern, or when the graph is acyclic, the problem is again solvable in $\mathcal{O}(|E|m)$ time. When edits are allowed to arbitrary cyclic graphs, the problem becomes NP-complete, even on binary alphabets [Jain et al., RECOMB 2019]. These results hold even when edits are restricted to only substitutions. Despite the popularity of de Bruijn graphs in Computational Biology, the complexity of approximate pattern matching on de Bruijn graphs remained open. We investigate this problem and show that the properties that make de Bruijn graphs amenable to efficient exact pattern matching do not extend to approximate matching, even when restricted to the substitutions only case with alphabet size four. Specifically, we prove that determining the existence of a matching walk in a de Bruijn graph is NP-complete when substitutions are allowed to the graph. In addition, we demonstrate that an algorithm significantly faster than $\mathcal{O}(|E|m)$ is unlikely for de Bruijn graphs in the case where only substitutions are allowed to the pattern. This stands in contrast to pattern-to-text matching where exact matching is solvable in linear time, like on de Bruijn graphs, but approximate matching under substitutions is solvable in subquadratic $\tilde{O}(n\sqrt{m})$ time, where n is the text's length [Abrahamson, SIAM J. Computing 1987].

Keywords: Approximate pattern matching · Sequence alignment · de Bruijn graphs · Computational complexity

© The Author(s), under exclusive license to Springer Nature Switzerland AG 2022
I. Pe'er (Ed.): RECOMB 2022, LNBI 13278, pp. 263–278, 2022.
https://doi.org/10.1007/978-3-031-04749-7_16

1   Introduction

De Bruijn graphs are an essential tool in Computational Biology. Their role in de novo assembly spans back to the 1980s [40], and their application in assembly has been researched extensively since then [9,10,17,33,38,39,43,46]. More recently, de Bruijn graphs have been applied in metagenomics and in the representation of large collections of genomes [14,27,30,37,45] and for solving other problems such as read-error correction [32,35] and compression [8,24]. Due to the popularity of de Bruijn graphs in the modeling of sequencing data, an algorithm to efficiently find *walks* in a de Bruijn graph matching (or approximately matching) a given query pattern would be a significant advancement. For example, in metagenomics, such an algorithm could quickly detect the presence of a particular species within genetic material obtained from an environmental sample. Or, in the case of read-error correction, such an algorithm could be used to efficiently find the best mapping of reads onto a 'cleaned' reference de Bruijn graph with low-frequency k-mers removed [32]. To facilitate such tasks, several algorithms (often seed-and-extend type heuristics) and software tools have been developed that perform pattern matching on de Bruijn (and general) graphs [5,22,23,29,31,34,36,42].

The importance of pattern matching on labeled graphs in Computational Biology and other fields has caused a recent surge of interest in the theoretical aspects of this problem. In turn, this has led to many new fascinating algorithmic and computational complexity results. However, even with this improved understanding of the theory of pattern matching on labeled graphs, our knowledge is still lacking in many respects concerning specific, yet extremely relevant, graph classes. An overview of the current state of knowledge is provided in Table 1.

Table 1. The computational complexity of pattern matching on labeled graphs

	Exact matching	Approximate matching
Easy	Solvable in linear time	Solvable in $\mathcal{O}(|E|m)$ time
	• Wheeler graphs [16]	• DAGs: Substitutions/Edits to graph [29]
	(e.g. de Bruijn graphs,	• General graphs:
	NFAs for multiple strings)	Substitutions/Edits to pattern [6]
		• de Bruijn graphs: Substitutions to pattern
		-No strongly Sub-$\mathcal{O}(|E|m)$ alg. (this paper)
Hard	NO strongly sub-$\mathcal{O}(|E|m)$ Alg.	NP-Complete
	• General graphs [13,19]	• General graphs:
	(including DAGs with	Substitutions/Edits to vertex labels [6,26]
	total degree ≤ 3)	• de Bruijn graphs:
		Substitutions to vertex labels (this paper)

For general graphs, we can consider exact and approximate matching. For exact matching, conditional lower-bounds based on the Strong Exponential Time

Hypothesis (SETH), and other conjectures in circuit complexity, indicate that an $\mathcal{O}(|E|m^{1-\varepsilon} + |E|^{1-\varepsilon}m)$ time algorithm with any constant $\varepsilon > 0$, for a graph with $|E|$ edges and a pattern of length m, is highly unlikely (as is the ability to shave more than a constant number of logarithmic factors from the $\mathcal{O}(|E|m)$ time complexity) [13,19]. These results hold for even very restricted types of graphs, for example, DAGs with maximum total degree three and binary alphabets. For approximate matching, when edits are only allowed in the pattern, the problem is solvable in $\mathcal{O}(|E|m)$ time [6]. If edits are also permitted in the graph, but the graph is a DAG, matching can be done in the same time complexity [29]. However, the problem becomes NP-complete when edits are allowed in arbitrary cyclic graphs. This was originally proven in [6] for large alphabets and more recently proven for binary alphabets in [26]. These results hold even when edits are restricted to only substitutions. The distinction between modifications to the graph and modifications to the pattern is important as these two problems are fundamentally different. When changes are made to cyclic graphs the same modification can be encountered multiple times while matching a pattern with no additional cost (see Section 3.1 in [26] for a detailed discussion). Furthermore, algorithmic solutions appearing in [29,36,42] are for the case where modifications are performed only to the pattern.

De Bruijn graphs are interesting from a theoretical perspective. Many graphs allow for extending Burrows-Wheeler Transformation (BWT) based techniques for efficient pattern matching. Sufficient conditions for doing this are captured by the definition of Wheeler graphs, introduced in [16], and further studied in [3,4,12,15,20]. De Bruijn graphs are themselves Wheeler graphs, hence on a de Bruijn graph exact pattern matching is solvable in linear time. However, the complexity of approximate matching in de Bruijn graphs when permitting modifications to the graph or modifications to the pattern remained open [26].

We make two important contributions (see Table 1). First, we prove that for de Bruijn graphs, despite exact matching being solvable in linear time, the approximate matching problem with vertex label substitutions is NP-complete. Second, we prove that a strongly subquadratic time algorithm for the approximate pattern matching problem on de Bruijn graphs, where substitutions are only allowed in the pattern, is not possible under SETH. This confirms the optimality of the known quadratic time algorithms when considering polynomial factors. To the best of our knowledge, these are the first such results for any type of Wheeler graph. Note that pattern-to-text matching (under substitutions) can be solved in sub-quadratic $\tilde{\mathcal{O}}(n\sqrt{m})$ time, where n is the text's length [2].

1.1   Technical Background and Our Results

Notation for Edges: For a directed edge from a vertex u to a vertex v we will use the notation (u, v). Additionally, we will refer to u as the *tail* of (u, v), and v as the *head* of (u, v).

Walks Versus Paths: A distinction must be made between the concept of a *walk* and a *path* in a graph. A walk is a sequence of vertices v_1, v_2, ..., v_t such that for each $i \in [1, t-1]$, $(v_i, v_{i+1}) \in E$. Vertices can be repeated in a walk. A path is a walk where vertices are not repeated. The length of a walk is defined as the number of edges in the walk, $t-1$, or equivalently one less than the number of vertices in the sequence (counted with multiplicity). This work will be concerning the existence of walks.

Induced Subgraphs: An induced subgraph of a graph $G = (V, E)$ consists of a subset of vertices $V' \subseteq V$, and all edges $(u, v) \in E$ such that $u, v \in V'$. This is in contrast to an arbitrary subgraph of G, where an edge can be omitted from the subgraph, even if both of its incident vertices are included.

de Bruijn Graphs: An *order-k* full de Bruijn graph is a compact representation of all k-mers (strings of length k) from an alphabet Σ of size σ. It consists of σ^k vertices, each corresponding to a unique k-mer (which we call as its *implicit label*) in Σ^k. There is a directed edge from each vertex with implicit label $s_1 s_2 ... s_k \in \Sigma^k$ to the σ vertices with implicit labels $s_2 s_3 ... s_k \alpha$, $\alpha \in \Sigma$. We will work with induced subgraphs of full de Bruijn graphs in this paper. We assign to every vertex v a label $L(v) \in \Sigma$, such that the implicit label of v is $L(u_1)L(u_2)...L(u_{k-1})L(v)$ where $u_1, u_2, ..., u_{k-1}, v$ is any length $k-1$ walk ending at v. This is equivalent to the notion of a de Bruijn graph constructed from k-mers commonly used in Computational Biology.

Strings and Matching: For a string S of length n indexed from 1 to n, we use $S[i]$ to denote the ith symbol in S. We use $S[i, j]$ to denote the substring $S[i]S[i+1]...S[j]$. If $j < i$, then we take $S[i, j]$ as the empty string. As mentioned above, we will consider every vertex v as labeled with a single symbol $L(v) \in \Sigma$. A pattern $P[1, m]$ matches a walk $v_1, v_2, ..., v_m$ iff $P[i] = L(v_i)$ for every $i \in [1, m]$.

With these definitions in hand, we can formally define our first problem.

Problem 1 (Approximate matching with vertex label substitutions). Given a vertex labeled graph $D = (V, E)$ with alphabet Σ of size σ, pattern $P[1, m]$, and integer $\delta \geq 0$, determine if there exists a walk in D matching P after at most δ substitutions to the vertex labels.

Theorem 1. *Problem 1 is NP-complete on de Bruijn graphs with $\sigma = 4$.*

Theorem 1 is proven in Sect. 2. Intuitively, our reduction transforms a general directed graph into a de Bruijn that maintains key topological properties related to the existence of walks. The distinct problem of approximately matching a pattern to a *path* in a de Bruijn graph was shown to be NP-complete in [31]. As mentioned by the authors of that work, the techniques used there do not appear to be easily adaptable to the problem for walks. Our approach uses edge transformations more closely inspired by those used in [28] for proving hardness on the paired de Bruijn sound cycle problem.

Problem 2 (Approximate matching with substitutions within the pattern). Given a vertex labeled graph $D = (V, E)$ with alphabet Σ of size σ, pattern $P[1, m]$, and integer $\delta \geq 0$, determine if there exists a walk in D matching P after at most δ substitutions to the symbols in P.

For Problem 2 we provide a hardness result based on SETH, which is frequently used for establishing conditional optimality of polynomial time algorithms [1,7,13,18,19,25]. We refer the reader to [44] for the definition of SETH and for the reduction to the Orthogonal Vectors problem (OV), which is utilized to prove Theorem 2.

Theorem 2. *Conditioned on SETH, for all constants $\varepsilon > 0$, there does not exist an $\mathcal{O}(|E|m^{1-\varepsilon} + |E|^{1-\varepsilon}m)$ time algorithm for Problem 2 on de Bruijn graphs with $\sigma = 4$.*

Note that the order of the de Bruijn graphs used in ours proofs are $\Theta(\log^2 |V|)$ for Theorem 1 and $\Theta(\log |V|)$ for Theorem 2.

2 NP-Completeness of Problem 1 on de Bruijn Graphs

Our proof of NP-completeness uses a reduction from the Hamiltonian Cycle Problem on directed graphs, which is the problem of deciding if there exists a cycle through a directed graph that visits every vertex exactly once. It was proven NP-complete even when restricted to directed graphs where the number of edges is linear in the number of vertices [41]. To present the reduction, we introduce the concept of *merging* two vertices. To merge vertices u and v, we create a new vertex w. We then take all edges with either u or v as their head and make w their new head. Next, we take all edges with either u or v as their tail and make w their new tail. This makes the edges (u, v) and (v, u) (if they existed) into self-loops for w. If two self-loops are formed, we delete one of them. Finally, we delete the original vertices u and v.

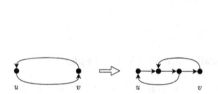

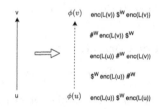

Fig. 1. Gadget to remove cycles of length 2 from the initial input graph.

Fig. 2. The transformation from edges to paths used in our reduction.

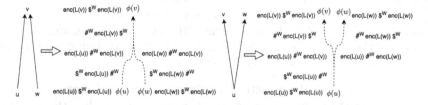

Fig. 3. Vertices with the same implicit label are merged while transforming D to D', causing edges with shared vertices to become paths with shared vertices.

2.1   Reduction

We start with an instance of the Hamiltonian cycle problem on a directed graph where the number of edges is linear in the number of vertices. We can assume there are no self-loops or vertices with in-degree or out-degree zero. To simplify the proof, we first eliminate any cycles of length 2 using the gadget in Fig. 1. We denote the resulting graph as $D = (V, E)$ and let $n = |V|$. We assign each vertex $v \in V$ a unique integer $L(v) \in [0, n-1]$. Let $\ell = \lceil \log n \rceil$, $\mathrm{bin}(i)$ be the standard binary encoding of i using ℓ bits and $\Sigma = \{\$, \#, 0, 1\}$. Define $\mathrm{enc}(i) = (0^{2\ell}1)^{2\ell}\mathrm{bin}(i)$, $W = |\mathrm{enc}(i)|$, and $k = 3W$.

We construct a new (de Bruijn) graph $D' = (V', E')$ as follows: Initially D' is the empty graph. For $i = 0, 1, \ldots, n-1$, for each edge $(u, v) \in E$ where $L(v) = i$, create a new path whose concatenation of vertex labels is $\#^W\mathrm{enc}(i)\$^W\mathrm{enc}(i)$. The vertex u will correspond with a new vertex $\phi(u)$ at the start of this path, and the vertex v will correspond with a new vertex $\phi(v)$ at the end of this path. The vertex $\phi(v)$ has the implicit label $\mathrm{enc}(L(v))\$^W\mathrm{enc}(L(v))$. The vertex $\phi(u)$ is temporarily assigned the implicit label $\mathrm{enc}(L(u))\$^W\mathrm{enc}(L(u))$. See Fig. 2. We call vertices with implicit labels of the form $\mathrm{enc}(L(\cdot))\$^W\mathrm{enc}(L(\cdot))$ *marked vertices*. We use the notation $\phi((u, v))$ to denote the path created when applying this transformation to $(u, v) \in E$. After the path $\phi((u, v))$ is created, vertices in V' having the same implicit label are merged, and parallel edges are deleted (Fig. 3). See Fig. 4 for a complete example. Finally, let $\delta = 2\ell(n-1)$ and

$$P = \#^W\mathrm{enc}(0)\$^W\mathrm{enc}(0)\#^W\mathrm{enc}(1)\$^W\mathrm{enc}(1)\#^W \ldots$$
$$\#^W\mathrm{enc}(n-1)\$^W\mathrm{enc}(n-1)\#^W\mathrm{enc}(0)\$^W\mathrm{enc}(0).$$

We will show that there exists a walk in D' matching P with at most δ vertex label substitutions iff D contains a Hamiltonian cycle.

Proof of Correctness

Lemma 1. *The graph D' constructed as above is a de Bruijn graph.*

Proof (Overview). Three properties must be proven: (i) Implicit labels are unique, meaning for every implicit label at most one vertex is assigned that label; (ii) No edges are missing, i.e., if the implicit label of $y \in V'$ is $S\alpha$ for some string $S[1, k-1]$ and symbol $\alpha \in \Sigma$, and there exists a vertex $x \in V'$ with implicit label $\beta S[1, k-1]$ for some symbol $\beta \in \Sigma$, then $(x, y) \in E'$; (iii) Implicit labels are well-defined, in that every walk of length $k-1$ ending at a vertex $x \in V'$ matches the same string (the implicit label of x); The most involved of these is proving property (ii), which requires analyzing several cases. The complete proof is given in the full version [21].   □

The correctness of the reduction remains to be shown. Lemmas 2–4 establish useful structural properties of D', Lemma 5 proves that the existence of a Hamiltonian Cycle in D implies an approximate matching in D', and Lemmas 6–9 demonstrate the converse.

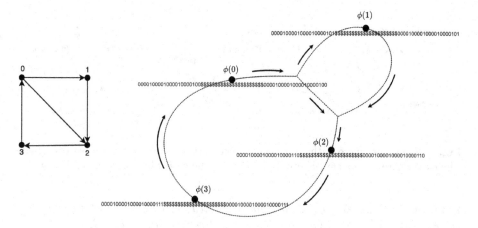

Fig. 4. (Left) A graph before the reduction is applied to it. (Right) The transformed graph. Implicit labels for marked vertices are shown and the path directions are annotated by arrows beside each path.

Lemma 2. *Any walk between two marked vertices $\phi(u)$ and $\phi(v)$ containing no additional marked vertices has length $4W$. Hence, we can conclude any such walk is a path.*

Proof (Overview). This is proven using induction on the number of edges transformed. It is shown that for every vertex, a key property regarding the distances to its closest marked vertices continues to hold after vertices on any newly created path are merged. We defer the complete proof to the full version [21]. □

Lemma 3. *For $(u_1, v_1), (u_2, v_2) \in E$, unless $u_1 = u_2$ or $v_1 = v_2$, $\phi((u_1, v_1))$ and $\phi((u_2, v_2))$ share no vertices.*

Proof. In the case where $\{u_1, v_1\} \cap \{u_2, v_2\} = \emptyset$ (Fig. 5 left), every implicit vertex label in $\phi((u_1, v_1))$ contains $\text{enc}(L(u_1))$ or $\text{enc}(L(v_1))$ (or both), and contains neither $\text{enc}(L(u_2))$ nor $\text{enc}(L(v_2))$. Similarly, every implicit vertex label in $\phi((u_2, v_2))$ contains $\text{enc}(L(u_2))$ or $\text{enc}(L(v_2))$ (or both) and contains neither $\text{enc}(L(u_1))$ nor $\text{enc}(L(v_1))$. This implies that none of the implicit labels match between the two paths, thus no vertices are merged. In the case where $v_1 = u_2$ and $u_1 \neq v_2$ (Fig. 5, right), the implicit labels of vertices $\phi((u_1, v_1))$ not containing $\text{enc}(L(u_1))$ have $\#$ symbols in different positions than implicit labels of vertices in $\phi((u_2, v_2))$ not containing $\text{enc}(L(v_2))$, and, since $v_1 \neq v_2$, cannot match the implicit labels of vertices in $\phi((u_2, v_2))$ containing $\text{enc}(L(v_2))$. Vertices in $\phi((u_1, v_1))$ with implicit labels containing $\text{enc}(L(u_1))$ have $\#$ symbols in different positions than implicit labels of vertices in $\phi((u_2, v_2))$ not containing $\text{enc}(L(u_2))$, and, since $u_1 \neq u_2$, cannot match the implicit labels of vertices in $\phi((u_2, v_2))$ containing $\text{enc}(L(u_2))$. The case $u_1 = v_2$ and $u_2 \neq v_1$ is symmetric. The case $u_1 = v_2$ and $v_1 = u_2$ cannot happen since, by the use of our gadget in Fig. 1, D cannot contain the edges (u_1, v_1) and (v_1, u_1). □

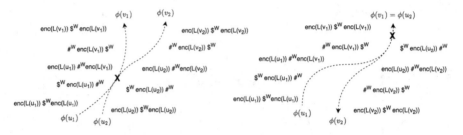

Fig. 5. Examples where paths between marked vertex cannot share any vertex: (Left) The case where $\{u_1, v_1\} \cap \{u_2, v_2\} = \emptyset$. (Right) The case where $v_1 = u_2$ and $u_1 \neq v_2$.

Lemma 4. *There exists a path from a marked vertex $\phi(u) \in V'$ to a marked vertex $\phi(v) \in V'$ containing no other marked vertices iff $(u, v) \in E$.*

Proof (Overview). It is clear from construction that if $(u, v) \in D$, then such a path exists in D'. In the other direction, we utilize Lemmas 2 and 3 to show that such a path existing without a corresponding edge would create a contradiction. The complete proof is provided in the full version [21].   □

Lemma 5. *If D has a Hamiltonian cycle, then P can be matched in D' with at most δ substitutions to vertex labels of D'.*

Proof. To obtain a matching walk, follow the cycle corresponding to a solution in D starting with the marked vertex in V' corresponding to the vertex in V with label 0. By Lemma 4, each edge traversed in D corresponds to a path in D'. While traversing these paths, modify the vertex labels in D' corresponding to the substrings bin(i) to match P. Assuming no conflicting substitutions are needed, this requires at most $2\ell(n-1)$ substitutions.

It remains to be shown that no conflicting label substitutions will be necessary. Consider the edges $(u_1, v_1), (u_2, v_2) \in E$ used in the Hamiltonian cycle in D. We will never have $u_1 = u_2$ or $v_1 = v_2$. Hence, by Lemma 3, the sets of vertices on the paths $\phi((u_1, v_1))$ and $\phi((u_2, v_2))$ are disjoint.   □

Lemma 6. *If P can be matched in D' with at most δ substitutions to vertex labels of D', then all \$'s in P are matched with non-substituted \$'s in D' and all #'s in P are matched with non-substituted #'s in D'. Consequently, we can assume the only substitutions are to the vertex labels corresponding to bin(i)'s within enc(i)'s.*

Proof (Overview). We establish the existence of a long, non-branching path for every marked vertex that can be traversed at most once when matching P. This, combined with maximal paths of, \$, #, and 0/1-symbols, all being of length W, makes it so that 'shifting' P to match a portion of D forces the shift to occur throughout the walk traversed while matching P. Utilizing the large Hamming distance between shifted instances of two encodings, we can then show that not matching all non-0/1 symbols requires more than δ substitutions. The complete proof is provided in the full version [21].   □

Post-substitution to vertex labels, we will refer to a vertex as marked if there exists a walk ending at it that matches a string of the form $enc(L(u))\$^W enc(L(u))$, $u \in V$. Note that this definition does not require all length $k - 1$ walks ending at such a vertex to match the same string.

Lemma 7. *If P can be matched in D' with at most δ substitutions to vertex labels of D', then no additional marked vertices are created due to vertex substitutions.*

Proof. Pre-substitution, only marked vertices have implicit labels of the form $S_1\$^W S_2$ where S_1 and S_2 contain no $\$$ symbols. Hence, the only way that a vertex could have a walk ending at it that matches a pattern of that form post-substitution is if either it was originally a marked vertex, or some non-0/1-symbols were substituted in D'. However, by Lemma 6 the latter case cannot happen, and only originally marked vertices have walks ending at them matching strings of the form $S_1\$^W S_2$ post-substitution. □

Lemma 8. *If P can be matched in D' with at most δ substitutions to vertex labels of D', then each originally marked vertex in D' is visited exactly once, except for an originally marked vertex at the end of a path matching $enc(0)\$^W enc(0)$ that is visited twice.*

Proof. First, we show that all marked vertices, except the one with implicit label $enc(0)\$^W enc(0)$, are visited at most once. Pre-substitution, a marked vertex with implicit label $enc(i)\$^W enc(i)$ is at the end of a unique, branchless path of length W matching $enc(i)$. By Lemma 6, the only substitutions to this path made while matching P are substitutions making it match $enc(i')$, $i' \neq i$. If this path were modified to match $enc(i')$, $i' > 0$, then the only way the marked vertex could be visited twice while matching P is if after traversing the path, another path matching $\$^W$ is taken back to the start of this $enc(i')$ path. However, any edges leaving this marked vertex are labeled with $\#$, making this impossible. By similar reasoning, the path matching $enc(0)$ ending at a marked vertex is visited at most twice. We now show that each marked vertex is visited at least once. Suppose some marked vertex is not visited. By Lemma 7, no additional marked vertices are created. Hence, a marked vertex ending a path matching $enc(i)$, $i > 0$ is visited at least twice, or a marked vertex ending a path matching $enc(0)$ is visited at least three times, a contradiction. □

Lemma 9. *If P can be matched in D' with at most δ substitutions to vertex labels of D', then D has a Hamiltonian cycle.*

Proof. By Lemma 4, the paths between marked vertices traversed while matching with P correspond to edges between vertices in D. Combined with marked vertices being visited exactly once from Lemma 8 (except the marked vertex ending a path matching $enc(0)$), the walk matched by P in D' corresponds to a Hamiltonian cycle through D beginning and ending at the vertex labeled 0. □

This completes the proof of Theorem 1. To see that $k = \Theta(\log^2 |V'|)$, first recall that $|V|$ is the number of vertices in the original graph, where we assumed $|E| = \mathcal{O}(|V|)$. At most $4W|E| = \mathcal{O}(k|V|)$ vertices are created in the reduction. Also, the proof of Lemma 6 establishes that there is a unique set of at least $\Theta(k)$ vertices for every marked vertex, each one corresponding to a vertex in the original graph. Combining, we have that $|V'| = \Theta(k|V|)$. By construction, $k = \Theta(\log^2 |V|)$, and since $|V'| = \Theta(k|V|)$, $k = \Theta(\log^2 |V'|)$ as well.

3 Hardness for Problem 2 on de Bruijn Graphs

Reduction. The Orthogonal Vectors Problem is defined as follows: given two sets of binary vectors $A, B \subseteq \{0,1\}^d$ where $|A| = |B| = N$, determine whether there exists vectors $a \in A$ and $b \in B$ such that their inner product is zero. Conditioned on SETH, a standard reduction shows that this cannot be solved in time $d^{\Theta(1)} N^{2-\varepsilon}$ for any constant $\varepsilon > 0$ [44].

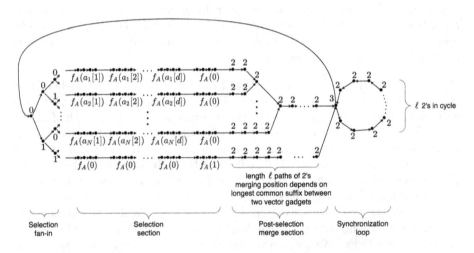

Fig. 6. An illustration of the reduction from OV to Problem 2.

Let the given instance of OV consist of $A, B \subseteq \{0,1\}^d$ where $|A| = |B| = N = 2^m$ for some natural number m. Hence, we have $\lceil \log(N+1) \rceil = \log N + 1$. This will ease computation later. We also assume that $d > \log N$. This is reasonable, as if $d \leq \log N$, then $|A|$ and $|B|$ would contain either all vectors of length d or repetitions.

We will next provide a formal description of the graph D our reduction creates from the set $A = \{a_1, a_2, ..., a_N\}$ and the pattern P it creates from the set $B = \{b_1, b_2, ..., b_N\}$. The reader may find Fig. 6 helpful. The graph will consist of four sections. We name these according to their function in the reduction: the Selection fan-in, the Selection section, the Post-Selection merge section, and the Synchronization loop.

We start with the Selection fan-in. Let 2^c be the smallest power of 2 such that $2^c \geq N + 1$. The Selection fan-in consists of a complete binary tree with 2^c leaves where all paths are directed away from the root. The root is labeled 0 and the children of every node are labeled 0 and 1, respectively.

The Selection section consists of $N + 1$ paths. We first define the mappings f_A and f_B from $\{0, 1\}$ to sequences of length four as $f_A(0) = 1100$, $f_A(1) = 1111$, $f_B(0) = 0110$, $f_B(1) = 0000$. These mappings have the property that $d_H(f_A(0), f_B(0)) = d_H(f_A(0), f_B(1)) = d_H(f_A(1), f_B(0)) = 2$ and $d_H(f_A(1), f_B(1)) = 4$, where $d_H(x, y)$ is the Hamming distance between strings x and y. We make the ith path for $1 \leq i \leq N$ a path of $4(d + 1)$ vertices with labels matching the string $f_A(a_i[1])f_A(a_i[2])...f_A(a_i[d])f_A(0)$. We make the $(N + 1)^{th}$ path have $4(d + 1)$ vertices and match the string $f_A(0)^d f_A(1)$. Let s_i denoted the start vertex of path i. We arbitrarily choose $N + 1$ leaves, $l_1, l_2,..., l_{N+1}$, from the Selection fan-in and add the edges (l_i, s_i) for $1 \leq i \leq N + 1$.

We define the implicit label size as $k = \lceil \log(N + 1) \rceil + 4(d + 1)$ and $\ell = k - 1$. To construct the Post-selection merge section, we start with $N + 1$ length $\ell - 1$ paths, each matching the string 2^ℓ. For every path in the Selection section, we add an edge from the last vertex in the path to one of the paths matching 2^ℓ. This is done so that every path matching 2^ℓ in the Post-selection merge section is connected to exactly one path from the Selection section. Next, we merge two vertices if they have the same implicit label. This is repeated until all vertices in the Post-selection merge section have a unique implicit label.

To construct the Synchronization loop we create a directed cycle with $\ell + 1 = k$ vertices. One of these is labeled with the symbol 3, and the rest with the symbol 2. Edges from each ending vertex in the Post-selection Merge section to the vertex labeled 3 are then added. A final edge from the vertex labeled 3 to the root of the binary tree in the Selection fan-in completes the graph, which we denote as D.

Let $t = 5d + \lceil \log(N + 1) \rceil$. To complete the reduction, we make the pattern

$$P = (2^\ell 3)^t \ 2^{\lceil \log(N+1) \rceil} f_B(b_1[1]) f_B(b_1[2]) \ldots f_B(b_1[d]) f_B(1)$$
$$(2^\ell 3)^t \ 2^{\lceil \log(N+1) \rceil} \ f_B(b_2[1]) f_B(b_2[2]) \ldots f_B(b_2[d]) f_B(1)$$
$$\cdots$$
$$(2^\ell 3)^t \ 2^{\lceil \log(N+1) \rceil} f_B(b_N[1]) f_B(b_N[2]) \ldots f_B(b_N[d]) f_B(1)$$

and the maximum number of allowed substitutions $\delta = N \lceil \log_2(N + 1) \rceil + 2(d + 1) + (2d + 4)(N - 1)$.

We call substrings in P of the form $f_B(b_i[1]) f_B(b_i[2]) \ldots f_B(b_i[d]) f_B(1)$ and paths in D matching strings of the form $f_A(a_i[1]) f_A(a_i[2])...f_A(a_i[d]) f_A(0)$ vector gadgets. Note that $|E| = \mathcal{O}(dN)$ and $m = |P| = \mathcal{O}(d^2 N)$. Hence, an algorithm for approximate matching running in time $\mathcal{O}(m|E|^{1-\varepsilon} + m^{1-\varepsilon}|E|)$ for some $\varepsilon > 0$ would imply an algorithm for OV running in time $d^{\Theta(1)} N^{2-\varepsilon}$. This implies that once the correctness of the reduction has been established, Theorem 2 follows.

3.1 Proof of Correctness

Proofs of Lemma 10 and Lemma 11 are provided in the full version [21].

Lemma 10. *The graph D is a de Bruijn graph.*

Lemma 11. *In an optimal solution, 3's in P are matched with 3's in D.*

Lemma 12. *In an optimal solution, vector gadgets in P are matched with vector gadgets in D.*

Proof. Suppose otherwise. By Lemma 11, this can only occur if some vector gadget in P is matched against the Synchronization loop. This requires at least $4(d + 1)$ substitutions. We can instead match the $\lceil \log(N + 1) \rceil$ 2's preceding the vector gadget in P with the Selection fan-in and the vector gadget in P with the $(N + 1)^{th}$ path in the Selection section. Due to $d_H(f_A(0), f_B(0)) = d_H(f_A(0), f_B(1)) = 2$ and $d(f_A(1), f_B(1)) = 4$, this requires $\lceil \log(N+1) \rceil + 2d + 4$ substitutions in P. Since, $\log N < d < 2d$ we have $\log N < 2d - 1$. Using that N is some power of 2, $\lceil \log(N + 1) \rceil + 2d + 4 = \log N + 1 + 2d + 4 < 4d + 4$. Hence, the cost decreases by matching the vector gadget in P to a vector gadget in D instead.   □

Lemma 13. *If there exists a vector $a \in A$ and $b \in B$ such that $a \cdot b = 0$, then P can be matched to D with at most δ substitutions.*

Proof. Match the vector gadget for b in P with the vector gadget for a in the Selection section of D. This costs $2(d + 1)$ substitutions. Match the remaining $N - 1$ vector gadgets in P with the $(N + 1)^{th}$ path in the Selection section, requiring $(2d+4)(N-1)$ substitutions in total. The total number of substitutions of 2's in P to match the Selection fan-in is $N \lceil \log(N + 1) \rceil$. Adding these, the total number of substitutions is exactly δ. The synchronization loop can be used for matching all additional symbols in P without any further substitutions.   □

Lemma 14. *If P can be matched in D with at most δ substitutions, then there exists vectors $a \in A$ and $b \in B$ such $a \cdot b = 0$.*

Proof. By Lemma 12, we can assume vector gadgets in P are only matched against vector gadgets in D. Suppose that there does not exist a pair of orthogonal vectors $a \in A$ and $b \in B$. Then, which ever vector gadget in D we choose to match a vector gadget in P to, matching the vector gadget requires at least $2d+4$ substitutions. Hence, the total cost is at least $(2d + 4)N + N \lceil \log(N + 1) \rceil > \delta$, proving the contrapositive of Lemma 14.   □

4 Discussion

We leave open several interesting problems. An NP-completeness proof for Problem 1 on de Bruijn graphs when $k = \mathcal{O}(\log n)$ and the alphabet size is constant is still needed. Additionally, we need to extend these hardness results to when substitutions are allowed in both the graph and the pattern, and when insertions and deletions in some form are allowed in the graph and (or) the pattern. It seems unlikely that adding more types of edit operations would make the problems computationally easier, and we conjecture these variants are NP-complete on de Bruijn graphs as well. It also needs to be determined whether Problem 1 is NP-complete on de Bruijn graphs with binary alphabets, or whether the SETH-based hardness results hold for Problem 2 on binary alphabets. A practical question is whether these problems are hard for small δ values on de Bruijn graphs (the problem for general graphs was proven to $W[2]$ hard in terms of δ in [11]). In applications, the allowed error thresholds are quite small. Clearly, the problems are slice-wise-polynomial with respect to δ, i.e., for a constant δ it is solvable in polynomial time via brute force, but are they fixed-parameter-tractable in δ? The reduction presented here (as well as the reductions in [6,26]) is based on the Hamiltonian cycle problem, where a large δ value is used. This makes the existence of such a fixed-parameter-tractable algorithm a distinct possibility.

Acknowledgement. This research is supported in part by the U.S. National Science Foundation (NSF) grants CCF-1704552, CCF-1816027, CCF-2112643, and CCF-2146003.

References

1. Abboud, A., Backurs, A., Hansen, T.D., Williams, V.V., Zamir, O.: Subtree isomorphism revisited. ACM Trans. Algorithms **14**(3), 27:1—27:23 (2018). https://doi.org/10.1145/3093239
2. Abrahamson, K.R.: Generalized string matching. SIAM J. Comput. **16**(6), 1039–1051 (1987). https://doi.org/10.1137/0216067
3. Alanko, J., D'Agostino, G., Policriti, A., Prezza, N.: Wheeler languages. CoRR abs/2002.10303 (2020). https://arxiv.org/abs/2002.10303
4. Alanko, J.N., Gagie, T., Navarro, G., Benkner, L.S.: Tunneling on wheeler graphs. In: Data Compression Conference, DCC 2019, Snowbird, UT, USA, -26–29 March 2019. pp. 122–131 (2019). https://doi.org/10.1109/DCC.2019.00020
5. Almodaresi, F., Sarkar, H., Srivastava, A., Patro, R.: A space and time-efficient index for the compacted colored de Bruijn graph. Bioinform **34**(13), i169–i177 (2018). https://doi.org/10.1093/bioinformatics/bty292
6. Amir, A., Lewenstein, M., Lewenstein, N.: Pattern matching in hypertext. J. Algorithms **35**(1), 82–99 (2000). https://doi.org/10.1006/jagm.1999.1063
7. Backurs, A., Indyk, P.: Which regular expression patterns are hard to match? In: IEEE 57th Annual Symposium on Foundations of Computer Science, FOCS 2016, 9–11 October 2016, pp. 457–466. Hyatt Regency, New Brunswick (2016). https://doi.org/10.1109/FOCS.2016.56
8. Benoit, G., et al.: Reference-free compression of high throughput sequencing data with a probabilistic de Bruijn graph. BMC Bioinform. **16**, 288:1–288:14 (2015). https://doi.org/10.1186/s12859-015-0709-7

9. Chikhi, R., Limasset, A., Jackman, S., Simpson, J.T., Medvedev, P.: On the representation of de Bruijn graphs. J. Comput. Biol. **22**(5), 336–352 (2015). https://doi.org/10.1089/cmb.2014.0160

10. Chikhi, R., Rizk, G.: Space-efficient and exact de Bruijn graph representation based on a bloom filter. Algorithms Mol. Biol. **8**, 22 (2013). https://doi.org/10.1186/1748-7188-8-22

11. Dondi, R., Mauri, G., Zoppis, I.: Complexity issues of string to graph approximate matching. In: Leporati, A., Martín-Vide, C., Shapira, D., Zandron, C. (eds.) LATA 2020. LNCS, vol. 12038, pp. 248–259. Springer, Cham (2020). https://doi.org/10.1007/978-3-030-40608-0_17

12. Egidi, L., Louza, F.A., Manzini, G.: Space efficient merging of de Bruijn graphs and wheeler graphs. CoRR abs/2009.03675 (2020). https://arxiv.org/abs/2009.03675

13. Equi, M., Grossi, R., Mäkinen, V., Tomescu, A.I.: On the complexity of string matching for graphs. In: 46th International Colloquium on Automata, Languages, and Programming, ICALP 2019, 9–12 July 2019, Patras, Greece. pp. 55:1–55:15 (2019). https://doi.org/10.4230/LIPIcs.ICALP.2019.55

14. Flick, P., Jain, C., Pan, T., Aluru, S.: Reprint of "a parallel connectivity algorithm for de Bruijn graphs in metagenomic applications". Parallel Comput. **70**, 54–65 (2017). https://doi.org/10.1016/j.parco.2017.09.002

15. Gagie, T.: r-indexing wheeler graphs. CoRR abs/2101.12341 (2021). https://arxiv.org/abs/2101.12341

16. Gagie, T., Manzini, G., Sirén, J.: Wheeler graphs: a framework for BWT-based data structures. Theor. Comput. Sci. **698**, 67–78 (2017). https://doi.org/10.1016/j.tcs.2017.06.016

17. Georganas, E., Buluç, A., Chapman, J., Oliker, L., Rokhsar, D., Yelick, K.A.: Parallel de Bruijn graph construction and traversal for de novo genome assembly. In: International Conference for High Performance Computing, Networking, Storage and Analysis, SC 2014, New Orleans, LA, USA, -16–21 November 2014. pp. 437–448 (2014). https://doi.org/10.1109/SC.2014.41

18. Gibney, D.: An efficient elastic-degenerate text index? not likely. In: Boucher, C., Thankachan, S.V. (eds.) SPIRE 2020. LNCS, vol. 12303, pp. 76–88. Springer, Cham (2020). https://doi.org/10.1007/978-3-030-59212-7_6

19. Gibney, D., Hoppenworth, G., Thankachan, S.V.: Simple reductions from formula-sat to pattern matching on labeled graphs and subtree isomorphism. In: 4th Symposium on Simplicity in Algorithms, SOSA 2021, Virtual Conference, 11–12 January 2021. pp. 232–242 (2021). https://doi.org/10.1137/1.9781611976496.26

20. Gibney, D., Thankachan, S.V.: On the hardness and inapproximability of recognizing wheeler graphs. In: 27th Annual European Symposium on Algorithms, ESA 2019, 9–11 September 2019, Munich/Garching, Germany. pp. 51:1–51:16 (2019). https://doi.org/10.4230/LIPIcs.ESA.2019.51

21. Gibney, D., Thankachan, S.V., Aluru, S.: The complexity of approximate pattern matching on de Bruijn graphs (2022)

22. Heydari, M., Miclotte, G., de Peer, Y.V., Fostier, J.: Browniealigner: accurate alignment of illumina sequencing data to de Bruijn graphs. BMC Bioinform. **19**(1), 311:1–311:10 (2018). https://doi.org/10.1186/s12859-018-2319-7

23. Holley, G., Peterlongo, P.: Blastgraph: intensive approximate pattern matching in string graphs and de-Bruijn graphs. In: PSC 2012 (2012)

24. Holley, G., Wittler, R., Stoye, J., Hach, F.: Dynamic alignment-free and reference-free read compression. J. Comput. Biol. **25**(7), 825–836 (2018), https://doi.org/10.1089/cmb.2018.0068

25. Hoppenworth, G., Bentley, J.W., Gibney, D., Thankachan, S.V.: The fine-grained complexity of median and center string problems under edit distance. In: 28th Annual European Symposium on Algorithms, ESA 2020, 7–9 September 2020, Pisa, Italy (Virtual Conference). pp. 61:1–61:19 (2020), https://doi.org/10.4230/LIPIcs.ESA.2020.61

26. Jain, C., Zhang, H., Gao, Yu., Aluru, S.: On the complexity of sequence to graph alignment. In: Cowen, L.J. (ed.) RECOMB 2019. LNCS, vol. 11467, pp. 85–100. Springer, Cham (2019). https://doi.org/10.1007/978-3-030-17083-7_6

27. Kamal, M.S., Parvin, S., Ashour, A.S., Shi, F., Dey, N.: De-Bruijn graph with MapReduce framework towards metagenomic data classification. Int. J. Inf. Technol. 9(1), 59–75 (2017)

28. Kapun, E., Tsarev, F.: On NP-hardness of the paired de Bruijn sound cycle problem. In: Darling, A., Stoye, J. (eds.) WABI 2013. LNCS, vol. 8126, pp. 59–69. Springer, Heidelberg (2013). https://doi.org/10.1007/978-3-642-40453-5_6

29. Kavya, V.N.S., Tayal, K., Srinivasan, R., Sivadasan, N.: Sequence alignment on directed graphs. J. Comput. Biol. 26(1), 53–67 (2019). https://doi.org/10.1089/cmb.2017.0264

30. Li, D., Liu, C., Luo, R., Sadakane, K., Lam, T.W.: MEGAHIT: an ultra-fast single-node solution for large and complex metagenomics assembly via succinct de Bruijn graph. Bioinformatics 31(10), 1674–1676 (2015). https://doi.org/10.1093/bioinformatics/btv033

31. Limasset, A., Cazaux, B., Rivals, E., Peterlongo, P.: Read mapping on de Bruijn graphs. BMC Bioinform. 17, 237 (2016). https://doi.org/10.1186/s12859-016-1103-9

32. Limasset, A., Flot, J., Peterlongo, P.: Toward perfect reads: self-correction of short reads via mapping on de Bruijn graphs. Bioinformatics 36(2), 651 (2020). https://doi.org/10.1093/bioinformatics/btz548

33. Lin, Y., Shen, M.W., Yuan, J., Chaisson, M., Pevzner, P.A.: Assembly of long error-prone reads using de Bruijn graphs. In: Proceedings of the Research in Computational Molecular Biology - 20th Annual Conference, RECOMB 2016, Santa Monica, CA, USA, 17–21 April 2016, p. 265 (2016). https://link.springer.com/content/pdf/bbm%3A978-3-319-31957-5%2F1.pdf

34. Liu, B., Guo, H., Brudno, M., Wang, Y.: deBGA: read alignment with de Bruijn graph-based seed and extension. Bioinformatics 32(21), 3224–3232 (2016). https://doi.org/10.1093/bioinformatics/btw371

35. Morisse, P., Lecroq, T., Lefebvre, A.: Hybrid correction of highly noisy long reads using a variable-order de Bruijn graph. Bioinformatics 34(24), 4213–4222 (2018). https://doi.org/10.1093/bioinformatics/bty521

36. Navarro, G.: Improved approximate pattern matching on hypertext. Theor. Comput. Sci. 237(1–2), 455–463 (2000). doi.org/10.1016/S0304-3975(99)00333--3

37. Pell, J., Hintze, A., Canino-Koning, R., Howe, A., Tiedje, J.M., Brown, C.T.: Scaling metagenome sequence assembly with probabilistic de Bruijn graphs. Proc. Natl. Acad. Sci. USA 109(33), 13272–13277 (2012). https://doi.org/10.1073/pnas.1121464109

38. Peng, Y., Leung, H.C.M., Yiu, S., Chin, F.Y.L.: IDBA - a practical iterative de Bruijn graph de novo assembler. In: Proceedings of the 14th Annual International Conference on Research in Computational Molecular Biology, RECOMB 2010, Lisbon, Portugal, 25–28 April 2010. pp. 426–440 (2010). https://doi.org/10.1007/978-3-642-12683-3_28

39. Peng, Y., Leung, H.C.M., Yiu, S., Lv, M., Zhu, X., Chin, F.Y.L.: IDBA-tran: a more robust de novo de Bruijn graph assembler for transcriptomes with uneven expression levels. Bioinformatics **29**(13), 326–334 (2013). https://doi.org/10.1093/bioinformatics/btt219

40. Pevzner, P.A.: 1-tuple DNA sequencing: computer analysis. J. Biomol. Struc. Dyn. **7**(1), 63–73 (1989)

41. Plesník, J.: The np-completeness of the hamiltonian cycle problem in planar digraphs with degree bound two. Inf. Process. Lett. **8**(4), 199–201 (1979). doi.org/10.1016/0020-0190(79)90023--1

42. Rautiainen, M., Marschall, T.: Aligning sequences to general graphs in o (v+ me) time. bioRxiv p. 216127 (2017)

43. Ren, X., et al.: Evaluating de Bruijn graph assemblers on 454 transcriptomic data. PLoS ONE **7**(12), e51188 (2012)

44. Williams, V.V.: Hardness of easy problems: basing hardness on popular conjectures such as the strong exponential time hypothesis (invited talk). In: 10th International Symposium on Parameterized and Exact Computation, IPEC 2015, 16–18 September 2015, Patras, Greece. pp. 17–29 (2015). https://doi.org/10.4230/LIPIcs.IPEC.2015.17

45. Ye, Y., Tang, H.: Utilizing de Bruijn graph of metagenome assembly for meta-transcriptome analysis. Bioinformatics **32**(7), 1001–1008 (2016). https://doi.org/10.1093/bioinformatics/btv510

46. Zerbino, D.R., Birney, E.: Velvet: algorithms for de novo short read assembly using de Bruijn graphs. Genome Res. **18**(5), 821–829 (2008)

ProTranslator: Zero-Shot Protein Function Prediction Using Textual Description

Hanwen Xu[1] and Sheng Wang[2]([✉])

[1] Department of Automation, Tsinghua University, Beijing, China
[2] Paul G. Allen School of Computer Science & Engineering, University of Washington, Seattle, WA, USA
swang@cs.washington.edu

Abstract. Accurately finding proteins and genes that have a certain function is the prerequisite for a broad range of biomedical applications. Despite the encouraging progress of existing computational approaches in protein function prediction, it remains challenging to annotate proteins to a novel function that is not collected in the Gene Ontology and does not have any annotated proteins. This limitation, a "side effect" from the widely-used multi-label classification problem setting of protein function prediction, hampers the progress of studying new pathways and biological processes, and further slows down research in various biomedical areas. Here, we tackle this problem by annotating proteins to a function only based on its textual description so that we don't need to know any associated proteins for this function. The key idea of our method ProTranslator is to redefine protein function prediction as a machine translation problem, which translates the description word sequence of a function to the amino acid sequence of a protein. We can then transfer annotations from functions that have similar textual description to annotate a novel function. We observed substantial improvement in annotating novel functions and sparsely annotated functions on CAFA3, SwissProt and GOA datasets. We further demonstrated how our method accurately predicted gene members for a given pathway in Reactome, KEGG and MSigDB only based on the pathway description. Finally, we showed how ProTranslator enabled us to generate the textual description instead of the function label for a set of proteins, providing a new scheme for protein function prediction. We envision ProTranslator will give rise to a protein function "search engine" that returns a list of proteins based on the free text queried by the user.

Keywords: Protein function prediction · Natural language processing · Systems biology

Availability: https://github.com/HanwenXuTHU/ProTranslator/

1 Introduction

Accurately identifying protein functions serves as the basis for studying a wide range of biomedical problems [1–4], such as cell cycle regulation [5], neuronal morphogenesis

I. Pe'er (Ed.): RECOMB 2022, LNBI 13278, pp. 279–294, 2022.
https://doi.org/10.1007/978-3-031-04749-7_17

[6], signal transduction [7] and drug discovery [8, 9]. Experimentally testing the functions of millions of proteins across tens of thousands of functions is impractical. As a result, many computational approaches have been proposed to predict protein functions according to protein domains [1, 10], protein motifs [11, 12], protein sequence [13–17], protein-protein interactions [18–20], protein text description [21], and protein structures [22, 23]. These features have also been integrated to jointly perform the prediction [24–27].

The standard problem setting of protein function prediction is to form it as a multi-label classification problem, where the input is the feature vector of a protein and the output is a set of functions predefined as controlled vocabulary in the Gene Ontology (GO) [28]. This problem setting enables protein function prediction to easily incorporate new machine learning techniques in the feature extraction or the classification component, but inevitably restricts the predicted function to be within the set of controlled vocabulary. As a result, existing methods are not able to classify proteins into functions that are not in the GO and do not have any annotated proteins. clusDCA is able to classify proteins to the function that do not have any annotated proteins [18], but it still requires that function to be within the GO graph. This limitation substantially hinders the progress towards understanding new molecular functions and biological processes, further slowing down research in downstream applications.

We aim to develop an algorithm that enables us to classify proteins into any function that does not have any annotated proteins and is not in GO. The only information we need for that function is a textual description, which could be a few sentences describing this function. The key idea of our method is to embed descriptions of all GO functions into the same low-dimensional space, where similar functions are co-located. When we need to annotate a new function that is not in GO, we will project this new function in this low-dimensional space based on its textual description and then transfer annotations from other GO functions. To embed the textual description, we used large-scale language model PubMedBert [29], which is pre-trained on millions of scientific papers and obtained the state-of-art performance on specialized biomedicine tasks. We then embed proteins by integrating protein sequence, protein textual description and protein-protein interaction network. Finally, we learnt a linear transformation from protein embedding space to function embedding space according to known annotations.

We validated our method on CAFA3 [2], SwissProt [30], and GOA [28] datasets and observed substantial improvement on functions that do not have any annotations and functions that are sparsely annotated. We further demonstrated how our method could predict gene members of pathways in Reactome [31], KEGG [32], MsigDB [33] by only using the pathway description without seeing any specific gene belonging to it. Finally, we showed how to generate sentences that can best describe the function of a set of given proteins based on our method. We envision that our method enables us to build a "search engine" for function prediction, where users only need to provide a few keywords or sentences to describe the function that they want to annotate and then our system will return the associated proteins and genes for this function.

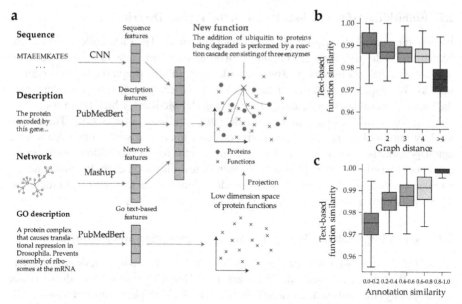

Fig. 1. **a,** Flowchart of ProTranslator. ProTranslator embeds proteins into the low-dimensional space by integrating protein sequence, description and network features. It then embeds GO terms according to the textual description. GO terms are then projected into the protein embedding space according to known annotations. To annotate a new function, ProTranslator will project the new function into this low-dimensional space according to its textual description and annotate it using nearby proteins. **b, c,** Box plots comparing text-based GO similarity for GO terms that have different distances on the graph (**b**) and annotation similarity (**c**).

2   Methods

2.1   Problem Definition

Previous studies modelled the protein function prediction as a multi-label classification problem. They used features of proteins, denoted as X, to predict a subset Y of predefined functions $Y_0 = \{y_1, y_2, ..., y_{z_0}\}$ where z_0 denotes the number of predefined functions. For each function y_j in Y_0, it has annotated protein feature set X_j collected for the model training. However, this modelling approach restricts the scope of protein prediction research to the function within the controlled vocabulary Y_0. To tackle this problem, we redefine this problem by considering the potential novel function set without any annotated protein collected. The novel function set is defined as $U_0 = \{u_1, u_2, ..., u_{z_1}\}$ and z_1 denotes the number of novel functions. There is also a new protein feature set X_j^U for each u_j and $X_j^U \notin \{X_1, ..., X_{z_0}\}$. This problem is defined as: with only protein features $\{X_1, ..., X_{z_0}\}$ and their annotations $\{Y_1, Y_2, ..., Y_{z_0}\}$ seen before, the prediction method should learn to classify a new protein feature set X_j^U into a novel function u_j.

2.2 Embedding GO Functions Based on the Textual Description

We use the textual description to embed GO functions. They are collected from the "definition" field from the Gene Ontology. In order to embed a new function to the same low-dimensional space only based on textual description, we disregard other information, such as GO graph and protein annotations, when embedding GO functions. We obtain the vector representation for each GO term using PubMedBert, which is pre-trained on both the PubMed's abstracts and PubMedCentral's full-text articles [29]. The corpus used by PubMedBert is best aligned with our task in comparison to other pre-trained language models. To obtain a fixed-size feature vector for GO definitions with various numbers of words, We average the last hidden states' output on each dimension across all tokens (words or subwords) to obtain the low-dimensional text representations. The final representation vector for each GO term is d_{Bert} dimensions.

2.3 Embedding Proteins Based on Sequence, Description and Network

To embed proteins, we consider three widely-used features: sequence, description and network. We followed the state-of-the-art approach DeepGOPlus to extract the sequence features using convolution neural networks (CNN) [14]. Multiple 1-d convolution kernels with different sizes are used in the first layer and the size ranges from R_D to R_U with step T, which results in e different sizes. The number of filters of each size is set to d_0. Then a max pooling layer is used to extract information across kernels.

$$FS_t = f_1([\omega_{1,t}*x_{1:t}^s, \omega_{1,t}*x_{2:1+t}^s, ..., \omega_{1,t}*x_{L-t+1:L}^s]), \tag{1}$$

where $\omega_{1,t}$ represents the 1-d convolution kernel in the first layer with the window size t and $x^s \in \mathbb{R}^{n \times 21 \times L}$ represents the input sequence one-hot encodings. n denotes the number of proteins. f_1 denotes the max pooling layer with the kernel size $L - t + 1$. We then concatenate different $FS_t \in \mathbb{R}^{n \times d_0}$ together as the sequence features $FS \in \mathbb{R}^{n \times d_{seq}}$.

$$FS = concat(FS_{t_1}, FS_{t_2}, ...FS_{t_e}), t_1 \le t_2 \le ... \le t_e \le L. \tag{2}$$

We embed the protein descriptions similarly to the process of embedding the textual description of GO functions. The description of each protein was obtained from GeneCards [34, 35]. Each protein is then represented by a low-dimensional vector $FD \in \mathbb{R}^{d_{Bert}}$. The gene network data $FN \in \mathbb{R}^{d_{Mashup}}$ is provided by the pre-trained Mashup representations of each protein according to their topology in multiple protein-protein interaction networks [20]. Then we add one-depth fully connected layers to reshape each kind of feature (FS, FD and FN) and set the output dimension to h. The fully connected layers are denoted as $Layer_{FCN,1}$, $Layer_{FCN,2}$, $Layer_{FCN,3}$:

$$M_S, M_D, M_N = Layer_{FCN,1}(FS), Layer_{FCN,2}(FD), Layer_{FCN,3}(FN). \tag{3}$$

M_S, M_D, M_N denote the processed features vector of protein sequence, description and network respectively, where M_S, M_D, M_N are all $n \times h$ matrices. Then we concatenate them together and denote it as M_{SDN}, which is $n \times 3h$.

$$M_{SDN} = concat(M_s, M_D, M_N). \tag{4}$$

2.4 Protein Function Prediction Based on GO Embeddings and Protein Embeddings

Our model predicts the protein function by projecting GO terms and proteins into the same low-dimensional space. Let $FT \in \mathbb{R}^{z \times d_{Bert}}$ be the representation vectors of the protein function text data, where z represents the number of protein functions. Let B be the binary label matrix, which is n by z. $B_{i,j} = 1$ only if protein i has the function j. The binary cross-entropy loss function is defined as:

$$
L = \sum_{i=1}^{n} \sum_{j=1}^{z} \Big[-B_{i,j} \times log\Big(1/1 + exp\Big(M_{SDN} W (FT)^{T}\Big)\Big)\Big)
$$
$$
- (1 - B_{i,j}) \times log\big(exp\big(M_{SDN} W (FT)^{T}\big)/\big(1 + exp\big(M_{SDN} W (FT)^{T}\big)\Big]
$$

(5)

where $W \in R^{3h \times d_{Bert}}$ are learnable parameters. During the prediction process, we can annotate a new protein using the following equation:

$$
p_j = 1/\Big(1 + exp\Big(m_{SDN} W (FT_j)^{T}\Big)\Big),
$$

(6)

where we use p_j to represent the probability of the new protein has the function j and m_{SDN} denotes this protein's concatenated features extracted by the method in Sect. 2.3. Since our method utilizes the textual description of a new function to classify proteins, it is able to annotate a new function even if it is not annotated to any protein in the training data.

2.5 Annotate Novel Functions, Sparse Functions and Gene Sets to Pathways

As for the terms seen in the training, we could combine the similarity based prediction method using DiamondScore [14, 36] to enhance ProTranslator, which is denoted as Pro-Translator + DiamondScore. The previous prediction score $S^{j}_{ProTranslator}$ for function j is the output of the deep learning model. Therefore we redefine the overall prediction score of ProTranslator + DiamondScore on function j as:

$$
S^{j}_{ProTanslator+DiamondScore} = \alpha \times S^{j}_{DiamondScore} + (1 - \alpha) \times S^{j}_{ProTranslator},
$$

(7)

The DiamondScore is calculated as:

$$
S^{j}_{DiamondScore} = \frac{\sum_{x^s \in E} bitscore\big(x^s_q, x^s\big) \times I(j \in J_{x^s})}{\sum_{x^s \in E} bitscore\big(x^s_q, x^s\big)},
$$

(8)

where x^s_q is the query sequence and E is the similar sequences set. $J_{x^{s_i}}$ is the annotations set of proteins with sequence feature x^s. I is the identity function and $bitscore$ is the sequence similarity score predicted by BLAST [10].

2.6 Text Generation by the Protein Sequence Features

We develop a model to generate the description of proteins from the sequence features based on the Transformer architecture [37]. For each GO function, we average all the one hot encodings of the sequences of its samples as the input in the text generation model. We still leverage the convolutional kernels in DeepGOCNN [14] to process the sequence features. Then we add the main Transformer architecture. Since the DeepGOCNN model discards the positional information when setting the max pooling layer to the maximum size, we remove the positional encodings at the encoder stack bottoms in Transformer. The multi-head self-attention could be written as:

$$MultiHead\left(M_{SDN,l}, M_{SDN,l}, M_{SDN,l}\right) = concat(head_1, ..., head_o)$$

$$head_i = softmax\left(\frac{M_{SDN,l}A_i^Q(M_{SDN,l}A_i^K)^T}{\sqrt{d_k}}\right)M_{SDN,l}A_i^V, \qquad (9)$$

where $M_{SDN,l}$ represents the input of the l_{th} sub-layer. A_i^Q, A_i^K, A_i^V are learnable parameters. To make the optimization process more stable, we adopt the pre-layer normalization in the Transformer [38]:

$$M_{SDN,l+1} = M_{SDN,l} + \mathcal{F}\left(LN\left(M_{SDN,l}; \theta_l\right)\right), \qquad (10)$$

where SDN_l and SDN_{l+1} represent the l_{th} sub-layer's input and output. LN denotes the layer normalization and θ_l represents the parameters of the sub-layer $\mathcal{F}$ in the encoder or decoder.

3 Experimental Setup

3.1 Calculating Similarities Between GO Functions

We calculated three kinds of similarities between two GO functions: text-based similarity, GO graph-based similarity and annotation-based similarity. The text-based similarity is calculated using the cosine similarity between the representation vectors of their textual descriptions. We calculated the GO graph-based similarity using the shortest distance between two GO functions on the GO graph, which is built based on "is_a" and "part_of" relationships. We calculated the annotation-based similarity by using the cosine similarity between the binary annotation vectors of two GO functions. The binary annotation vector $Annt^j \subseteq \mathbb{R}^z$ of a GO function j is defined as $Annt_i^j = 1$ if function j is i or one of i s ancestor in the GO hierarchy otherwise $Annt_i^j = 0$.

3.2 Datasets and Evaluation

We used the Gene Ontology (GO) that was released on June 16, 2021. The descriptions in the 'def' field were used as the textual description. We considered three datasets: CAFA3 [2], SwissProt [30], and GOA(Human) [28]. The preprocessed CAFA3 challenge dataset and SwissProt dataset were obtained from the online data files provided by DeepGOPlus.

The pre-trained gene network features for humans were downloaded from STRING database v9.1 [39]. We collected the gene descriptions from GeneCards [34, 35]. The CAFA3 dataset was released in September, 2016. We selected the proteins from the intersection of the CAFA3 dataset, the gene network features file and the gene description file and 11,679 proteins were finally selected. The SwissProt dataset was published in January, 2016 and we finally selected 5,889 proteins. The annotations were propagated according to the hierarchical structure of GO based on "is_a" and "part_of" relationships. We collected the annotations of the GOA(Human) dataset from the Gene Ontology Consortium website. The annotation file was generated on May 1, 2021. We leveraged 3-fold cross-validiation to evaluate these datasets and selected 10% of the leaf nodes in the GO graph as the novel functions in the zero-shot setting and excluded their annotations in the training dataset. We investigated the performance of ProTranslator and current state-of-art methods on annotating sparse functions with proteins less than 20 using the same GOA(Human) dataset and additional GOA(Mouse) dataset. The 3-fold cross-validiation was adopted. We calculated the area under the receiver of characteristic curve (AUROC) [40] of our model on the novel and sparse functions. In the text generation, we used the bilingual evaluation understudy (BLEU) [41] score as the metric. The BLEU score was computed first between segments of generated texts and references and then averaged over them.

To classify genes into pathways, we collected the Reactome [31] and KEGG [32] pathways description and gene sets. We finally obtained 2,007 and 264 pathways in Reactome and KEGG, the average gene number in each pathway is 4.6 and 22.6 respectively. We also collected the Molecular Signatures Database (MSigDB) [33] for pathway prediction. The text in "DESCRIPTION_FULL" was selected as the text data. We evaluated our approach on pathway C2, which has the most complete textual description. There were 3,704 pathways and the average number of genes for each pathway is 3.7 in pathway C2. In each pathway, the genes in both the genesets and STRING database for humans were selected for evaluation. In the text generation part, we leveraged the GOA(Human) datasets. We select 70% functions in GO data as the training functions and 30% function as validation functions.

The input length L of a protein was set to 2000 in annotating functions. We set the range of convolutional kernel size R_D and R_U to 8 and 128, and the step T was 8. Then we could get $e = 16$ different sizes of kernels. We set d_0 to 512 and therefore d_{seq} was 8192. The dimension of PubMedBert representations d_{Bert} is 768. The dimension of Mashup representations $d_{Mashup} = 800$ and $d_{Mashup} = 1000$ for GOA(Human) and GOA(Mouse) datasets. The dimension h was set to 1500. When combining the similarity based prediction method, ProTranslator used the setting of DeepGOPlus, and $\alpha = 0.68$, $\alpha = 0.63$, $\alpha = 0.46$ for BP, MF and CC. In the Transformer architecture, we used 6 encoder layers and 6 decoder layers. We set the hidden dimension of the Transformer to 512. The attention layer heads number o was 8 and d_k was 64. We also used the warm-up stage during the training process. Here we set the warm-up steps to 2000. We used the greedy decode strategy in the inference process.

3.3 Comparison Approaches

We compared our method to two comparison approaches that have the same model architecture while replacing the combined features with sequence features only. In one comparison approach, we replaced the PubMedBert embedded text vectors with Term Frequency–Inverse Document Frequency (TF-IDF) embeddings to investigate the influence of text embedding methods. In the other comparison approach, the text vectors were replaced with the ontology network vectors to make comparison with the representations of topological features. We named these two comparison approaches 'tf-idf' and 'Graph-based'. To investigate how our method performed compared with current state-of-art methods when annotating sparsely annotated functions, we selected DeepGOPlus as the comparison approaches since it has been shown to exceed multiple benchmarks in the previous research [14]. DeepGOPlus used the latest released 'alphas' when considering similarity based predictions, which are 0.68, 0.63 and 0.46 for BP, MF and CC.

4 Results

4.1 Gene Ontology Term Description Similarity Reflects Function Annotation Similarity

The key idea of our method is to annotate a novel function by transferring annotations from other functions that have similar textual description. Therefore, we first examined the correlation between the text-based GO term similarity and the annotation-based GO term similarity (see **Experimental Setup**). We observed a strong correlation between these two similarities, indicating that GO terms with similar textual description tend to have similar protein annotations (Fig. 1c). We next compared the text-based similarity with GO graph-based similarity. We found that terms that are close on the graph have much higher textual similarity (Fig. 1b). Since previous work has demonstrated how GO graph can be used to assist function annotation, especially for sparsely annotated functions [18, 42, 43], the strong consistency between text-based GO similarity and GO graph-based similarity further raises our confidence that textual description can be used to enhance protein function prediction.

4.2 ProTranslator Enables Protein Function Prediction in the Zero-Shot Setting

We next sought to examine whether ProTranslator can classify proteins in the zero-shot setting where the test function does not have any annotated proteins in the training data. We summarized the results of ProTranslator on three GO domains of biological process (BP), molecular function (MF) and cellular component (CC) across three datasets in Fig. 2. To simulate the zero-shot setting, we held out all protein annotations of test functions from the training data. We first compared ProTranslator to TF-IDF, which models the textual description using a frequency-based vector space model, and observed at least 13%, 11%, 14% improvements on BP, MF, CC domains on three datasets by using ProTranslator, indicating the superior performance of embedding text description using large-scale pre-trained language models. We then compared ProTranslator to a graph-based approach, which embeds the GO graph structure to annotate novel functions. ProTranslator also outperformed this graph-based approach by a large margin, which demonstrates the advantage of using text data against GO graphs to annotate novel functions. More importantly, the graph-based approach requires the function to be within the GO graph, whereas ProTranslator supports the annotation of any function only based on a short text description.

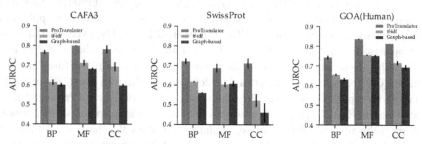

Fig. 2. Performance of ProTranslator in the zero-shot setting. Bar Plots comparing the AUROC of ProTranslator, tf-idf and the graph-based approach on CAFA3, SwissProt, GOA(Human). We held out all protein annotations of test functions from the training data.

4.3 ProTranslator Obtains Substantial Improvement in the Few-Shot Setting

After confirming the superior performance of ProTranslator in the zero-shot setting, we next investigated whether ProTranslator can achieve better performance in the few-shot setting where each test GO term only has very few annotated proteins. We compared Pro-Translator to the state-of-the-art approach DeepGOPlus [14] on GOA Human and Mouse datasets (Fig. 3). Similar to DeepGOPlus, we also incorporated the protein sequence feature into our model to improve the performance and denoted it as ProTranslator + DiamondScore. We observed substantial improvement of ProTranslator over DeepGO-Plus when each GO term only has a limited number of annotation proteins between 1 to 20. We further found that the improvement of our method is larger for terms that have fewer annotated proteins, indicating the more prominent performance of our method in annotating new functions.

To further understand the effect of sequence-based features, we excluded the sequence-based DiamondScore from both our method and DeepGOPlus. DeepGOCNN [14] is the implementation of DeepGOPlus without using sequence-based Diamond-Score. We observed a decreased performance for both our method and DeepGOPlus. Nevertheless, ProTranslator still outperforms DeepGOCNN with a large margin. More-over, ProTranslator without DiamondScore also outperforms DeepGOPlus on functions that have less than 10 annotated proteins on GOA(Human), again indicating the prominent performance of using textual description for protein function prediction.

4.4 ProTranslator Annotated Genes to Pathways by Only Using the Pathway Description

After observing the superior performance of ProTranslator on functions collected in the Gene Ontology, we next evaluate ProTranslator on a more challenging setting of classifying genes into a pathway without knowing any genes in that pathway. Since ProTranslator could annotate any GO term as long as its textual description is available, we hypothesized that ProTranslator could also predict genes of a given pathway by only using the description of that pathway. Specifically, we trained ProTranslator using the function annotation and text description of Gene Ontology and then applied this model to pathways in Reactome, KEGG and MSigDB. Notably, even though graph-based approaches, such as clusDCA, are able to annotate GO terms that do not have any proteins, they cannot be applied to these pathways as they require the functions to be within the GO graph. In contrast, our method does not have this restriction, as it only relies on a short description of pathways. We summarized the performance of ProTranslator on Reactome, KEGG and MSigDB pathway in Fig. 4a–c. To avoid potential data leakage between pathways and GO terms, we excluded pathways that have more than 90% shared genes with an existing GO term. We examined the performance of our method at different AUROC thresholds and found that our method could annotate 81% of Reactome pathways with AUROC larger than 0.85 and 84% of KEGG pathways with AUROC larger than 0.75, demonstrating the accuracy of annotating genes to pathways and functions that are not collected in the GO.

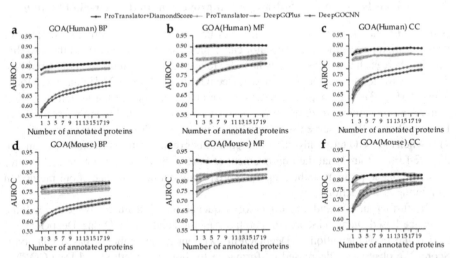

Fig. 3. Performance of ProTranslator on sparsely annotated functions. Plots comparing the AUROC of using ProTrainslator with sequence feature (ProTranslator + DiamondScore), Pro-Translator, DeepGOPlus and DeepGOCNN on annotating functions with number of annotated proteins from 1 to 20 in the training data in GOA(Human) (a–c) and GOA(Mouse) (d–f).

To better understand the prominent performance of ProTranslator on classifying genes into these functions, we further investigated how the text information determined the performance of ProTranslator. We observed that the pathway descriptions of Reactome and KEGG dataset were closer to the GO descriptions than those of MSigDB in the embedding space (Fig. 5a), which explains the more prominent performance of ProTranslator on Reactome and KEGG than MSigDB. Then we plotted the three datasets separately and colored each pathway using its AUROC during the cross-validation (Fig. 5b–d). We observed a clear pattern that the pathway whose textual description is closer to GO description tends to have a higher AUROC. This again indicates the substantial contribution of textual description in classifying genes into pathways and shows that the performance of our method depends on the quality of the textual description.

4.5   ProTranslator Generates Text Description for a Gene Set

The superior performance of ProTranslator comes from its novel setting of modeling the protein function prediction as a machine translation problem. We have extensively validated how to find associated proteins for a given function. Here, we aim to explore whether we can also generate the functional textual description for a set of proteins. For a given set of proteins, we used the average feature representations of them as input and then generated a novel textual description using ProTranslator. We evaluated this method on the GOA (Human) dataset by comparing the generated textual description to the ground truth curated GO term description and obtained a 0.26 BLEU. To avoid potential data leakage, we excluded the test term that has more than 0.5 Jaccard annotation-based similarity with any training GO term. By further examining the generated text, we observed that many of them are highly consistent with the curated GO term description (Table 1), suggesting the possibility of using our method to automatically expand GO and curate new functions.

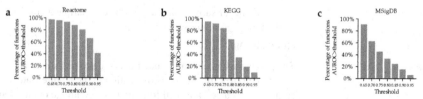

Fig. 4. Classifying genes into pathways only based on pathway description. a–c, Bar plots showing the percentage of pathways with AUROC greater than different thresholds (x-axis) on Reactome (a) and KEGG (b) and MSigDB (c). We didn't see any genes of these pathways in the training stage. Pathways that have many overlapped genes with an existing GO term are excluded.

4.6   Ablation Experiment

ProTranslator integrated protein sequence, protein description and protein-protein network as features to embed proteins. To understand the contribution of each kind of feature, we conducted ablation studies on the GOA(Human) dataset to explore how the performances would change with different feature combinations (Fig. 6). We found that each of these three components makes an important contribution for the function annotation. ProTranslator could achieve the best performance with all the three features together

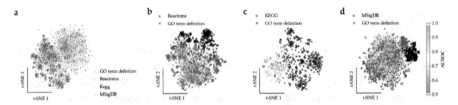

Fig. 5. Visualization of the joint embedding space of pathways and GO terms based on textual description. a, t-SNE plot showing the embedding space of GO terms and pathways in Reactome, KEGG, and MSigDB dataset. 1500 randomly selected GO terms are shown here. GO terms and pathways are embedded using their textual descriptions. **b, c, d,** t-SNE plots show the co-embedding space of Reactome (b), KEGG (c), MSigDB (d) and GO term. Each pathway is collected by its AUROC during the cross-validation. 500 randomly selected GO terms are shown in each plot.

Table 1. GO term description generated by our method according to the annotated proteins. The nearest text refers to the text of the training GO term that is closest to the test GO term.

	GO:0032588
Generated text	The lipid bilayer surrounding a vesicle transporting substances between the trans - golgi network and other parts of the cell
Nearest text in the training	The network of interconnected tubular and cisternal structures located within the Golgi apparatus on the side distal to the endoplasmic reticulum, from which secretory vesicles emerge. The trans-Golgi network is important in the later stages of protein secretion where it is thought to play a key role in the sorting and targeting of secreted proteins to the correct destination
Ground truth text	The lipid bilayer surrounding any of the compartments that make up the trans - golgi network
	GO:0048738
Generated text	The process whose specific outcome is the progression of a cardiac cell over time, from its formation to the mature state. a cardiac cell is a cell that will form part of the cardiac organ of an individual
Nearest text in the training	The process in which a relatively unspecialized cell acquires the specialized structural and/or functional features of a cell that will form part of the cardiac organ of an individual
Ground truth text	The process whose specific outcome is the progression of cardiac muscle over time, from its formation to the mature structure

and the AUROC was lowest with only the sequence feature. This observation verified that ProTranslator could be applied to cases where the protein sequence, description and network data were not all available.

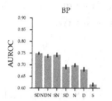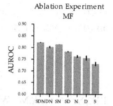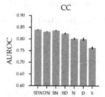

Fig. 6. Ablations studies on contributions of different features. Bar plots comparing the AUROC of using different feature combinations. S, D and N denote the protein sequence, description and network respectively. The tick label under each bar implies the combination, e.g., SN means using protein sequence and network features.

5 Conclusion and Discussion

We have presented ProTranslator, a text based protein function prediction framework. Through experiments on predicting GO terms in zero-shot and few-shot settings, we have verified that our approach was able to annotate novel functions by only using textual descriptions. We have further successfully applied ProTranslator to annotate genes to pathways from Reactome, KEGG and MsigDB [31–33] using only the pathway description. We observed that the performance of ProTranslator was better for functions that have text descriptions similar to GO term descriptions. Finally, we have demonstrated how our method can be used to generate novel textual descriptions for a given set of genes, offering the possibility to automatically curate new GO terms.

Despite the novelty and prominent performance of our method, there are still a few limitations of our method. Firstly, textual description is required to annotate the new function, which could be difficult to get for an under-studied new function. We plan to incorporate genomics of these functions into our framework to supplement the text information. We will also provide interactive interfaces for users to modify their text based on the annotations provided by our method. Secondly, the AUROCs of these novel functions are relatively lower compared to AUROCs of functions that have many annotations. Predicting for densely annotated functions is known to be less challenging [18]. Although AUROC values are not very high, our method can be used to narrow down the candidate proteins for a given new function, thus substantially reducing the experimental and other validation efforts.

This work is inspired by the decade-long attempts to automatically curate Gene Ontology, including NeXO [44] and CliXO [45]. The key difference between us and these pioneering works is that they reconstructed the hierarchical structure and gene clusters in the GO, whereas we generate the textual description of terms. These textual descriptions play a key role in scientific communication and collaborations and their curation is often most labor-intensive. Our method complements these existing efforts by using a novel natural language processing perspective and fills in an important gap towards automating GO curation. Another line of related works is automatically generating the term name for a set of genes or proteins [46–48]. Compared to these approaches, we generate a free text that contains a few sentences, which are more informative than a simple term name. Moreover, these existing approaches restricted the generated term to be a known phrase in the existing literature, whereas the text generated by our method is de novo, thus offering the unique description to a novel function or pathway.

References

1. Radivojac, P., Clark, W.T., Oron, T.R., Schnoes, A.M., Wittkop, T., Sokolov, A., et al.: A large-scale evaluation of computational protein function prediction. Nat. Methods **10**, 221–227 (2013)
2. Zhou, N., Jiang, Y., Bergquist, T.R., Lee, A.J., Kacsoh, B.Z., Crocker, A.W., et al.: The CAFA challenge reports improved protein function prediction and new functional annotations for hundreds of genes through experimental screens. Genome Biol. **20**, 244 (2019)
3. Jiang, Y., Oron, T.R., Clark, W.T., Bankapur, A.R., D'Andrea, D., Lepore, R., et al.: An expanded evaluation of protein function prediction methods shows an improvement in accuracy. Genome Biol. **17**, 184 (2016)
4. Friedberg, I., Radivojac, P.: Community-wide evaluation of computational function prediction. Methods Mol. Biol. **1446**, 133–146 (2017)
5. Dick, F.A., Rubin, S.M.: Molecular mechanisms underlying RB protein function. Nat. Rev. Mol. Cell Biol. **14**, 297–306 (2013)
6. Freixo, F., Martinez Delgado, P., Manso, Y., Sánchez-Huertas, C., Lacasa, C., Soriano, E., et al.: NEK7 regulates dendrite morphogenesis in neurons via Eg5-dependent microtubule stabilization. Nat. Commun. **9**, 2330 (2018)
7. Pierri, C.L.: SARS-CoV-2 spike protein: flexibility as a new target for fighting infection. Sig. Transduct. Target Ther. **5**, 254 (2020)
8. Menche, J., Sharma, A., Kitsak, M., Ghiassian, S.D., Vidal, M., Loscalzo, J., et al.: Disease networks. Uncovering disease-disease relationships through the incomplete interactome. Science **347**, 1257601 (2015)
9. Cheng, F., Kovács, I.A., Barabási, A.-L.: Network-based prediction of drug combinations. Nat. Commun. **10**, 1197 (2019)
10. Altschul, S.F., Madden, T.L., Schäffer, A.A., Zhang, J., Zhang, Z., Miller, W., et al.: Gapped BLAST and PSI-BLAST: a new generation of protein database search programs. Nucleic Acids Res. **25**, 3389–3402 (1997)
11. Jones, P., Binns, D., Chang, H.-Y., Fraser, M., Li, W., McAnulla, C., et al.: InterProScan 5: genome-scale protein function classification. Bioinformatics **30**, 1236–1240 (2014)
12. Zohra Smaili, F., Tian, S., Roy, A., Alazmi, M., Arold, S.T., Mukherjee, S., et al.: QAUST: protein function prediction using structure similarity, protein interaction, and functional motifs. Genomics Proteomics Bioinform. (2021). https://doi.org/10.1016/j.gpb.2021.02.001
13. Kulmanov, M., Khan, M.A., Hoehndorf, R., Wren, J.: DeepGO: predicting protein functions from sequence and interactions using a deep ontology-aware classifier. Bioinformatics **34**, 660–668 (2018)
14. Kulmanov M, Hoehndorf R. DeepGOPlus: improved protein function prediction from sequence. Bioinformatics (2021). https://doi.org/10.1093/bioinformatics/btaa763
15. Fa, R., Cozzetto, D., Wan, C., Jones, D.T.: Predicting human protein function with multi-task deep neural networks. PLoS ONE **13**, e0198216 (2018)
16. You, R., Zhang, Z., Xiong, Y., Sun, F., Mamitsuka, H., Zhu, S.: GOLabeler: improving sequence-based large-scale protein function prediction by learning to rank. Bioinformatics **34**, 2465–2473 (2018)
17. Strodthoff, N., Wagner, P., Wenzel, M., Samek, W.: UDSMProt: universal deep sequence models for protein classification. Bioinformatics **36**, 2401–2409 (2020)

18. Wang, S., Cho, H., Zhai, C., Berger, B., Peng, J.: Exploiting ontology graph for predicting sparsely annotated gene function. Bioinformatics **31**, i357–i364 (2015)
19. Vazquez, A., Flammini, A., Maritan, A., Vespignani, A.: Global protein function prediction from protein-protein interaction networks. Nat. Biotechnol. **21**, 697–700 (2003)
20. Cho, H., Berger, B., Peng, J.: Compact integration of multi-network topology for functional analysis of genes. Cell Syst. **3**, 540–548.e5 (2016)
21. You, R., Huang, X., Zhu, S.: DeepText2GO: improving large-scale protein function prediction with deep semantic text representation. Methods **145**, 82–90 (2018)
22. Yang, J., Yan, R., Roy, A., Xu, D., Poisson, J., Zhang, Y.: The I-TASSER suite: protein structure and function prediction. Nat. Methods **12**, 7–8 (2014)
23. Whisstock, J.C., Lesk, A.M.: Prediction of protein function from protein sequence and structure. Q. Rev. Biophys. **36**, 307–340 (2003)
24. Borgwardt, K.M., Ong, C.S., Schönauer, S., Vishwanathan, S.V.N., Smola, A.J., Kriegel, H.-P.: Protein function prediction via graph kernels. Bioinformatics **21**(Suppl 1), i47-56 (2005)
25. Zhang, C., Freddolino, P.L., Zhang, Y.: COFACTOR: improved protein function prediction by combining structure, sequence and protein–protein interaction information. Nucleic Acids Res. **45**, W291–W299 (2017)
26. You, R., Yao, S., Xiong, Y., Huang, X., Sun, F., Mamitsuka, H., et al.: NetGO: improving large-scale protein function prediction with massive network information. Nucleic Acids Res. **47**, W379–W387 (2019)
27. Yao, S., You, R., Wang, S., Xiong, Y., Huang, X., Zhu, S.: NetGO 2.0: improving large-scale protein function prediction with massive sequence, text, domain, family and network information. Nucleic Acids Res. **49**, W469–475 (2021)
28. Ashburner, M., Ball, C.A., Blake, J.A., Botstein, D., Butler, H., Cherry, J.M., et al.: Gene ontology: tool for the unification of biology. The Gene Ontology Consortium. Nat. Genet. **25**, 25–29 (2000)
29. Gu, Y., Tinn, R., Cheng, H., Lucas, M., Usuyama, N., Liu, X., et al.: Domain-specific language model pretraining for biomedical natural language processing. ACM Trans. Comput. Healthcare **3**, 1–23 (2021)
30. Boeckmann, B., Bairoch, A., Apweiler, R., Blatter, M.-C., Estreicher, A., Gasteiger, E., et al.: The SWISS-PROT protein knowledgebase and its supplement TrEMBL in 2003. Nucleic Acids Res. **31**, 365–370 (2003)
31. Fabregat, A., Jupe, S., Matthews, L., Sidiropoulos, K., Gillespie, M., Garapati, P., et al.: The Reactome pathway knowledgebase. Nucleic Acids Res. **46**, D649–D655 (2018)
32. Kanehisa, M., Goto, S.: KEGG: kyoto encyclopedia of genes and genomes. Nucleic Acids Res. **28**, 27–30 (2000)
33. Liberzon, A., Subramanian, A., Pinchback, R., Thorvaldsdóttir, H., Tamayo, P., Mesirov, J.P.: Molecular signatures database (MSigDB) 3.0. Bioinformatics **27**, 1739–1740 (2011)
34. Rebhan, M., Chalifa-Caspi, V., Prilusky, J., Lancet, D.: GeneCards: integrating information about genes, proteins and diseases. Trends Genet. **13**, 163 (1997)
35. Rebhan, M., Chalifa-Caspi, V., Prilusky, J., Lancet, D.: GeneCards: a novel functional genomics compendium with automated data mining and query reformulation support. Bioinformatics **14**, 656–664 (1998)
36. Buchfink, B., Xie, C., Huson, D.H.: Fast and sensitive protein alignment using DIAMOND. Nat. Methods **12**, 59–60 (2015)
37. Vaswani, A., Shazeer, N., Parmar, N.: Attention is all you need. In: 31st Conference on Neural Information Processing Systems (NIPS 2017) (2017)
38. Xiong, R., Yang, Y., He, D., Zheng, K., Zheng, S., Xing, C., et al.: On layer normalization in the transformer architecture. In: Proceedings of the 37th International Conference on Machine Learning (2020)

39. Franceschini, A., Szklarczyk, D., Frankild, S., Kuhn, M., Simonovic, M., Roth, A., et al.: STRING v9.1: protein-protein interaction networks, with increased coverage and integration. Nucleic Acids Res. **41,** D808–815 (2013)

40. Zou, K.H., O'Malley, A.J., Mauri, L.: Receiver-operating characteristic analysis for evaluating diagnostic tests and predictive models. Circulation **115,** 654–657 (2007)

41. Papineni, K., Roukos, S., Ward, T., Zhu, W.-J., Bleu: a method for automatic evaluation of machine translation. In: Proceedings of the 40th Annual Meeting of the Association for Computational Linguistics (2002)

42. Yu, G., Fu, G., Wang, J., Zhao, Y.: NewGOA: predicting new GO annotations of proteins by Bi-random walks on a hybrid graph. IEEE/ACM Trans. Comput. Biol. Bioinform. **15,** 1390–1402 (2018)

43. Zhao, Y., Fu, G., Wang, J., Guo, M., Yu, G.: Gene function prediction based on gene ontology hierarchy preserving hashing. Genomics **111,** 334–342 (2019)

44. Dutkowski, J., Kramer, M., Surma, M.A., Balakrishnan, R., Cherry, J.M., Krogan, N.J., et al.: A gene ontology inferred from molecular networks. Nat. Biotechnol. **31,** 38–45 (2013)

45. Kramer, M., Dutkowski, J., Yu, M., Bafna, V., Ideker, T.: Inferring gene ontologies from pairwise similarity data. Bioinformatics **30,** i34-42 (2014)

46. Wang, S., Ma, J., Fong, S., Rensi, S., Han, J., Peng, J., et al.: Deep functional synthesis: a machine learning approach to gene functional enrichment. bioRxiv 2019:824086. https://doi.org/10.1101/824086

47. Wang, S., Ma, J., Yu, M.K., Zheng, F., Huang, E.W., Han, J., et al.: Annotating gene sets by mining large literature collections with protein networks. Pac. Symp. Biocomput. **23,** 602–613 (2018)

48. Zhang, Y., Chen, Q., Zhang, Y., Wei, Z., Gao, Y., Peng, J., et al.: Automatic term name generation for gene ontology: task and dataset. In: Findings of the Association for Computational Linguistics: EMNLP 2020 (2020)

Short Papers

Single-Cell Multi-omic Velocity Infers Dynamic and Decoupled Gene Regulation

Chen Li[1], Maria Virgilio[1,2], Kathleen L. Collins[2,3,4], and Joshua D. Welch[1,5(✉)]

[1] Department of Computational Medicine and Bioinformatics,
University of Michigan, Ann Arbor, MI, USA
[2] Department of Cellular and Molecular Biology, University of Michigan, Ann Arbor, MI, USA
[3] Department of Microbiology and Immunology, University of Michigan, Ann Arbor, MI, USA
[4] Department of Internal Medicine, University of Michigan, Ann Arbor, MI, USA
[5] Department of Computer Science and Engineering, University of Michigan, Ann Arbor, MI, USA
welchjd@umich.edu

Abstract

Computational approaches can leverage single-cell snapshots to infer sequential gene expression changes during developmental processes. For example, cell trajectory inference algorithms use pairwise cell similarities to map cells onto a "pseudotime" axis corresponding to predicted developmental progress. However, trajectory inference based on similarity cannot predict the directions or relative rates of cellular transitions. Methods for inferring RNA velocity [1,2] address these limitations by fitting a system of differential equations that describes the directions and rates of transcriptional changes using spliced and unspliced transcript counts. Crucially, the dynamical model of RNA velocity [1] also infers a latent time value for each cell, providing a mechanistic means of reconstructing the order of gene expression changes during cell differentiation.

Single-cell multi-omic measurements provide an opportunity to incorporate epigenomic data into mechanistic models of trancription. The epigenome and transcriptome both change during cellular differentiation, and thus the temporal snapshots in single-cell multi-omic datasets potentially reveal the interplay among these molecular layers. For example, if epigenomic lineage priming occurs at a particular genomic locus, single-cell multi-omic data could reveal a significant time lag between chromatin remodeling of a gene and its transcription. Similarly, observing the dynamic changes in both the expression of a transcription factor and the chromatin accessibility of putative binding sites could reveal their temporal relationship.

Existing RNA velocity models assume that the transcription rate of a gene is uniform throughout the induction phase of gene expression. However, epigenomic changes play a key role in regulating gene expression, such as tightening or loosening the chromatin compaction of promoter and enhancer regions. For example, a transition from euchromatin to heterochromatin significantly reduces the rate of transcription at that locus, because transcriptional machinery cannot

© The Author(s), under exclusive license to Springer Nature Switzerland AG 2022
I. Pe'er (Ed.): RECOMB 2022, LNBI 13278, pp. 297–299, 2022.
https://doi.org/10.1007/978-3-031-04749-7_18

access the DNA. Therefore, a more realistic model would reflect the influence of enhancer and promoter chromatin accessibility on transcription rate.

MultiVelo describes the process of gene expression as a system of three ordinary differential equations (ODEs) characterized by a set of switch time and rate parameters. The time-varying levels of chromatin accessibility, unspliced pre-mRNA, and spliced mature mRNA are related by ODEs describing the rates of chromatin opening and closing, RNA transcription, splicing, and degradation or nuclear export. We assume that chromatin opening rapidly leads to full accessibility and similarly that chromatin closing rapidly leads to full inaccessibility. Each gene has distinct rate parameters describing its unique kinetics. We assume that the transcription rate is proportional to the chromatin accessibility and thus is time-varying, and we model the distinct phases or states that a cell traverses as its time advances. There are two states each for chromatin accessibility and RNA: chromatin opening, chromatin closing, transcriptional induction, and transcriptional repression. We hypothesize that multiple orders of events are possible: chromatin closing can occur either before or after transcriptional repression begins. We refer to the first ordering as Model 1 and the second ordering as Model 2. In addition, we refer to the time interval when chromatin opens before transcription initiates as *priming* and the interval in which chromatin accessibility and gene expression move in opposite directions as *decoupling*.

When being applied to real biological datasets, MultiVelo accurately fits the observed chromatin accessibility, unspliced pre-mRNA, and spliced mRNA counts in 10X Multiome embryonic mouse brain. MultiVelo identifies many clear examples of genes that are best described by either Model 1 or Model 2. Statistics of fitted genes show that most of the highly variable genes possess both induction and repression phases (a complete trajectory), and for genes that only have partial trajectories, induction-only phase portraits appear more often than repression-only. MultiVelo also identifies epigenomic priming and decoupling in the mouse brain dataset and predicts the four possible states for each cell in each gene.

We used MultiVelo to quantify epigenomic priming in SHARE-seq [3] data from mouse hair follicle. MultiVelo correctly identifies direction of differentiation from TACs to IRS and hair shaft cells, consistent with the diffusion map analysis reported in the initial paper. It also faithfully captures the priming state in genes that show a clear time delay visually across modalities, such as *Wnt3*.

Not only can MultiVelo perform well on mono-lineage or bi-lineage differentiation systems, but also in complex multi-lineage cell populations like human hematopoietic stem and progenitor cells (HSPC). We found that incorporating chromatin information significantly improves the local consistency and biological accuracy of predicted cell directions. Moreover, we found that as with the mouse brain dataset, Model 2 genes in the HSPC dataset are significantly enriched for gene ontology terms related to the cell cycle.

Lastly, we applied MultiVelo to a 10X Multiome dataset from developing human cortex [4]. As with the embryonic mouse brain dataset, MultiVelo inferred velocity vectors consistent with known patterns of brain cell development.

We again identified clear examples of both Model 1 and Model 2 genes. We used the inferred global latent time to study the relationship between the expression of a transcription factor (TF) and the accessibility of its binding sites and observed that the time of the highest RNA expression of the TF consistently preceded the time of corresponding high accessibility of downstream targets. We also collected a list of single-nucleotide polymorphisms (SNPs) associated with brain disorders and distinguished three major groups based on whether their maximum accessibility occurred early or late in latent time and before or after the expression of the linked gene.

In this work, we presented MultiVelo, a computational approach for inferring epigenomic regulation of gene expression from single-cell multi-omic datasets. We extend the dynamical RNA velocity model to incorporate chromatin to more accurately predict the past and future state of each cell, jointly infer the instantaneous rate of induction or repression for each modality, and determine the extent of coupling or time lag between modalities. MultiVelo uses a probabilistic latent variable model to estimate the switch time and rate parameters of gene regulation, providing a quantitative summary of the temporal relationship between epigenomic and transcriptomic changes. We expect that MultiVelo will provide fundamental insights into the mechanisms by which epigenomic changes regulate gene expression during cell fate transitions.

The full version of this manuscript is available on bioRxiv (https://www.biorxiv.org/content/10.1101/2021.12.13.472472v1).

References

1. Bergen, V., Lange, M., Peidli, S., Wolf, F.A., Theis, F.J.: Generalizing RNA velocity to transient cell states through dynamical modeling. Nat. Biotechnol. **38**, 1408–1414 (2020). https://doi.org/10.1038/s41587-020-0591-3
2. La Manno, G., et al.: RNA velocity of single cells. Nature **560**, 494–498 (2018). https://doi.org/10.1038/s41586-018-0414-6
3. Ma, S., et al.: Chromatin potential identified by shared single-cell profiling of RNA and chromatin. Cell **183**, 1103–1116 (2020). https://doi.org/10.1016/j.cell.2020.09.056
4. Trevino, A.E., et al.: Chromatin and gene-regulatory dynamics of the developing human cerebral cortex at single-cell resolution. Cell **184**(19), 5053–5069.e23 (2021). https://doi.org/10.1016/j.cell.2021.07.039

Ultrafast and Interpretable Single-Cell 3D Genome Analysis with Fast-Higashi

Ruochi Zhang[✉], Tianming Zhou, and Jian Ma[✉]

Computational Biology Department, School of Computer Science,
Carnegie Mellon University, Pittsburgh, PA 15213, USA
ruochiz@andrew.cmu.edu, jianma@cs.cmu.edu

The advent of high-throughput whole-genome mapping methods for the three-dimensional (3D) genome organization such as Hi-C has revealed distinct features of chromatin folding in various scales within the cell nucleus. These multiscale 3D genome features collectively contribute to vital genome functions such as transcription. However, the variation of 3D genome features and their functional significance in single cells remain poorly understood. The recent advances of single-cell Hi-C (scHi-C) technologies have provided us with unprecedented opportunities to probe chromatin interactions at single-cell resolution. These new technologies and datasets have the promise to unveil the connections between genome structure and function in single cells for a wide range of biological contexts. Unfortunately, existing scHi-C analysis methods are hindered by the technical noise, the sparseness of the data, and the complex chromatin interaction patterns. The lack of computational scalability and interpretation in the current methods poses further challenges for large-scale scHi-C analysis.

To directly address these important challenges, here we develop Fast-Higashi, a novel interpretable and scalable framework for embedding and integrative analysis of scHi-C data. Our proposed Fast-Higashi algorithm jointly produces embeddings and meta-interactions (analogous to the definition of metagenes in scRNA-seq analysis) for a given scHi-C dataset. As shown in Fig. 1a, the input of Fast-Higashi consists of a collection of scHi-C contact maps. In our Fast-Higashi formulation, every contact map can be approximated by a weighted sum of a set of meta-interactions. The weights are then decomposed into the product of single cell embeddings, a chromosome-specific transformation matrix, and meta-interaction-specific bin weights. To mitigate the sparseness of the scHi-C contact maps and improve the model robustness, we develop a new partial random walk with restart ("Partial RWR") algorithm and efficiently incorporate it into the optimization procedure of Fast-Higashi (Fig. 1b). Crucially, Fast-Higashi simultaneously learns the underlying meta-interactions and the cell embeddings.

Applications to several scHi-C datasets from cell lines and complex brain tissues demonstrate that Fast-Higashi is able to generate overall comparable or even better embeddings than existing methods, but is also >40× faster than the neural network-based baseline, enabling ultrafast analysis of scHi-C datasets. Moreover, Fast-Higashi is able to infer critical chromatin meta-interactions that define cell types with strong connections to cell type-specific transcription.

R. Zhang and T. Zhou—These two authors contributed equally.

I. Pe'er (Ed.): RECOMB 2022, LNBI 13278, pp. 300–301, 2022.
https://doi.org/10.1007/978-3-031-04749-7_19

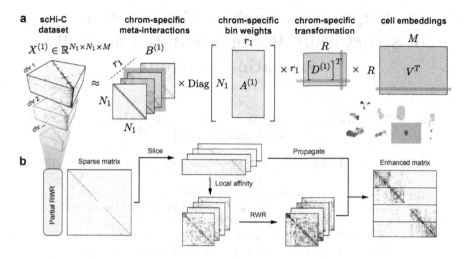

Fig. 1. Overview of Fast-Higashi. **a.** Workflow of the Fast-Higashi algorithm. Given an input scHi-C dataset of k chromosomes, Fast-Higashi models it as k 3-way tensors $X^{(c)}$ where the first two dimensions correspond to genomic bins and the last dimension corresponds to the single cells. Fast-Higashi then decomposes the tensors $X^{(c)}$ into four factors: a set of meta-interactions ($B^{(c)}$), a genomic bin weights indicating importance for each bin ($A^{(c)}$), the cell embedding matrix V that is shared across all chromosomes, and a chromosome-specific transformation matrix $D^{(c)}$ that transforms the shared cell embeddings into chromosome specific ones. **b.** Workflow of the partial random walk with restart (Partial RWR) algorithm. The Partial RWR algorithm is integrated into the tensor decomposition framework. When calculating the decomposed factors for frontal slices of the tensor $X^{(c)}$, the corresponding slices would be imputed through Partial RWR first. The impute process includes the calculation of local affinity, standard RWR algorithm, and information propagation using both sliced tensor and RWR imputed affinity matrix.

Together, Fast-Higashi is a powerful and scalable framework for the analysis of large-scale scHi-C datasets, with the distinct advantage that it can jointly unveil the finer cell types from complex tissues and infer interpretable meta-interactions that provide biological insights. As the development of scHi-C related technologies expands rapidly, Fast-Higashi has the potential to become an essential method in the toolbox of single-cell epigenomic analysis to greatly enhance the integrative analysis of 3D genome organization, genome functions, and cellular phenotypes at single-cell resolution for a wide range of biological applications.

Link to the bioRxiv preprint: https://www.biorxiv.org/content/10.1101/2022.04.18.488683v1.

DiffDomain Enables Identification of Structurally Reorganized Topologically Associating Domains

Dunming Hua[1], Ming Gu[1], Yanyi Du[1], Li Qi[2], Xiangjun Du[1], Zhidong Bai[3], Xiaopeng Zhu[4], and Dechao Tian[1(✉)]

[1] School of Public Health (Shenzhen), Sun Yat-sen University, Shenzhen 510275, Guangdong, China
tiandch@mail.sysu.edu.cn
[2] Chongqing Municipal Center for Disease Control and Prevention, Chongqing 400042, China
[3] KLASMOE and School of Mathematics and Statistics, Northeast Normal University, Changchun 130024, Jilin, China
[4] Computational Biology Department, School of Computer Science, Carnegie Mellon University, Pittsburgh, PA 15213, USA

The recent development of mapping technologies, such as Hi-C, that probes the 3D genome organization reveals that a chromosome is divided into topologically associating domains (TADs). TADs are genomic regions where chromatin loci are more frequently interacting with chromatin loci from the same TADs rather than from other TADs. TADs are functional units for transcriptional regulation, such as constraining interactions between enhancers and promoters. Although earliest studies have found that TADs are stable, multiple recent studies found large-scale TAD reorganization in diseases, cell differentiation, somatic cellular reprogramming, neuronal cell types, species, and individual cells. Thus, it is important to identify reorganized TADs through comparative analysis to improve understanding of the functional relevance of 3D genome organization, a priority of current work in the field. However, random perturbations and variations in numbers of mapped reads pose computational challenges for identifying reorganized TADs. Existing methods are highly dependent on changes at TAD boundaries, challenged by low-resolution Hi-C data, and inferior in controlling false positives. Importantly, methods using emerging single-cell Hi-C data are under-explored.

To fill these gaps, we develop DiffDomain, an algorithm leveraging high-dimensional random matrix theory to identify structurally reorganized TADs between a pair of biological conditions. The inputs are two Hi-C contact matrices of a given TAD region (Fig. 1A). The core of DiffDomain is that it formulates the problem as a hypothesis testing problem. Intuitively, if a TAD is not structurally reorganized, a properly normalized difference matrix D (Fig. 1B-D) would resemble a Wigner random matrix (null hypothesis). P value is computed using the asymptotic distribution of the largest singular value of D (Fig. 1E). DiffDomain identifies a reorganized TAD if the P value is less than 0.05. DiffDomain

---

D. Hua, M. Gu and Y. Du—These three authors contributed equally.

I. Pe'er (Ed.): RECOMB 2022, LNBI 13278, pp. 302–303, 2022.
https://doi.org/10.1007/978-3-031-04749-7_20

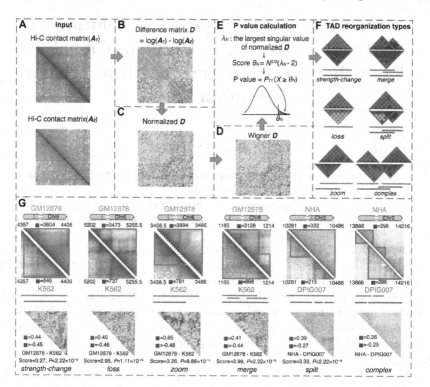

Fig. 1. Workflow of DiffDomain. (**A**) Input of DiffDomain are Hi-C contact matrices (A_1 and A_2) of a TAD. (**B**) Difference between log-transformed A_1 and A_2, denoted as D. (**C**) D is normalized by standardizing each of its k-off diagonal parts. (**D**) D is transformed by dividing $\sqrt{N}$. Under null hypothesis, it is a Wigner random matrix. (**E**) P value is calculated based on the fact that θ_N, normalized largest singular value of D, follows Tracy-Widom distribution with β_1 (denoted as T_1 distribution). A TAD is identified as a reorganized TAD if P value ≤ 0.05. (**F**) Reorganized TADs are classified into six subtypes. (**G**) Example of reorganized TADs. *Top:* Hi-C data in conditions 1 and 2; *Bottom:* normalized differential matrix D.

also classifies identified reorganized TADs into six subtypes to aid biological analyses and interpretations (Fig. 1F). A few examples are shown in Fig. 1G.

Application to real Hi-C data reveals that DiffDomain outperforms alternative methods in false-positive rates (at least 41-fold decrease on average) and accuracy in identifying truly reorganized TADs (at least 1.75 times increase). We demonstrate the robustness of DiffDomain and its biological applications by analyzing Hi-C data from different human cell types and disease states. It also works on ultra-sparse pseudo-bulk Hi-C data created from single-cell Hi-C data.

In summary, DiffDomain is a statistically sound method for identifying structurally reorganized TADs. DiffDomain could be a valuable method of an essential toolkit for emerging comparative TAD analysis in health and disease and other biological conditions.

Link to the bioRxiv preprint: doi: https://doi.org/.

Joint Inference of Repeated Evolutionary Trajectories and Patterns of Clonal Exclusivity or Co-occurrence from Tumor Mutation Trees

Xiang Ge Luo[1,2], Jack Kuipers[1,2], and Niko Beerenwinkel[1,2](✉)

[1] Department of Biosystems Science and Engineering, ETH Zurich, Basel,
Switzerland
niko.beerenwinkel@bsse.ethz.ch
[2] SIB Swiss Institute of Bioinformatics, Basel, Switzerland

Cancer progression is an evolutionary process shaped by both deterministic and stochastic forces. Recent advances in multi-region sequencing, single-cell sequencing, and phylogenetic tree inference empower more precise reconstruction of the mutational history of each tumor. At the same time, the increased resolution also highlights the extensive diversity across tumors and patients. Resolving the interactions among mutations and recovering the recurrent evolutionary processes may offer greater opportunities for designing successful therapeutic strategies.

Many cancer mutations do not appear independently of each other but rather exhibit patterns of clonal exclusivity and co-occurrence. Existing cancer progression models that address the problem of inferring the temporal order of mutations and predicting tumor progression can only handle binary genotypes derived from cross-sectional bulk sequencing data. As such, they have not been designed for tree-structured data as they cannot capture the subclonal structure within a tumor nor utilize the existing order information from the tumor phylogenies. Other methods that infer repeated mutational trajectories directly from phylogenetic trees are mostly based on pairwise ancestor-descendant relationships or consensus trees. These methods do not explicitly utilize the pairwise frequencies of mutational events between subclones and thus cannot account for clonal exclusivity.

Here, we present a novel probabilistic framework, called TreeMHN, for joint inference of repeated evolutionary trajectories and patterns of clonal exclusivity or co-occurrence from a cohort of intra-tumor phylogenetic trees. We denote each subclone in a tree by the mutational pathway that runs from the root to the node where the subclone is attached and assume an underlying tree-generating process: the waiting times of the subclones are exponentially-distributed random variables parameterized by a Mutual Hazard Network (MHN). The diagonal entries of an MHN represent the baseline rates of mutations, and the off-diagonal entries correspond to the positive (co-occurring), negative (exclusive), or zero (no) effects of a mutation on its downstream mutations. The observed tree structures are jointly determined by the waiting times of subclones and an independent sampling time. The marginal probability of observing a tree is

I. Pe'er (Ed.): RECOMB 2022, LNBI 13278, pp. 304–305, 2022.
https://doi.org/10.1007/978-3-031-04749-7_21

equal to the probability that all observed mutational events happen before the sampling event, and all unobserved events that could happen next did not happen before the sampling event. Given an MHN, we can generate heterogeneous mutation trees. Conversely, given a set of trees for a patient cohort, we can estimate the underlying MHN. After estimating the MHN, we can also compute the probability and the expected waiting time of different evolutionary trajectories.

To ensure the scalability of TreeMHN, we provide two inference methods based on maximum likelihood estimation (MLE) and a hybrid Monte Carlo expectation-maximization (MC-EM) algorithm. With our efficient parameter estimation procedure, it is the maximum tree size, rather than the total number of mutations, which is typically much larger, that limits the computation time of TreeMHN. Through simulation experiments we demonstrated the superior performance of TreeMHN over alternative methods in estimating both patterns of clonal exclusivity or co-occurrence and the associated probability distribution of the trajectories. Notably, TreeMHN does not rely on any particular phylogenetic method and is robust against uncertainty in the phylogenetic trees.

We have applied our method to a real dataset containing single-cell mutation trees for 123 acute myeloid leukemia patient samples. The inferred patterns of clonal exclusivity are highly consistent with and enrich known statistical and biological findings. Beyond detecting the existence of pairwise interactions, TreeMHN estimates the baseline rates of mutational events, directional strengths of the enabling or inhibiting effects, as well as the probabilistic temporal ordering among mutations. Moreover, conditioned on an observed tree structure and the estimated MHN, we can predict the next most probable mutational event. Given its flexibility, TreeMHN can be applied to tree data of different cancer types and the results may be useful for further experimental studies as well as treatment designs.

The link to the full preprint of this paper is https://doi.org/10.1101/2021. 11.04.467347. The R package TreeMHN is open-source and available at https:// github.com/cbg-ethz/TreeMHN.

Fast and Optimal Sequence-to-Graph Alignment Guided by Seeds

Pesho Ivanov$^{(\boxtimes)}$, Benjamin Bichsel, and Martin Vechev

Department of Computer Science, ETH Zurich, Zürich, Switzerland
{pesho.ivanov,benjamin.bichsel,martin.vechev}@inf.ethz.ch

Abstract. We present a novel A* *seed heuristic* that enables fast and optimal sequence-to-graph alignment, guaranteed to minimize the edit distance of the alignment assuming non-negative edit costs.

We phrase optimal alignment as a shortest path problem and solve it by instantiating the A* algorithm with our seed heuristic. The seed heuristic first extracts non-overlapping substrings (*seeds*) from the read, finds exact seed *matches* in the reference, marks preceding reference positions by *crumbs*, and uses the crumbs to direct the A* search. The key idea is to punish paths for the absence of foreseeable seed matches. We prove admissibility of the seed heuristic, thus guaranteeing alignment optimality.

Our implementation extends the free and open source aligner and demonstrates that the seed heuristic outperforms all state-of-the-art optimal aligners including GRAPHALIGNER, VARGAS, PASGAL, and the prefix heuristic previously employed by ASTARIX. Specifically, we achieve a consistent speedup of $>60\times$ on both short Illumina reads and long HiFi reads (up to 25 kbp), on both the *E. coli* linear reference genome (1 Mbp) and the MHC variant graph (5 Mbp). Our speedup is enabled by the seed heuristic consistently skipping $>99.99\%$ of the table cells that optimal aligners based on dynamic programming compute.

ASTARIX Aligner and Evaluations: https://github.com/eth-sri/astarix.

Full Paper: https://www.biorxiv.org/content/10.1101/2021.11.05.467 453.

Keywords: Genome graph · Optimal alignment · Semi-global alignment · Edit distance · Shortest path · Long reads · A* algorithm · Seed heuristic

1 Introduction

Alignment of reads to a reference genome is an essential and early step in most bioinformatics pipelines. While linear references have been used traditionally, an increasing interest is directed towards graph references capable of representing biological variation [1]. Specifically, a *sequence-to-graph* alignment is a base-to-base correspondence between a given read and a walk in the graph. As sequencing errors and biological variation result in inexact read alignments, edit distance

© The Author(s), under exclusive license to Springer Nature Switzerland AG 2022
I. Pe'er (Ed.): RECOMB 2022, LNBI 13278, pp. 306–325, 2022.
https://doi.org/10.1007/978-3-031-04749-7_22

is the most common metric that alignment algorithms optimize in order to find the most probable read origin in the reference.

Suboptimal Alignment. In the last decades, approximate and alignment-free methods satisfied the demand for faster algorithms which process huge volumes of genetic data [2]. *Seed-and-extend* is arguably the most popular paradigm in read alignment [3–5]. First, substrings (called *seeds* or *kmers*) of the read are extracted, then aligned to the reference, and finally prospective matching locations are *extended* on both sides to align the full read.

While such a heuristic may produce acceptable alignments in many cases, it fundamentally does not provide quality guarantees, resulting in suboptimal alignment accuracy. In contrast, we demonstrate in this work that seeds can benefit optimal alignment as well.

Key Challenges in Optimal Alignment. Finding optimal alignments is desirable but expensive in the worst case, requiring $\mathcal{O}(Nm)$ time [6], for graph size N and read length m. Unfortunately, most optimal sequence-to-graph aligners rely on dynamic programming (DP) and always reach this worst-case asymptotic runtime. Such aligners include VARGAS [7], PASGAL [8], GRAPHALIGNER [9], HGA [10], and VG [1], which use bit-level optimizations and parallelization to increase their throughput.

In contrast, ASTARIX [11] follows the promising direction of using a heuristic to avoid worst-case runtime on realistic data. To this end, ASTARIX rephrases the task of alignment as a shortest-path problem in an *alignment graph* extended by a *trie index*, and solves it using the A* algorithm instantiated with a problem-specific prefix heuristic. Importantly, its choice of heuristic only affects performance, not optimality. Unlike DP-based algorithms, this prefix heuristic allows scaling sublinearly with the reference size, substantially increasing performance on large genomes. However, it can only efficiently align reads of limited length.

This Work: Optimal Alignment for Short and Long Reads. In this work, we address the key challenge of scaling to long HiFi reads, while retaining the superior scaling of ASTARIX in the size of the reference graph. To this end, we instantiate the A* algorithm with a novel seed heuristic, which outperforms existing optimal aligners on both short and long HiFi reads. Specifically, the seed heuristic utilizes information from the whole read to narrowly direct the A* search by placing *crumbs* on graph nodes which lead up to a *seed match*, i.e., an exact match of a substring of the read.

Overall, the contributions presented in this work are:

1. A novel A* seed heuristic that exploits information from the whole read to quickly align it to a general graphs reference.
2. An optimality proof showing that the seed heuristic always finds an alignment with minimal edit distance.
3. An implementation of the seed heuristic as part of the ASTARIX aligner.

$$(\langle u,i \rangle, \langle v,i+1 \rangle, q[i], \Delta_{\text{match}}) \in E_{\mathsf{a}}^q \quad \text{if } (u,v,\ell) \in E_r, \ell = q[i] \qquad \text{(match)}$$
$$(\langle u,i \rangle, \langle v,i+1 \rangle, q[i], \Delta_{\text{subst}}) \in E_{\mathsf{a}}^q \quad \text{if } (u,v,\ell) \in E_r, \ell \neq q[i] \quad \text{(substitution)}$$
$$(\langle u,i \rangle, \langle v,i \quad \rangle, \varepsilon, \ \Delta_{\text{del}} \ ) \in E_{\mathsf{a}}^q \quad \text{if } (u,v,\ell) \in E_r \qquad \text{(deletion)}$$
$$(\langle u,i \rangle, \langle u,i+1 \rangle, q[i], \Delta_{\text{ins}} \ ) \in E_{\mathsf{a}}^q \qquad\qquad\qquad\qquad \text{(insertion)},$$

Fig. 1. Formal definition of alignment graph edges $E_{\mathsf{a}}^q \subseteq V_{\mathsf{a}}^q \times V_{\mathsf{a}}^q \times \Sigma_\varepsilon \times \mathbb{R}_{\geq 0}$. Here, $u, v \in V_r$, $0 \leq i < |q|$, $\ell \in \Sigma$, and ε represents the empty string, indicating that letter ℓ was deleted.

4. An extensive evaluation of our approach, showing that we align both short Illumina reads and long HiFi reads to both linear and graph references $\geq 60\times$ faster than existing optimal aligners.
5. A demonstration of superior empirical runtime scaling in the reference size N: $N^{0.46}$ on Illumina reads and $N^{0.11}$ on HiFi reads.

2   Prerequisites

We define the task of alignment as a shortest path problem (Sect. 2.1) to be solved using the A* algorithm (Sect. 2.2).

2.1   Problem Statement: Alignment as Shortest Path

In the following, we formalize the task of optimally aligning a read to a reference graph in terms of finding a shortest path in an *alignment graph*. Our discussion closely follows [11, §2] and is in line with [12].

Reference Graph. A reference graph $G_r = (V_r, E_r)$ encodes a collection of references to be considered when aligning a read. Its directed edges $E_r \subseteq V_r \times V_r \times \Sigma$ are labeled by nucleotide letters from $\Sigma = \{\mathsf{A}, \mathsf{C}, \mathsf{G}, \mathsf{T}\}$, hence any walk π_r in G_r spells a string $\sigma(\pi_r) \in \Sigma^*$.

An alignment of a read $q \in \Sigma^*$ to a reference graph G_r consists of (i) a walk π_r in G_r and (ii) a sequence of edits (matches, substitutions, deletions, and insertions) that transform $\sigma(\pi_r)$ to q. An alignment is *optimal* if it minimizes the sum of edit costs for a given real-valued cost model $\Delta = (\Delta_{\text{match}}, \Delta_{\text{subst}}, \Delta_{\text{del}}, \Delta_{\text{ins}})$. Throughout this work, we assume that edit costs are non-negative—a pre-requisite for the correctness of A*. Further, we assume that $\Delta_{\text{match}} \leq \Delta_{\text{subst}}, \Delta_{\text{ins}}, \Delta_{\text{del}}$—a prerequisite for the correctness of our heuristic. We note that our approach naturally works for cyclic reference graphs.

Alignment Graph. In order to formalize optimal alignment as a shortest path finding problem, we rely on an *alignment graph* $G_{\mathsf{a}}^q = (V_{\mathsf{a}}^q, E_{\mathsf{a}}^q)$. Its nodes V_{a}^q are *states* of the form $\langle v, i \rangle$, where $v \in V_r$ is a node in the reference graph and $i \in \{0, \ldots, |q|\}$ corresponds to a position in the read q. Its edges E_{a}^q are selected such that any path π_{a} in G_{a}^q from $\langle u,0 \rangle$ to $\langle v,i \rangle$ corresponds to an alignment of the first i letters of q to G_r. Further, the edges are weighted, which allows us to define an *optimal alignment* of a read $q \in \Sigma^*$ as a shortest path π_a in G_a^q from $\langle u, 0 \rangle$ to $\langle v, |q| \rangle$, for any $u, v \in V_r$. Figure 1 formally defines the edges E_a^q.

2.2   A* Algorithm for Finding a Shortest Path

The A* algorithm is a shortest path algorithm that generalizes Dijkstra's algorithm by directing the search towards the target. Given a weighted graph $G = (V, E)$, the A* algorithm finds a shortest path from sources $S \subseteq V$ to targets $T \subseteq V$. To prioritize paths that lead to a target, it relies on an admissible heuristic function $h: V \to \mathbb{R}_{\geq 0}$, where $h(v)$ estimates the remaining length of the shortest path from a given node $v \in V$ to a target $t \in T$.

Algorithm. In a nutshell, the A* algorithm maintains a set of *explored* nodes, initialized by all possible starting nodes S. It then iteratively *expands* the explored state v with lowest estimated total cost $f(v)$ by exploring all its neighbors. Here, $f(v) := g(v) + h(v)$, where $g(v)$ is the distance from $s \in S$ to v, and $h(v)$ is the estimated distance from v to $t \in T$. When the A* algorithm expands a target node $t \in T$, it reconstructs the path leading to t and returns it. For completeness, Appendix A.1 provides an implementation of A*.

Admissibility. The A* algorithm is guaranteed to find a shortest path if its heuristic h provides a lower bound on the distance to the closest target, often referred to as h being *admissible* or optimistic.

Further, the performance of the A* algorithm relies critically on the choice of h. Specifically, it is crucial to have low estimates for the optimal paths but also to have high estimates for suboptimal paths.

Discussion. To summarize, we use the A* algorithm to find a shortest path from $\langle u, 0 \rangle$ to $\langle v, |q| \rangle$ in G_a^q. To guarantee optimality, its heuristic function $h\langle v, i \rangle$ must provide a lower bound on the shortest distance from state $\langle v, i \rangle$ to a terminal state of the form $\langle w, |q| \rangle$. Equivalently, $h\langle v, i \rangle$ should lower bound the minimal cost of aligning $q[i:]$ to G_r starting from v, where $q[i:]$ denotes the suffix of q starting at position i (0-indexed). The key challenge is thus finding a heuristic that is not only admissible but also yields favorable performance.

3   Seed Heuristic

We instantiate the A* algorithm with a novel, domain-specific *seed heuristic* which allows to quickly align reads to a general reference graph. We first intuitively explain the seed heuristic and showcase it on a simple example (Sect. 3.1). Then, we formally define the heuristic and prove its admissibility (Sect. 3.2). Finally, we adapt our approach to rely on a trie, which leads to a critical speedup (Sect. 3.3).

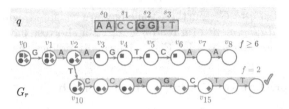

Fig. 2. A toy overview example using the seed heuristic to align a read q to a reference graph G_r. The read is split into four colored seeds, where their corresponding crumbs are shown inside reference graph nodes as symbols with matching color. The optimal alignment is highlighted as a green path ending with a tick (✓) and includes one substitution (T→A) and one deletion (C). (Color figure online)

3.1 Overview

Figure 2 showcases the seed heuristic on an overview example. It shows a read q to be aligned to a reference graph G_r. Our goal is to find an optimal alignment starting from an arbitrary node $v \in G_r$. For simplicity of the exposition, we assume unit edit costs $\Delta = (0, 1, 1, 1)$, which we generalize in Sect. 3.2.

Intuition. The A* search requires us to provide a lower bound of the remaining path cost from a state $\langle v, i \rangle$ to a target state. Clearly, to align the whole query, each of the remaining seeds (i.e. at or after position i in q) has to be eventually aligned. The intuition underlying the seed heuristic is to punish the state for the absence of any foreseeable match of each remaining seed. Notice that the order of the seeds is not directly taken into account.

In order to quickly check if a seed s can lead to a match, we follow a procedure similar to the one used by Hansel and Gretel who were placing breadcrumbs to find their trail back home. Before aligning a query, we will precompute all *crumbs* from all seeds so that not finding a crumb for a seed s on node v indicates that seed s could not be matched exactly before the query is fully aligned continuing from v. This way, assuming that a shortest path includes many seed matches, the crumbs will direct the A* search along with it.

If a crumb from an expected seed is missing in node v, its corresponding seed s could not possibly be aligned exactly and this will incur the cost of at least one substitution, insertion, or deletion. Assuming unit edit costs, $h\langle v, i \rangle$ yields a lower bound on the cost for aligning $q[i:]$ starting from v by simply returning the number of missing expected crumbs in v.

Crumbs Precomputation Example. Figure 2 shows four seeds as colored sections of length 2 each, and represents their corresponding crumbs as ■, ▶, ● and ◆, respectively. Four crumbs are expected if we start at v_2, but ■ is missing, so $h\langle v_2, 0 \rangle = 1$. Analogously, if we reach v_2 after aligning one letter from the read, we expect 3 crumbs (except ■), and we find them all in v_2, so $h\langle v_2, 1 \rangle = 0$. To precompute the crumbs for each seed, we first find all positions in G_r from which the seed aligns exactly. Figure 2 shows these exact matches as colored sections

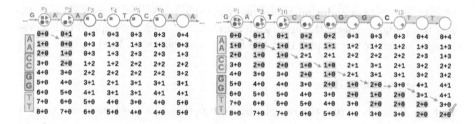

Fig. 3. Exploration of G_a^q, searching for a shortest path from the first to the last row using the seed heuristic. The table entry in the ith row (zero indexed) below node v shows $g\langle v, i \rangle + h\langle v, i \rangle$, where $g\langle v, i \rangle$ is the shortest distance from any starting state $\langle u, 0 \rangle$ to $\langle v, i \rangle$. States that may (Depending on how the A* algorithm handles tie-braking, different sets of states could be explored. For simplicity, we show all states that *could potentially* be explored) be expanded by the A* algorithm are highlighted in pink , and the rest of the states are shown for completeness even though they are never expanded. The shortest path corresponding to the best alignment is shown with green arrows ($\rightarrow$). (Color figure online)

of G_r. Then, from each match we traverse G_r backwards and add crumbs to nodes that could lead to these matches. For example, because seed CC can be matched starting in node v_{10}, crumbs ▶ are placed on all nodes leading up to v_{10}. Similarly, seed AA has two exact matches, one starting in node v_0 and one starting in node v_6. However, we only add crumbs ■ to nodes v_0, v_1, and v_3–v_6, but not to node v_2. This is because v_2 is (i) strictly after the beginning of the match of AA at v_1 and (ii) too far before the match of AA at v_6. Specifically, any alignment starting from node v_2 and still matching AA at v_6 would induce an overall cost of 4 (it would require deleting the 4 letters A, G, T, and C). Even without a crumb ■ on v_2, our heuristic provides a lower bound on the cost of such an alignment: it never estimates a cost of more than 4, the number of seeds.

Guiding the Search Example. Figure 3 demonstrates how $h\langle v, i \rangle$ guides the A* algorithm towards the shortest path by showing which states may be expanded when using the seed heuristic. Specifically, the unique optimal alignment in Fig. 2 starts from node v_1, continues to v_2, and then proceeds through node v_{10} (instead of v_3).

While the seed heuristic initially explores all states of the form $\langle v, 0 \rangle$ (we discuss in Sect. 3.3 how to avoid this by using a trie), it skips expanding any state that involves nodes v_3–v_8. This improvement is possible because all these explored states are penalized by the seed heuristic by at least 3, while the shortest path of cost 2 will be found before considering states on nodes v_3–v_8. Here, the heuristic function accurately predicts that expanding v_{10} may eventually lead to an exact alignment of seeds CC , GG and TT , while expanding v_3 may not lead to an alignment of either seed. In particular, the seed heuristic is not misled by the short-term benefit of correctly matching A in v_2, and instead provides a long-term recommendation based on the whole read. Thus, even though the walk to v_3 aligns exactly the first two letters of q, A* does not expand v_3 because the seed heuristic guarantees that the future cost will be at least 3.

3.2 Formal Definition

Next, we formally define the seed heuristic function $h\langle v, i \rangle$. Overall, we want to ensure that $h\langle v, i \rangle$ is admissible, i.e., that it is a lower bound on the cost of a shortest path from $\langle v, i \rangle$ to some $\langle w, |q| \rangle$ in G_a^q.

Seeds. We split read $q \in \Sigma^*$ into a set *Seeds* of non-overlapping seeds $s_0, \ldots, s_{|Seeds|-1} \in \Sigma^*$. For simplicity, in this work we ensure that all seeds have the same length and are consecutive, i.e., we split q into substrings $s_0 \cdot s_1 \cdots s_{|Seeds|-1} \cdot t$, where all s_j are seeds of length k and we ignore the suffix t of q, which is shorter than k. We note that our approach could be trivially generalized to seeds of different lengths or non-consecutive seeds as long as they do not overlap. An interesting future work item is investigating how different choices of seeds affect the performance of our approach, and selecting seeds accordingly.

Matches. For each seed $s \in Seeds$, we locate all nodes $u \in M(s)$ in the reference graph that can be the start of an exact match of s:

$$M(s) := \{u \in V_r \mid \exists \text{walk } \pi \text{ starting from } u \in G_r \text{ and spelling } \sigma(\pi) = s\}.$$

To compute $M(s)$ efficiently, we leverage the trie introduced in Sect. 3.3.

Crumbs. For seed s_j starting at position i in q, we place crumbs on all nodes $u \in V_r$ which can reach a node $v \in M(s_j)$ using less than $i + n_{del}$ edges:

$$C(s) := \{u \in V_r \mid \exists v \in M(s): dist(u, v) < i + n_{del}\},$$

where $dist(u, v)$ is the length of a shortest walk from u to v.

Later in this section, we will select n_{del} to ensure that if an alignment uses more than n_{del} deletions, its cost must be so high that the heuristic function is trivially admissible.

To compute $C(s)$ efficiently, we can traverse the reference graph backwards from each $v \in M(s)$ by a backward breadth-first-search (BFS).

Heuristic. Let $Seeds_{\geq i}$ be the set of seeds that start at or after position i of the read, formally defined by $Seeds_{\geq i} := \{s_j \mid \lceil i/k \rceil \leq j < |Seeds|\}$. This allows us to define the number of expected but missing crumbs in state $\langle v, i \rangle$ as $misses\langle v, i \rangle := \left| \{v \notin C(s) \mid s \in Seeds_{\geq i}\} \right|$. Finally, we define the seed heuristic as

$$h\langle v, i \rangle = (|q| - i) \cdot \Delta_{match} + misses\langle v, i \rangle \cdot \delta_{min}, \tag{1}$$

$$\text{for } \delta_{min} = \min(\Delta_{subst} - \Delta_{match}, \Delta_{del}, \Delta_{ins} - \Delta_{match}), \tag{2}$$

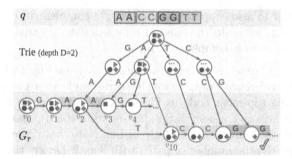

Fig. 4. Reference graph from Fig. 2, extended by a trie of depth $D = 2$. For simplicity, the reverse-complement reference graph and parts marked by "..." are omitted.

Intuitively, Eq. (1) reflects that the cost of aligning each remaining letter from $q[i:]$ is at least Δ_{match}. In addition, every inexact alignment of a seed induces an additional cost of at least δ_{min}. Specifically, every substitution costs Δ_{subst} but requires one less match; every deletion costs Δ_{del}; and every insertion costs Δ_{ins} but also requires one less match.

We note that $h\langle v, i\rangle$ implicitly also depends on the reference graph G_r, the read q, the set of seeds, and the edit costs Δ.

In order for an alignment with at least n_{del} deletions to have a cost so high that the heuristic function is trivially admissible, we ensure $n_{\mathrm{del}} \cdot \Delta_{\mathrm{del}} \geq h\langle v, i\rangle$ by defining

$$n_{\mathrm{del}} := \left\lceil \frac{|q| \cdot \Delta_{\mathrm{match}} + |Seeds| \cdot \delta_{min}}{\Delta_{\mathrm{del}}} \right\rceil. \tag{3}$$

In Theorem 1, we show that $h\langle v, i\rangle$ is admissible, ensuring that our heuristic yields optimal alignments.

Theorem 1 (Admissibility). *The seed heuristic $h\langle v, i\rangle$ is admissible.*

Proof. We provide a proof for Theorem 1 in Appendix A.2.   □

3.3   Trie Index

Considering all nodes $v \in V_r$ as possible starting points for the alignment means that the A* algorithm would explore all states of the form $\langle v, 0\rangle$, which immediately induces a high overhead of $|V_r|$. In line with previous works [11,13], we avoid this overhead by complementing the reference graph with a trie index to produce a new graph $G_r^+ = (V_r^+, E_r^+)$, where V_r^+ is the union of the reference graph nodes V_r and the new trie vertices, and E_r^+ is the union of E_r, the trie edges, and edges connecting the trie leafs with reference nodes. Note that constructing this trie index is a one-time pre-processing step that can be reused for multiple queries.

Since we want to also support aligning reverse-complement reads by starting from the trie *root*, we build the trie not only from the original reference graph and also from its reverse-complement.

Intuition. Figure 4 extends the reference graph G_r from Fig. 2 with a trie. Here, any path in the reference graph uniquely corresponds to a path starting from the trie *root* (the top-most node in Fig. 4). Thus, in order to find an optimal alignment, it suffices to consider paths starting from the trie *root*, by using state $\langle root, 0 \rangle$ as the only source for the A* algorithm. Note that if the reference graph branches frequently, the number of paths with length D may rise exponentially, leading to an exponential number of trie leaves. To counteract this exponential growth, we can select D logarithmically small, as $\log_4 N$.

For a more thorough introduction to the trie and its construction, see [11]. Importantly, our placement of crumbs (Sect. 3.2) generalizes directly to reference graphs extended with a trie (see also Fig. 4).

Reusing the Trie to Find Seed Matches. As a second usage of the trie, we can also exploit it to efficiently locate all matches $M(s)$ of a given seed s. In order to find all nodes where a seed match begins, we align (without errors) $\bar{s}$, the reverse-complement of s. To this end, we follow all paths spelling $\bar{s}$ starting from the *root*—the final nodes of these paths then correspond to nodes in $M(s)$. We ensure that the seed length $|s|$ is not shorter than the trie depth D, so that matching all letters in $\bar{s}$ ensures that we eventually transition from a trie leaf to the reference graph.

Optimization: Skip Crumbs on the Trie. Generally, we aim to place as few crumbs as possible, in order to both reduce precomputation time and avoid misleading the A* algorithm by unnecessary crumbs. In the following, we introduce an optimization to avoid placing crumbs on trie nodes that are "too close" to the match of their corresponding seed so they cannot lead to an optimal alignment.

Specifically, when traversing the reference graph backwards to place crumbs for a match of seed s starting at node w, we may "climb" from a reference graph node u to a trie node u' backwards through an edge that otherwise leads from the trie to the reference. Assuming s starts at position i in the read, we have already established that we can only consider nodes u that can reach w with less than $i + n_{del}$ edges (see Sect. 3.2). Here, we observe that it is sufficient to only climb into the trie from nodes u that can reach w using more than $i - n_{ins} - D$ edges, for

$$n_{ins} := \left\lceil \frac{|q| \cdot \Delta_{\text{match}} + |Seeds| \cdot \delta_{min}}{\Delta_{\text{ins}}} \right\rceil. \tag{4}$$

We define n_{ins} analogously to n_{del} to ensure that n_{ins} insertions will induce a cost that is always higher than $h\langle u, i \rangle$. We note that we can only avoid climbing into the trie if all paths from u to w are too short, in particular the longest one.

The following Lemma 1 shows that this optimization preserves optimality.

Lemma 1 (Admissibility when skipping crumbs). *The seed heuristic remains admissible when crumbs are skipped in the trie.*

Proof. We provide a proof for Lemma 1 in Appendix A.2.   □

In order to efficiently identify all nodes u that can reach w by using more than $i - D - n_{\text{ins}}$ edges (among all nodes at a backward-distance at most $i + n_{\text{del}}$ from w), we use topological sorting: considering only nodes at a backward-distance at most $i + n_{\text{del}}$ from w, the length of a longest path from a node v to w is (i) ∞ if v lies on a cycle and (ii) computable from nodes closer to w otherwise.

4   Evaluations

In the following, we demonstrate that our approach aligns faster than existing optimal aligners due to its superior scaling. Specifically, we address the following research questions:

Q1 What speedup can the seed heuristic achieve?
Q2 How does the seed heuristic scale with reference size?
Q3 How does the seed heuristic scale with read length?

The modes of operation which we analyze include both short (Illumina) and long (HiFi) reads to be aligned on both linear and graph references.

4.1   Seed Heuristic Implementation

Both the seed heuristic and the prefix heuristic reuse the same free and open source C++ codebase of the ASTARIX aligner [11]. It includes a simple implementation of a graph and trie data structure which is not optimized for memory usage. In order to easily align reverse complement reads, the reverse complement of the graph is stored alongside its straight version. The shortest path algorithm only constructs explored states explicitly, so most states remain implicitly defined and do not cause computational burden.

Both heuristics benefit from a default optimization in ASTARIX called *greedy matching* [11, Section 4.2] which skips adding a state to the A* queue when only one edge is outgoing from a state and the upcoming read and reference letters match.

4.2   Setting

All experiments were executed on a commodity machine with an Intel Core i7-6700 CPU @ 3.40 GHz processor, using a memory limit of 20 GB and a single thread. We note that while multiple tools support parallelization when aligning a single read, all tools can be trivially parallelized to align multiple reads in parallel.

Compared Aligners. We compare the novel seed heuristic to prefix heuristic (both heuristics are implemented in ASTARIX), GRAPHALIGNER, PASGAL, and VARGAS. We provide the versions and commands used to run all aligners and read simulators in Appendix A.3. We note that we do not compare to VG [1] and HGA [10] since the optimal alignment functionality of VG is meant for debugging purposes and has been shown to be inferior to other aligners [10, Tab. 4], and HGA makes use of parallelization and a GPU but has been shown to be superseded in the single CPU regime [10, Fig. 9]. PASGAL and VARGAS are compiled with AVX2 support. We execute the prefix heuristic with the default lookup depth $d = 5$.

Seed Heuristic Parameters. The choice of parameters for the seed heuristic influences its performance. Increasing the trie depth increases its number of nodes, but decreases the average out-degree of its leaves. We set the trie depth for all experiments to $D = 14 \approx \log_4 N$.

Shorter seeds are more likely to be matched perfectly by an optimal alignment, as they contain less letters that could be subject to edits. Thus, shorter seeds can tolerate higher error rate, but at the cost of slower precomputation due to a higher total number of matches, and slower alignment due to more off-track matches. In our experiments, we use seed lengths of $k = 25$ for Illumina reads and $k = 150$ for HiFi reads.

Data. We aligned reads to two different reference graphs: a linear *E. coli* genome (strain: K-12 substr. MG1655, ASM584v2) [14], with length of 4 641 652 bp (approx. 4.7 Mbp), and a variant graph with the Major Histocompatibility Complex (MHC), of size 5 318 019bp (approx. 5 Mbp), taken from the evaluations of PASGAL [8]. Additionally, we extracted a path from MHC in order to create a linear reference MHC-linear of length 4 956 648bp which covers approx. 93% of the original graph. Because of input graph format restrictions, we execute GRAPHALIGNER, VARGAS and PASGAL only on linear references in FASTA format (*E. coli* and the MHC-linear), while we execute the seed heuristic and the prefix heuristic on the original references (*E. coli* and MHC). This yields an underestimation of the speedup of the seed heuristic, as we expect the performance on MHC-linear to be strictly better than on the whole MHC graph.

To generate both short Illumina and long HiFi reads, we relied on two tools. We generated short single-end 200bp Illumina MSv3 reads using ART simulator [15]. We generated long HiFi reads using the script RANDOMREADS.SH[1] with sequencing lengths 5–25 kbp and error rates 0.3%, which are typical for HiFi reads.

Edit Costs. We execute ASTARIX with edit costs typical to the corresponding sequencing technology: $\Delta = (0, 1, 5, 5)$ for Illumina reads and $\Delta = (0, 1, 1, 1)$ for HiFi reads. As the performance of DP-based tools is independent of edit costs, we are using the respective default edits costs when executing GRAPHALIGNER, PASGAL and VARGAS.

Metrics. We report the performance of aligners in terms of runtime per one thousand aligned base pairs [s/kbp]. Since we measured runtime end-to-end (including loading reference graphs and reads from disk, and building the trie

---

[1] https://github.com/BioInfoTools/BBMap/blob/master/sh/randomreads.sh

Table 1. Runtime and memory comparison of optimal aligners. Simulated Illumina and HiFi reads are aligned to linear *E. coli* and graph MHC references. The runtime of the seed heuristic is expressed as absolute time per aligned kbp, while the other aligners are compared to the seed heuristic at a fold change. Additionally, the fraction of explored states is shown for the seed heuristic and the prefix heuristic.

	Illumina		HiFi		
Tool	*E. coli*	MHC	*E. coli*	MHC	
Seeds heuristic	0.019	0.041	0.001	0.002	s/kbp
(this work)	2.4	2.6	2.4	1.7	GB (max used)
	99.9996	99.9981	99.9989	99.9984	% skipped states
Prefix heuristic	269x	180x	n/a	n/a	x slowdown
	7.7	9.6	>20	>20	
	99.9501	99.9501	n/a	n/a	
GRAPHALIGNER	424x	212x	118x	64x	
	0.2	0.2	3.6	3.4	
VARGAS	133x	67x	1 413x	705x	
	<0.1	<0.1	7.3	7.3	
PASGAL	263x	130x	1 367x	736x	
	0.6	0.6	0.6	0.6	

index for ASTARIX), we ensured that alignment time dominates the total runtime by providing sufficiently many reads to each aligner. In order to prevent excessive runtimes for slower tools, we used a different number of reads for each tool and explicitly report them for each experiment.

Since shortest path approaches skip considerable parts of the computation performed by aligners based on dynamic programming, the commonly used Giga Cell Updates Per Second (GCUPS) metric is not adequate for measuring performance in our context.

We measured used memory by max_rss (Maximum Resident Set Size) from Snakemake[2].

We do not report accuracy or number of unaligned reads, as all evaluated tools align all reads with guaranteed optimality according to edit distance. We note that VARGAS reports a warning that some of its alignments are not optimal—we ignore this warning and focus on its performance.

---

[2] https://snakemake.readthedocs.io/en/stable/.

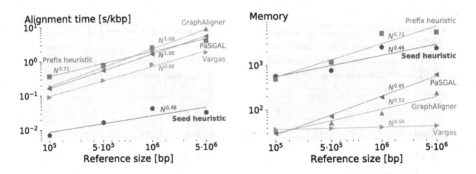

Fig. 5. Performance degradation with reference size for **Illumina reads**. Log-log plots of total alignment time (left) and memory usage (right) show the scaling difference between aligners.

4.3   Q1: Speedup of the Seed Heuristic

Table 1 shows that the seed heuristic achieves a speedup of at least 60 times compared to all considered aligners, across all regimes of operation: both Illumina and HiFi reads aligned on *E. coli* and MHC references.

In the Illumina experiments, the seed heuristic is given 100k reads, while the other tools are given 1000 reads. In the HiFi experiments, the seed heuristic is given reads that cover the reference 10 times, and the other tools are given reads of coverage 0.1.

The key reason for the speedup of the seed heuristic is that on all four experiments, it skips $\geq 99.99\%$ of the Nm states computed by the DP approaches of GRAPHALIGNER, PASGAL, and VARGAS. This fraction accounts for both the explored states during the A* algorithm, and the number of crumbs added to nodes during precomputation for each read.

The prefix heuristic exceeded the available memory on HiFi reads, as it is not designed for long reads.

4.4   Q2: Scaling with Reference Size

In order to study the scaling of the aligners in terms of the reference size, we extracted prefixes of increasing length from MHC-linear. We then generated reads from each prefix, and ran all tools on all prefixes with the corresponding reads.

Illumina Reads. Figure 5 shows the runtime scaling and memory usage for Illumina reads. The seed heuristic was provided with 10k reads, while other tools were provided with 1k reads. The runtime of GRAPHALIGNER, PASGAL and VARGAS grow approximately linearly with the reference length, whereas the runtime of the seed heuristic grows roughly with $\sqrt[3]{N}$, where N is the reference

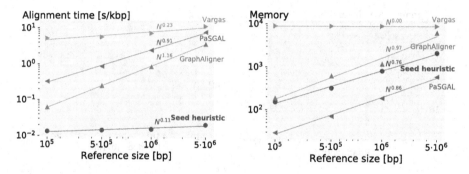

Fig. 6. Performance degradation with reference size for **HiFi reads**. Log-log plots of total alignment time (left) and memory usage (right) show the scaling difference between aligners. Linear best fits correspond to polynomials of varying degree.

size. Even on relatively small graphs like MHC, the speedup of the seed heuristic reaches 200 times. Note that the scaling of the prefix heuristic is substantially worse than the seed heuristic since the 200 bp reads are outside of its operational capabilities.

HiFi Reads. Figure 6 shows the runtime scaling and memory usage for HiFi reads. The respective total lengths of all aligned reads are 5 Mbp for the seed heuristic, 500 kbp for GRAPHALIGNER, and 100 kbp for VARGAS and PASGAL. We do not show the prefix heuristic, since it explores too many states and runs out of memory. Crucially, we observe that the runtime of the seed heuristic is almost independent of the reference size, growing as $N^{0.11}$. We believe this improved trend compared to short reads is because the seed heuristic obtains better guidance on long reads, as it can leverage information from the whole read.

For both Illumina and HiFi reads, we observe near-linear scaling for PASGAL and GRAPHALIGNER as expected from the theoretical $\mathcal{O}(Nm)$ runtime of the DP approaches. We conjecture that the runtime of VARGAS for long reads is dominated by the dependence from the read length, which is why on HiFi reads we observe better than linear runtime dependency on N but very large runtime. The current alignment bottleneck of ASTARIX-SEEDS is its memory usage, which is distributed between remembering crumbs and holding a queue of explored states.

4.5   Q3: Scaling with Read Length

Figure 7 shows the runtime and memory scaling with increasing length of aligned HiFi reads on MHC reference. Here we used reads with a total length of 100 Mbp for the seed heuristic and 2 Mbp for all other aligners.

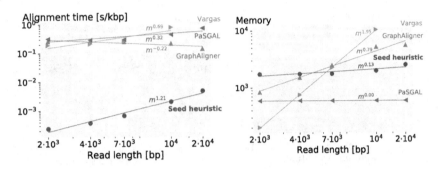

Fig. 7. Performance degradation with HiFi read length. Log-log plots of total alignment time (left) and memory usage (right) show the scaling difference between aligners.

The scaling of the seed heuristic in terms of read length is slightly worse than that of other aligners. However, this is compensated by its superior scaling in terms of reference size (see Sect. 4.4), leading to an overall better absolute runtime. We note that the memory usage of the seed heuristic does not heavily depend on the read length and for reads longer than 10 kbp, it is superior to GRAPHALIGNER and VARGAS.

5   Conclusion

We have presented an optimal read aligner based on the A* algorithm instantiated with a novel seed heuristic which guides the search by preferring crumbs on nodes that lead towards optimal alignments even for long reads.

The memory usage is currently limiting the application of ASTARIX for bigger references due to the size of the trie index. A remaining challenge is designing a heuristic function able to handle not only long but also noisier reads, such as the uncorrected PacBio reads that may reach 20% of mistakes. Possible improvements of the seed heuristic may include inexact matching of seeds, careful choice of seed positions, and accounting for the seeds order.

Acknowledgements. We want to thank the anonymous reviewers for the valuable feedback. The first author is grateful for the Vipassana meditation which inspired the idea to punish paths for the absence of foreseeable seed matches.

A   Appendix

Algorithm 1. A* algorithm

1: **function** A*$(G$: Graph, S: Sources, T: Targets, h: Heuristic function)
2:    $f \leftarrow Map(default = \infty)$: Nodes $\rightarrow \mathbb{R}_{\geq 0}$ ▷ Map nodes from G to priorities
3:    $Q \leftarrow MinPriorityQueue(priority = f)$ ▷ Priorities according to f
4:    **for all** $s \in S$ **do**
5:        $f[s] \leftarrow 0.0$
6:        $Q.push(s)$ ▷ Initially, explore all $s \in S$
7:    **while** $Q \neq \emptyset$ **do**
8:        $curr \leftarrow Q.pop()$ ▷ Get state with minimal f to be expanded
9:        **if** $curr \in T$ **then**
10:           **return** BACKTRACKPATH($curr$)   ▷ Reconstruct a walk to $curr$
11:          **for all** $(curr, next, cost) \in G.outgoingEdges(curr)$ **do**
12:             $\hat{f}_{next} \leftarrow f[curr] + cost + h(next)$ ▷ Candidate value for $f[next]$
13:             **if** $\hat{f}_{next} < f[next]$ **then**
14:                $f[next] \leftarrow \hat{f}_{next}$
15:                $Q.push(next)$ ▷ Explore state $next$
16:       **assert** $False$ ▷ Cannot happen if T is reachable from S

A.1   A* Algorithm

Algorithm 1 shows an implementation of the A* algorithm, taken from [11, §A.1]. We omit the implementation of BACKTRACKPATH for simplicity.

A.2   Proofs

In the following, we provide proofs for Theorem 1 and Lemma 1, restated here for convenience.

Theorem 1 (Admissibility). *The seed heuristic $h\langle v, i \rangle$ is admissible.*

Proof. Let A be an optimal alignment of $q[i:]$ starting from $v \in G_r$. We will prove that the cost of A is at least $h\langle v, i \rangle$.

If A contains at least n_{del} deletions, its cost is at least $n_{del} \cdot \Delta_{del}$, which is at least $|q| \cdot \Delta_{match} + |Seeds| \cdot \delta_{min}$ by plugging in n_{del} from Eq. (3). This is an upper bound for $h\langle v, i \rangle$, which we observe after maximizing $h\langle v, i \rangle$ by substituting $i = 1$ and $misses = |Seeds|$ into Eq. (1), which concludes the proof in this case.

Otherwise, A contains less than n_{del} deletions. If we interpret A as a path in G_a^q, we first observe that A must spell $q[i:]$. Thus, A must in particular also contain all seeds $s_j \in Seeds_{\geq i}$ as substrings. We then split A into *subalignments* $A_{-1}, A_0, \ldots, A_p$, selected such that $A_0, \ldots, A_{p-1}$ spell the seeds $s_j \in Seeds_{\geq i}$, and A_{-1} and A_p spell the prefix and suffix of $q[i:]$ which do not cover any full seed.

This ensures that we can compute a lower bound on the cost of A as follows:

$$\text{cost}(A) = \sum_{j=-1}^{p} \text{cost}(A_k) \tag{5}$$

$$\geq \sum_{j=-1}^{p} |\sigma(A_k)| \cdot \Delta_{\text{match}} + \sum_{j=0}^{p-1} \left\{ \begin{array}{ll} 0 & \text{if } v \in C\left(s_{\lceil i/k \rceil + j}\right) \\ \delta_{min} & \text{if } v \notin C\left(s_{\lceil i/k \rceil + j}\right) \end{array} \right\} \tag{6}$$

$$= (|q| - i) \cdot \Delta_{\text{match}} + \left| \{ v \notin C(s) \mid s \in Seeds_{\geq i} \} \right| \cdot \delta_{min} \tag{7}$$

$$= h\langle v, i \rangle \tag{8}$$

Here, Eq. (5) follows from our decomposition of A. If we ignore the right-hand side in Eq. (6) (right of "+"), the inequality follows because matching all letters is the cheapest method to align any string. The right-hand side follows from a more precise analysis for subalignments A_k that spell a seed $s_{\lceil i/k \rceil + j}$ without a corresponding crumb in v. The absence of such a crumb indicates that no exact match of $s_{\lceil i/k \rceil + j}$ in G_r can be reached within less than $i + n_{\text{del}}$ steps from v. However, because A contains less than n_{del} deletions, A_k must start within less than $i + n_{\text{del}}$ steps from v. Thus, A_k does not align $s_{\lceil i/k \rceil + j}$ exactly, meaning that it introduces a cost of at least δ_{min}.

Equation (7) follows from observing that $A_{-1}, \ldots, A_p$ have a total length of $|q| - i$, and observing that the right-hand sum adds up δ_{min} for every expected but missing crumb. Finally, Eq. (8) follows from our definition of $h\langle v, i \rangle$, concluding the proof. □

Lemma 1 (Admissibility when skipping crumbs). *The seed heuristic remains admissible when crumbs are skipped in the trie.*

Proof. Consider a reference graph with a match of seed s starting in node w. Now, consider a node v that cannot reach w using more than $i - D - n_{\text{ins}}$ edges. We can then show that a trie node v' with a path to v does not require a crumb for the match of s in node w.

Specifically, any path from *root* through nodes v' and v to node w has total length greater or equal $i - n_{\text{ins}}$. Thus, matching s at w requires at least n_{ins} insertions. Hence, the cost of such a path is at least $n_{\text{ins}} \cdot \Delta_{\text{ins}} = |q| \cdot \Delta_{\text{match}} + n \cdot \delta_{min}$. Observing that this is an upper bound for $h\langle v, i \rangle$ concludes the proof. □

A.3   Versions, Commands, Parameters for Running all Evaluated Approaches

In the following, we provide details on how we executed the newest versions of the tools discussed in Sect. 4:

Executing AStarix
Obtained from https://github.com/eth-sri/astarix
Seed heuristic
Command   `astarix align-optimal -D 14 -a astar-seeds -seeds_len l -f reads.fq -g`
                                               `graph.gfa >output`
Prefix heuristic
Command   `astarix align-optimal -D 14 -a astar-prefix -d 5 -f reads.fq -g`
                                             `graph.gfa >output`

For aligning Illumina reads, `astarix` is used with additional `-M 0 -S 1 -G 5` and for HiFi reads with `-M 0 -S 1 -G 1` which better match the error rate profiles for these technologies.

Executing Other Tools
Vargas
Obtained from   https://github.com/langmead-lab/vargas (v0.2, commit `b1ad5d9`)
Command   `vargas align -g graph.gdef -U reads.fq -ete`
Comment   `-ete` stands for end to end alignment; default is 1 thread
PaSGAL
Obtained from   https://github.com/ParBLiSS/PaSGAL (commit `9948629`)
Command   `PaSGAL -q reads.fq -r graph.vg -m vg -o output -t 1`
Comment   Compiled with AVX2
GraphAligner
Obtained from   https://github.com/maickrau/GraphAligner (v1.0.13, commit `02c8e26`)
Command   `GraphAligner -seeds-first-full-rows 64 -b 10000 -t 1 -f reads.fq -g`
                      `graph.gfa -a alignments.gaf >output` (commit `9948629`)
Comment   `-seeds-first-full-rows` forces the search from all possible reference positions instead of using seeds; `-b 10000` sets a high alignment bandwidth; these two parameters are necessary for an optimal alignment according to the author and developer of the tool

Simulating Reads
Illumina
           `art_illumina -ss MSv3 -sam -i graph.fasta -c N -l 200 -o dir -rnd_seed 42`
HiFi
           `randomreads.sh -Xmx1g build=1 ow=t seed=1 ref=graph.fa illuminanames=t addslash=t pacbio=t pbmin=0.003 pbmax=0.003 paired=f gaussianlength=t minlength=5000 midlength=13000 maxlen=25000 out=reads.fq`
Comment   `BBMapcoverage`, https://github.com/BioInfoTools/BBMap/blob/master/sh/randomreads.sh (commit: a9ceda0)

A.4   Notations

Table 2 summarizes the notational conventions used in this work.

Table 2. Notational conventions.

Object	Notation
Queries	$Q = \{q_i | q_i \in \Sigma^m\}$
Read	$q \in Q$
Length	$m := |q| \in \mathbb{N}$
Position in read	$q[i] \in \Sigma,\ i \in \{0, \ldots, m-1\}$
Reference graph	$G_r = (V_r, E_r)$
Size	$|G_r| := |V_r| + |E_r| \in \mathbb{N}$
Nodes	$u, v \in V_r$
Number of nodes	$N := |V_r| \in \mathbb{N}$
Edges	$e \in E_r := V_r \times V_r \times \Sigma$
Edge letter	$\ell \in \Sigma$
Reference graph with a trie	$G_r^+ = (V_r^+, E_r^+)$
Trie depth	$D \in \mathbb{N}_{>0}$
Alignment graph	$G_a^q = (V_a^q, E_a^q)$
State	$\langle u, i \rangle \in V_a^q := V \times \{0, \ldots, m\}$
Edges	$(\langle u, i \rangle, \langle v, j \rangle, \ell, w) \in E_a^q \subseteq V_a^q \times V_a^q \times \Sigma_\varepsilon \times \mathbb{R}_{\geq 0},\ \Sigma_\varepsilon = \Sigma \cup \{\varepsilon\}$
Edge cost	$w \in \mathbb{R}_{\geq 0}$
Alignment	$\pi \in E_e^*$ and $\sigma(\pi) = q$
Alignment cost	$\mathrm{cost}(\pi) \in \mathbb{R}_{\geq 0}$
Seed heuristic	$h\langle u, i \rangle$
State	$\langle u, i \rangle$
Seed length	k
Maximum number of deletions	n_{del}
Maximum number of insertions	n_{ins}
In all graphs	$G = (V, E) \in \{G_r, G_e, G_a^q\}$
Walk	$\pi \in G : \pi \in E^*$
Walk spelling	$\sigma(\pi) \in \Sigma^*$
Path	A walk without repeating nodes
A*	$A^*(G, S, T, h)$
Graph	$G = (V, E)$
Nodes	$u, v \in V$
Edges	$e \in E \subseteq V \times V \times \mathbb{R}_{\geq 0}$
Source states	$S \subseteq V$
Target states	$T \subseteq V$
Heuristic function	$h : V \to \mathbb{R}_{\geq 0}$
Minimum cost to a target	$h^*(u)$
Explored state	A state pushed to the queue of Algorithm 1
Expanded state	A state popped from the queue of Algorithm 1

References

1. Garrison, E., et al.: Variation graph toolkit improves read mapping by representing genetic variation in the reference. Nat. Biotechnol. **36**(9), 875–879 (2018)
2. Kucherov, G.: Evolution of biosequence search algorithms: a brief survey. Bioinformatics **35**(19), 3547–3552 (2019)
3. Altschul, S.F., Gish, W., Miller, W., Myers, E.W., Lipman, D.J.: Basic local alignment search tool. J. Mol. Biol. **215**(3), 403–410 (1990)

4. Langmead, B., Salzberg, S.L.: Fast gapped-read alignment with Bowtie 2. Nature Methods (2012)
5. Li, H., Durbin, R.: Fast and accurate short read alignment with Burrows-Wheeler transform. Bioinformatics **25**(14), 1754–1760 (2009)
6. Equi, M., Grossi, R., Mäkinen, V., Tomescu, A., et al.: On the complexity of string matching for graphs. In: 46th International Colloquium on Automata, Languages, and Programming (ICALP 2019), Schloss Dagstuhl-Leibniz-Zentrum für Informatik (2019)
7. Darby, C.A., Gaddipati, R., Schatz, M.C., Langmead, B.: Vargas: heuristic-free alignment for assessing linear and graph read aligners. Bioinformatics **36**(12), 3712–3718 (2020)
8. Jain, C., Misra, S., Zhang, H., Dilthey, A., Aluru, S.: Accelerating sequence alignment to graphs. In: International Parallel and Distributed Processing Symposium (IPDPS) (2019). ISSN 1530-2075
9. Rautiainen, M., Mäkinen, V., Marschall, T.: Bit-parallel sequence-to-graph alignment. Bioinformatics **35**(19), 3599–3607 (2019)
10. Feng, Z., Luo, Q.: Accelerating sequence-to-graph alignment on heterogeneous processors. In: 50th International Conference on Parallel Processing, pp. 1–10 (2021)
11. Ivanov, P., Bichsel, B., Mustafa, H., Kahles, A., Rätsch, G., Vechev, M.: ASTARIX: fast and optimal sequence-to-graph alignment. In: Schwartz, R. (ed.) RECOMB 2020. LNCS, vol. 12074, pp. 104–119. Springer, Cham (2020). https://doi.org/10.1007/978-3-030-45257-5_7
12. Rautiainen, M., Marschall, T.: Aligning sequences to general graphs in $O(V+mE)$ time. Bioinformatics (2017, preprint)
13. Dox, G., Fostier, J.: Efficient algorithms for pairwise sequence alignment on graphs. Master's thesis, Ghent University (2018)
14. Howe, K.L., et al.: Ensembl Genomes 2020-enabling non-vertebrate genomic research. Nucleic Acids Res. **48**, D689–D695 (2020)
15. Huang, W., Li, L., Myers, J.R., Marth, G.T.: ART: a next-generation sequencing read simulator. Bioinformatics **28**(4), 593–594 (2011)

CLMB: Deep Contrastive Learning for Robust Metagenomic Binning

Pengfei Zhang[1,2], Zhengyuan Jiang[1,2], Yixuan Wang[1,4], and Yu Li[1,3(✉)]

[1] Department of Computer Science and Engineering, CUHK,
Hong Kong SAR, China
`liyu@cse.cuhk.edu.hk`
[2] University of Science and Technology of China, Hefei, Anhui, China
[3] The CUHK Shenzhen Research Institute, Hi-Tech Park, Nanshan,
Shenzhen 518057, China
[4] Department of Mathematics, HIT, Weihai 264209, China

Abstract. The reconstruction of microbial genomes from large metagenomic datasets is a critical procedure for finding uncultivated microbial populations and defining their microbial functional roles. To achieve that, we need to perform metagenomic binning, clustering the assembled contigs into draft genomes. Despite the existing computational tools, most of them neglect one important property of the metagenomic data, that is, the noise. To further improve the metagenomic binning step and reconstruct better metagenomes, we propose a deep Contrastive Learning framework for Metagenome Binning (CLMB), which can efficiently eliminate the disturbance of noise and produce more stable and robust results. Essentially, instead of denoising the data explicitly, we add simulated noise to the training data and force the deep learning model to produce similar and stable representations for both the noise-free data and the distorted data. Consequently, the trained model will be robust to noise and handle it implicitly during usage. CLMB outperforms the previous state-of-the-art binning methods significantly, recovering the most near-complete genomes on almost all the benchmarking datasets (up to 17% more reconstructed genomes compared to the second-best method). It also improves the performance of bin refinement, reconstructing 8–22 more high-quality genomes and 15–32 more middle-quality genomes more than the second-best result. Impressively, in addition to being compatible with the binning refiner, single CLMB even recovers on average 15 more HQ genomes than the refiner of VAMB and Maxbin on the benchmarking datasets. On a real mother-infant microbiome dataset with 110 samples, CLMB is scalable and practical to recover 365 high-quality and middle-quality genomes (including 21 new ones), providing insights into the microbiome transmission. CLMB is open-source and available at https://github.com/zpf0117b/CLMB/.

Keywords: Metagenomic binning · Contrastive learning · Deep learning · Noise · Sequence clustering

© The Author(s), under exclusive license to Springer Nature Switzerland AG 2022
I. Pe'er (Ed.): RECOMB 2022, LNBI 13278, pp. 326–348, 2022.
https://doi.org/10.1007/978-3-031-04749-7_23

1   Introduction

Studies of microbial communities are increasingly dependent on high-throughput, whole-genome shotgun sequencing datasets [1,2]. General studies assemble short sequence reads obtained from metagenome sequencing into longer sequence fragments (contigs), and subsequently group them into genomes by metagenome binning [3,4]. Metagenome binning is a crucial step in recovering the genomes, which therefore provides access to uncultivated microbial populations and understanding their microbial functional roles.

In recent years, we have witnessed great progress in metagenome binning. Firstly, the composition and the abundance of each contig are proved useful for binning [5,6]. Secondly, several programs have been developed for fully automated binning procedures, which leverage both composition and abundance as features. MetaBAT [7], MetaBAT2 [8], CONCOCT [5], and Maxbin2 [9] utilize the composition and abundance information and take the metagenome binning as the clustering task. VAMB [10] performs dimensionality reduction, encoding the data using VAE first and subsequently conducting the clustering task. Thirdly, a new approach 'multi-split' is developed and achieves great performance [10,11]. It gathers contigs from all the samples and calculates the abundance among samples, clustering them into bins and splitting the bins by sample.

Earlier works on metagenomics binning achieved good performance by applying different strategies for clustering. However, they ignored the potential factors in real-world conditions that influence the quality of metagenomic short reads, such as the low total biomass of microbial-derived genomes in clinical isolates [12] and the imperfect genomic sequencing process, for example, base substitutions, insertions, and deletions [13]. As a consequence of the factors, metagenomic sequences are susceptible to the noise issue, such as contamination noise and alignment noise [12]. The potential noise can influence the quality of metagenomics sequences, and therefore make it difficult to distinguish whether certain contigs come from the same type of or different bacterial genomes, impacting the correctness of the formed draft genomes. Furthermore, all of the existing binners are restricted by data volume.

To learn a high-quality draft genome for each bacterium, we design a novel deep Contrastive Learning algorithm for Metagenomic Binning (CLMB) to handle the noise (Fig. 1). The basic idea of the CLMB module is that, since the noise of the real dataset is hard to detect, we add simulated noise to the data and force the trained model to be robust to them. Essentially, instead of denoising the data explicitly, we add simulated noise to the training data and ask the deep learning model to produce similar and stable representations for both the noise-free data and the distorted data. Consequently, the trained model will be robust to noise and handle it implicitly during usage. By effectively tackling the noise in the metagenomics data using the contrastive deep learning framework [14,15], we can group pairs of contigs that originate from the same type of bacteria together while dividing contigs from different species to different bins. Moreover, CLMB performs data augmentation before training and take the augmented data as training data. Unlike other binners, CLMB uses the augmented data, instead of the raw data, for training. Therefore, the data volume for training is largely increased, which improves the representation of the deep learning model and prevents overfitting. CLMB also keeps the 'multi-split' approach, which combines the contigs of all the samples for binning, because the contrastive deep learning benefits more from a larger data size [14].

On the CAMI2 Toy Human Microbiome Project Dataset [16], CLMB outperforms the previous state-of-the-art binning methods significantly, recovering the most near-complete genomes on almost all the benchmarking datasets. Specifically, CLMB reconstructs up to 17% more near-complete genomes compared to the second-best method. We then investigate the recovered genomes under different criteria and find that more information contained in data contributes to the binning performance of CLMB. By involving CLMB, the performance of bin refinement is improved, reconstructing 8–22 more high-quality genomes and 15–32 more middle-quality genomes more than the second-best result. Binning refiner with CLMB and VAMB [10] achieves the best performance than any other binners. Impressively, in addition to being compatible with the binning refiner, single CLMB even recovers on average 15 more HQ genomes than the refiner of VAMB and Maxbin on the benchmarking datasets. Furthermore, CLMB is applied to a real mother-infant microbiome dataset with 110 samples and recovers 365 high-quality and middle-quality genomes, including 21 new ones. As a crucial step for metagenomic research, the genome recovered by CLMB provides insights into the microbiome transmission.

Our contributions in this paper are summarized as follows:

- We propose a new metagenomic binner, CLMB, based on deep contrastive learning. It is the first binner that can effectively handle the noise in the metagenomic data. By implicitly modeling the noise using contrastive learning, our method can learn stable and robust representations for the contigs, thus leading to better binning results.
- We propose a novel data augmentation approach for metagenomic binning under the contrastive learning framework. Experiments suggest that it can indeed help us model the noise implicitly.
- We carefully evaluate the contribution of different properties and features to metagenomic binning using our method, including the sequence encoding, dimension, abundance, etc. We also show how our method can be combined with other binners to further improve the binning step. It can guide the users to achieve a better binning result.

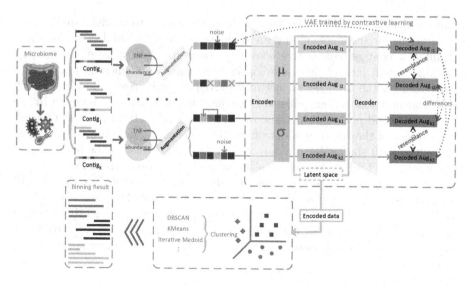

Fig. 1. Overview of CLMB workflow. CLMB takes contigs from sampled microbiome as inputs. Then, the abundances and the per-sequence tetranucleotide frequencies (TNF) are calculated, concatenated, and subsequently augmented to a pair of distorted data. All the augmented data are passed through VAE to train it with contrastive learning. After training, the concatenated features of each contig are passed through VAE to obtain the encoded data in the latent space as the representation. Finally, a general clustering algorithm can be applied to the representations to obtain binning results.

2   Methods

The key idea of CLMB is to involve explicitly modeled noise it in the data, to learn effective contig representations, and to pull together the representations of functionally similar contigs, while pushing apart dissimilar contigs. We achieve the goal with deep contrastive learning.

The CLMB pipeline is shown in Fig. 1. The inputs of CLMB are the contigs assembled from sequencing reads. For each contig, the abundances and the per-sequence tetranucleotide frequencies (TNF) are respectively calculated and transformed to numerical vectors of s-dimensional and 103-dimensional, denoted A_{in} and T_{in} (Methods C.1 in Appendix, s denotes the number of samples), both of which were concatenated as the input feature, denoted $concat(A_{in}, T_{in})$. Given the feature, we simulate noise in different forms, such as Gaussian noise and random mask, and add the noise to it, resulting in slightly distorted feature as the augmented data. Specifically, for each contig, two random augmented data are generated based on the feature data (Sect. 2.1) and used to train a neural network with contrastive learning, *i.e.*, contrasting the training pair of each contig between each other and against other data pairs [14]. As for the neural network model, we select the variational autoencoder (Sect. 2.2), due to its capability of learning smooth latent state representations of the input data [17,18]. When training the VAE model (Sect. 2.4), we force the model to produce similar representations for the augmented data of the same contig while distinct for those of different contigs (contrastive learning). More specifically, by discriminating the augmented data of the same contig from massive augmented data of the other contigs, the deep neural network (VAE) parameterizes a locally smooth nonlinear function f_θ that pulls together multiple distortions of a contig in the latent space and pushes away those of the other contigs. Intuitively, as the representations of the augmented data from the same contig are pulled together by f_θ, contigs with similar feature data can be pulled together in the latent space, which are more likely to be placed in the same cluster. After contrastive learning, $concat(A_{in}, T_{in})$ of each contig can be encoded by the trained VAE to the mean of their denoised distributions in the latent space (Sect. 2.5). The mean data of the contigs are the representations that we learn, which are subsequently clustered with the common clustering algorithms (*e.g.*, minibatch k-means [19], DBSCAN [20], iterative medoid clustering algorithm [7,10])[1] and put into respective bins (Sect. 2.5).

2.1 Data Augmentation

Data augmentation is essentially the process of modeling the noise explicitly. Any noise in real-world conditions that influence the quality of metagenomic short reads might result in the implicit change of feature data. For example, base deletion during genomic sequencing causes a statistical error of the tetramer frequencies and consequently the distortion of TNFs. Therefore, we perform data augmentation to the feature data for interpretability and effectiveness. We design three augmentation approaches for three noise cases, considering the real-life metagenoimc sequencing and data analytic pipeline.

1. Gaussian noise. It simulates the unexpected noise in metagenomic sequences. Assuming the features conform to Gaussian distribution with mean μ and variance σ^2, the noise obtained by sampling the Gaussian distribution $N(0, \sigma^2)$ and scaled in 0.15μ is added to the feature data.
2. Random mask. This simulates undetected read mapping during the assembly. Each dimension of the feature data might be masked with 0.01 probability.
3. Random shift. This kind of noise covers the imperfect genomic sequencing process. Two dimensions, i and j, of the feature data are chosen, and the number $f[i]$ on dimension i turns into $\frac{9f[i]}{10}$ while the number $f[j]$ on dimension j turns into $f[j] + \frac{f[i]}{10}$. The total percentage of chosen pairs of dimension is 0.01.

For each data augmentation during training, Gaussian noise and one of the other approaches are selected to generate training pairs for the feature data of each contig. After this, a minibatch of N contigs generates the augmented data with size $2N$.

---

[1] Minibatch k-means and DBSCAN are implemented by scikit-learn: https://scikit-learn.org. Iterative medoid clustering algorithm are implemented by [10]: https://github.com/RasmussenLab/vamb/blob/master/doc/tutorial.ipynb.

2.2   Architecture of the VAE

We employ the VAE architecture constructed in [10]. For a minibatch of N contigs, augmented data with size $2 * N$ are passed through the VAE module. Each $(s + 103)$-dimensional vector, generated from the augmentation of $concat(A_{in}, T_{in})$, is firstly passed through two fully connected layers with batch normalization [21] and dropout (P = 0.2) [22], termed the encoding layers, parameterizing function f_e. The output of the last layer, with N_h dimension, is then passed to two different fully connected layers with N_h dimensions, termed the μ and σ layers, parameterizing function f_μ and f_σ, respectively. The latent layer, l, is obtained by sampling the Gaussian distribution using the μ and σ layers as parameters, i.e., $l_i \sim N(\mu_i, \sigma_i)$ for each neuron $i = 1, 2, ..., N_h$. The sampled latent representation is then passed through the decoding layers, with the same size as the encoding layers except for arranged in a reverse order, parameterizing function f_d. Followed by the last decoding layer is a fully connected layer of $s + 103$ dimensions with function f_s parameterized, in which the vector is splitted into two output vectors of dimension s and 103, A_{out} and T_{out}, as the output abundance and TNFs, respectively. We use linear activation for the μ layer, softplus activation for the σ layer, and leaky rectified linear activation [23] for the other layers.

2.3   Loss Function

The loss function of CLMB is a trade-off for three goals:

1. The decoded data should be similar to the input data, which is a requirement of training autoencoder;
2. The Gaussian distribution dependent on the μ and σ layers for sampling is constrained by a prior $N(0, I)$, which is the prerequisite of VAE [17,18].
3. The decoded data for the augmented data of the same contig are as similar as possible, while those of different contigs are as dissimilar as possible, which is the terminal condition of contrastive learning [14].

To satisfy the first goal, we have

$$L_1 = w_A \sum ln(A_{out} + 10^{-9}) \cdot A_{in} + w_T \sum (T_{out} - T_{in})^2, \qquad (1)$$

where the w_A and w_T are the weighting terms. We use cross-entropy to penalize the abundance bias and the sum of squared errors to penalize the TNFs bias.

To satisfy the second goal, we have

$$L_2 = -\sum \frac{1}{2}(1 + ln(\sigma) - \mu^2 - \sigma) \qquad [24, 14]. \qquad (2)$$

We use the Kullback-Leibler divergence to penalize the deviance from this distribution.

To satisfy the third goal, we investigate the structure of each minibatch of $2 * N$ (distorted) augmented data, which are obtained by performing data augmentation to $\{concat(A_{in}, T_{in})_k\}_{k=1}^N$ of N contigs. All the data are passed through the VAE module, and we denote the output data from the decoding layer as $X = \{x_k \in R^{s+103}\}_{k=1}^{2N}$. For a pair of positive data x_i and x_j (derived from the feature data of the same contig), the other $2*N-2$ samples are treated as negatives. To distinguish the positive pair from the negatives, we define the cosine distance between two vectors $cos(x_i, x_j) = \frac{x_i^T \cdot x_j}{||x_i|| \cdot ||x_j||}$ and use the normalized temperature-scaled cross-entropy loss:

$$l_{i,j} = -log \frac{e^{\frac{cos(x_i, x_j)}{\tau}}}{\sum_{s=1, s \neq i}^{2N} e^{\frac{cos(x_i, x_s)}{\tau}}}, \tag{3}$$

where the temperature τ is a parameter we can tune. Note that $l(i,j)$ is asymmetrical. Suppose all the pairs $X = \{x_k \in R^{s+103}\}_{k=1}^{2N}$ are put in an order, in which x_{2k-1} and x_{2k} denote a pair of positive data, the summed-up loss within this minibatch is:

$$L_3 = \frac{1}{2N} \sum_{k=1}^N (l_{2k-1,2k} + l_{2k,2k-1}). \tag{4}$$

Finally, the combined loss function is

$$LOSS = L_1 + w_2 L_2 + w_3 L_3. \tag{5}$$

The weighting terms are set as $w_A = 0.85 ln(s)^{-1}$ and $w_T = 0.15/103$ (in accordance with [10]), $\tau = 0.1$, $w_2 = \frac{L_{1(0)}/L_{2(0)}}{2.75 \times 10^5 N_h}$, $w_3 = 0.15 L_{1(0)}/L_{3(0)}$, where $L_{1(0)}, L_{2(0)}, L_{3(0)}$ indicate the value of L_1, L_2, L_3 at the first epoch and are initially set to 1.

2.4   Training with Contrastive Learning

Here, we have modelled the noise explicitly, constructed the architecture, and defined the loss function we should optimize. The contrastive learning algorithm for training process will force the architecture to be robust to the noise we modelled. The pseudocode for training is presented in Algorithm 1.

As shown in Algorithm 1, in each training epoch, the contigs are randomly separated to several minibatches. The augmented data of each minibatch are put into VAE for training. The loss function is determined after L_1, L_2, L_3 are calculated. We train VAE by optimizing $LOSS$ using the Adam optimizer [25] and using one Monte Carlo sample of the Gaussian latent representation.

Algorithm 1 trains VAE by discriminating the data in sampled minibatch. However, due to insufficient memory capacity (either of CPU or GPU), a limited proportion of data are sampled to a minibatch, which might lead to a problem that the VAE fits well with the data in the minibatch rather than the whole dataset. Therefore, contrastive learning can benefit from shuffled, larger batch size and more epochs for training [14]. We train the model with minibatches of 4096 contigs for 600 epochs.

Algorithm 1. The contrastive learning algorithm for training VAE

Input: batchsize N, constant parameter τ, structure of $f_e, f_\mu, f_\sigma, f_d, f_s$, feature data $concat(A_{in}, T_{in})$

1: **for** sampled minibatch $\{concat(A_{in}, T_{in})_k\}_{k=1}^N$ **do**
2: select one data augmentation form pair with augmentation functions t_1, t_2;
3: **for all** $k \in \{1, 2, ..., k\}$ **do**
4:    $Aug_{2k-1} = t_1(concat(A_{in}, T_{in})_k)$; $Aug_{2k} = t_2(concat(A_{in}, T_{in})_k)$ #Augmentation
5:    $\mu_{2k-1} = f_\mu(f_e(Aug_{2k-1}))$; $\mu_{2k} = f_\mu(f_e(Aug_{2k}))$
6:    $\sigma_{2k-1} = f_\sigma(f_e(Aug_{2k-1}))$; $\sigma_{2k} = f_\sigma(f_e(Aug_{2k}))$
7:    sample l_{2k-1}, l_{2k} from the multivariate gaussian distribution $N(\mu_{2k-1}, \sigma_{2k-1})$, $N(\mu_{2k}, \sigma_{2k})$ respectively.   #Representation
8:    $x_{2k-1} = f_d(l_{2k-1})$; $x_{2k} = f_d(l_{2k})$ #Projection
9:    $A_{out_{2k-1}}, T_{out_{2k-1}} \quad = \quad f_s(x_{2k-1})$; $A_{out_{2k}}, T_{out_{2k}} = f_s(x_{2k})$ #Splitting
10:   $L_1 = w_A \sum_{k=1}^{2N} ln(A_{out_k} + 10^{-9}) \cdot A_{in_{\lfloor \frac{k+1}{2} \rfloor}} + w_T \sum(T_{out} - T_{in_{\lfloor \frac{k+1}{2} \rfloor}})^2$
11:   $L_2 = -\sum_{k=1}^{2N} \frac{1}{2}(1 + ln(\sigma_k) - \mu_k^2 - \sigma_k)$
12:   $L_3 = \frac{1}{2N} \sum_{k=1}^{N}(l_{2k-1,2k} + l_{2k,2k-1})$, where $l_{i,j}$ is defined in Equation 3
13:   **if** in the first epoch and $w_2 = w_3 = 1$ **then**
14:     calculate w_2, w_3 based on the value of L_1, L_2, L_3
15:   $LOSS = L_1 + w_2 L_2 + w_3 L_3$
16:   update networks $f_e, f_\mu, f_\sigma, f_d, f_s$ to minimize LOSS
17: **return** encoding structure f_e, f_μ

2.5 Productive Model

After training, we define the productive function $f_\theta(x) = f_\mu(f_e(x))$, i.e., the mapping parameterized by the encoder layers connected with the μ layers. Therefore, given the feature data $concat(A_{in}, T_{in})$ of a contig, we obtain the representations $f_\mu(f_e(concat(A_{in}, T_{in})))$ by passing the data through the encoder layers and the μ layers. Once we obtain the representations of all the contigs, we cluster them with the common clustering algorithms (*e.g.*, minibatch k-means [19], DBSCAN [20]). We find that the iterative medoid clustering algorithm developed by [10] is the state-of-art clustering algorithm specifically for metagenome binning (Fig. 7 in Appendix). After clustering, contigs in the same cluster are put into the same bin. Moreover, for the multisplit workflow, the contigs in the same bin should also be separated based on their source samples [10].

3 Results

3.1 Datasets and Evaluation Metrics

Datasets. To show the performance of CLMB, we use the benchmarking datasets, which are five synthetic datasets from the CAMI2 Toy Human Microbiome Project Dataset [16]: Airways (10 samples), Gastrointestinal (GI, 10 samples), Oral (10 samples), Skin (10 samples), and Urogenital (Urog, 9 samples)[2]. For each dataset, contigs $< 2,000$ base pairs are discarded. We obtain the abundance data in numpy[3] format from the website of [10][4], which are calculated using $jgi\_summarize\_bam\_contig\_depths$, implemented by [8] on BAM files created with bwa-mem [26] and sorted with samtools [27].

Evaluation Metrics. We adopt the evaluation metrics for taxonomic binning defined in [16] as done in previous work [8,10]. After the bins are obtained, we match each bin with each reference genome. We define the number of nucleotides in the genome covered by contigs from the bin as true positives (TP); the number of nucleotides from other genomes covered by contigs in the bin as the false positives (FP); the number of nucleotides in the genome covered by contigs in the dataset, but not by any contig in the bin as the false negatives (FN). Then, $Precision = \frac{TP}{TP+FP}$ and $Recall = \frac{TP}{TP+FN}$ are calculated. All the CAMI2 datasets have taxonomy files with the definition of strain, species, and genus taxonomic levels.

3.2 CLMB Recovers More Near-Complete Genomes on Most Benchmarking Datasets

We ran CLMB on the five CAMI2 datasets. For each dataset, the augmented data serve as training data, while the original feature data serve as testing data. Therefore, CLMB obtains a specific encoding function f_θ parameterized by VAE for each dataset. In addition, as the data augmentation is performed several times during training, CLMB has a larger data volume for training than the input data volume.

We also benchmarked VAMB [10], MetaBAT2 [7] and Maxbin2 [9] on the five benchmarking datasets for comparison. We evaluated the binning performance by the number of recovered Near-Complete (NC, $recall > 90\%$ and $precision > 95\%$) genomes as the previous works [10,16,28]. Firstly, CLMB reconstructed 4–21 more NC genomes at the strain level over the second-best binners on three of the five benchmarking datasets (Airways, GI, Urog), and equivalent NC strains to VAMB on Skin and Oral datasets (Fig. 2a and Table 2 in Appendix). Secondly, the increased performance of CLMB relative to MetaBAT2 and Maxbin2 is very significant. Moreover, the increased performance of CLMB to VAMB is positively correlated with the difficulty of the CAMI2 datasets (which is defined as the reciprocal of the Shannon entropy of the datasets[5] because higher Shannon entropy indicates more information contained in the dataset and lower difficulty for

---

[2] You can get the whole package data from https://data.cami-challenge.org/participate, or get the contigs and calculated abundance from https://codeocean.com/capsule/1017583/tree/v1.

[3] https://numpy.org.

[4] https://codeocean.com/capsule/1017583/tree/v1.

[5] The Shannon entropy of the five datasets are calculated by [10] on their Supplementary Table 1.

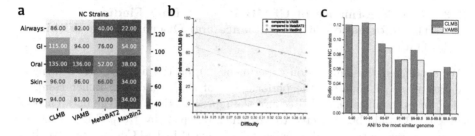

Fig. 2. Performance comparison on benchmarking datasets. a. Number of NC strains recovered from the five benchmarking datasets for CLMB, VAMB, MetaBAT2, and MaxBin2. **b.** The linear fitting and 95% confidence interval of the difficulty of the dataset and the increased number of NC strains recovered by CLMB relative to VAMB (Pearson correlation coefficient = 0.85), MetaBAT2 (Pearson correlation coefficient = −0.77) and MaxBin2 (Pearson correlation coefficient = −0.75). The difficulty is defined as the reciprocal of the Shannon entropy (see Supplementary Table 1 from [10]) of the dataset and is always positive. **c.** The ratio of recovered NC genomes to total reference genomes (which is regarded as ideally recoverable genomes), divided by the ANI to the most similar reference genomes across CAMI2 datasets. CLMB, pink; VAMB, yellow. (Color figure online)

binning.) (Fig. 2b). That indicates that our method indeed resolves the bottleneck of the other methods when the dataset becomes more noisy and difficult. More specifically, CLMB reconstructed more NC strains for most datasets compared to MetaBAT2 and Maxbin2. Compared to VAMB, CLMB reconstructed more NC strains for high-difficulty datasets and approximately equivalent NC strains for low-difficulty datasets. Thirdly, CLMB reconstructed on average 10% more species under any criteria for the GI and Urog datasets, and 8% more species under stricter criteria (e.g., *Recall* > 0.90) for the Airways and Skin datasets. However, if loosening the criterion (e.g., *Recall* > 0.70), CLMB reconstructed 1%–5% fewer species on Airways and Skin datasets than VAMB, which had similar performance to CLMB on the Oral dataset with VAMB 0.5% better across all the criteria except for *Recall* > 0.99 (Table 3 in Appendix). At the genus level, CLMB outperformed VAMB on datasets Airways, GI, Oral, Skin under stricter criteria, but on the contrary under looser criteria. On the Urog dataset, CLMB was the second-best binner, recovering approximately 10% fewer genus than MetaBAT2 (Table 4 in Appendix).

We further mapped the recovered genomes to reference genomes and counted the average nucleotide identity (ANI) between each reference genome. Ideally, all the reference genomes are recovered after the sequencing, assembly, and binning process, which is, however, extremely hard in real-world conditions. For each reference genome, we found the most similar genome and counted the ANI between them. The NC genomes recovered by CLMB can be mapped to 6% of all reference genomes having > 99.9% ANI to the most similar genome (Fig. 2c). Moreover, compared to VAMB, the NC genomes recovered by CLMB were mapped to more reference genomes across all the intervals of ANI except for 99.5%–99.9% ANI.

3.3 The Performance of CLMB Benefits from Finding the Information of Resemblance and Discrimination Within Data

We conducted the data fusion experiment [29] on the five CAMI2 challenging datasets, *i.e.*, comparing the performance of the abundance, k-mer composition, or both concatenated. Because the representation of all the data encoded by CLMB would be projected to 32-dimension space by f_e, f_μ (Fig. 1), we also projected raw data to 32-dimension space using Principal Components Analysis (PCA) [30], termed as 'projected data', to avoid the clustering results affected by different dimensions. We tested the number of NC strains produced by binning with raw data, projected data, and CLMB-encoded data in the data fusion experiment, respectively (Fig. 3).

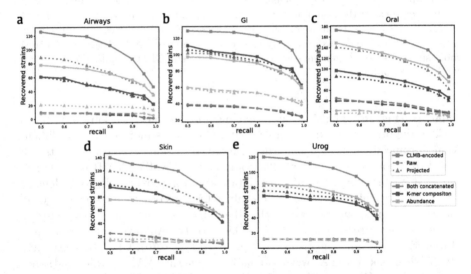

Fig. 3. Results of data fusion experiments. Fusion test of 5 benchmarking datasets for CLMB, precision = 0.95, recall range from 0.5 to 0.99. Color: Abundance (Yellow), k-mer composition (Purple), both concatenated (Green) Linestyle: Raw data (Round), Projected data (triangle), CLMB-encoded data (square). (Color figure online)

On datasets Airways, Oral, Skin, and Urog, the raw data of both concatenated did not achieve better results than the raw data of single abundance or single k-mer composition, but the projected data of both concatenated yielded 5%–700% more genomes than that of single data. This interesting result proved that the dimension of input data did affect the clustering and binning result, and more information contained in the concatenated data was beneficial to the clustering result after eliminating the variation of dimensionality. On dataset GI, the raw data of both concatenated achieved worse results than the raw data of single abundance, but the projected data of both concatenated yielded worse results than the single k-mer composition. This might stem from the information conflict between k-mer composition and abundance. With contrastive learning, the three CLMB-encoded data recovered 3–12 times more NC genomes than the corresponding raw data. Moreover, the CLMB-encoded data of both concatenated and abundance also recovered on average 19% and 189% more than the projected data ones, although the CLMB-encoded data of k-mer composition had similar performance to the projected data. Most importantly, the CLMB-encoded data of both concatenated achieved the best performance across all the datasets, recovering on average 17% more genomes more than the second-best results, no matter what performance the raw data or projected data of both concatenated achieved.

We also visualized the raw data and CLMB-encoded data on dataset Skin, using t-SNE [31] (Fig. 4). Firstly, the CLMB-encoded data of both concatenated appeared to have genomes more clearly separated than any other cases. Figure 4a, 4c, and 4e showed that, more information contained in both concatenated contributed little to the cluster separation, which is similar to the result of the data fusion experiment. However, Fig. 4b, 4d, and 4f showed that, the CLMB-encoded data of both concatenated appeared to have genomes more clearly separated than any other cases. It suggests that CLMB leverages the information within data to achieve better performance.

Furthermore, the performance of CLMB-encoded data of both concatenated was dependent on the number of selected samples (which decided the dimension of the abundance) (Fig. 8 in Appendix). Another experiment tested the effect of different k (2–5) for encoding k-mers composition, and showed that CLMB could k = 4 assist to select the best kmer for or second performance on all the datasets (Fig. 9 in Appendix).

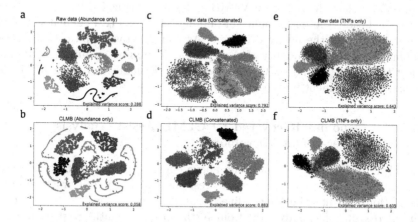

Fig. 4. T-SNE visualiztion of data fusion experiments on Skin dataset. We randomly selected 10 of 15 strains with maximum contigs from the CAMI2 Skin dataset. Each point represents a contig from that strain, and points with same color means originating from same strain (*i.e.*, the same reference genome). **a, b.** Raw data (a) and CLMB-encoded data (b) of abundance. **c, d.** Raw data (c) and CLMB-encoded data (d) of both concatenated. **e, f.** Raw data (e) and CLMB-encoded data (f) of k-mer composition.

3.4   The Performance of the Ensemble Binning Is Improved by Involving CLMB

The ensemble binning refinement method is popular after draft metagenome binning because they combine bins from multiple programs. To show that CLMB is compatible with the ensemble binning tool, we ran MetaWRAP bin-refinement [32,33] on the five CAMI2 challenging datasets by involving CLMB. Because MetaWRAP bin refiner used CheckM [34] to assess the quality of recovered genomes, we here evaluated the performance by the number of recovered high-quality (HQ, *completeness* > 90% and *contamination* < 5%) genomes or middle-quality (MQ, 50% < *completeness* < 90% and *contamination* < 5%) genomes as the previous works [35,36]. The bin refiner of two binners usually outperformed single binner, and the refiner of CLMB and VAMB performed best, recovering 8–22 more HQ genomes and 15–32 more MQ genomes than the second-best method. We also found that the refiner of CLMB and Maxbin2 outperformed that of VAMB and Maxbin2 on four of five datasets (Fig. 5a, b). Moreover, CLMB and VAMB agreed on over a half of the HQ genomes and MQ genomes, but CLMB recovered more unique HQ genomes on average (Fig. 5c, d).

Notice that the CheckM results are not equivalent to the benchmarking results for each binner, which is due to different evaluation methods. We then revisited the benchmarking experiments except for evaluating the performance by the number of recovered HQ genomes and MQ genomes. On datasets GI, Oral, and Urog, CLMB recovered 21–22 more HQ genomes or 6–18 more MQ genomes than VAMB, which had similar performance to CLMB on Airways and better performance than CLMB on Skin (Fig. 5e, f). Impressively, on datasets Airways, GI, Oral, and Urog, single CLMB even recovered on average 15 more HQ genomes than the refiner of VAMB and Maxbin (Fig. 5a, b).

In conclusion, the performance of binning refiner is highly dependent on the performance of all the involved binners. As many metagenomics studies screen the bins based on their quality after metagenome binning for future analysis, we expect that more HQ and MQ genomes can be distinguished using CLMB and the binning refinement methods.

3.5   The Genomes Recovered by CLMB Assist Analysis for Mother-Infant Microbiome

Experiment Datasets. Unlike the above experiments on synthetic datasets, we apply CLMB to real-world data to test the scalability and practicability in this section. We use the longitudinally sampled microbiome of mother-infant pairs across multiple body sites from birth up to 4 months postpartum from [35], which are available at the NCBI Sequence Read Archive (SRA) [37] under BioProject number PRJNA352475 and SRA accession number SRP100409. We select 10 mother-infant pairs with 110 samples and 496342 contigs in total for this experiment.

We ran CLMB on the dataset with default parameters. We recovered 373 (HQ+MQ) genomes, in which there are 24 new-found species consisting of 30

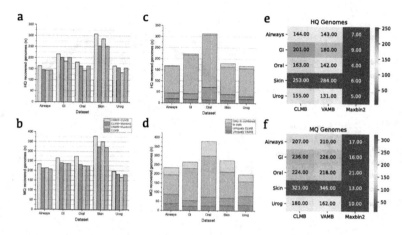

Fig. 5. Quality assessment of genomes recovered by binners. a, b. The number of high-quality (a) and middle-quality (b) genomes obtained using MetaWRAP binning refinement tool. We used the binning result from 1) CLMB and VAMB (light cyan), 2) CLMB and MaxBin2 (purple), and 3) VAMB and MaxBin2 (green). The number of high-quality (a) and middle-quality (b) genomes recovered by a single CLMB (pink) is used for comparison. **c, d.** The source of the results of the MetaWRAP binning refinement tool. We investigated the number of HQ (c) and MQ (d) genomes uniquely from one of the two binners (dark pink, medium pink), found in both binners (light pink), and the number of genomes that were not HQ (c) or MQ (d) in any binner but were regenerated as HQ (c) or MQ (d) in the binning refinement output (lightest pink). **e, f.** Number of HQ (e) and MQ (f) genomes recovered by single CLMB, single VAMB, and single MaxBin2. (Color figure online)

bins. CLMB recover more bins compared to VAMB on this dataset (Fig. 6b). We then reconstructed the phylogeny of all (HQ+MQ) genomes and obtained the unrooted tree [38], which are annotated with the metadata file (Fig. 6a). The new-found species, as annotated, are more from samples of mothers. We also found that the microbiome of the infants shared more species. For example, 11 stool samples from 5 infants share strain *Escherichia coli*, and 8 samples collected from stool and tongue dorsum of 4 infants contain strain *Rothia sp902373285* across all ages. On the contrary, few species are shared among mothers in the tree. Moreover, the range of species reconstructed in mothers' samples overlaps little with the range of species reconstructed in infants' samples. More than half of the bins are recovered from stool samples, probably because of the larger sequencing files obtained from stool samples than those obtained from samples of other sources (human body sites). We then counted the newly exclusive species of the 10 infants. We found that the proportion of exclusive species has largely changed as they grew up (Fig. 6c). At the age of 4 months, the proportions of exclusive species are within a small range, indicating most infants contained 20%–30% exclusive species found in their microbiome.

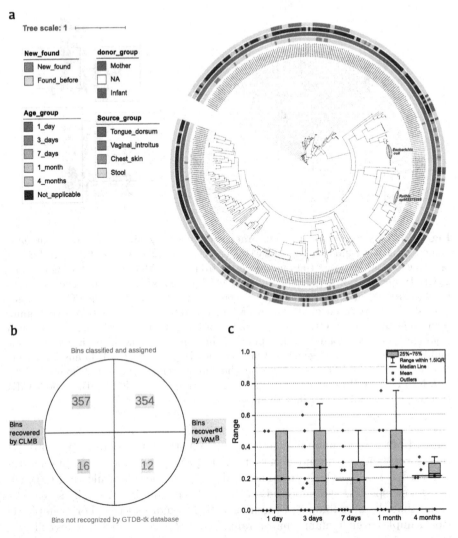

Fig. 6. Metagenomic analysis on mother-infant microbiome. a. Cladogram of species tree of all the 365 bins generated. The annotation rings, from inner to outer: 1) the bins of new-found strains (green) or discovered before (light pink) in [35]; 2) the sample is donated by mother (red) or infant (blue); 3) the age of infant donor, 1 day (onion green), 3 days (dark green), 7 days (olive green), 1 month (cyan) or 4 months (yellow-green). Not applicable (dark) for mother donors; 4) which human body site the sample is collected from, tongue dorsum (pink), vaginal introitus (lighter pink), chest skin (vermeil), or stool (light cyan). The bins classified as strain *Escherichia coli* and strain *Rothia sp902373285* are marked red. **b.** Some bins that are recovered by CLMB/VAMB are classified as strains by GTDB-tk, while the others are unknown strains in GTDB-tk database. **c.** The ratio of exclusive species to the total number of species in infants' microbiome. The samples, which obtain 0 strain, are not considered. (Color figure online)

4   Discussions

Here, by conciously handling the noise occured in metagenome research, we show improvements on both benchmarking datasets and real-world datasets, at similar time cost (Table 1 in Appendix). The improvements, as we have shown, benefit not only from the dimensionality reduction, but also from the model trained by the contrastive learning framework and its robustness to noise. Furthermore, experiments and applications on real-world datasets demonstrate the scalability and practicability of CLMB.

From the algorithm perspective, CLMB can handle the numerical data that potentially contain error [39], which is not limited to metagenome binning. CLMB is promising to handle noise, a significant factor that interferes the data precision. Therefore, we believe that our findings can inspire not only the field of metagenomics [40], but also other related fields, like structural and functional fields [41–44].

5   Appendix

A   Figures

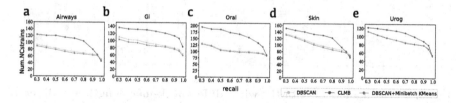

Fig. 7. Performance of different clustering algorithms based on five datasets. Orange: DBSCAN Algorithm. Green: Exclude the outlier using DBSCAN first and cluster the others points using minibatch k-means algorithm. Red: Iterative medoid algorithm, which is developed by [10] and used by CLMB. (Color figure online)

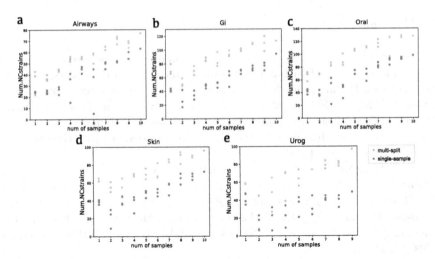

Fig. 8. Performance of CLMB with different samples. For any given number of samples, samples were randomly drawn 3 times and executed independently. For "single-sample", all the samples were run independently. We note that for increasing number of samples, the random subsets chosen is not independent, due to only having 9 (Urog) or 10 (Airways, GI, Skin, Oral) samples in total. Orange: Multi-split workflow of CLMB, Green: Single sample workflow of CLMB. (Color figure online)

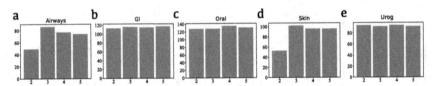

Fig. 9. Performance of CLMB with different k-mer length on different datasets. It is assessed by the number of reconstructed NC strains. The performance varies among the datasets.

B   Tables

Table 1. Number of genomes at the strain level reconstructed with a precision of at least 95%

Dataset	CLMB (GPU)	VAMB (GPU)	MetaBAT2 (CPU)	MaxBin2 (CPU)
Airways	40	35	44	545
Gastrointestinal	24	29	18	342
Oral	38	43	40	852
Skin	37	37	45	950
Urogenital	15	26	11	103
Mother-infant	145	100	N/A	N/A

Table 2. Number of genomes at the strain level reconstructed with a precision of at least 95%

Dataset	Binner	RECALL						
		0.50	0.60	0.70	0.80	0.90	0.95	0.99
CAMI2 Airways	MaxBin2	42	39	38	33	23	17	13
	MetaBAT2	80	72	66	56	40	30	18
	VAMB	125	**123**	**120**	**113**	79	60	41
	CLMB	**126**	121	119	106	**86**	**65**	**46**
CAMI2 GI	MaxBin2	64	63	63	60	53	50	45
	MetaBAT2	99	97	94	87	76	68	58
	VAMB	121	120	118	113	100	91	77
	CLMB	**129**	**128**	**127**	**123**	**115**	**105**	**85**
CAMI2 Oral	MaxBin2	64	61	55	46	39	31	21
	MetaBAT2	88	86	84	79	73	58	38
	VAMB	**181**	**174**	**166**	**152**	**135**	113	81
	CLMB	173	169	164	151	**135**	**114**	**84**
CAMI2 Skin	MaxBin2	56	53	50	46	34	30	27
	MetaBAT2	106	98	93	76	65	53	42
	VAMB	139	**133**	**129**	116	**97**	80	63
	CLMB	**140**	130	126	**119**	96	**81**	**69**
CAMI2 Urog	MaxBin2	37	36	36	35	34	29	26
	MetaBAT2	77	74	71	70	69	61	44
	VAMB	118	114	109	101	89	74	50
	CLMB	**120**	**118**	**111**	**105**	**94**	**83**	**56**

Table 3. Number of genomes at the species level reconstructed with a precision of at least 95%

Dataset	Binner	RECALL						
		0.50	0.60	0.70	0.80	0.90	0.95	0.99
CAMI2 Airways	MaxBin2	41	38	37	32	22	16	12
	MetaBAT2	76	69	63	53	38	28	17
	VAMB	**98**	**97**	**95**	**90**	61	45	27
	CLMB	95	92	91	85	**66**	**47**	**30**
CAMI2 GI	MaxBin2	59	58	58	55	51	48	44
	MetaBAT2	91	89	87	81	74	66	57
	VAMB	89	88	88	85	80	74	63
	CLMB	**101**	**100**	**99**	**96**	**92**	**85**	**71**
CAMI2 Oral	MaxBin2	63	60	54	46	39	31	21
	MetaBAT2	87	85	83	78	72	57	38
	VAMB	**129**	**126**	**124**	**116**	**103**	**84**	58
	CLMB	123	122	119	111	101	83	**59**
CAMI2 Skin	MaxBin2	56	53	50	46	34	30	27
	MetaBAT2	100	92	88	73	63	52	42
	VAMB	107	**103**	**100**	87	69	59	48
	CLMB	**108**	101	99	**94**	**75**	**64**	**56**
CAMI2 Urog	MaxBin2	34	33	33	32	31	26	24
	MetaBAT2	66	64	62	61	60	54	39
	VAMB	69	69	67	64	59	53	39
	CLMB	**74**	**74**	**71**	**68**	**64**	**60**	**43**

Table 4. Number of genomes at the genus level reconstructed with a precision of at least 95%

Dataset	Binner	RECALL						
		0.50	0.60	0.70	0.80	0.90	0.95	0.99
CAMI2 Airways	MaxBin2	30	28	27	23	16	11	9
	MetaBAT2	48	42	38	31	23	16	9
	VAMB	**52**	**51**	**50**	**49**	33	19	8
	CLMB	51	50	49	46	**36**	**23**	**12**
CAMI2 GI	MaxBin2	38	37	37	35	32	31	29
	MetaBAT2	**56**	**54**	**53**	48	42	37	34
	VAMB	47	46	46	45	43	38	34
	CLMB	50	50	50	**49**	**46**	**43**	**40**
CAMI2 Oral	MaxBin2	42	41	40	37	32	25	18
	MetaBAT2	55	54	52	50	47	41	28
	VAMB	**66**	**63**	**63**	**61**	**54**	**47**	34
	CLMB	62	62	61	59	53	45	**37**
CAMI2 Skin	MaxBin2	46	44	41	38	30	27	24
	MetaBAT2	**64**	**61**	**61**	52	46	39	33
	VAMB	58	58	56	50	44	37	31
	CLMB	57	55	55	**54**	**48**	**41**	**36**
CAMI2 Urog	MaxBin2	28	28	28	27	26	23	21
	MetaBAT2	**35**	**34**	**33**	**32**	**32**	**29**	**23**
	VAMB	29	29	29	26	24	22	18
	CLMB	33	33	31	30	29	27	21

C   Methods

In this section, we show the methods and experiments in our research.

C.1   Feature Calculation of TNFs and Abundance

We use the same approach to calculate TNFs and abundance as the previous work [10]. For each contig, we count the frequencies of each tetramer with definite bases, and, to satisfy statistical constraints, project them into a 103-dimensional independent orthonormal space to obtain TNFs [6]. As a result, the TNFs for each contig are a 103-dimensional numerical vector. We also count the number of individual reads mapped to each contig. More specifically, a read mapped to n contigs counts $1/n$ towards each. The read counts are normalized by sequence length and total number of mapped reads, which generates the abundance value in reads per kilobase sequence per million mapped reads (RPKM). The resulted abundance for each contig is a s-dimensional numerical vector, where s is the number of samples. TNFs are normalized by z-scaling each tetranucleotide across the sequences, and abundance are normalized across samples.

C.2   Benchmarking

CLMB and VAMB [10] were run with default parameters with multi-split enabled. MetaBAT2 [8] was run with setting minClsSize = 1 and other parameters as default. MaxBin2 [9] was run with default parameters. The benchmarking results were calculated using benchmark.py script implemented by [10]. The mapping of the recovered genomes to the reference genomes was the intermediate result[6] of benchmark.py script. FastANI [45] with default parameters was used to calculate ANI between the reference genomes. For the binning refinement experiment, we use metaWRAP bin_refinement API [32,33] with parameters –c 50 and –x 10, indicating we keep the genomes qualifying *completeness* > 50% and *contamination* < 10%. The completeness and contamination of the genomes recovered by the bins are calculated using CheckM [34] with default parameters. We use the pipeline integrated in MetaGEM [11] for binning refinement experiment.

C.3   Data Fusion Experiment

We define the feature data as the raw data, and obtained the projected data by projecting the feature data to 32-dimension space using PCA. For the CLMB-encoded data, we obtained them by encoding the feature data to 32-dimension space with the deep contrastive learning framework. We assess the performance of these data by clustering them with the iterative medoid clustering and obtained the benchmarking results. All the experiments on CAMI2 datasets were run with default parameters with multi-split enabled, and the experiments on MetaHIT datasets was run with default parameters with multi-split disabled. For comparison to other clustering methods, we use MiniBatchKMeans (n_clusters = 750, batch_size = 4096, max_iter = 25, init_size = 20000, reassignment_ratio = 0.02) and DBSCAN (eps = 0.35, min_samples = 2) implemented by scikit-learn.

C.4   Binning of the Mother-Infant Transmission Dataset

We downloaded the sequencing datasets of selected mother-infant pairs (marked as 10001, 10002, 10003, 10005, 10006, 10007, 10008, 10009, 10015, 10019) using SRA Toolkit and filtered them based on quality using fastp [46]. Then, we assembled the short sequence reads into contigs using MEGAHIT [47,48] and mapped the reads to the contigs using kallisto [49] in order to speed up this process for large datasets. The coabundance across samples can be subsequently calculated using kallisto quantification algorithm. With the assemblies and coabundances, we ran CLMB with default parameters and multi-split enabled. Then, we splited the fasta file into bins based on the result of clustering using create_fasta.py script. CheckM [34] on lineage specific workflow with default parameters was applied to the resulting bins to calculate the completeness and contamination, and only those with sufficient quality (*completeness* ≥ 50%,

---

[6] The variable **recprecof** in class **Binning**.

contamination ≤ 5%) were considered for further analysis. Then, we use GTDB-tk [38] on for taxonomic assignment of each bins and phylogeny inference. We visualized the tree with iTOL [50].

References

1. Van Dijk, E.L., Auger, H., Jaszczyszyn, Y., Thermes, C.T.: years of next-generation sequencing technology. Trends Genet. **6**, 9 (2014)
2. Tringe, S., Rubin, E.: Metagenomics: DNA sequencing of environmental samples. Nat. Rev. Genet. **6**, 805–814 (2005)
3. Quince, C., Walker, A., Simpson, J., et al.: Shotgun metagenomics, from sampling to analysis. Nat. Biotechnol. **35**, 833–844 (2017)
4. Miller, J.R., Koren, S., Sutton, G.: Assembly algorithms for next-generation sequencing data. Genomics **95**, 315–327 (2010)
5. Alneberg, J., Bjarnason, B., de Bruijn, I., et al.: Binning metagenomic contigs by coverage and composition. Nat. Methods **11**, 1144–1146 (2014)
6. Kislyuk, A., Bhatnagar, S., Dushoff, J., et al.: Unsupervised statistical clustering of environmental shotgun sequences. BMC Bioinform. **10**, 1–16 (2009)
7. Kang, D.D., Froula, J., Egan, R., Wang, Z.: MetaBAT: an efficient tool for accurately reconstructing single genomes from complex microbial communities. PeerJ **3**, e1165 (2015)
8. Kang, D.D., et al.: MetaBAT2: an adaptive binning algorithm for robust and efficient genome reconstruction from metagenome assemblies. PeerJ **7**, e7359 (2019)
9. Wu, Y.-W., Simmons, B.A., Singer, S.W.: MaxBin 2.0: an automated binning algorithm to recover genomes from multiple metagenomic datasets. Bioinformatics **32**, 15 (2016)
10. Nissen, J.N., Johansen, J., Allese, R.L., et al.: Improved metagenome binning and assembly using deep variational autoencoders. Nat. Biotechnol. **39**, 555–560 (2021)
11. Zorrilla, F., Buric, F., Patil, K.R., Zelezniak, A.: metaGEM: reconstruction of genome scale metabolic models directly from metagenomes. Nucleic Acids Res. **49**(21), e126–e126 (2021)
12. van Belkum, A., Burnham, C.D., Rossen, J.W.A., et al.: Innovative and rapid antimicrobial susceptibility testing systems. Nat. Rev. Microbiol. **18**, 299–311 (2020)
13. Fischer-Hwang, I., Ochoa, I., Weissman, T., et al.: Denoising of aligned genomic data. Sci. Rep. 15067 (2019)
14. Hinton, T.C., Kornblith, S., Norouzi, M., Hinton, G.: A simple framework for contrastive learning of visual representations. In: ICML (2020)
15. Han, W., et al.: Self-supervised contrastive learning for integrative single cell RNA-seq data analysis. bioRxiv (2021)
16. Sczyrba, A., Hofmann, P., Belmann, P., et al.: Critical assessment of metagenome interpretation-a benchmark of metagenomics software. Nat. Methods **14**, 1063–1071 (2017)
17. Kingma, D.P., Welling, M.: Auto-encoding variational bayes. Arxiv (2014). https://arxiv.org/abs/1312.6114
18. Rezende, D.J., Mohamed, S., Wierstra, D.: Stochastic backpropagation and approximate inference in deep generative models. Proc. Mach. Learn. Res. 1278–1286 (2014)

19. Sculley, D.: Web-scale k-means clustering. In: Proceedings of 19th International Conference on World Wide Web, pp. 1177–1178 (2010)
20. Ester, M., Kriegel, H.-P., Sander, J., Xu, X.: A density-based algorithm for discovering clusters in large spatial databases with noise. In: KDD 1996 Proceedings (1996)
21. Ioffe, S., Szegedy, C.: Batch normalization: accelerating deep network training by reducing internal covariate shift. Arxiv (2015). https://arxiv.org/abs/1502.03167
22. Hinton, G.E., Srivastava, N., Krizhevsky, A., Sutskever, I., Salakhutdinov, R.R.: Improving neural networks by preventing co-adaptation of feature detectors. Arxiv (2012). https://arxiv.org/pdf/1207.0580.pdf
23. Maas, A.L., Maas, A.L., Hannun, A.Y., Ng, A.Y.: Rectifier nonlinearities improve neural network acoustic models. Arxiv (2013). https://arxiv.org/pdf/1207.0580.pdf
24. Doersch, C.: Tutorial on variational autoencoders (2021). https://arxiv.org/abs/1606.05908
25. Kingma, D.P., Ba, J.L.: Adam: a method for stochastic optimization. Arxiv (2017). https://arxiv.org/abs/1412.6980
26. Li, H., Durbin, R.: Fast and accurate short read alignment with burrows-wheeler transform. Bioinformatics **25**, 1754–1760 (2009)
27. Li, H., et al.: The sequence alignment/map format and samtools. Bioinformatics **25**, 2078–2079 (2009)
28. Bowers, R.M., et al.: Minimum information about a single amplified genome (MISAG) and a metagenome-assembled genome (MIMAG) of bacteria and archaea. Nat. Biotechnol. **35**, 725–731 (2017)
29. Haghighat, M., Abdel-Mottaleb, M., Alhalabi, W.: Discriminant correlation analysis: real-time feature level fusion for multimodal biometric recognition. IEEE Trans. Inf. Forensics Secur. **11**, 1984–1996 (2016)
30. Jolliffe, I.T., Cadima, J.: Principal component analysis: a review and recent developments. Philos. Trans. Ser. A Math. Phys. Eng. Sci. **374**, 20150202 (2016)
31. van der Maaten, L., Hinton, G.: Visualizing data using t-SNE. J. Mach. Learn. Res. **9**, 2579–2605 (2008)
32. Uritskiy, G.V., DiRuggiero, J., Taylor, J.: MetaWRAP-a flexible pipeline for genome-resolved metagenomic data analysis. Microbiome 158 (2018)
33. Song, W.Z., Thomas, T.: Binning_refiner: improving genome bins through the combination of different binning programs. Bioinformatics **33**, 1873–1875 (2017)
34. Parks, D.H., Imelfort, M., Skennerton, C.T., Hugenholtz, P., Tyson, G.W.: CheckM: assessing the quality of microbial genomes recovered from isolates, single cells, and metagenomes. Genome Res. **25**, 1043–1055 (2015)
35. Ferretti, P., et al.: Mother-to-infant microbial transmission from different body sites shapes the developing infant gut microbiome. Cell Host Microbe **24**, 133–145.e5 (2018)
36. Pasolli, E., et al.: Extensive unexplored human microbiome diversity revealed by over 150,000 genomes from metagenomes spanning age, geography, and lifestyle. Cell **176**, 649–662 (2019)
37. Leinonen, R., et al.: The sequence read archive. Nucleic Acids Res. **39**, D19–D21 (2011)
38. Chaumeil, P.-A., Mussig, A.J., Hugenholtz, P., Parks, D.H.: GTDB-Tk: a toolkit to classify genomes with the genome taxonomy database. Bioinformatics **36**, 1925–1927 (2020)

39. Li, Y., et al.: DLBI: deep learning guided Bayesian inference for structure reconstruction of super-resolution fluorescence microscopy. Bioinformatics ISMB **34**(13), i284–i294 (2018)

40. Li, Y., et al.: HMD-ARG: hierarchical multi-task deep learning for annotating antibiotic resistance genes. Microbiome **9**, 1–12 (2021)

41. Li, Y., et al.: Deep learning in bioinformatics: introduction, application, and perspective in the big data era. Methods **166**, 4–21 (2019)

42. Chen, X., Li, Y., Umarov, R., Gao, X., Song, L.: RNA secondary structure prediction by learning unrolled algorithms. In: International Conference on Learning Representations 2020 (2020)

43. Li, H., et al.: Modern deep learning in bioinformatics. J. Mol. Cell Biol. **12**, 823–827 (2020)

44. Wei, J., Chen, S., Zong, L., Gao, X., Li, Y.: Protein-RNA interaction prediction with deep learning: structure matters. arXiv preprint arXiv:2107.12243 (2021)

45. Jain, C., Rodriguez-R, L.M., Phillippy, A.M., et al.: High throughput ANI analysis of 90k prokaryotic genomes reveals clear species boundaries. Nat. Commun. 5114 (2018)

46. Chen, S., Zhou, Y., Chen, Y., Gu. J.: fastp: an ultra-fast all-in-one FASTQ preprocessor. Bioinformatics **34**, i884–i890 (2018)

47. Li, D., Liu, C.-M., Luo, R., Sadakane, K., Lam, T.-W.M.: An ultra-fast single-node solution for large and complex metagenomics assembly via succinct de bruijn graph. Bioinformatics **31**(10), 1674–1676 (2015)

48. Li, D., et al.: MEGAHIT v1.0: a fast and scalable metagenome assembler driven by advanced methodologies and community practices. Methods (2016)

49. Bray, N.L., Pimentel, H., Melsted, P., Pachter, L.: Near-optimal probabilistic RNA-seq quantification. Nat. Biotechnol. **34**, 525–527 (2016)

50. Letunic, I., Bork, P.: Interactive tree of life (iTOL) v5: an online tool for phylogenetic tree display and annotation. Nucleic Acids Res. **49**, W293–W296 (2021)

Unsupervised Cell Functional Annotation for Single-Cell RNA-Seq

Dongshunyi Li[1], Jun Ding[2], and Ziv Bar-Joseph[1,3]($\boxtimes$)

[1] Computational Biology Department, School of Computer Science,
Carnegie Mellon University, Pittsburgh 15213, USA
`zivbj@cs.cmu.edu`
[2] Meakins-Christie Laboratories, Department of Medicine,
McGill University Health Centre, Montreal, Quebec H4A 3J1, Canada
[3] Machine Learning Department, School of Computer Science,
Carnegie Mellon University, Pittsburgh 15213, USA

Introduction. One of the first steps in the analysis of single cell RNA-Sequencing data (scRNA-Seq) is the assignment of cell types. While a number of supervised methods have been developed for this, in most cases such assignment is performed by first clustering cells in low-dimensional space and then manually annotating each cluster using known marker genes or cluster specific differentially expressed genes [1]. Several clustering methods have been developed and used for scRNA-Seq data. However, to date, these methods have only relied on the observed expression data [2–5]. There are several additional complementary datasets that can be used to improve clustering and reduce noise related grouping. Specifically, gene sets [6] have been compiled to characterize many processes, pathways and conditions. While the exact processes or functions that are activated in specific cells or clusters are unknown, we can use these sets to guide the grouping of cells by placing more emphasis on co-expression of genes in known sets when clustering single cell data. Since cells of the same type likely share many of the biological processes, such design can both, improve the identification of good clusters and help in annotating them based on the function of the sets associated with each cluster.

Here we introduce UNIFAN (**Un**supervised Single-cell **F**unctional **An**notation) to simultaneously cluster and annotate cells with known biological processes (including pathways). For each cell, we first infer its gene set activity scores based on the co-expression of genes in known gene sets. Next, UNIFAN clusters cells by using the learned gene set activity scores and a reduced dimension representation of the expression of genes in the cell. The gene set activity scores are used by an "annotator" to guide the clustering such that cells sharing similar biological processes are more likely to be grouped together. This design allows us to use prior knowledge about gene membership to guide the dimension reduction and cluster assignment. Gene sets selected as predictive by the annotator, in turn, provide useful annotations for each cell cluster. To allow the selection of marker genes, we also added a set of most variable genes selected using Seuratv3 [3] as features of the annotator. See [7] for the specific loss

I. Pe'er (Ed.): RECOMB 2022, LNBI 13278, pp. 349–352, 2022.
https://doi.org/10.1007/978-3-031-04749-7_24

functions we use, the model training process and how hyperparameter values are selected.

Results. We first evaluated if UNIFAN can accurately cluster cells and reveal key pathways and cellular functions activated in cells assigned to different clusters. For this, we used a scRNA-Seq dataset [8], composed of 25,185 cells and 19,404 genes. UNIFAN clusters correspond well to the manually annotated cell types (ARI: 0.81, NMI: 0.77). To annotate cell clusters, we examined the coefficients assigned by the "annotator" to different gene sets for each cluster. We found that UNIFAN correctly assigns gene sets based on the type of cells. For cluster 0, the set "GOBP POSITIVE REGULATION OF T CELL RECEPTOR SIGNALING PATHWAY" is assigned a large weight and this cluster is annotated as CD4+ T cells in the original paper. For cluster 5 (which mainly contains CD8+ T cells), one of the top scoring sets is "REACTOME NEF MEDIATED CD8 DOWN REGULATION". Cluster 3 and 6 correspond to classical monocyte (cMonocyte) and non-classical monocyte (ncMonocyte), respectively. While UNIFAN assigns processes related to "antigen presentation" and "inflammation" to both clusters, the process related to wound healing "GOBP REGULATION OF INFLAMMATORY RESPONSE TO WOUNDING" only appears in cluster 6. One of the main differences between ncMonocyte and cMonocyte is their role in wound healing [9] and so such assignment helps with correctly annotating these clusters. In addition to the gene sets, we also evaluated genes highly weighted by the annotator by comparing them to known cell type marker sets [6], which are not used in model learning. The most enriched cell type marker sets for each cluster correspond well to the true labels, indicating that UNIFAN can indeed identify the marker genes for each cell type (cluster). We observed similar results for the other datasets as shown in [7].

We also compared UNIFAN's clustering performance with prior methods proposed for clustering scRNA-Seq data using several datasets. The number of cells in these datasets ranges from 366 (Aorta in Tabula Muris [10]) to 96,282 ("Atlas lung" dataset [11]) and so they can provide a good representation of current scRNA-Seq datasets. The methods we compared to included two graph-based methods Leiden clustering [2] and Seuratv3 [3], a kernel-based method SIMLR [4] and a deep-learning based method DESC [5]. For each dataset, we ran each method ten times using different initializations. The results show that, for all datasets, UNIFAN outperforms all other methods (e.g., average ARI of UNIFAN and the best performing prior method on "HuBMAP Spleen": UNIFAN-0.75, DESC-0.31; on "Tabula Muris": UNIFAN-0.70, SIMLR-0.53). The large improvement may result from the ability of UNIFAN's to focus on the more relevant sets of co-expressed genes rather than on co-expression that may results from noise. We observe the same results when using other metrics and we also evaluate the different parts of UNIFAN to determine which input and processing contributes the most to its success. See [7] for details.

Discussion. We presented UNIFAN which improves both clustering and cluster annotations by using a large collection of gene sets [6]. UNIFAN infers gene set activity scores and uses them to regularize the clustering of cells. Such design improves the ability to identify biologically meaningful co-expressed genes and to use these to group cells. In addition to leading to improved clustering, UNIFAN also assigns a subset of the gene sets to clusters which can help characterize their cell types.

We compared UNIFAN to several popular methods for clustering scRNA-Seq data using datasets spanning a large number of organs from both human and mouse. As we show, UNIFAN consistently outperforms other methods across these datasets. We also analyzed the gene sets selected by UNIFAN for various clusters and demonstrated that they match well with the known cell types.

Analysis of UNIFAN identified the annotator and the gene sets and genes it uses as the main sources for the improvement. The fact that adding variable genes as input improves performance is likely the result of the fact that current gene sets, while very useful, are incomplete. We may still be missing from current collections sets of genes characterizing some less known biological processes. In such cases, the selected genes capture groupings that are missed by the known gene sets.

Code Availability: https://github.com/doraadong/UNIFAN.

Full-Text Preprint: https://www.biorxiv.org/content/10.1101/2021.11.20.469410v1.

References

1. Abdelaal, T., et al.: A comparison of automatic cell identification methods for single-cell RNA sequencing data. Genome Biol. **20**, 1–19 (2019). https://doi.org/10.1186/s13059-019-1795-z
2. Traag, V.A., Waltman, L., Van Eck, N.J.: From Louvain to leiden: guaranteeing well-connected communities. Sci. Rep. **9**, 1–12 (2019). https://doi.org/10.1038/s41598-019-41695-z
3. Stuart, T., et al.: Comprehensive integration of single-cell data. Cell **177**, 1888–1902 (2019). https://doi.org/10.1016/j.cell.2019.05.031
4. Wang, B., Zhu, J., Pierson, E., Ramazzotti, D., Batzoglou, S.: Visualization and analysis of single-cell RNA-SEQ data by kernel-based similarity learning. Nat. Methods **14**, 414–416 (2017). https://doi.org/10.1038/nmeth.4207
5. Li, X., et al.: Deep learning enables accurate clustering with batch effect removal in single-cell RNA-SEQ analysis. Nat. Commun. **11**, 1–14 (2020). https://doi.org/10.1038/s41467-020-15851-3
6. Subramanian, A., et al.: Gene set enrichment analysis: a knowledge-based approach for interpreting genome-wide expression profiles. Proc. Natl. Acad. Sci. USA **102**, 15545–15550 (2005). https://www.pnas.org/content/102/43/15545. https://www.pnas.org/content/102/43/15545.full.pdf
7. Li, D., Ding, J., Bar-Joseph, Z.: Unsupervised cell functional annotation for single-cell RNA-SEQ. bioRxiv (2021). https://www.biorxiv.org/content/early/2021/11/21/2021.11.20.469410. https://www.biorxiv.org/content/early/2021/11/21/2021.11.20.469410.full.pdf
8. Van Der Wijst, M.G., et al.: Single-cell RNA sequencing identifies celltype-specific CIS-EQTLS and co-expression QTLS. Nat. Genet **50**, 493–497 (2018). https://doi.org/10.1038/s41588-018-0089-9

9. Schmidl, C., et al.: Transcription and enhancer profiling in human monocyte subsets. Blood **123**, e90–e99 (2014). https://doi.org/10.1182/blood-2013-02-484188

10. Consortium, T.M., et al.: Single-cell transcriptomics of 20 mouse organs creates a tabula muris. Nature **562**, 367–372 (2018). https://doi.org/10.1038/s41586-018-0590-4

11. Adams, T.S., et al.: Single-cell RNA-seq reveals ectopic and aberrant lung-resident cell populations in idiopathic pulmonary fibrosis. Sci. Adv. **6**, eaba1983 (2020). https://www.science.org/doi/abs/10.1126/sciadv.aba1983. https://www.science.org/doi/pdf/10.1126/sciadv.aba1983

A Novel Matrix Factorization Model for Interpreting Single-Cell Gene Expression from Biologically Heterogeneous Data

Kun Qian[1], Shiwei Fu[2], Hongwei Li[1], and Wei Vivian Li[2]([✉]) [iD]

[1] School of Mathematics and Physics, China University of Geosciences,
Wuhan, Hubei 430074, China
[2] Department of Biostatistics and Epidemiology, Rutgers,
The State University of New Jersey, Piscataway 08854, USA
vivian.li@rutgers.edu

Single-cell RNA sequencing (scRNA-seq) technologies enable gene expression measurement at a single-cell resolution, and have opened a new frontier to understand animal development, physiology, and disease-associated molecular mechanisms [1,2]. Rapid advances of scRNA-seq technologies have resulted in the generation of large-scale single-cell gene expression datasets from different platforms in different laboratories, using samples that span a broad range of species, tissue types, and experimental conditions. The increasing number of scRNA-seq datasets emphasizes the need for integrative biological analysis to help assess and interpret similarities and differences between single-cell samples and to obtain in-depth insights into the underlying biological systems. A fundamental goal in integrative scRNA-seq data analysis is to jointly define cell clusters, obtain their functional interpretation and annotation, and identify differentially activated biological pathways in distinct cell types and biological conditions. However, a key challenge for achieving this goal is the heterogeneity present in single-cell gene expression data. As expression data from different sources are associated with various types of technical effects [3], expression patterns of biological interest need to be discerned from cell-specific and sample-specific effects in order to compare single-cell transcriptomes across samples and biological contexts. In addition to technical variability, genuine cellular heterogeneity is present in different cell types and cell states with distinct behaviors and functions, and in response to different perturbations.

To help remove the batch effects emerging from scRNA-seq data generated by different sequencing platforms or library-preparation protocols, several batch correction methods, including mnnCorrect, BBKNN, and BEER, have been developed. However, batch correction methods assume that the differences between the single-cell samples are purely technical and non-biological, and thus are not appropriate for analyzing biologically different scRNA-seq datasets, such as tissue biopsy data from different patients or data of the same tissue type from related species. In practice, there are multiple integration methods that have been used to analyze single-cell gene expression data from biologically heterogeneous sources. For example, Seurat matches cell states across samples by identifying the so-called anchor cells in a lower-dimensional space constructed

I. Pe'er (Ed.): RECOMB 2022, LNBI 13278, pp. 353–355, 2022.
https://doi.org/10.1007/978-3-031-04749-7_25

with canonical correlation analysis. Similarly, Scanorama matches cell clusters by identifying mutual nearest neighbors in a lower-dimensional space constructed with randomized singular value decomposition. scMerge performs clustering in each sample, matches clusters across samples, and then uses control genes to correct for inter-sample variation. In addition, LIGER identifies both shared and dataset-specific metagene factors to enable integration of multiple single-cell samples.

Even though the above methods have been shown useful in batch-effect removal and integrative analysis of multiple single-cell samples, they do not account for the situation where heterogeneous samples come from distinct biological conditions (e.g., different experimental groups or disease phases), and thus may compromise the results. To address this challenge, we propose a novel method named scINSIGHT to jointly model and interpret gene expression patterns in single-cell samples from biologically heterogeneous sources. scINSIGHT uses a new model based on non-negative matrix factorization (NMF) [4] to decompose gene expression patterns of distinct cell types and biological conditions. Compared with existing tools, scINSIGHT has the following advantages: (1) it explicitly models coordinated gene expression patterns that are common among or unique to biological conditions, enabling the decomposition of common and condition-specific gene modules from high-dimensional gene expression data; (2) it achieves precise identification of cell populations across single-cell samples, using common gene modules that capture cellular identities; (3) it enables efficient comparison between samples and biological conditions based on cellular compositions and module expression; (4) it discovers sparse and directly interpretable module expression patterns to assist functional annotation. We evaluated the performance of scINSIGHT in both simulation and real data studies, both of which demonstrated its accuracy and effectiveness for interpreting single cell gene expression from biologically heterogeneous data.

We benchmarked the performance of scINSIGHT in both simulation and real data studies, in comparison with analysis without integration or with six alternative integration methods. Using the ground truth information in simulation as a reference, we confirmed scINSIGHT's ability to accurately decompose common and condition-specific gene modules, and to precisely identify cellular identities based on the inferred expression of common gene modules. In the three real data applications, scINSIGHT repeatedly demonstrated its effectiveness to analyze, compare, and interpret single-cell gene expression data across samples and biological conditions. Based on its identified cell clusters and decomposed gene modules, scINSIGHT is able to discover T cell states associated with response to immunotherapy in melanoma patients, B cell types associated with disease phase of COVID-19 patients, and dermal cell populations for murine skin wound healing. In addition, scINSIGHT consistently showed higher accuracy and interpretability than the other methods in the above real data studies.

The scINSIGHT R package is freely available at https://github.com/Vivianstats/scINSIGHT and https://cran.r-project.org/web/packages/scINSIGHT/index.html. The full article is available at https://doi.org/10.1186/s13059-022-02649-3.

References

1. Luecken, M.D., Theis, F.J.: Current best practices in single-cell RNA-SEQ analysis: a tutorial. Mol. Syst. Biol. **15**(6), e8746 (2019)
2. Potter, S.S.: Single-cell RNA sequencing for the study of development, physiology and disease. Nat. Rev. Nephrol. **14**(8), 479–492, e8746 (2018)
3. Brennecke, P., et al.: Accounting for technical noise in single-cell RNA-SEQ experiments. Nat. Methods **10**(11), 1093 (2013)
4. Lee, D.D., Seung, H.S.: Learning the parts of objects by non-negative matrix factorization. Nature **401**(6755), 788–791 (1999)

Tractable and Expressive Generative Models of Genetic Variation Data

Meihua Dang[1][(✉)], Anji Liu[1], Xinzhu Wei[1], Sriram Sankararaman[1,2,3], and Guy Van den Broeck[1]

[1] Department of Computer Science, UCLA, Los Angeles, USA
{mhdang,liuanji,aprilwei,sriram,guyvdb}@cs.ucla.edu
[2] Department of Human Genetics, School of Medicine, UCLA, Los Angeles, USA
[3] Department of Computational Medicine, School of Medicine, UCLA, Los Angeles, USA

Generative models of genetic sequence data play a central role in population genomics. By modeling dependencies across individuals and sites, these models have empowered genomic analyses such as genotype imputation, haplotype phasing, and ancestry inference. Such models also form the basis for programs that simulate artificial genomes (AGs) that, in turn, have played a critical role in testing evolutionary hypothesis, inferring population genetic models, validating empirical results, and benchmarking methods. The ability to accurately and efficiently simulate AGs has been important. Classical probabilistic models based on hidden Markov model (HMM) are tractable in computing likelihoods and thus widely applied in genotype imputation, but they are not accurate enough in modeling dependencies. While recently popularized deep learning methods such as deep generative adversarial networks (GANs), variational autoencoders (VAEs), and restricted Boltzmann machines (RBMs) are more expressive than HMMs, they are limited in computing exact probabilistic likelihoods and challenging to train.

We propose a class of probabilistic models that can both give us those tractability advantages and keep high expressiveness. To model the distribution over a sequence of variants, we propose a class of latent variable models where each hidden variable is associated with a SNP and the hidden variables are connected via a tree-structured graphical model. This model, termed the *hidden Chow-Liu tree* (HCLT), generalizes previously proposed HMMs. Although HMMs also associate each hidden random variable with a SNP, the hidden variables are related by a chain (a special type of tree) with the restriction that the only edges are present between consecutive SNPs along the genome. By allowing for more general tree structures, the HCLT model can potentially capture long-range correlations or linkage-disequilibrium (LD) among SNPs. While the HCLT model is more expressive than HMMs, it is unclear if such a model can be efficiently learned from data. A second contribution of our work is an affirmative answer to this question by representing HCLTs as Probabilistic Circuits (PCs), a large class of probabilistic models encoded using circuit representations. PCs have been shown to permit tractable inference tasks (e.g., marginal likelihood

S. Sankararaman and G. Van den Broeck—Equal contribution.

computation) which are beyond the reach of most deep generative models. The representation of HCLTs as PCs enables us to leverage recent advances in deep learning such as stochastic learning algorithms and the use of GPUs to enable efficient parameter estimation and inference. HCLT can scale to around 10,000 SNPs and 5,000 genotypes and learning converges in less than 2 h, and single SNP imputation for all sites on such model takes around 5 s. We also leverage the framework of PCs to explore more restrictive models including Markov models.

Finally, we perform extensive experiments to show that HCLTs generate more accurate AGs relative to more restrictive models (fully factorized models and Markov models) suggesting that the structure encoded by the HCLT captures dependencies in genetic variation data. More interestingly, we find that HCLTs as well as deep generative models (GANs and RBMs) preserve LD structure among SNPs. When trained on a subset of individuals from the 1000 Genomes Project (1KGP) across 805 SNPs that are distributed across the genome (and chosen to capture global population structure) as well as a second dataset of 10K SNPs from a contiguous region on chromosome 15, HCLTs greatly improved over existing methods. Compared to HMMs, averaged log-likelihoods of HCLTs improved from −438 to −389 on the 805 SNPs setting and from −633 to −357 on 10K SNPs setting, while results are evaluated on a distinct set of individuals not used in training. We also evaluate the AGs generated by different models by comparing the PCA plots, allele frequencies, and linkage disequilibrium (LD) patterns and observe that the AGs generated by the HCLTs are substantially closer to the patterns observed in real data. In comparison to the next-best method, GANs, the Wasserstein 2D distances between the PCA representations of real versus generated individuals are 0.0015 with a 62.5% improvement and 0.0029 with a 55.4% improvement on 805 and 10K data respectively. The R-squared correlations between real and generated LDs are 0.99 and 0.95, while RBMs achieve 0.98 and 0.95 respectively. Our results suggest that the increased expressivity of HCLTs leads to more accurate models of genetic variation. Furthermore, recent advances in learning and inference enabled by PCs allows us to fully exploit this increased capacity.

CONCERT: Genome-Wide Prediction of Sequence Elements That Modulate DNA Replication Timing

Yang Yang, Yuchuan Wang, Yang Zhang, and Jian Ma[✉]

Computational Biology Department, School of Computer Science,
Carnegie Mellon University, Pittsburgh, USA
jianma@cs.cmu.edu

Proper control of the replication timing (RT) program is of vital importance to maintain genome integrity. However, the genome-wide sequence determinants regulating RT remain mostly unclear. A major algorithmic challenge is to delineate a series of potential sequence determinants in shaping the RT programs over large-scale genomic domains.

Here, we develop a new computational method, named CONCERT, to simultaneously predict RT from sequence features and identify genomic sequence elements that modulate RT in a genome-wide manner. As shown in Fig. 1, CONCERT integrates two functionally cooperative modules, a selector, which performs importance estimation-based sampling of the genomic sequences to detect predictive elements, and a predictor, which incorporates the bi-directional recurrent neural networks and the self-attention mechanism to perform selective learning of long-range spatial dependencies across genomic loci. We apply CONCERT to predict RT in mouse embryonic stem cells (mESCs) and multiple human cell types with high accuracy. In particular, each of the five early replication control elements (ERCEs) in mESCs that were experimentally validated through CRISPR-mediated deletions in a recent study by Sima et al. *Cell* (2019) can be reliably identified by our method. Furthermore, by applying to multiple human cell types, CONCERT reveals conserved and cell type-specific sequence elements that may play key roles in RT regulation and maintaining nuclear organization. The identified important genomic loci show novel connections with different types of genomic and epigenomic features. Notably, CONCERT also identifies sequence elements that may harbor under-explored roles for RT regulation. The results suggest that the dependencies between RT and DNA sequences are likely to only exist for a limited number of genomic loci, with variations across different cell types. Our new method shows the potential for prioritizing experimental characterizations of possible sequence determinants of RT.

Together, CONCERT is a generic interpretable machine learning framework for predicting large-scale functional genomic profiles based on sequence features and provides new insights into the potential sequence determinants of the RT program.

Link to the bioRxiv preprint: https://www.biorxiv.org/content/10.1101/2022.04.21.488684v1.

Y. Wang and Y. Zhang—These two authors contributed equally.

© The Author(s), under exclusive license to Springer Nature Switzerland AG 2022
I. Pe'er (Ed.): RECOMB 2022, LNBI 13278, pp. 358–359, 2022.
https://doi.org/10.1007/978-3-031-04749-7_27

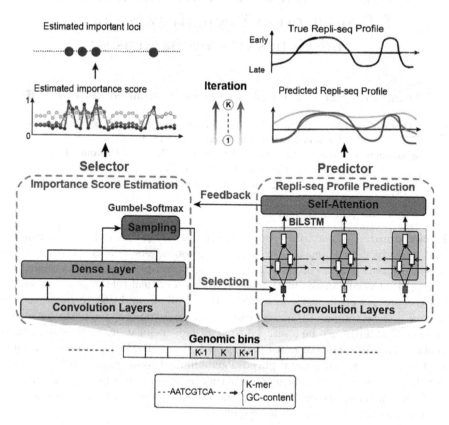

Fig. 1. Overview of the CONCERT model. There are two connected primary functional cooperative modules, the selector and the predictor, which are trained jointly within one framework in modeling long-range spatial dependencies across genomic loci, detecting predictive genomic loci, and learning context-aware feature presentations of the genomic sequences. The input to both the selector and the predictor contains DNA sequence features of the genomic loci within each context window, including K-mer frequency-based features with dimension reduction and GC profile-based features. The output includes both predicted genomic signals and locus-wise estimated importance scores. The estimated scores can be further processed by filtering and local peak detection to delineate predictive sequence elements that are important for the genomic signal prediction. The selector uses the Gumbel-Softmax trick to perform importance estimation-based genomic loci sampling based on the sequence features. The normalized importance scores of the genomic loci estimated by the selector are shared with the predictor. The predictor is mainly constructed with the convolution layers, the BiLSTM (bi-directional long-short term memory neural networks) layer, and the self-attention layer. The predictor integrates the sequence features and the estimated locus-wise importance scores as input to perform spatial dependency learning and RT profile prediction across the genomic loci.

CORSID Enables *de novo* Identification of Transcription Regulatory Sequences and Genes in Coronaviruses

Chuanyi Zhang[1], Palash Sashittal[2], and Mohammed El-Kebir[2]

[1] Department of Electrical and Computer Engineering, University of Illinois at Urbana-Champaign, Urbana, IL 61801, USA
[2] Department of Computer Science, University of Illinois at Urbana-Champaign, Urbana, IL 61801, USA
melkebir@illinois.edu

Background. Coronaviruses are comprised of a single-stranded RNA genome that is ready to be translated by the host ribosome. As their genomes are comprised of many genes, coronaviruses rely on transcription regulatory sequences (TRSs) that occur upstream of genes and play a critical role in gene expression via the process of discontinuous transcription [3]. In addition to being crucial for our understanding of the regulation and expression of coronavirus genes, we propose that TRSs can be leveraged to identify gene locations in the coronavirus genome. However, current motif finding tools (e.g., MEME [1]) are not designed for finding TRSs, and general-purpose gene-finding tools (e.g., Prodigal [2]) are not designed to leverage the genomic structure of coronaviruses, specifically the TRS sites located upstream of the genes in the genome, nor are they able to directly identify these regulatory sequences.

Methods. Firstly, we introduce the TRS IDENTIFICATION (TRS-ID) problem of identifying TRS given prescribed gene locations. Genomes of coronaviruses can be split at the start of the first open reading frame (ORF) into two subsequences: the leader region and the body region, containing TRS-L and TRS-Bs respectively. To find TRSs in these regions, we align all candidate regions preceding each gene and the leader region, and find the best *TRS alignment*. The TRS alignment is a special multiple sequence alignment that contains no gaps in the aligned TRS-L and no internal gaps in other aligned TRS-Bs for genes, as template switching occurs due to complementary base pairing between them. The total matching score between each TRS-B $\mathbf{a}_1, \ldots, \mathbf{a}_n$ and the TRS-L $\mathbf{a}_0$ is the score $s(A)$ of the TRS alignment $A = [\mathbf{a}_0, \ldots, \mathbf{a}_n]^\top$, and the largest contiguous subsequence $\mathbf{c}(A)$ of the leader sequence in the TRS alignment is referred to as the *core sequence*. We introduce CORSID-A, a polynomial time algorithm.

Problem 1 (TRS IDENTIFICATION(TRS-ID)). Given sequences $\mathbf{w}_0, \ldots, \mathbf{w}_n$, core-sequence length $\omega > 0$ and score threshold $\tau > 0$, find a TRS alignment

---

C. Zhang and P. Sashittal—Joint first authorship.

© The Author(s), under exclusive license to Springer Nature Switzerland AG 2022
I. Pe'er (Ed.): RECOMB 2022, LNBI 13278, pp. 360–362, 2022.
https://doi.org/10.1007/978-3-031-04749-7_28

$A = [\mathbf{a}_0, \ldots, \mathbf{a}_n]^\top$ such that (i) $\mathbf{a}_i$ corresponds to a subsequence in $\mathbf{w}_i$ for all $i \in \{0, \ldots, n\}$, (ii) the core sequence $\mathbf{c}(A)$ has length at least ω, (iii) the minimum score $s_{\min}(A)$ is at least τ, and (iv) the alignment has maximum score $s(A)$.

We also formulate the TRS AND GENE IDENTIFICATION (TRS-GENE-ID) problem of simultaneously identifying TRS sites and gene locations in unannotated coronavirus genomes. Without prescribed genes, ORFs split the genome into many intervals. By associating putative TRS-Bs in each interval with the immediate downstream ORF, we define the induced gene set of a TRS alignment. Using the fact that genomes of coronaviruses are highly compact, we require the induced gene set to be non-overlapping and cover most of the genome. This corresponds to finding a maximum weight independent set on the interval graph of all ORFs, which we do in polynomial time using CORSID.

Results. We show that CORSID-A outperforms existing motif-based methods in identifying TRS sites in coronaviruses (Fig. 1b). CORSID outperforms state-of-the-art gene finding methods in finding genes in coronavirus genomes, achieving a higher precision and recall (Fig. 1b). CORSID is the first method to perform accurate and simultaneous identification of TRS sites and genes in coronavirus genomes without the use of any prior information.

Preprint: https://doi.org/10.1101/2021.11.10.468129

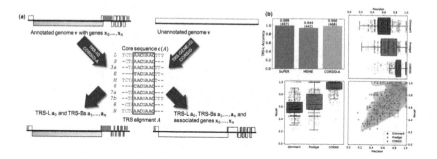

Fig. 1. (a) An example of the TRS alignment and both CORSID-A (left) and CORSID (right) use it to identify TRS and additionally genes. (b) Top left: CORSID-A finds TRS-L more accurately than other motif finding methods. Top right and lower left: CORSID achieves higher precision and recall than other gene identifiers.

References

1. Bailey, T.L., et al.: MEME SUITE: tools for motif discovery and searching. Nucleic Acids Res. **37**(Suppl. 2), W202–W208 (2009)
2. Hyatt, D., Chen, G.L., LoCascio, P.F., Land, M.L., Larimer, F.W., Hauser, L.J.: Prodigal: prokaryotic gene recognition and translation initiation site identification. BMC Bioinform. **11**(1), 1–11 (2010)
3. Sola, I., Moreno, J.L., Zúñiga, S., Alonso, S., Enjuanes, L.: Role of nucleotides immediately flanking the transcription-regulating sequence core in coronavirus subgenomic mRNA synthesis. J. Virol. **79**(4), 2506–2516 (2005)

Learning Probabilistic Protein-DNA Recognition Codes from DNA-Binding Specificities Using Structural Mappings

Joshua L. Wetzel[1], Kaiqian Zhang[1], and Mona Singh[1,2(✉)]

[1] Lewis-Sigler Institute for Integrative Genomics, Princeton University, Princeton, USA
[2] Department of Computer Science, Princeton University, Princeton, USA
mona@cs.princeton.edu

Summary. Knowledge of how proteins interact with DNA is essential for understanding gene regulation. While DNA-binding specificities for thousands of transcription factors (TFs) have been determined, the specific amino acid-base interactions comprising their structural interfaces are largely unknown. This lack of resolution hampers attempts to leverage these data in order to predict specificities for uncharacterized TFs or TFs mutated in disease. Here we introduce rCLAMPS (Recognition Code Learning via Automated Mapping of Protein-DNA Structural interfaces), a probabilistic approach that uses DNA-binding specificities for TFs from the same structural family to simultaneously infer both which nucleotide positions are contacted by particular amino acids within the TF as well as a recognition code that relates each base-contacting amino acid to nucleotide preferences at the DNA positions it contacts. We apply rCLAMPS to homeodomains, the second largest family of TFs in metazoans.

Methods. rCLAMPS takes as input a corpus of DNA-binding specificities (represented as position weight matrices, or PWMs) for a set of proteins from the same DNA-binding family, along with co-complex structural data for that family. Prior to running the procedure, analogous positions across the proteins are known; however, the positions within the PWM that are contacted by these amino acids are not known. rCLAMPS uses a Gibbs sampling approach to simultaneously infer which positions within each PWM are contacted by these amino acids and a set of pairwise amino acid-to-base contact energy parameters that combine linearly to form a protein-DNA recognition code for the protein family, as detailed in Fig. 1.

Results. We apply rCLAMPS to a diverse set of 763 naturally occurring and synthetic homeodomain proteins along with their DNA-binding specificities. First, we show that the linear recognition code it learns can predict *de novo* DNA-binding specificities for TFs with accuracy comparable to that of state-of-the-art combinatorial predictors based on random forests, inferring over 91% of PWM columns correctly on held out TFs. Next, we show that the inferred

I. Pe'er (Ed.): RECOMB 2022, LNBI 13278, pp. 363–365, 2022.
https://doi.org/10.1007/978-3-031-04749-7_29

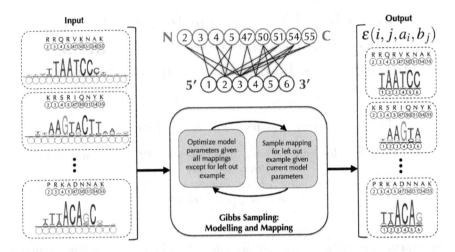

Fig. 1. Overview of rCLAMPS. (Top, middle) Our approach first analyzes protein-DNA co-complex structural data for a TF family to determine commonly observed pairwise contacts between positions in the protein (orange circles) and positions within DNA (blue circles) that together comprise a structural interface or "canonical" contact map. Here we show such a contact map for the homeodomain TF family, with protein positions corresponding to match states in Pfam homeodomain model PF00046. **(Left)** Given a set of TFs and their corresponding DNA-binding specificities as PWMs, the positions (and amino acids) within each TF that interact with DNA are known (orange circles and amino acids above), but initially the positions within the PWMs that are contacted by these amino acids are not known (dotted blue circles). **(Middle, bottom)** We use a Gibbs sampling approach to map the PWM positions to DNA positions within the contact map wherein base preferences at each nucleotide position are described in terms of additive amino acid-to-base contacts energies. **(Right)** After Gibbs sampling is complete, we have a mapping of each TF-PWM pair to the TF family contact map, along with a linear recognition code for the TF family that consists of pairwise energy estimates for each amino-to-base pairing in each of the (i, j) amino acid-to-nucleotide position pairs in the contact map. (Color figure online)

amino acid-nucleotide contacts allow for accurate transfer of specificity information from wildtype to mutant TFs at single base position resolution, enabling improved inferences of DNA-binding specificities for mutant TFs beyond that of *de novo* prediction. Finally, we show the set of contacts learned implies a multiple PWM alignment that is in better agreement with ground truth than that produced by a state-of-the-art multiple PWM alignment method that does not consider protein sequence information.

Conclusion. We describe a novel probabilistic framework rCLAMPS that enables fully automated analyses of large compendia of TF DNA-binding specificities to jointly discover mappings of underlying sets of amino acid-to-base contacts and structure-aware TF family-wide recognition codes. Our approach is an important step towards automatically uncovering the determinants of protein-DNA specificity from large compendia of DNA-binding specificities, and inferring the altered functionalities of TFs mutated in disease. Software implementing rCLAMPS is freely available at https://github.com/jlwetzel-slab/rCLAMPS. A complete manuscript describing this work can be accessed at: https://www.biorxiv.org/content/10.1101/2022.01.31.477772v1.

Uncertainty Quantification Using Subsampling for Assembly-Free Estimates of Genomic Distance and Phylogenetic Relationships

Eleonora Rachtman[1], Shahab Sarmashghi[2], Vineet Bafna[3],
and Siavash Mirarab[2(✉)]

[1] Bioinformatics and Systems Biology Graduate Program, UC San Diego,
La Jolla, CA, USA
[2] Department of Electrical and Computer Engineering, UC San Diego,
La Jolla, CA, USA
smirarab@ucsd.edu
[3] Department of Computer Science and Engineering, UC San Diego,
La Jolla, CA, USA

Keywords: Genomic distance · Uncertainty quantification · Genome
skimming · Phylogenetic uncertainty · Data subsampling ·
Assembly-free and alignment-free distance calculation · k-mer-based
distance calculation

Extended Abstract

Computing the distance between two genomes, often without using alignments or even access to assembled sequences, is fundamental to many downstream analyses. For example, biodiversity research increasingly relies on low-coverage whole-genome sequencing to identify samples and their evolutionary relationships using assembly-free and alignment-free methods that estimate genomic distances from these genome skims [1]. However, the use of these alignment-free methods, including in the fast-growing field of genome skimming [2,5], is hampered by a major methodological gap. While accurate methods (many k-mer-based) for assembly-free distance calculation exist, robust methods for measuring the uncertainty of estimated distances and downstream analyses such as phylogenetic inference have not been sufficiently developed.

Our results show that bootstrapping (i.e., *re*sampling – sampling with replacement), which is the standard non-parametric method of measuring phylogenetic uncertainty, is not a suitable solution for k-mer-based distance estimators because resampling data can violate the fundamental assumptions of these methods. Building on a rich but underutilized literature in statistics [3], we propose using subsampling (i.e., with no replacement) for estimating distance uncertainty. Subsampling, unlike bootstrapping, artificially increases the variance of the estimator distribution, necessitating an extra correction.

I. Pe'er (Ed.): RECOMB 2022, LNBI 13278, pp. 366–368, 2022.
https://doi.org/10.1007/978-3-031-04749-7_30

Our proposed correction procedure computes a distribution of distances for each pair of genomes or genome skims. More formally, input is a set of N genome skims S_i, each with n reads (can be extended to allow n to change). We choose a constant $\alpha < 1$ (default $\alpha = 9/10$) to set $b = n^\alpha$ noting $b \to \infty$ and $b/n \to 0$ as $n \to \infty$. We perform m (user-provided) rounds of subsampling. In each round r, we subsample b reads uniformly at random for each skim i and compute distances between these subsampled skims, giving us an estimate d^r_{ij} for each pair i, j of skims. These distances need to be next corrected and used for estimating a tree per replicate. For getting the final tree, we can use either the *main* estimates with no sub-sampling, $\hat{d}_{ij}$, or the related quantity $\overline{d_{ij}}$, defined as the average distance (*mean* estimate) across all m sub-sample replicates. Let d_{ij} be the true distance. Under certain assumptions, $\sqrt{n^\alpha/n}(d^r_{ij} - \hat{d}_{ij})$, $\sqrt{n^\alpha/n}(d^r_{ij} - \overline{d_{ij}})$, $(\hat{d}_{ij} - d_{ij})$, and $(\overline{d_{ij}} - d_{ij})$ all asymptotically converge to the same distribution. Accordingly, we consider two expressions for correcting distance. First, when the final tree is inferred from main estimates, we center all the subsampled distances around zero, apply the correction, and then center them back around the main estimate:

$$y^r_{ij} = \sqrt{\frac{n^\alpha}{n}}(d^r_{ij} - \overline{d_{ij}}) + \hat{d}_{ij} \qquad (1)$$

The alternative is to use the extended majority rule (i.e., greedy) consensus of the m replicate trees as the final tree. Since this final tree does not refer to the main distances, we have no reason to use $\hat{d}_{ij}$ in the correction and instead use:

$$x^r_{ij} = \sqrt{\frac{n^\alpha}{n}}(d^r_{ij} - \overline{d_{ij}}) + \overline{d_{ij}} \qquad (2)$$

This procedure can be easily extended to assemblies by subsampling k-mers.

In simulations and on real data, we show that our subsampling procedure paired with k-mer-based distance estimation method Skmer [4] produces reliable support values from both genome and genome skims. We evaluate the method under conditions with model misspecification and show that while support is not always fully calibrated, it is predictive and distinguishes correct and incorrect branches. The software is available publicly at https://github.com/shahab-sarmashghi/Skmer. Raw data and summary of results are deposited in https://github.com/noraracht/subsample_support_scripts. Complete version of the manuscript is available at http://dx.doi.org/10.2139/ssrn.3986497.

References

1. Bohmann, K., Mirarab, S., Bafna, V., Gilbert, M.T.P.: Beyond DNA barcoding: the unrealized potential of genome skim data in sample identification. Mol. Ecol. **29**(14), 2521–2534 (2020). https://doi.org/10.1111/mec.15507
2. Coissac, E., Hollingsworth, P.M., Lavergne, S., Taberlet, P.: From barcodes to genomes: extending the concept of DNA barcoding. Mol. Ecol.. **25**(7), 1423–1428 (2016). https://doi.org/10.1111/mec.13549

3. Politis, D.N., Romano, J.P., Wolf, M.: Subsampling. In: Michalos, A.C. (ed.) Encyclopedia of Quality of Life and Well-Being Research. Springer, Dordrecht (1999). https://doi.org/10.1007/978-94-007-0753-5_2909
4. Sarmashghi, S.: Skmer: assembly-free and alignment-free sample identification using genome skims. Genome Biol. **20**, 1–20 (2019))
5. Weitemier, K., et al.: Hyb-Seq: combining target enrichment and genome skimming for plant phylogenomics. Appl. Plant Sci. **2**(9), 1400042 (2014). https://doi.org/10.3732/apps.1400042

SOPHIE: Viral Outbreak Investigation and Transmission History Reconstruction in a Joint Phylogenetic and Network Theory Framework

Pavel Skums[1(✉)], Fatemeh Mohebbi[1], Vyacheslav Tsyvina[1], Pelin Icer[2], Sumathi Ramachandran[3], and Yury Khudyakov[3]

[1] Georgia State University, Atlanta, GA, USA
pskums@gsu.edu
[2] ETH Zurich, Basel, Switzerland
[3] Centers for Disease Control and Prevention, Atlanta, GA, USA

Abstract. Reconstruction of transmission networks from viral genomes sampled from infected individuals is a major computational problem of genomic epidemiology. For this problem, we propose a maximum likelihood framework SOPHIE (SOcial and PHilogenetic Investigation of Epidemics) based on the integration of phylogenetic and random graph models. SOPHIE is scalable, accounts for intra-host diversity and accurately infers transmissions without case-specific epidemiological data.

Keywords: Genomic epidemiology · Transmission network · Maximum likelihood inference

1 Introduction

Advances of sequencing technologies have a profound effect on epidemiology and virology. In particular, genomic epidemiology is becoming a major methodology for investigation of outbreaks and surveillance of transmission dynamics [1].

The hallmark of viruses as species is an extremely high genomic diversity originating from their error-prone replication. First generation of genomic epidemiology methods largely ignored intra-host viral diversity, but later studies demonstrated that taking it into account greatly enhances the predictive power of transmission inference algorithms [2,3]. Despite the significant progress achieved with the appearance of the next generation of transmission inference method, a number of computational, modelling and algorithmic challenges still need to be addressed. This includes development of scalable methodology based on maximum likelihood or Bayesian rather than maximum parsimony approach; problems with utilization of case-specific epidemiological information; accounting for non-independence of transmission events.

P.S. was supported by the NIH grant 1R01EB025022 and by the NSF grant 2047828.

I. Pe'er (Ed.): RECOMB 2022, LNBI 13278, pp. 369–370, 2022.
https://doi.org/10.1007/978-3-031-04749-7_31

2   Methods

We propose to address aforementioned challenges by integrating two components: the evolutionary relationships between viral genomes represented by their phylogenies and the expected structural properties of inter-host social networks. Frequently cited properties of social contact networks include power law degree distribution, small diameter, modularity and presence of hubs. All of them are reflected by network vertex degrees. Thus, we model social networks as random graphs with given expected degree distributions (EDDs). The goal is to find transmission networks that are consistent with observed genomic data and have the highest probability to be subnetworks of random contact networks.

This methodology is implemented within a maximum likelihood framework SOPHIE (SOcial and PHilogenetic Investigation of Epidemics). SOPHIE samples from the joint distribution of phylogeny ancestral traits defining transmission networks, estimates the probabilities that sampled networks are subgraphs of a random contact network and summarize them accordingly into the consensus network. This approach is scalable, accounts for intra-host diversity and accurately infers transmissions without case-specific epidemiological data.

3   Results

We applied SOPHIE to synthetic data simulated under different epidemiological and evolutionary scenarios, as well as to experimental data from epidemiologically curated HCV outbreaks. The experiments confirm the effectiveness of the new methodology. We demonstrated that the proposed approach is capable of achieving a substantial accuracy improvement over state-of-the-art parsimony-based phylogenetic methods, while retaining their scalability and speed.

4   Disclaimer

The conclusions in this report do not necessarily reflect the official position of the Centers for Disease Control and Prevention. Experimental data were used as approved by the Institutional Review Board of the CDC (protocol 7270.0).

References

1. Armstrong, G.L., et al.: Pathogen genomics in public health. N. Engl. J. Med. **381**(26), 2569–2580 (2019)
2. Skums, P., et al.: Quentin: reconstruction of disease transmissions from viral quasispecies genomic data. Bioinformatics **34**(1), 163–170 (2017)
3. Wymant, C., et al.: Phyloscanner: inferring transmission from within-and between-host pathogen genetic diversity. Mol. Biol. Evol. **35**(3), 719–733 (2017)

Identifying Systematic Variation at the Single-Cell Level by Leveraging Low-Resolution Population-Level Data

Elior Rahmani[1]([✉]), Michael I. Jordan[1,2], and Nir Yosef[1,3,4,5]([✉])

[1] Department of Electrical Engineering and Computer Science,
University of California, Berkeley, Berkeley, CA, USA
{erahmani,niryosef}@berkeley.edu, jordan@cs.berkeley.edu
[2] Department of Statistics, University of California, Berkeley, Berkeley CA, USA
[3] Center for Computational Biology, University of California, Berkeley, Berkeley, CA, USA
[4] Chan-Zuckerberg Biohub, San Francisco, CA, USA
[5] Ragon Institute of MGH, MIT, and Harvard, Cambridge, MA, USA

Abstract. A major limitation in single-cell genomics is a lack of ability to conduct cost-effective population-level studies. As a result, much of the current research in single-cell genomics focuses on biological processes that are broadly conserved across individuals, such as cellular organization and tissue development. This limitation prevents us from studying the etiology of experimental or clinical conditions that may be inconsistent across individuals owing to molecular variation and a wide range of effects in the population. In order to address this gap, we developed "kernel of integrated single cells" (Keris), a novel model-based framework to inform the analysis of single-cell gene expression data with population-level effects of a condition of interest. By inferring cell-type-specific moments and their variation across conditions using large tissue-level bulk data representing a population, Keris allows us to generate testable hypotheses at the single-cell level that would otherwise require collecting single-cell data from a large number of donors. Within the Keris framework, we show how the combination of low-resolution, large bulk data with small but high-resolution single-cell data enables the identification of changes in cell-subtype compositions and the characterization of subpopulations of cells that are affected by a condition of interest. Using Keris we estimate linear and non-linear age-associated changes in cell-type expression in large bulk peripheral blood mononuclear cells (PBMC) data. Combining with three independent single-cell PBMC datasets, we demonstrate that Keris can identify changes in cell-subtype composition with age and capture cell-type-specific subpopulations of senescent cells. This demonstrates the promise of enhancing single-cell data with population-level information to study compositional changes and to profile condition-affected subpopulations of cells, and provides a potential resource of targets for future clinical interventions.

A preprint of the full paper is available at https://www.biorxiv.org/content/10.1101/2022.01.27.478115v1.

© The Author(s), under exclusive license to Springer Nature Switzerland AG 2022
I. Pe'er (Ed.): RECOMB 2022, LNBI 13278, p. 371, 2022.
https://doi.org/10.1007/978-3-031-04749-7_32

Belayer: Modeling Discrete and Continuous Spatial Variation in Gene Expression from Spatially Resolved Transcriptomics

Cong Ma, Uthsav Chitra, Shirley Zhang, and Benjamin J. Raphael[✉]

Department of Computer Science, Princeton University, Princeton, NJ 08544, USA
braphael@princeton.edu

Motivation: Spatially resolved transcriptomics (SRT) technologies simultaneously measure gene expression and spatial location of cells in a 2D tissue slice, enabling the study of spatial patterns of gene expression. Current approaches to model spatial variation in gene expression assume either that gene expression is determined by discrete cell types or that gene expression varies continuously across a tissue slice. However, neither of these modeling assumptions adequately represent spatial variation in gene expression: the first assumption ignores continuous variation within a spatial region containing cells of the same type, while the second assumption does not allow for discontinuous changes in expression, e.g., due to a sharp change in cell type composition.

Results: We propose a model of spatial patterns in gene expression that incorporates both discontinuous and continuous spatial variation in gene expression. Specifically, we model the expression of a gene in a *layered tissue* slice as a piecewise linear function of a single spatial coordinate with potential discontinuities at tissue layer boundaries. We formulate the problem of inferring tissue layer boundaries and gene expression functions for all genes simultaneously. We derive a dynamic programming algorithm to find the optimal boundaries of layers when these boundaries are lines parallel to a coordinate axis. We generalize this algorithm to arbitrarily curved tissue layers by transforming the tissue geometry to be axis-aligned using the theory of conformal maps from complex analysis. We implement these algorithms in a method called Belayer. Applying Belayer to simulated data and to spatial transcriptomics data from the human dorsolateral prefrontal cortex, we demonstrate that Belayer achieves both higher accuracy in clustering tissue layers compared to state-of-the-art SRT clustering methods, and higher accuracy in identifying cortical layer marker genes compared to commonly-used methods for identification of spatially varying genes.

Cong Ma and Uthsav Chitra contributed equally and author order was decided by a coin toss.

I. Pe'er (Ed.): RECOMB 2022, LNBI 13278, pp. 372–373, 2022.
https://doi.org/10.1007/978-3-031-04749-7_33

Availability: Software is available at www.github.com/raphael-group/belayer.

Preprint: A preprint of the full manuscript is available at https://doi.org/10.1101/2022.02.05.479261.

Lossless Indexing with Counting de Bruijn Graphs

Mikhail Karasikov[1,2,3] , Harun Mustafa[1,2,3] , Gunnar Rätsch[1,2,3,4,5(✉)] ,
and André Kahles[1,2,3(✉)]

[1] Department of Computer Science, ETH Zurich, Zurich, Switzerland
Gunnar.Ratsch@ratschlab.org, andre.kahles@inf.ethz.ch
[2] Biomedical Informatics Research, University Hospital Zurich, Zurich, Switzerland
[3] Swiss Institute of Bioinformatics, Lausanne, Switzerland
[4] Associate Faculty in the Department of Biology, ETH Zurich, Zurich, Switzerland
[5] ETH AI Center, ETH Zurich, Zurich, Switzerland

Motivation: High-throughput sequencing data is rapidly accumulating in public repositories. Making this resource accessible for interactive analysis at scale requires efficient approaches for its storage and indexing. There have recently been remarkable advances in solving the experiment discovery problem and building compressed representations of annotated de Bruijn graphs where k-mer sets can be efficiently indexed and interactively queried [1–6]. However, approaches for representing and retrieving other quantitative attributes such as gene expression or genome positions in a general manner have yet to be developed.

Methods: In this work, we propose the concept of *Counting de Bruijn graphs* generalizing the notion of *annotated* (or *colored*) *de Bruijn graphs*. Counting de Bruijn graphs supplement each node-label relation with one or many attributes (e.g., a k-mer abundance or its positions in a genome). To represent them, we first observe that many schemes for the representation of compressed binary matrices already support the rank operation on the columns or rows, which can be used to define an inherent indexing of any additional quantitative attributes. Based on this property, we generalize these schemes and introduce a new approach for representing non-binary sparse matrices in compressed data structures (Fig. 1). Finally, we notice that relation attributes are often easily predictable from a node's local neighborhood in the graph. Notable examples are genome positions shifting by 1 for neighboring nodes in the graph, or expression counts that are often similar in neighboring nodes. We exploit this regularity of graph annotations and apply an invertible delta-like coding to achieve better compression.

Results: We show that Counting de Bruijn graphs can index the abundance of each k-mer in 2,652 human RNA-Seq read sets in a representation that is over 8× smaller and yet faster to query compared to state-of-the-art bioinformatics

© The Author(s), under exclusive license to Springer Nature Switzerland AG 2022
I. Pe'er (Ed.): RECOMB 2022, LNBI 13278, pp. 374–376, 2022.
https://doi.org/10.1007/978-3-031-04749-7_34

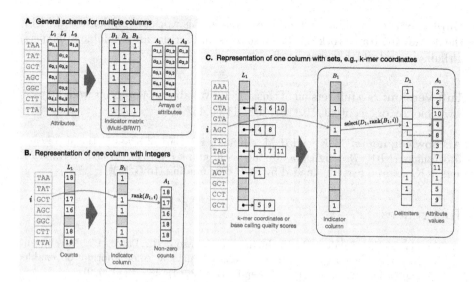

Fig. 1. The proposed representation of sparse matrices in compressed form. **Panel A:** General scheme for sparse matrices with abstract attributes, where the non-assigned attributes are eliminated by an indicator binary matrix stored in a compressed representation (e.g., Multi-BRWT) supporting the rank operation on its columns to enable the access to the corresponding attribute for any given cell of the matrix. These attributes are stored separately, typically in a form of compressed arrays. **Panel B:** The scheme applied to a single column with integer values (e.g., k-mer counts) and the query algorithm (e.g., the count of k-mer i is retrieved as $A_1[\text{rank}(B_1, i)]$). Empty cells in grey represent zeros. **Panel C:** The scheme applied to a single column where each cell is a set of numbers, or a tuple (e.g., representing k-mer coordinates). The "zero" attributes (empty sets) are eliminated with an indicator bitmap and the non-empty sets are encoded in an array that holds all numbers and a delimiting bitmap.

tools. It also takes less time and memory to construct. Furthermore, Counting de Bruijn graphs with positional annotations losslessly represent entire reads in indexes on average 27% smaller than the input compressed with *gzip -9* for human Illumina RNA-Seq and 57% smaller for PacBio HiFi sequencing of viral samples. A complete joint searchable index of all viral PacBio SMRT reads from NCBI's SRA (152,884 read sets, 875 Gbp) takes only 178 GB. Finally, on the full RefSeq collection, we generate a lossless and fully queryable index that is 4.4× smaller compared to the respective *MegaBLAST* index. The techniques proposed in this work naturally complement existing methods and tools employing de Bruijn graphs and significantly broaden their applicability: from indexing k-mer counts and genome positions to implementing novel sequence alignment algorithms on top of highly compressed and fully searchable graph-based sequence indexes.

Implementation: The methods presented in this work are implemented within the `MetaGraph` framework [4]. The resources and scripts are available at https://github.com/ratschlab/counting_dbg.

Full Version: See full version of this article available as a preprint from https://www.biorxiv.org/content/10.1101/2021.11.09.467907.

Acknowledgments. M. K. and H. M. are funded as part of Swiss National Research Programme (NRP) 75 "Big Data" by the SNSF grant #407540_167331. M. K., H. M., and A. K. are also partially funded by ETH core funding (to G. R.).

References

1. Almodaresi, F., Pandey, P., Ferdman, M., Johnson, R., Patro, R.: An efficient, scalable, and exact representation of high-dimensional color information enabled using de Bruijn graph search. J. Comput. Biol. **27**(4), 485–499 (2020)
2. Bradley, P., Den Bakker, H.C., Rocha, E.P., McVean, G., Iqbal, Z.: Ultrafast search of all deposited bacterial and viral genomic data. Nat. Biotechnol. **37**(2), 152–159 (2019)
3. Danciu, D., Karasikov, M., Mustafa, H., Kahles, A., Rätsch, G.: Topology-based sparsification of graph annotations. Bioinformatics **37**(Suppl._1), i169–i176 (2021). https://doi.org/10.1093/bioinformatics/btab330
4. Karasikov, M., et al.: Metagraph: indexing and analysing nucleotide archives at petabase-scale. bioRxiv (2020). https://doi.org/10.1101/2020.10.01.322164. https://www.biorxiv.org/content/early/2020/11/03/2020.10.01.322164
5. Marchet, C., Iqbal, Z., Gautheret, D., Salson, M., Chikhi, R.: REINDEER: efficient indexing of k-mer presence and abundance in sequencing datasets. Bioinformatics **36**(Suppl._1), i177–i185 (2020). https://doi.org/10.1093/bioinformatics/btaa487
6. Pandey, P., Almodaresi, F., Bender, M.A., Ferdman, M., Johnson, R., Patro, R.: Mantis: a fast, small, and exact large-scale sequence-search index. Cell Syst. **7**(2), 201–207.e4 (2018)

Uncovering Hidden Assembly Artifacts: When Unitigs are not Safe and Bidirected Graphs are not Helpful (ABSTRACT)

Amatur Rahman[1]([✉]) and Paul Medvedev[1,2,3]([✉])

[1] Department of Computer Science and Engineering, Penn State, University Park, USA
{aur1111,pzm11}@psu.edu
[2] Department of Biochemistry and Molecular Biology, Penn State, University Park, USA
[3] Center for Computational Biology and Bioinformatics, Penn State, University Park, USA

Summary: Recent assemblies by the T2T and VGP consortia have achieved significant accuracy but required a tremendous amount of effort and resources. More typical assembly efforts, on the other hand, still suffer both from mis-assemblies (joining sequences that should not be adjacent) and from under-assemblies (not joining sequences that should be adjacent). To better understand the common algorithm-driven causes of these limitations, we investigated the unitig algorithm, which is a core algorithm at the heart of most assemblers. The unitig algorithms returns, roughly speaking, all non-branching paths in the assembly graph. It is already known that the unitig algorithm contributes to under-assembly [1,2] and can trivially create mis-assemblies when there are sequencing errors. However, it is widely assumed that if it were not for sequencing errors, unitigs would always be safe (i.e. substrings of the sequenced genome). In this work, we prove that, contrary to popular belief, even when there are no sequencing errors, unitigs are not always safe.

The unitig algorithm also needs to account for the fact that the strand from which a read is sequenced is unknown. Most assemblers do so via two common approaches to constructing the de Bruijn graph (dBG). In one, every k-mer is "doubled" prior to constructing the dBG, i.e. for every k-mer in the input, both it and its reverse complement is added to the dBG. In the other approach, edges are given two instead of one orientation, thereby capturing the way that double-stranded strings can overlap. This results in a bidirected dBG. In this work, we prove that when the bidirected graph is used to model double-strandedness, the unitig algorithm under-assembles by failing to merge the two halves of palindromes. To the best of our knowledge, this paper is the first to theoretically predict the existence of these assembler artifacts and confirm and measure the extent of their occurrence in practice.

A full version of this paper is available as a preprint https://doi.org/10.1101/2022.01.20.477068.

I. Pe'er (Ed.): RECOMB 2022, LNBI 13278, pp. 377–379, 2022.
https://doi.org/10.1007/978-3-031-04749-7_35

Unitigs are not Safe: One of our two main theorems (Theorem 1 in the full paper [3]) gives necessary and sufficient conditions for a unitig to be safe in the directed dBG constructed from error-free reads. The technical nature of the theorem requires that we only give an informal presentation of it here. To formalize the concept of mis-assembly, consider a set of reads and the set of their constituent k-mers K. A *sequenced segment* is defined as a maximal substring of the genome whose constituent k-mers all belong to K. A set of reads therefore induces a set S of sequenced segments. Figure 1 illustrates the concept. Given S, we say that a unitig w in the dBG constructed from S is *unsafe* iff it is not a subwalk of a walk that corresponds to a string in S. This definition of an unsafe unitig captures the notion of a potential mis-assembly, as the unitig is not present in the sequenced part of the genome.

Our Theorem 1 shows that w is unsafe iff for all $S \in \mathcal{S}$, one of four cases hold. We use Fig. 2 to illustrate this, focusing on the unitig $w = ACT \rightarrow CTT \rightarrow TTG$. Consider a walk g corresponding to a sequenced segment. Clearly, if g does not contain any k-mer from w (e.g. the green segment), then g cannot make w safe (i.e. g does not contain w as a subwalk). If g starts in the middle of w and does not visit its own starting vertex again (e.g. the red segment), then g cannot make w safe. Similarly, if g ends in the middle of w and did not visit its own ending vertex previously (i.e. the blue segment), then g cannot make w safe. If g starts and ends in the middle of w, with the ending vertex to the right of the starting vertex, and contains each of those vertices exactly twice (e.g. the orange segment), then g cannot make w safe. Figure 2 shows the situation when all the sequenced segments fall into these four cases, making w is unsafe in spite of it being a unitig. Our Theorem 1 proves the other direction as well, i.e. that under all other conditions, g makes w safe.

```
ACGTACTTAA ACG GGGAA TACC GCCTA
| | | | | | | | | | | | | | | | | | | | | | | | |
ACGTA | | | | | | | GGGAA | | | | GCCTA
| | | TACTT | | | | | | | | GGAAT | | | | | | | |
| | | | | CTAAA | | | | | | | | | | TACCG | | | |
```

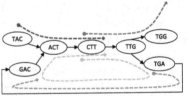

Fig. 1. Illustration of sequenced segments. The black text shows the reference genome and the sever red sequences are reads aligned to the reference. The green boxes highlight the resulting sequenced segments when $k = 3$. (Color figure online)

Fig. 2. Illustration of the three cases of our Theorem 1. $\mathcal{S}$ has three segments, each of which is marked by a dashed line starting at a dot and ending at a diamond.

Bidirected Graphs Result in Under-Assembly: The second of our two main theorems states that naively using the bidirected graph actually contributes to under-assembly, compared to the doubled graph. We prove there is a bijection between maximal unitigs in the doubled dBG and in the bidirected dBG, except that palindromic unitigs in the doubled dBG are split in half in the bidirected dBG. To do so, we have given proper definitions for things like walks and unitigs in the context of bidirected graphs. Previous papers used these concepts somewhat informally; when definitions were given, they worked in the context of that paper but failed to have more general desired properties.

Experimental Results: Using experimental simulations, we confirmed the presence of these theoretically-predicted artifacts in real genomes and popular assemblers. We simulated error-free reads from the T2T human reference chromosome 1 and confirmed that the unitigs that were unsafe were exactly the unitigs that satisfied the conditions of Theorem 1. We also demonstrated that real assemblers like SPAdes and MEGAHIT output unsafe contigs that closely match the unsafe unitigs that our theory predicts. We constructed both doubled and bidirected de Bruijn graphs for varying k-mer size from human chromosome 21 to verify the validity of our second theorem. Since long palindromes are rare in real genomes, we artificially introduced many palindromic regions in the reference genome (human chromosome 4), and observed the palindrome splitting artifact in the assembled contigs generated by SPAdes and MEGAHIT.

References

1. Tomescu, A.I., Medvedev, P.: Safe and complete contig assembly via omnitigs. In: Singh, M. (ed.) RECOMB 2016. LNCS, vol. 9649, pp. 152–163. Springer, Cham (2016). https://doi.org/10.1007/978-3-319-31957-5_11
2. Cairo, M., Khan, S., Rizzi, R., Schmidt, S., Tomescu, A.I., Zirondelli, E.C.: The hydrostructure: a Universal Framework for Safe and Complete Algorithms for Genome Assembly, arXiv preprint arXiv:2011.12635 (2020)
3. Rahman, A., Medvedev, P.: Uncovering hidden assembly artifacts: when unitigs are not safe and bidirected graphs are not helpful, bioRxiv (2022)

Mapping Single-Cell Transcriptomes to Copy Number Evolutionary Trees

Pedro F. Ferreira[1,2], Jack Kuipers[1,2(✉)], and Niko Beerenwinkel[1,2(✉)]

[1] Department of Biosystems Science and Engineering, ETH Zurich,
Basel, Switzerland
{jack.kuipers,niko.beerenwinkel}@bsse.ethz.ch
[2] SIB Swiss Institute of Bioinformatics, Basel, Switzerland

Cancer is an evolutionary process in which cells accumulate mutations and form different subclones with potentially varying phenotypes. The diversity of cell states that make up a tumor is caused by genomic and epigenomic changes, as well as interactions with its microenvironment. The resulting intra-tumor heterogeneity is a major cause of treatment failure and relapse. In particular, the manner in which different genomic changes, as opposed to other factors, contribute to cell-specific states is highly relevant in treatment decision making, but remains an open question.

In the absence of reliable and scalable experimental platforms to obtain joint RNA and copy number aberration (CNA) profiles from the same single cells, computational approaches can be used to match cells derived from the same sample but sequenced on separate platforms. Because single-cell CNA profiles can be grouped into subclones, the most natural approach is to assign single-cell transcriptomes to CNA subclones by assuming that gene expression is proportional to copy number. However, this naive approach ignores the underlying evolutionary history of the cancer cells. The phylogenetic relationships among tumor clones can be inferred by a number of computational methods, including for single-cell CNAs. Furthermore, cells within the same CNA subclone may have different expression profiles and form different groups of cell states as a result of other somatic mutations, epigenetic events, or environmental interactions. While totally unsupervised clustering and trajectory inference methods for single-cell RNA-sequencing (scRNA-seq) data may find spurious structure that does not respect the underlying subclonal structure, fully supervised methods are over-restrictive in forcing the clustering structure to be driven only by CNA subclones.

Here, we present SCATrEx (Single-Cell Augmentation of CNA Trees with gene Expression), a Bayesian model designed to find a hierarchical clustering structure in scRNA-seq data on top of a CNA tree. SCATrEx uses a novel nested tree structure based on the tree-structured stick breaking process to augment a given CNA tree with nodes corresponding to non-CNA events, such as undetected mutations or epigenetic changes with consequences at the gene expression level. Specifically, we model each node in the known CNA tree as a tree-structured stick breaking process containing unobserved nodes with different gene expression states. This enables a joint analysis of CNAs and gene

I. Pe'er (Ed.): RECOMB 2022, LNBI 13278, pp. 380–381, 2022.
https://doi.org/10.1007/978-3-031-04749-7_36

expression at single-cell level. In this model, the gene expression of a cell is proportional to the copy number profile of the subclone it belongs to and influenced by node-specific cell state factors that are propagated through the augmented tree. Additional structured noise factors contribute to gene expression variability that does not follow the tree structure. SCATrEx is a generative model of the raw read counts, and so avoids issues related to normalization and pre-processing that may result in removal of the signal of interest.

We employ a search-and-score approach to learn the augmented tree structure and its parameters by maximizing the marginal data log-likelihood. At each step, we propose either local changes to the tree structure or parameter updates, and we approximate the new marginal log-likelihood by computing a variational lower bound. To alleviate the computational cost of performing this approximation, we use a fully factorized variational distribution. Our model is fully continuous, enabling the use of automatic differentiation to optimize the variational lower bound. Our implementation can run on GPUs to further decrease runtime.

We compared SCATrEx with fully unsupervised approaches to reconstruct the clusters of cells in the data as simulated by an augmented CNA tree, as well as with cell-to-clone assignment methods. Our simulations demonstrate that SCATrEx's joint approach is accurate in both clustering and cell-to-clone assignment, and robust to various levels of noise in the data. In a real data example of a mouse xenograft of triple-negative breast cancer, we obtained an augmented tree and inferred gene expression states for the observed clones as well as for clone-specific populations of cells. Even though these populations had the same copy number states, they contained different expression levels for genes that are associated with immune escape in cancer.

This paper is available at https://doi.org/10.1101/2021.11.04.467244. The SCATrEx Python package is open-source and available at https://github.com/cbg-ethz/SCATrEx.

ImmunoTyper-SR: A Novel Computational Approach for Genotyping Immunoglobulin Heavy Chain Variable Genes Using Short Read Data

Michael Ford[1(✉)], Ananth Hari[1,3], Oscar Rodriguez[4], Junyan Xu[1],
Justin Lack[2], Cihan Oguz[2], Yu Zhang[2], Sarah Weber[2], Mary Magliocco[2],
Jason Barnett[2], Sandhya Xirasagar[2], Smilee Samuel[2], Luisa Imberti[6],
Paolo Bonfanti[7], Andrea Biondi[7], Clifton L. Dalgard[5], Stephen Chanock[1],
Lindsey Rosen[2], Steven Holland[2], Helen Su[2], Luigi Notarangelo[2],
SPS NIAID COVID Consortium, Uzi Vishkin[3], Corey Watson[4],
and S. Cenk Sahinalp[1(✉)]

[1] National Cancer Institute, NIH, Bethesda, MD, USA
{mike.ford,cenk.sahinalp}@nih.gov
[2] National Institute of Allergy and Infectious Diseases, NIH, Bethesda, MD, USA
[3] Department of Electrical Engineering, University of Maryland, College Park,
MD, USA
[4] Department of Biochemistry and Molecular Genetics, University of Louisville,
Louisville, KY, USA
[5] Uniformed Services University of the Health Sciences, Bethesda, MD, USA
[6] Diagnostic Department, ASST Spedali Civili di Brescia, Brescia, Italy
[7] University of Milano-Bicocca-Fondazione MBBM, Monza, Italy

Introduction. The human immunoglobulin heavy chain (*IGH*) locus on chromosome 14 includes more than 40 functional copies of the variable gene (*IGHV*). These combine with the joining genes, diversity genes, constant genes and immunoglobulin light chains to code for antibodies (Ab) and B cell receptors, forming a critical part of the adaptive immune system due to their role in identifying and neutralizing pathogenic invaders. Despite known associations with clinical phenotypes, such as infectious disease and vaccine response [1,2], autoimmune/inflammatory conditions [3,4], and cancer [5], our understanding of population-level genetic diversity in the *IGH* locus and its contribution to Ab function in disease remains limited [6,7], due primarily to the complexity of the locus. *IGH* is known to contain many large structural variants, including segmental duplications, large insertions and deletions, and other copy number variants (CNVs) [8–10]. Of primary interest are the *IGHV* genes due to their important role in defining epitope structure. The genes are short (mean of 291 bp) and have highly similar sequence, within and across different genes. To date, there has been only one published computational pipeline for germline *IGHV* genotyping using short read WGS data [11,12], which remain the most ubiquitous and available genome sequencing data type.

© The Author(s), under exclusive license to Springer Nature Switzerland AG 2022
I. Pe'er (Ed.): RECOMB 2022, LNBI 13278, pp. 382–384, 2022.
https://doi.org/10.1007/978-3-031-04749-7_37

We present ImmunoTyper-SR, a novel computational approach for genotype and CNV analysis of functional germline *IGHV* genes using short read WGS data, using a database of known *IGHV* sequences as a reference. ImmunoTyper-SR is based on a novel combinatorial optimization formulation that aims to minimize the total edit distance between the reads and their assigned alleles while maintaining additional constraints on the number and distribution of reads across each allele identified. ImmunoTyper-SR is able to produce accurate *IGHV* allele and CNV calls using short read WGS with moderate coverage, and was tested on 12 individuals with diverse genetic backgrounds from the 1000 Genomes Project [13] (1kGP), using independently-generated targeted long read-based *IGH* assemblies as ground truth [10]. We also used ImmunoTyper-SR to investigate associations between *IGHV* genotypes, type I IFN autoantibodies and COVID-19 disease severity using WGS data from a cohort of 542 individuals from the NIAID COVID Consortium ("NIAID cohort").

Methods. ImmunoTyper-SR functions in three main steps; first, by recruiting reads that are likely to overlap with *IGHV* sequence from the input mapped WGS BAM file; second, by assigning potential alleles as candidates for every recruited read by mapping to an allele sequence database; and third, by assigning reads to their best matched allele using a novel integer linear programming (ILP) formulation.

The ILP formulation has an objective function that minimizes the total edit distance between all reads and their assigned allele sequence, while allowing reads to be discarded. It also contains constraints that ensure the reads will be assigned in a way that is consistent with expected read depth and variance, while allowing for multiple copies to be called for any given allele.

Results. We tested ImmunoTyper-SR calls from WGS sequences for 12 1kGP individuals against independently generated *IGH* assemblies. The results demonstrate that ImmunoTyper-SR is the first short read WGS-based *IGHV* genotyping tool with allele-level granularity, with a mean precision and recall values of 83.7% and 80% respectively for identification of each allele sequence and its copy number exactly - significantly higher than the only other previously published comparable method. Many of the miscalls are due to highly similar alleles; if calls are allowed to be at most 3 bp from the ground truth allele, the mean F-score increases to 87.9%.

We also applied ImmunoTyper-SR to a set of genome sequences from 542 COVID patients, nine of whom had been sequenced twice. We found genotypes from these subjects had a 92.3% mean Jaccard similarity score between paired samples, when ignoring CNVs, which demonstrates ImmunoTyper's accuracy in allele calls. This dataset was also used to investigate genetic associations to anti-interferon type I (anti-IFN) antibodies, which have been previously associated with COVID severity. We found two alleles had associations with p-values less than 0.05, however we are unable to draw any conclusions about anti-IFN associations, due to the alleles being very rare (present in at most 10 individuals), and the case/control ratio in the dataset being heavily skewed.

As the first short read WGS-based *IGHV* genotyping tool with allele-level granularity, ImmunoTyper-SR opens the door to applying the power of the largest WGS datasets available to uncover the mysteries of one of the least understood loci in the human genome.

References

1. Yeung, Y.A., et al.: Germline-encoded neutralization of a staphylococcus aureus virulence factor by the human antibody repertoire. Nat. Commun. **7**, 1–14 (2016)
2. Lee, J.H., et al.: Vaccine genetics of IGHV1-2 VRC01-class broadly neutralizing antibody precursor naïve human b cells. NPJ Vaccines **6**, 1–12 (2021)
3. Johnson, T.A., et al.: Association of an IGHV3-66 gene variant with Kawasaki disease. J. Hum. Genet. **66**, 475–489 (2021)
4. Parks, T., et al.: Association between a common immunoglobulin heavy chain allele and rheumatic heart disease risk in Oceania. Nat. Commun. **8**, 1–10 (2017)
5. Cui, M., et al.: Immunoglobulin expression in cancer cells and its critical roles in tumorigenesis. Front. Immunol. **12**, 893 (2021)
6. Watson, C.T., Glanville, J., Marasco, W.A.: The individual and population genetics of antibody immunity. Front. Immunol. **38**, 459–470 (2017)
7. Collins, A.M., Yaari, G., Shepherd, A.J., Lees, W., Watson, C.T.: Germline immunoglobulin genes: disease susceptibility genes hidden in plain sight? Curr. Opin. Syst. Biol. **24**, 100–108 (2020). https://doi.org/10.1016/j.coisb.2020.10.011
8. Watson, C.T., et al.: Complete haplotype sequence of the human immunoglobulin heavy-chain variable, diversity, and joining genes and characterization of allelic and copy-number variation. Am. J. Hum. Genet. **92**, 530–546 (2013). https://doi.org/10.1016/j.ajhg.2013.03.004
9. Gadala-Maria, D., et al.: Identification of subject-specific immunoglobulin alleles from expressed repertoire sequencing data. Front. Immunol. **10**, 129 (2019)
10. Rodriguez, O.L., et al.: A Novel Framework for Characterizing Genomic Haplotype Diversity in the Human Immunoglobulin Heavy Chain Locus. Front. Immunol. **11**, 1–16 (2020)
11. Luo, S., Yu, J.A., Song, Y.S.: Estimating copy number and allelic variation at the immunoglobulin heavy chain locus using short reads. PLoS Comput. Biol. **12**, 1–21 (2016)
12. Luo, S., Jane, A.Y., Li, H., Song, Y.S.: Worldwide genetic variation of the IGHV and TRBV immune receptor gene families in humans. Life Sci. Alliance **2** (2019)
13. 1000-Genomes-Project-Consortium et al.: A global reference for human genetic variation. Nature **526**, 68 (2015)

AutoComplete: Deep Learning-Based Phenotype Imputation for Large-Scale Biomedical Data

Ulzee An[1]([⊠]), Na Cai[2], Andy Dahl[3], and Sriram Sankararaman[1,4,5]

[1] Department of Computer Science, UCLA, Los Angeles, CA, USA
ulzee@ucla.edu, sriram@cs.ucla.edu
[2] Helmholtz Pioneer Campus, Helmholtz Zentrum München, Neuherberg, Germany
[3] Section of Genetic Medicine, University of Chicago, Chicago, IL, USA
[4] Department of Human Genetics, David Geffen School of Medicine, UCLA,
Los Angeles, CA, USA
[5] Department of Computational Medicine, David Geffen School of Medicine, UCLA,
Los Angeles, CA, USA

Biomedical datasets that aim to collect diverse phenotypic and genomic data across large numbers of individuals are plagued by the large fraction of missing data. The ability to accurately impute or "fill-in" missing entries in these datasets is critical to a number of downstream applications. A notable example of this problem arises in the context of the UK Biobank (UKBB) which contains thousands of phenotypes in conjunction with genetic data for $\approx$500,000 UK individuals. Here, the presence of phenotypic measurements across individuals vary drastically. Some traits exist for virtually all individuals (age, sex) while others are majority unreported (addictions, self-harm > 99%). As a result, missingness can substantially impact our ability to study clinically-relevant phenotypes or diseases. However, current imputation methods fall short in one or more aspects of being reliable or scalable in the domain of massive, highly incomplete, and heterogeneous biobank-scale data. Imputation methods have been developed which leverage the correlation between the values at missing features and the observed features (*i.e.*, the conditional distribution of observed entries given values of the observed features) to impute the missing features. While simple in principle, challenges to this approach arise from the presence of large numbers of features and observations in Biobank-scale data, the heterogeneous data types that are measured, and the complex structure to the patterns of missingness. Thus, imputation methods that can accurately impute heterogeneous data types in the presence of complex patterns of missingness while being scalable are needed.

We propose AutoComplete, a deep-learning based imputation method based on an auto-encoder architecture designed for highly incomplete biobank-scale phenotype data. The method can impute binary and continuous phenotypes while scaling with ease to incomplete datasets with half a million individuals and millions of entries. To handle the absence of ground truth in incomplete datasets, we developed a realistic procedure to simulate missing values as reflected in the real data by implementing *Copy-masking* which propagates missingness patterns already present in the data, from which AutoComplete learns imputation.

© The Author(s), under exclusive license to Springer Nature Switzerland AG 2022
I. Pe'er (Ed.): RECOMB 2022, LNBI 13278, pp. 385–386, 2022.
https://doi.org/10.1007/978-3-031-04749-7_38

Our approach is applicable to any complete or incomplete dataset and leverages all available samples and all observed features. Using conventional computational hardware, AutoComplete can scale to ≈300,000 individuals and ≈400 phenotypes from the UKBB with ease converging within ~6h.

To provide a realistic assessment, we evaluated AutoComplete on two cohorts of UKBB phenotypes: a set of 86 blood lab and cardiovascular phenotypes and a larger set of 372 phenotypes from an on-going study on mental health-related traits, where the majority of the phenotypes of interest are missing. Each dataset contains ≈300,000 unrelated individuals of white British ancestry. Across various settings of missingness, AutoComplete greatly improved imputation accuracy over existing methods. In comparison to the next-best method (SoftImpute), the average squared Pearson correlation coefficient (r^2) of AutoComplete improved by +17.5% over all phenotypes across all tested settings. For binary phenotypes, Precision-Recall also increased +7.2% on average and Receiver Operating Characteristic also increased +7.3%. AutoComplete improved r^2 by +30% over SoftImpute across mental health-related phenotypes while improving r^2 by +5% across cardiovascular phenotypes. Beyond the aggregate improvements, AutoComplete also improved imputation accuracy on clinically important phenotypes: improving r^2 for *psychotic.medicalhelp* from 0.05 to 0.22 (+440%, 530 cases and 2068 controls, 94% missing in data) and *anxiety.drugalcohol* from 0.13 to 0.19 (+46%, 2726 cases and 6061 controls, 89% missing in data). Our results illustrate the value of deep-learning based imputation where accurate imputation of highly missing phenotypes can substantially improve the power of downstream analyses.

Resistor: An Algorithm for Predicting Resistance Mutations Using Pareto Optimization over Multistate Protein Design and Mutational Signatures

Nathan Guerin[1], Teresa Kaserer[2(✉)], and Bruce R. Donald[1,3,4,5(✉)]

[1] Department of Computer Science, Duke University, Durham, USA
brd+recomb22@cs.duke.edu
[2] Institute of Pharmacy/Pharmaceutical Chemistry, University of Innsbruck, Innsbruck, Austria
teresa.kaserer@uibk.ac.at
[3] Department of Biochemistry, Duke University Medical Center, Durham, USA
[4] Department of Chemistry, Duke University, Durham, USA
[5] Department of Mathematics, Duke University, Durham, USA

RESISTOR optimizes four objectives to find the Pareto frontier of these resistance-causing criteria. The first two axes of optimization are provable approximations to the relative change in binding affinity (ΔK_a) of a drug and endogenous ligand upon a protein's mutation. These ΔK_a predictions are made by the provable thermodynamic- and ensemble-based multistate computational protein design algorithm K^* [3,5] after an initial sequence filter using the multistate design algorithm COMETS [4] from the computational protein design software OSPREY. By virtue of pruning using COMETS, RESISTOR inherits the empirical sublinearity characteristics of the COMETS sequence search, rendering RESISTOR, to our knowledge, the first provable structure-based resistance prediction algorithm that is sublinear in the size of the sequence space. This is important because the size of sequence space is exponential in the number of residue positions that can mutate to confer resistance. On the third axis RESISTOR uses empirically derived mutational signatures [1,6] to determine the probability that each single- or double-point mutation will occur in a given context. The fourth axis of optimization is over "mutational hotspots", *viz.* those locations in the protein where multiple amino substitutions are predicted to confer resistance. Some highlights of the algorithm are presented in Fig. 1.

As validation of the algorithm, we applied RESISTOR to a a number of kinase inhibitors used to treat lung adenocarcinoma and melanoma through inhibition of EGFR or BRAF kinase activity. In so doing, we searched over a set of 1257 sequences in EGFR and 1214 sequences in BRAF with an average conformation space size of $\sim 5.9 \times 10^{10}$, using experimental structures when available and docked complexes when not. This search generated a set of predicted resistance mutations that we compared to known resistance mutations in EGFR and report

Supported by the NIH (grants R01-GM078031, R01-GM118543, and R35-GM144042 to BRD) and the Austrian Science Fund (grant P34376-B to TK).

© The Author(s), under exclusive license to Springer Nature Switzerland AG 2022
I. Pe'er (Ed.): RECOMB 2022, LNBI 13278, pp. 387–389, 2022.
https://doi.org/10.1007/978-3-031-04749-7_39

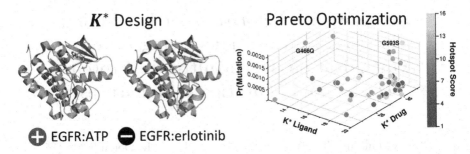

Fig. 1. Resistor finds the Pareto frontier from OSPREY positive and negative designs, mutational probabilities, and resistance hotspots. (left) To predict resistance mutations for EGFR (an example described in the manuscript) two structures are required as input to OSPREY to compute positive (+) and negative (−) design K^* scores. The structure for positive design is EGFR (green) bound to its endogenous ligand ATP (blue), for the negative design EGFR is bound to the drug erlotinib (pink). The goal of positive (resp. negative) design is to improve (resp. ablate) binding affinity. A mutation is a *resistance candidate* when its ratio of positive to negative K^* scores increases. (right) RESISTOR computes mutational probabilities using EGFR's coding DNA along with lung adenocarcinoma-specific trinucleotide mutational probabilities. It also generates a *hotspot score* representing how many different amino acids at a given location are resistance mutation candidates. Finally, it uses Pareto optimization on the positive and negative K^* scores, the mutational probabilities, and hotspot scores to identify the Pareto frontier of resistance mutations. (Color figure online).

herein that RESISTOR correctly identified eight clinically significant resistance mutations, including the "gatekeeper" T790M mutation to erlotinib and gefitinib and five known resistance mutations to osimertinib. This demonstrates that by exploiting the wealth of structural and sequence data available in the form of molecular structures and mutational signatures, RESISTOR is a general method for predicting resistance mutations that can be applied to a wide variety of cancer, antimicrobial, antiviral and antifungal drug targets.

The source code for RESISTOR is part of the OSPREY protein redesign software package available at https://github.com/donaldlab/OSPREY3. The full manuscript is available on bioRχiv at https://doi.org/10.1101/2022.01.18. 476733.

Competing Interests. BRD is a founder of Ten63 Therapeutics, Inc.

References

1. Alexandrov, L.B., et al.: Signatures of mutational processes in human cancer. Nature **500**(7463), 415–421 (2013)
2. Centers for Disease Control and Prevention: Antibiotic/antimicrobial resistance (2020). https://www.cdc.gov/drugresistance/index.html

3. Georgiev, I., Lilien, R.H., Donald, B.R.: The minimized dead-end elimination criterion and its application to protein redesign in a hybrid scoring and search algorithm for computing partition functions over molecular ensembles. J. Comput. Chem. **29**(10), 1527–1542 (2008)
4. Hallen, M.A., Donald, B.R.: Comets (constrained optimization of multistate energies by tree search): a provable and efficient protein design algorithm to optimize binding affinity and specificity with respect to sequence. J. Comput. Biol. **23**(5), 311–321 (2016)
5. Hallen, M.A., et al.: Osprey 3.0: open-source protein redesign for you, with powerful new features. J. Comput. Chem. **39**(30), 2494–2507 (2018)
6. Kaserer, T., Blagg, J.: Combining mutational signatures, clonal fitness, and drug affinity to define drug-specific resistance mutations in cancer. Cell Chem. Biol. **25**(11), 1359–1371 (2018)
7. Vasan, N., Baselga, J., Hyman, D.M.: A view on drug resistance in cancer. Nature **575**(7782), 299–309 (2019)

Ultra High Diversity Factorizable Libraries for Efficient Therapeutic Discovery

Zheng Dai[1](✉)(iD), Sachit D. Saksena[1](✉)(iD), Geraldine Horny[2],
Christine Banholzer[2], Stefan Ewert[2], and David K. Gifford[1](✉)(iD)

[1] Computer Science and Artificial Intelligence Laboratory,
Massachusetts Institute of Technology, Cambridge, MA 02139, USA
{zhengdai,sachit,gifford}@mit.edu
[2] Novartis Institutes for BioMedical Research (NIBR), Basel, Switzerland

The successful discovery of novel biological therapeutics by selection requires highly diverse libraries of candidate sequences that contain a high proportion of desirable candidates. Here we propose the use of computationally designed *factorizable libraries*, whose sequences are made of concatenated segments from smaller *segment libraries*, as a method of creating large libraries that meet an objective function at low cost.

Designing segment libraries that result in a factorizable library that meets an objective function is a computationally difficult task. We present a computational method we call Stochastically Annealed Product Spaces (SAPS), which optimizes segment libraries though iterative improvements with respect to an objective function to design a full length factorizable library. Key to our method is the *reverse kernel trick*, which allows us to efficiently evaluate an objective over the full factorizable library by casting the objective function as an inner product of feature vectors (see Fig. 1).

We show that SAPS outperforms five different benchmark sampling approaches on simulated datasets. We next apply SAPS to design factorizable libraries of the third complementarity determining region of antibody heavy chains (CDR-H3s). We show that this framework can generate factorized CDR-H3 segment libraries that, when joined combinatorially, contain $\sim 10^9$ unique sequences with highly specific and flexible design parameters. We compare these libraries to a randomized library and show that SAPS designed libraries are more diverse and more enriched in desirable sequences.

Applications of factorizable libraries include the discovery of biologics such as monoclonal antibody therapeutics [5], discovery of adeno-associated vectors (AAV) for gene therapy [1,8], T-cell receptor (TCR) discovery [2,4,7], and aptamer libraries [3,6].

Full Text Preprint: https://www.biorxiv.org/content/10.1101/2022.01.17.476670v1.

Data Availability: https://github.com/gifford-lab/FactorizableLibrary.

Z. Dai, S. D. Saksena and D. K. Gifford: Equal contribution.

© The Author(s), under exclusive license to Springer Nature Switzerland AG 2022
I. Pe'er (Ed.): RECOMB 2022, LNBI 13278, pp. 390–392, 2022.
https://doi.org/10.1007/978-3-031-04749-7_40

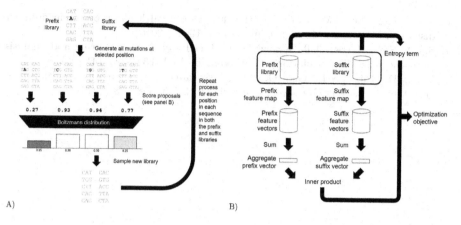

Fig. 1. Factorizable library evaluation and optimization. A) Optimization is achieved through iterative stochastic updates. An update step is performed by selecting a position in a sequence in one of the libraries and generating all possible mutations for that position. The mutated libraries are then scored, and then a Boltzmann distribution over the libraries is generated using the negated scores as energy values. The update is then sampled from the distribution. A full update sweep performs this for all positions in all sequences in both libraries. Multiple sweeps are done and the temperature of the Boltzmann distribution is lowered over time. For simplicity, the figure depicts this optimization on small DNA libraries. B) The score of the factorizable library is evaluated by mapping all the sequences in its prefix and suffix libraries to feature spaces. The feature vectors are then aggregated, and an inner product is taken between them, which by the distributive property produces the total score for the whole factorizable library. We refer to this as the reverse kernel trick, since this optimization requires expressing a "kernel function" that maps prefix suffix pairs to real values as an inner product. A position based entropy term is evaluated to quantify the diversity of sequences in the library, and a weighted sum of the two is then used to guide optimization.

Acknowledgements. This work was funded by NIH Grant R01 CA218094, and a gift from Schmidt Futures to D.K.G. The experimental work was funded by Novartis.

References

1. Bryant, D.H., et al.: Deep diversification of an AAV capsid protein by machine learning. Nat. Biotechnol. **39**, 691–696 (2021)
2. Holler, P.D., Holman, P.O., Shusta, E.V., O'Herrin, S., Wittrup, K.D., Kranz, D.M.: In vitro evolution of a t cell receptor with high affinity for peptide/mhc. Proc. Nat. Acad. Sci. **97**(10), 5387–5392 (2000). https://doi.org/10.1073/pnas.080078297, https://www.pnas.org/content/97/10/5387
3. Keefe, A.D., Pai, S., Ellington, A.: Aptamers as therapeutics. Nat. Rev. Drug Discov. **9**(7), 537–550 (2010)
4. Li, Y., et al.: Directed evolution of human t-cell receptors with picomolar affinities by phage display. Nat. Biotechnol. **23**(3), 349–354 (2005)
5. Liu, G., et al.: Antibody complementarity determining region design using high-capacity machine learning. Bioinformatics **36**(7), 2126–2133 (2020)

6. Maier, K.E., Levy, M.: From selection hits to clinical leads: progress in aptamer discovery. Mol. Ther. Methods Clin. Dev. **5**, 16014 (2016)
7. Smith, S.N., Harris, D.T., Kranz, D.M.: T cell receptor engineering and analysis using the yeast display platform. Methods Mol. Biol. **1319**, 95–141 (2015)
8. Wang, D., Tai, P.W.L., Gao, G.: Adeno-associated virus vector as a platform for gene therapy delivery. Nat. Rev. Drug Discov. **18**(5), 358–378 (2019)

Author Index